Disordered Solids

Structures and Processes

ETTORE MAJORANA INTERNATIONAL SCIENCE SERIES

Series Editor:

Antonino Zichichi

European Physical Society
Geneva, Switzerland

(PHYSICAL SCIENCES)

Recent volumes in the series:

Volume 36 **NEW TECHNIQUES FOR FUTURE ACCELERATORS II: RF and Microwave Systems**
Edited by M. Puglisi, S. Stipcich, and G. Torelli

Volume 37 **SPECTROSCOPY OF LIGHT AND HEAVY QUARKS**
Edited by Ugo Gastaldi, Robert Klapisch, and Frank Close

Volume 38 **MONTE CARLO TRANSPORT OF ELECTRONS AND PHOTONS**
Edited by Theodore M. Jenkins, Walter R. Nelson, and Alessandro Rindi

Volume 39 **NEW ASPECTS OF HIGH-ENERGY PROTON–PROTON COLLISIONS**
Edited by A. Ali

Volume 40 **DATA ANALYSIS IN ASTRONOMY III**
Edited by V. Di Gesù, L. Scarsi, P. Crane, J. H. Friedman, S. Levialdi, and M. C. Maccarone

Volume 41 **PROGRESS IN MICROEMULSIONS**
Edited by S. Martellucci and A. N. Chester

Volume 42 **DIGITAL SEISMOLOGY AND FINE MODELING OF THE LITHOSPHERE**
Edited by R. Cassinis, G. Nolet, and G. F. Panza

Volume 43 **NONSMOOTH OPTIMIZATION AND RELATED TOPICS**
Edited by F. H. Clarke, V. F. Dem'yanov, and F. Giannessi

Volume 44 **HEAVY FLAVOURS AND HIGH-ENERGY COLLISIONS IN THE 1–100 TeV RANGE**
Edited by A. Ali and L. Cifarelli

Volume 45 **FRACTALS' PHYSICAL ORIGIN AND PROPERTIES**
Edited by Luciano Pietronero

Volume 46 **DISORDERED SOLIDS: Structures and Processes**
Edited by Baldassare Di Bartolo

Disordered Solids
Structures and Processes

Edited by
Baldassare Di Bartolo
Boston College
Chestnut Hill, Massachusetts

Assitant Editors
Gönül Özen and John M. Collins
Boston College
Chestnut Hill, Massachusetts

Plenum Press • New York and London

Library of Congress Cataloging-in-Publication Data

Disordered solids : structures and processes / edited by Baldassare Di Bartolo.
p. cm. -- (Ettore Majorana international science series. Physical sciences ; v. 46)
Includes bibliographical references.
ISBN-13: 978-1-4684-5477-2 e-ISBN-13: 978-1-4684-5475-8
DOI: 10.1007/978-1-4684-5475-8
1. Order-disorder models. 2. Solids--Optical properties. 3. Semiconductors--Optical properties. 4. Glass--Optical properties. I. Di Bartolo, Baldassare. II. Series.
QC173.4.073D56 1990
530.4'1--dc20 89-49260
CIP

Proceedings of a course on Disordered Solids: Structures and Processes, held June 15–29, 1987, in Erice, Sicily, Italy

Softcover reprint of the hardcover 1st edition 1989
A Division of Plenum Publishing Corporation
233 Spring Street, New York, N.Y. 10013

"Has science promised happiness? I do not think so. It has promised the truth, and the question is to know if with truth we can achieve happiness."

Emile Zola

(Discourse to the students of Paris, 18 May, 1893)

PREFACE

This book presents an account of the course "Disordered Solids: Structures and Processes" held in Erice, Italy, from June 15 to 29, 1987. This meeting was organized by the International School of Atomic and Molecular Spectroscopy of the "Ettore Majorana" Centre for Scientific Culture.

The objective of this course was to present the advances in physical modelling, mathematical formalism and experimental techniques relevant to the interpretation of the structures of disordered solids and of the physical processes occurring therein.

Traditional solid-state physics treats solids as perfect crystals and takes great advantage of their symmetry, by means of such mathematical formalisms as the reciprocal lattice, the Brillouin zone, and the powerful tools of group theory. Even if in reality no solid is a perfect crystal, this theoretical approach has been of great usefulness in describing solids: deviations from perfect order have been treated as perturbations of the ideal model.

A new situation arises with truly disordered solids where any vestige of long range order has disappeared. The basic problem is that of describing these systems and gaining a scientific understanding of their physical properties without the mathematical formalism of traditional solid-state physics. While some of the old approaches may occasionally remain valid (e.g. chemical bonding approach for amorphous solids), the old ways will not do. Disorder is not a perturbation: with disorder, something basically new may be expected to appear. The challenge to theoretical physicists and to those who have to formulate and present these problems in didactical terms is evident: we must enlarge our view of the Solid State in order to encompass in its treatment disordered as well as periodic structures. This course was a response to such a challenge.

A total of 63 participants came from 40 laboratories and 17 different countries (Austria, Bangladesh, F.R. of Germany, France, India, Israel, Italy, Morocco, The Netherlands, P.R. of China, Poland, Portugal, Spain, Turkey, United Kingdom, U.S.A. and U.S.S.R.).

The secretaries of the course were Dr. J. Collins for the scientific aspects and Mr. J. Di Bartolo for the administrative aspects of the meeting.

Forty-seven lectures divided into 11 series were given. In addition 7 (one- or two-hour) "long" seminars and 19 "short" seminars were presented. Nine problem sessions were organized; the purpose of these sessions was to create a closer interaction between participants and lectures. During a problem session a lecturer did not introduce any new material, but explained in greater detail some points of the theory already presented, answered more questions, presented examples, etc.

A topic of general and great interest to physicists was also treated in a special session: "Elementary Particles: Discoveries and Perspectives."

Several participants presented overviews of the research projects of their laboratories and universities.

Two round table discussions were held. The first round-table discussion took place the third day of the meeting and had as subjects an evaluation of the work done in the first days of the course and the consideration of suggestions and proposals regarding the format of the lectures, the level of the presentations, etc. The second round-table discussion was held at the end of the course: the attendees had the opportunity of evaluating the work done during the entire meeting and to discuss various proposals for the next course of the International School of Atomic and Molecular Spectroscopy.

During the final session A. Ritter and L. McNeil presented a summary of the meeting and identified the major themes which appeared in various degrees throughout the course.

I wish to express my sincere gratitude to Drs. Gabriele and Savalli and to all the personnel of the "Ettore Majorana" Centre who contributed so much to create a congenial atmosphere for our meeting. I wish to acknowledge the sponsorship of the meeting by the Italian Ministry of Public Education, the Italian Ministry of Scientific and Technological Research, the Italian National Research Council, the Sicilian Regional Government, the European Physical Society and the Department of Physics of Boston College.

I would like to thank the members of the organizing committee (Professors Klingshirn and Silbey), the secretaries of the course (Dr. John Collins and Mr. John Di Bartolo), Ms. Gonul Ozen, Dr. Guzin Armagan and Mr. Daniel Di Bartolo who helped a lot with various aspects of the meeting. A warm "thank you" to Antonino La Francesca, "Nino"!

I would like to thank Ms. Patricia Vann of the Plenum Publishing Corporation for her encouragement and assistance during the preparation of this book. I wish to acknowledge the very patient and careful work done by the assistant editors of this book, Gonul Ozen and John Collins.

It was a privilege to direct this school and to meet so many fine people who came from many distant places to work together in the very friendly ambience provided by the Ettore Majorana Centre.

It is always a pleasure to return to my native Sicily, and to enjoy the warm hospitality of Erice. This is a magic place, made even more pleasurable by the scientific encounters that occur in the congenial atmosphere of this ancient village. We always long for these events and we look forward to many meetings in years to come.

B. Di Bartolo
Editor and Director of the International School of Atomic and Molecular Spectroscopy of the Ettore Majorana Centre

Erice, June 1987

CONTENTS

THE EVOLVING NOTION OF ORDER

G. Careri

Dipartimento di Fisica
Università di Roma I
00185 Roma, Italy

ABSTRACT

The notion of order in macroscopic physics has changed progressively during this century. This evolution is outlined in author's book on this subject, briefly reviewed here. Some major turning points (including the introduction of the internal field, the collective ground state and symmetry breaking) are considered in detail.

I. INTRODUCTION

The aim of these two lectures is to call the attention of an audience interested in disordered systems to the development of the notion of order which has taken place in physics in this century. Since Galileo this notion has been at the core of physics itself, and today we may say that the work of a physicist is essentially to look for order in that part of the world which is within his reach. As every graduate student already knows, the laws of both quantum and classical mechanics are appropriate space-time correlations among events observed in the laboratory, and even the invariance principles are correlations among physical laws, namely correlations of higher rank among the raw events themselves. Yet an explicit description of order in terms of these correlations in macroscopic systems is not easy, and it was only during the middle of this century that this goal was attained, at least not too far from equilibrium. I would like to say that my generation has been concerned with this unified expression of order in the frame of theoretical physics, much in the same way as the present generation is concerned with the description of disorder as it is possible today thanks to computer science.

In these two lectures I shall follow a small book that I wrote recently on these topics [1]. Therefore the present written contribution will be rather brief and it will be limited to a presentation of my book's contents, and to some remarks one the major turning points of our present notion of order in statistical physics and material science.

II. ORDER AND DISORDER IN MATTER

I have written my book for both undergraduate and graduate students of science departments interested in the description of material systems from the viewpoint of order and disorder. My concern was to cover the macroscopic description of matter that is typical of statistical physics, given the widely-observed fact that matter takes on macroscopic ordered structures which cannot generally be predicted by direct extrapolation from microscopic properties. For instance, if we start with only the notion of the dynamic laws governing movement of single molecules in a gas, it would be no simple task to predict the existence of a general phenomenon such as sound. Likewise, who would ever think that a crystal, with its typical spatial symmetries, would be formed when the temperature of an homogeneous and isotropic liquid composed of perfectly-spherical atoms is lowered? Thus in this area, a phenomenological description is perfectly appropriate because there is no general microscopic theory which can predict the macroscopic structure of any material system in any boundary conditions.

The level of complexity of the material system considered in my book is the one normally studied in atomic physics, in solid-state physics and in chemistry, where in principle all of the phenomenology can be traced back to the laws of non-relativistic quantum mechanics and of statistical mechanics. In other terms the energy of the system may be formally written as the sum of the kinetic energies of the nuclei and electrons and of the potential energy that contains all the electrical and magnetic interactions of the system, whether internal or external. It must be pointed out that none of the typical concepts of chemistry or of solid-state physics can emerge from the sole consideration of this reading of the system's total energy. As we know these concepts emerge only after approximation, and that is after the introduction of a proper model that can let the meaning of particular concept emerge. The same is true for the concepts of order and disorder, which are among the more general concepts that bridge the gap between the macroscopic and microscopic worlds, and are thus quite effective in describing the material systems at this level of complexity.

The book is divided into five chapters, the first being an area of contact with what the reader probably already knows about general physics. Instead of briefly summarizing statistical thermodynamics, I felt it would be wiser to recall several models to which statistical thermodynamics can be applied. Within the elementary context of statistical thermodynamics considered in this chapter, order and disorder appear to be two well-defined entities that merely complement each other. This position is drastically altered by the study of superfluids undertaken in the second chapter, because here order and disorder become two different ingredients in the description of the peculiar quantomechanic nature of the system.

Next, the more general question regarding the emergence of an order parameter as symmetry breaking in a material system is discussed in the third chapter. Starting from common phenomenology presented by a vast class of phase transitions in conditions of thermodynamic equilibrium, I have had to develop the notion of order parameter in more abstract form without the appropriate mathematical language. The phenomenology of phase transitions at equilibrium is the obvious reference point for the disorder order transitions outside equilibrium, considered in the fourth chapter. This subject is of particular interest in showing the new conceptual change required to understand the phenomenology of the laser, among other things. The laser is the prototype of an open physical system, and it contains all the necessary features for the statement of the induced order in the new terms of correlation among events. This new conceptual outlook was necessary to discuss the functional order of biology in the fifth chapter, which is seen as the extrapolation of the physical line of thought formed in the previous chapters.

As can be realized the conception of order in matter is progressively modified, so as to fit it to more and more complex models in more and more realistic conditions. The initial point of view is the outlook common to statistical thermodynamics, where disorder is measured by entropy and order is simply the complementary aspect. Later, the long-range order of superfluid helium is identified by the presence of a collective wave function connected with the fundamental energy state of the quantum system. This fact is then generalized to include even more complex material aggregates through the concept of order parameter, a quantity with few internal components that controls the spontaneous transitions toward order below a critical temperature far from absolute zero. However, the discussion of critical phenomena points out the need to consider the space- time fluctuations of the order parameter as well, and the statistical character of long-range order is then established with the introduction of a coherence length, within which there is a permanent statistical correlation in the distribution of matter. It is likewise through the order parameter that transitions toward order induced in open systems outside equilibrium are followed, and now we are faced with the alternative possibility of conceiving order as the space-time distribution correlation itself. The example of the laser clarifies this point, because here the elementary event can easily be identified in the emission of a light quantum, and the space-time correlation can either be viewed among emission events or in the distribution of excited atoms, and it is detected in the wave character of the coherent electromagnetic field. As a consequence, the statistical space-time correlation seems to be the natural evolution and merging of the two concepts of order and disorder. Furthermore, consideration of the systems outside equilibrium leads us to shift our attention from the basic elements of matter to the events taking place in processes involving matter. Therefore the closer the correlation among the events which give rise to these processes, the higher is the order of the material system.

III. CONCLUDING REMARKS

After the above outline of my book's contents let me consider what, in my opinion, are the major turning points in the development of the notion of order during this century.

- The first important change took place early in this century, when the ordering field introduced by Weiss and Van der Waals, was completed by the statistical distribution of populations on quantum energy levels (see box 1). As a later corollary along the same lines, we may recall the concept of negative temperature.

- A second relevant new outlook occurred after 1940, when Landau identified the superfluid component of liquid helium II in the collective ground state and disorder in the quasi-particle population of the excited states. This new outlook was completed in the sixties by Onsager and Feynman, who identified order in the coherent properties of a symmetric wave function (see box 2).

- The third turning point can be traced back to the Sixties, when in different fields of physics it was realized that the most ordered state was reached by symmetry breaking. This notion proved to be so general that it could be extended even to phenomena far from equilibrium (see box 3).

I hope the student will realize by now how deep the interdependence displayed by the notions of order and disorder is. For instance, molecular chaos follows detailed balance principle, and quasi-particles are described by wave functions; on the other hand, the more ordered state in a phase transition is reached thanks to a large-scale random fluctuation,

BOX 1

FROM STRUCTURES TO STATISTICAL DISTRIBUTIONS

- Ordered structures are created by an internal field
    - molecular aggregates (Van der Waals)
    - magnetic domains (Weiss)

- Generalization:

  order parameter η

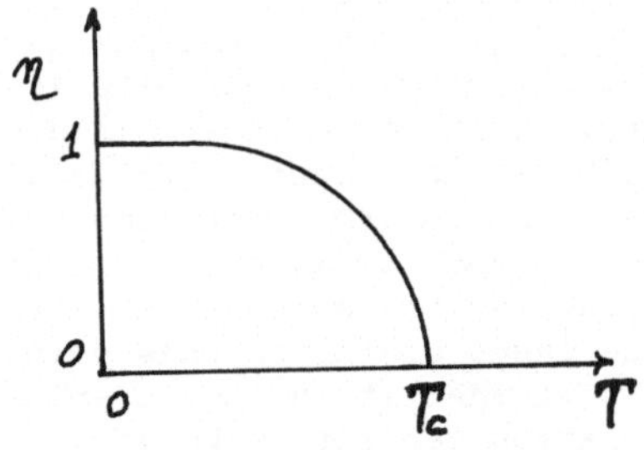

- Statistical thermodynamics on quantum levels E_i

$$A = -kT \ln Z$$

$$Z = \Sigma_i \exp(-E_i/kT)$$

$$S = -(\partial A/\partial T)_v$$

- Order and disorder in a perfect spin gas

lithium
fluoride

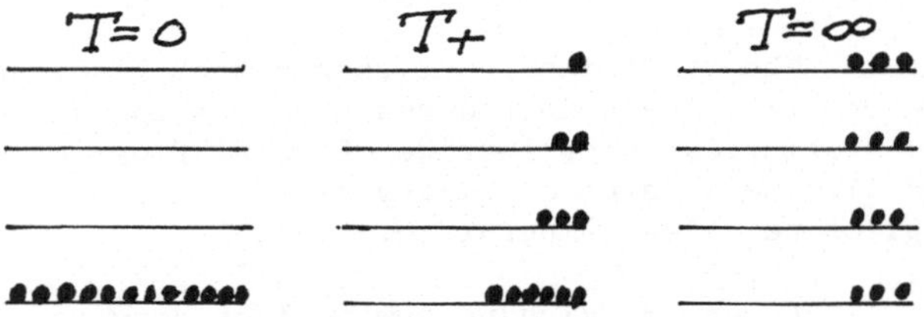

order
when only one level
is occupied

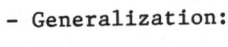

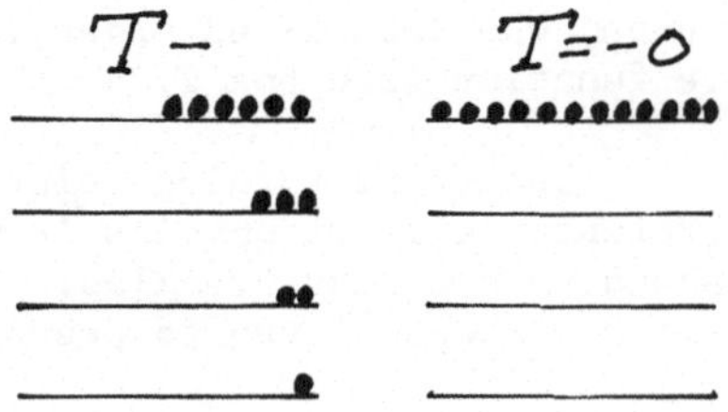

BOX 2

FROM COLLECTIVE QUANTUM LEVELS TO WAVE FUNCTIONS

- Liquid helium II : the most ordered liquid

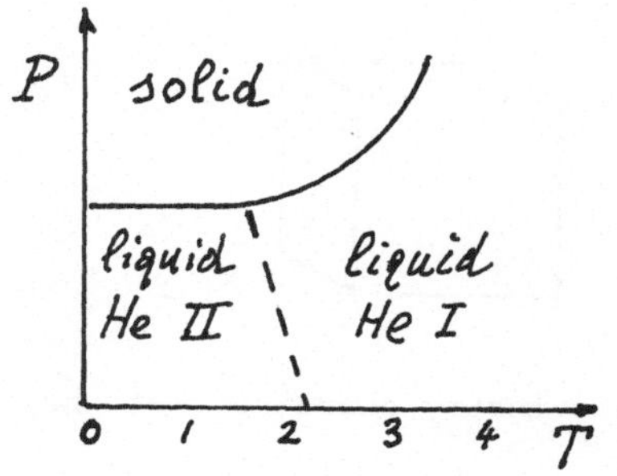

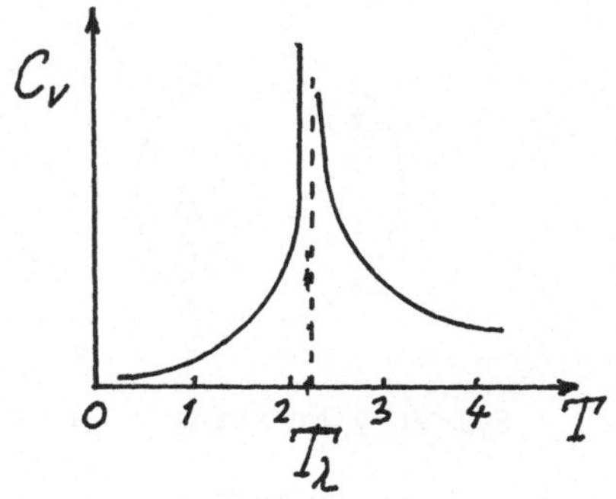

- Landau (1940-1950) : two fluids model

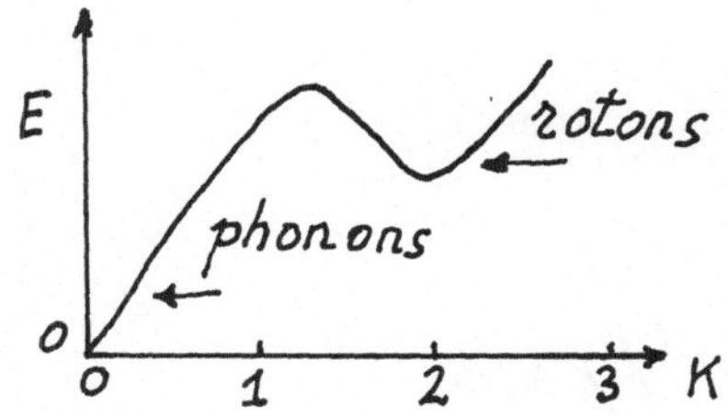

order = ground state

disorder = quasi-particles gas

- Onsager (1949) and Feynman (1954) : Bose wave function

order = Bose condensate

disorder = quasi particles + quantum vortices

- Exprimental evidence for collective Bose wave function

- quantized vortical lines
- superconductivity by electron pairs
- superfluidity of He^3 by spin pairs
- Josephson effects

BOX 3

ORDER AS SYMMETRY BREAKING

- Phase transitions :

disorder ——————→ order

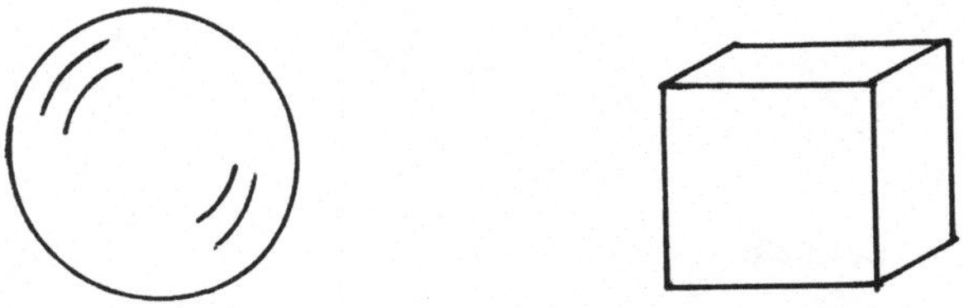

Symmetry breaking means to reduce symmetry characters of the systems, thus originating a new structure with a lower number of allowed invariant transformations.

- Analogies between laser and phase transitions :

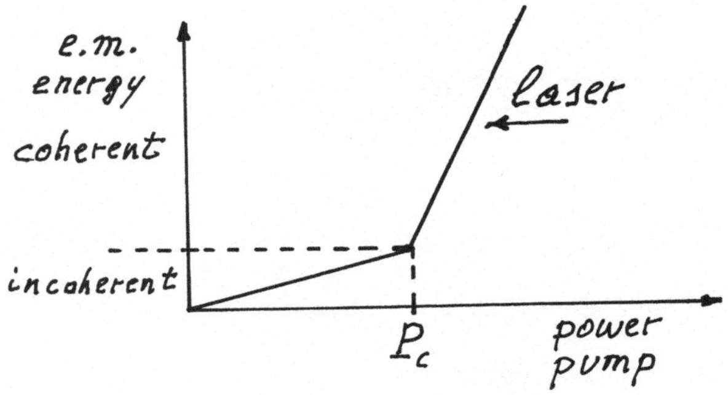

	order	disorder
Phase transitions	The internal field creates site correlations	Termal events destroy site correlations
Laser	The internal e.m. fields creates correlated emission	Random decay events destroy correlated events

which itself is a correlation among several fluctuation occuring at lower scales. For these reasons I believe that an understanding of the evolving notion of order can be beneficial today to students mainly interested in disordered states of matter.

REFERENCES

1. G. Careri, "Order and Disorder in Matter", Benjamin Cummings, Palo Alto, 1983.

PROBABILISTIC METHODS IN PHYSICS

N.G. van Kampen

University of Utrecht
Institute for Theoretical Physics
Utrecht, Netherlands

ABSTRACT

The fundamental concepts and methods of probability theory are reviewed. It is emphasized that probability calculus merely transforms one probability distribution into another. Hence for every application some a priori distribution must be defined on the basis of the underlying physics. The mere use of the word 'random' is not enough, particularly since that word is often used in the sense of 'irregular' or 'disordered'. The transition of a disordered system to a probabilistic (or stochastic) description is a fundamental problem, which cannot be solved by playing with words.

A stochastic process is an ensemble of functions of time, distinguished by a parameter for which a probability distribution is given. Alternatively it can be specified by a hierarchy of joint distributions for the values it takes at any set of time points $t_1, t_2, \dots t_r$. An important role is played in physics by the Markov processes, roughly described as processes having no memory (in analogy with the differential equations of deterministic physics). They are fully specified by the distribution of values at any one time, and the transition probability between two times. This transition probability obeys the master equation, in which the coefficients have a direct physical meaning.

The natural way of solving the master equation is provided by the expansion in a parameter (often the size of the system), which separates the macroscopic scale from the scale in which the fluctuations take place. The lowest term in the expansion reproduces the deterministic macroscopic rate equation. The next term describes the fluctuations in terms of the Fokker-Planck equation, whose solution is a time dependent Gaussian distribution. Higher terms provide corrections to this Gaussian behavior of the fluctuations. In this way the dilemma that arises from applying the Langevin approach to nonlinear systems is resolved.

A condition for the validity of this expansion is that the macroscopic equation is dissipative and tends to a stable equilibrium. If the macroscopic equation vanishes, the expansion tends to a Fokker-Planck equation with nonlinear coefficients. As a special example, the hopping of a charge carrier in an inhomogeneous medium with an external field is worked out. It leads to a diffusion equation in a rather unexpected form.

I. PROBABILITY THEORY

Definition. A 'random number' or 'random variable' is neither a number nor a variable. Its definition is composed of a given set (range or sample space) and a probability distribution over this set. The set may consist of discrete points labelled n, such as the six positions of a die and the energy levels of an atom; in that case the probability distribution attaches a nonnegative number p_n to each n with the only restriction $\sum_n p_n = 1$. Or the set may be an interval $a < x < b$; in that case the probability distribution is a nonnegative function $P(x)$ having the property $\int_a^b P(x)dx = 1$. If $P(x)$ is smooth it is called the probability density and $P(x)dx$ is the probability for the random number to lie in the interval $(x,x+dx)$. But $P(x)$ may also contain peaks; then a peak $b\delta(x-c)$ attaches a nonzero probability b to the single point c. The sample space can also consist of a combination of intervals and points, or may be a domain in a space of more dimensions. It can even be a more abstract set; then the probability distribution is defined as a nonnegative measure on that set, normalized to give a measure 1 to the total set. Important examples are

Poisson distribution. Range $n = 0,1,2,\ldots$;

$$p_n = \frac{\alpha^n}{n!}e^{-\alpha} \qquad (\alpha \text{ constant parameter}).$$

Lorentz or Cauchy distribution. Range $-\infty < x < \infty$;

$$P(x) = \frac{\gamma}{\pi}\frac{1}{x^2+\gamma^2} \qquad (\gamma \text{ positive constant}).$$

Gauss distribution. Range $-\infty < x < \infty$;

$$P(x) = (2\pi\sigma^2)^{-\frac{1}{2}} \exp[-\frac{(x-\mu)^2}{2\sigma^2}] \qquad (\sigma,\mu \text{ real}).$$

Multivariate Gauss distribution. Range $\mathbb{R}^r$;

$$P(x_1,x_2,\ldots,x_r) = C \exp[-\tfrac{1}{2}\sum_{i,j}^{r} A_{ij}\, x_i x_j + \sum_i^r B_i x_i].$$

A is a positive rxr matrix and B a constant vector. The normalizing constant C can be found by the reader.

Throughout this article we shall use either the continuous or the discrete notation, whichever is more convenient, but the results can easily be generalized. The average over a distribution $P(x)$ of a quantity $f(x)$ is

$$\langle f(x) \rangle = \int f(x)P(x)dx.$$

(In the mathematical literature often called expectation value of f(x) and written Ef(x)). In particular the <u>moments</u> are

$$\mu_m = \langle x^m \rangle = \int x^m P(x)dx.$$

A useful aid is the <u>characteristic function</u> of a distribution

$$G(k) = \langle e^{ikx} \rangle = \int e^{ikx} P(x)dx ,$$

where k is an auxiliary variable. It has the formal expansion

$$G(k) = \sum_{m=0}^{\infty} \frac{(ik)^m}{m!} \mu_m$$

and is therefore also called the 'moment generating function'.
If one expands its logarithm

$$\log G(k) = \sum_{m=1}^{\infty} \frac{(ik)^m}{m!} \kappa_m$$

the coefficients κ_m are called <u>cumulants</u>. They are combinations of the moments:

$$\kappa_1 = \mu_1 , \; \kappa_2 = \mu_2 - \mu_1^2 , \; \kappa_3 = \mu_3 - 3\mu_2\mu_1 + 2\mu_1^3 , \; \ldots$$

In particular $\kappa_2 = \langle x^2 \rangle - \langle x \rangle^2 = \langle (x-\langle x \rangle)^2 \rangle$ is called the <u>variance</u>. It is easily shown that for the Gauss distribution all moments beyond the second vanish, and that this is the only distribution with that property.

Multivariate distributions $P(x_1,x_2,\ldots,x_r)$ give rise to some additional concepts, which we formulate for r=2. The <u>marginal distributions</u> of $P(x_1,x_2)$ are

$$P_1(x_1) = \int P(x_1,x_2)dx_2 , \quad P_2(x_2) = \int P(x_1,x_2)dx_1 .$$

The <u>conditional distribution</u> of x_1, given the value of x_2, is

$$P(x_1|x_2) = \frac{P(x_1,x_2)}{P(x_2)} .$$

The moments $\langle x_1^{m_1} x_2^{m_2} \rangle$ are defined in the obvious way and the characteristic function is $G(k_1,k_2) = \langle \exp i(k_1x_1+k_2x_2) \rangle$. Each variable has its variance and their mutual <u>covariance</u> is

$$\mathrm{cov}(x_1,x_2) = \langle (x_1-\langle x_1 \rangle)(x_2-\langle x_2 \rangle) \rangle .$$

The random variables x_1, x_2 are called statistically independent if P factorizes: $P(x_1,x_2) = P_1(x_1)P_2(x_2)$. The factors P_1, P_2 are identical with the marginal distributions and also with the conditional ones. Necessary and sufficient for independence is that the joint c.f. $G(k_1,k_2)$ factorizes.

Statistical independence implies that the covariance is zero, but the converse is not true - unless one knows that the joint distribution is Gaussian. The word 'uncorrelated' actually means that the covariance vanishes, but is often loosely used to indicate statistical independence.

Transformation of variables. Suppose P(x) is given and one wants to know the probability Q(y) of a new variable $y = \psi(x)$. The general formula is

$$Q(y) = \int \sigma[y-\psi(x)]\, P(x)dx = \frac{P(x)}{|\psi'(x)|} \,.$$

If there are several values of x that give the same value y, say $x_\nu = x_\nu(y)$, then they all contribute to the delta function and one has

$$Q(y) = \sum_\nu \frac{P(x_\nu)}{|\psi'(x_\nu)|} \,.$$

Alternatively one may utilize the c.f. of Q,

$$\langle e^{iky} \rangle = \int e^{ik\psi(x)}\, P(x)dx \,.$$

The same transformation formulas apply when x stands for a set of variables $x_1, x_2, \ldots, x_r$ and y stands for another set $y_1, y_2, \ldots, y_s$. In particular if we have two variables x_1, x_2 and want to know the distribution of their sum $y = x_1 + x_2$ one finds

$$Q(y) = \int \delta(y-x_1-x_2)\, P(x_1,x_2)dx_1dx_2 = \int P(x_1,y-x_1)dx_1 \,.$$

This is the addition of random variables. If x_1 and x_2 are independent the distribution of their sum is the convolution of their individual distributions; the c.f. of their sum is the product of their individual characteristic functions.

All one ever does in probability calculus is transforming one distribution into another. Hence in every application a probability distribution must be given 'a priori'. In elementary problems, such as dice, this is usually done tacitly, but it does imply a physical assumption concerning the dice. In classical statistical mechanics the a priori probability is based on the ergodic theorem. In a crystal with random impurities one ought to specify it too, although that is not always done. The mere use of words like 'random' or 'totally random' is not enough.

This requirement is mitigated, however, by the law of large numbers or central limit theorem: the sum of a large number of independent variables, all having the same a priori distribution is approximately

Gaussian. This common one-variable distribution may be anything, provided that it has a finite variance σ^2. More precisely, the renormalized sum

$$y = \frac{x_1+x_2+\ldots+x_r-r\langle x\rangle}{\sqrt{r}}$$

has in the limit $r \to \infty$ the above Gaussian distribution (with $\mu=0$). This fact (together with the fact that it is so convenient in calculations) is the reason for the dominant role of the Gaussian or 'normal' distribution. The proof can be found, with any desired degree of mathematical sophistication, in the literature.

To demonstrate that the condition of finite σ^2 is needed one can easily show that the sum of Lorentzian variables is again Lorentzian and therefore never approaches a Gaussian. Distributions with this property are called 'stable' or 'Lévy distributions'. They can be fully characterized and have recently become popular because of their connection with scaling [1]. Unfortunately, apart from the Gaussian and the Lorentzian, they are rather awkward to handle.

A complex number $z = x+iy$ is random when x and y are random with a joint distribution $P(x,y)$. A random NxN matrix $||H_{ij}||$ is given by the joint distribution $P(H_{ij})$ of all its elements. A random hermitian matrix is given by a distribution of the real elements H_{ii} and the complex elements H_{ij} with $i < j$. Such matrices have been used in nuclear physics to describe Hamiltonians that are too complicated to treat in detail. The P is further specified by assuming that all these elements are statistically independent, and that P is invariant for orthogonal transformations of H. (These are precisely the assumptions of Maxwell's erroneous derivation of the velocity distribution law.) The eigenvalues $\lambda_1,\lambda_2,\ldots\lambda_N$ are functions of the H_{ij} and the problem is to find their joint distribution $Q(\lambda_1,\lambda_2,\ldots,\lambda_N)$.[2]

Another type of random matrix is more relevant for the present course. Consider a linear harmonic chain of N particles with fixed endpoints:

$$m_i\ddot{q}_i = \gamma_i(q_{i+1}-q_i) - \gamma_{i-1}(q_i-q_{i-1}) \,, \quad q_o = q_{N+1} = 0 \,.$$

When all masses m_i are equal the solution of these equations of motion amounts to diagonalizing the tridiagonal symmetric matrix

$$\begin{matrix} -(\gamma_o+\gamma_1) & \gamma_1 & 0 & 0 \\ \gamma_1 & -(\gamma_1+\gamma_2) & \gamma_2 & 0 \\ 0 & \gamma_2 & -(\gamma_2+\gamma_3) & \gamma_3 \\ \hline \end{matrix}$$

If the spring constants γ_1 ($i=0,1,2,\ldots,N$) are random the problem is again to find the distribution of the eigenvalues of this random matrix [3].

We have given a precise definition of random variables, random matrices, and random chains in terms of probability. Physicists like to visualize a probability distribution as an ensemble. However, the word 'random' is also used for 'irregular' or 'disordered'. This dual usage conceals the question: Is it justified to compute the properties of a disordered system by replacing it with a probabilistic ensemble? This question is usually answered by an appeal to 'self-averaging': A disordered solid may be regarded as a collection of small solid cubes, which constitute an ensemble in the flesh. This argument is only valid, however, if these cubes behave independently, that is, if all correlations have ranges short compared to the size of the sample observed in the laboratory. It does not apply to long-range properties such as percolation.

II. STOCHASTIC PROCESSES

Let x be a stochastic variable with distribution P(x). Every mathematical object that contains x as a parameter is also stochastic and may be visualized as an ensemble of similar objects, differing only by their individual values of x. The number of objects in the ensemble whose x lies in the interval x,x+dx is taken proportional to P(x)dx.

In particular, let f(t,x) be a family of functions of t, labeled by a parameter x. If x is a stochastic variable this family constitutes an ensemble of functions of t, and $y = f(t,x)$ is called a random function of t, or a stochastic process. One may compute averages, such as

$$\langle y(t_1)y(t_2)\ldots y(t_r)\rangle = \int f(t_1,x)f(t_2,x)\ldots f(t_r,x)P(x)dx \;.$$

The autocorrelation function of y is defined by

$$\kappa(t_1,t_2) = \langle y(t_1)y(t_2)\rangle - \langle y(t_1)\rangle\langle y(t_2)\rangle \;.$$

Alternatively one may look at y(t) as an infinite set of random variables, one for each t. The probability density of each of them is

$$P_1(y,t) = \int \delta[y-f(t,x)]\; P(x)dx \;.$$

The joint distribution of a set of r of these variables is

$$P_r(y_1,t_1;y_2,t_2;\ldots;y_r,t_r) = \int \prod_{i=1}^{r} \delta[y_i-f(t_i,x)] \;.\; P(x)dx \;.$$

There exists no joint probability for the entire infinite set, but the hierarchy of P_r for all r up to infinity enables one to compute all averages. This is an alternative description of a stochastic process, for some purposes more suitable than the ensemble picture. Kolmogorov's 'reconstruction theorem' states that both descriptions are equivalent.

A process is called stationary when all P_r remain the same if one shifts time by adding a constant τ to all $t_1,t_2,\ldots,t_r$. In particular $\langle y\rangle$

is independent of time and $\kappa(t_1,t_2) = \kappa(t_1-t_2)$. Moreover $\kappa(-\tau) = \kappa(\tau)$. Its Fourier transform (if it exists) is

$$S(f) = 2\int_{-\infty}^{\infty} \kappa(\tau)e^{2\pi i f\tau}\, d\tau = 4\int_{0}^{\infty} \kappa(\tau)\cos(2\pi f\tau)d\tau .$$

According to the Wiener-Khintchine theorem $S(f)$ is the intensity with which the wave of frequency f occurs in the process $y(t)$. This theorem is of fundamental importance, because what is usually measured (think of electronics) is the spectral density $S(f)$ of a current or a voltage, whereas theoretical derivations usual concern $\kappa(\tau)$.

A notorious example is 1/f noise. In semiconductors and metal films one has voltage or current fluctuations for which $S_f \sim 1/f$ over many decades. At high f this is cut off by noise of a more familiar kind, but at low f no limitation has yet been found, even down to $f \sim 10^{-3}$ Hertz. That implies that the autocorrelation function must extend over times of the order of an hour. No theory has yet been found that explains this [4].

A process is called Gaussian if each P_r, for fixed values of $t_1,t_2,\dots t_r$, is an r-variate Gaussian distribution of the random quantities $y_1,y_2,\dots,y_r$. A Gaussian process is uniquely determined by its mean $\langle y(t)\rangle$ and its autocorrelation function $\kappa(t_1,t_2)$. To see this we first define the renormalized process $\tilde{y}(t) = y(t) - \langle y(t)\rangle$, which is Gaussian with zero mean, and the same autocorrelation function $\kappa(t_1,t_2) = \langle\tilde{y}(t_1)\tilde{y}(t_2)\rangle$. Then use the following property of any multivariate Gauss distribution with zero mean:

$$\begin{aligned}
&\langle\tilde{y}(t_1)\tilde{y}(t_2)\dots\tilde{y}(t_{2q+1})\rangle = 0 \\
&\langle\tilde{y}(t_1)\tilde{y}(t_2)\dots\tilde{y}(t_{2q})\rangle = \sum \langle\tilde{y}(t_i)\tilde{y}(t_j)\rangle\langle\tilde{y}(t_k)\tilde{y}(t_\ell)\rangle \dots,
\end{aligned} \tag{1}$$

where $t_i,t_j,t_k,t_\ell,\dots$ stand for the $t_1,t_2,\dots$ taken pairwise and $\sum$ sums over all partitions in pairs. Thus all moments are expressed in κ, which proves the statement.

As an example take $\langle y(t)\rangle = 0$, and

$$\kappa(t_1,t_2) = \sigma^2 e^{-\gamma|t_1-t_2|} \qquad (\sigma^2,\gamma \text{ constants}) .$$

The resulting Gaussian process is stationary; it is called the Ornstein-Uhlenbeck process [5]. In the limit $\gamma \to \infty$, $\sigma^2 \to \infty$ with fixed $\sigma^2/\gamma = D$ the autocorrelation function becomes

$$\langle y(t_1)y(t_2)\rangle = 2D\,\delta(t_1-t_2)\ . \tag{2}$$

The resulting process is <u>Gaussian white noise</u>. This is not, however, a proper stochastic process (just as the delta function is not a proper function) and must therefore be handled with care.

III. MARKOV PROCESSES

In order to define them we first note that

$$P\left(y_r t_r | y_1,t_1;\ldots;y_{r-1},t_{r-1}\right) = \frac{P_r(y_1,t_1;\ldots;y_r,t_1)}{P_{r-1}(y_1,t_1;\ldots;y_{r-1},t_{r-1})}$$

is the conditional probability for y to have at t_r the value y_r, given that at $t_1,\ldots,t_{r-1}$ it has the values $y_1,\ldots,y_{r-1}$. Take a chronological sequence $t_1 < t_2 < \ldots < t_r$. A <u>Markov process</u> is characterized by the fact that this conditional probability depends only on the last preceding value y_{r-1} and not on the earlier ones:

$$P(y_r,t_r|y_1,t_1;\ldots;y_{r-1},t_{r-1}) = P(y_r,t_r|y_{r-1},t_{r-1}) \qquad (t_1 < t_2 <\ldots<t_r)\ .$$

Thus the value of y at any one time provides full information concerning the future probabilities; the earlier history is irrelevant.

We have given a precise definition in mathematical terms of a stochastic process, often simply called a 'process'. In physics, however, the word 'process' usually refers to a (time-dependent) physical phenomenon observed in the laboratory. The question whether such a <u>physical</u> process is Markovian or not can not be answered unless one specifies the variables one chooses to describe it. If a particle moves through a medium with randomly distributed scatterers its position $\underset{\sim}{r}(t)$ is not a Markov process, but its velocity $\underset{\sim}{v}(t)$ is, and therefore also the joint process $\{\underset{\sim}{r}(t),\underset{\sim}{v}(t)\}$. On the other hand, if a particle moves by hopping among traps its position $\underset{\sim}{r}(t)$ is Markovian, provided that the time spent between traps is negligible and that it resides in each trap so long that the heat motion of the lattice destroys those features of its wave function that retain information about how it arrived there. If it does not, one has to go back to the full Schrödinger equation, as in the theory of heavy fermions [6].

If one knows that a process is Markovian, its entire hierarchy $\{P_r\}$ can be constructed from the $P_1(y_1,t_1)$ and $P_2(y_1,t_1;y_2,t_2)$. For instance, for $t_1 < t_2 < t_3$

$$P_3(y_1,t_1;y_2,t_2;y_3,t_3) = P_2(y_1,t_1;y_2,t_2)\ P(y_3,t_3|y_2,t_2)\ .$$

But even P_1 and P_2 are not independent, but related by

$$P_1(y_2,t_2) = \int P(y_2,t_2|y_1,t_1)\ P_1(y_1,t_1)dy_1\ .$$

Hence it suffices to know P_1 at one particular time t_1. Finally P_2 itself obeys the Chapman-Kolmogorov identity. To derive this identity integrate the above expression for P_3 with respect to y_2:

$$P_2(y_1,t_1x;y_3,t_3) = \int P(y_3,t_3|y_2,t_2)\, P_2(y_2,t_2x;y_1,t_1)dy_2 .$$

Subsequently divide both members by $P_1(y_1,t_1)$, so that one obtains for $t_1 < t_2 < t_3$ the identity

$$P(y_3,t_3|y_1,t_1) = \int P(y_3,t_3|y_2,t_2)\, P(y_2,t_2|y_1,t_1)dy_2 . \tag{3}$$

This is an integral equation for the transition probability. It states that, in order to go from the initial value y_1 at t_1 to the later value y_3 at t_3, the process has to pass, at the intermediate time t_2, through some value y_2; and the probability for this to happen is the product of the probabilities for each step.

The Wiener process is a nonstationary Gaussian Markov process defined for $t > 0$, $-\infty < y < \infty$ by

$$P_1(y,0) = \delta(y) , \quad P(y,t|y_1,t_1) = \frac{1}{\sqrt{2\pi(t-t_1)}} \exp\left[-\frac{(y-y_1)^2}{2(t-t_1)}\right] .$$

This process describes the position of a diffusing particle.

The Ornstein-Uhlenbeck process, already mentioned, is defined for $-\infty < t < \infty$, $-\infty < y < \infty$ by

$$P_1(y) = \frac{1}{\sqrt{2\pi}} e^{-\frac{1}{2}y^2} \quad \text{and} \quad (\text{with } \tau = t-t_1 > 0)$$

$$P(y,t|y_1,t_1) = \frac{1}{\sqrt{2\pi(1-e^{-2\tau})}} \exp\left[-\frac{(y-y_1e^{-\tau})^2}{2(1-e^{-2\tau})}\right] .$$

It describes the velocity of a Brownian particle. Doob's theorem states that this is the only stationary Gaussian Markov process, apart from linear transformations of y and t [7].

The Poisson process describes a sequence of random events that occur independently from each other, such as the arrivals of α-particles from ^{238}U in my Geiger counter. The only parameter is the average number ν of events per unit time. In order to describe this sequence of dots on the time axis we call $n(t)$ the number of dots or events in the interval $(0,t]$. Then

$$P_1(n,0) = 0 , \quad P(n,t|n_1,t_1) = \frac{[\nu(t-t_1)]^{n-n_1}}{(n-n_1)!} e^{-\nu(t-t_1)} . \tag{4}$$

This process is not stationary, although the physical decay process is. The reason is that we could not formulate it as a process without introducing an arbitrary initial time at which to start counting. A formulation in terms of random sets of points is more elegant [8].

IV. THE MASTER EQUATION

Consider the transition probability (4). Of course for $t = t_1$ it reduces to δ_{n,n_1}. For a small time difference Δt one has

$$P(n,t_1+\Delta t|n_1,t_1) = \delta_{n,n_1}(1-\nu\Delta t) + \delta_{n,n_1+1}\,\nu\Delta t + O(\Delta t)^2 .$$

The first term refers to the possibility that no event occurs during Δt and the second to the possibility that one event occurred. The chance that more occur is of order $(\Delta t)^2$. This idea will now be used to derive from the Chapman-Kolmogorov equation the more useful master equation.

Write the CK equation for a discrete variable n,

$$P(n_3,t_3|n_1,t_1) = \sum_{n_2} P(n_3,t_3|n_2,t_2)\, P(n_2,t_2|n_1,t_1).$$

Take $t_3 = t_2 + \Delta t$; then one expects

$$P(n_3,t_2+\Delta t|n_1,t_1) = \delta_{n_3,n_1}(1-\nu\Delta t) + \Delta t\, W(n_3|n_2) + O(\Delta t)^2 . \tag{5}$$

The W here defined is the transition probability per unit time from n_2 to n_3. In general it may depend on t_2 but in many cases it doesn't, certainly not for stationary processes; for simplicity we omit this time dependence. The quantity ν must ensure conservation of probability, hence

$$\nu(n_2) = \sum_{n_3} W(n_3|n_2) .$$

Now substitute all this in the CK equation, divide by Δt, go to the limit $\Delta t \to 0$ and adjust the notation:

$$\partial_t P(n,t|n_1,t_1) = \sum_{n'} \{W(n|n')P(n',t|n_1,t_1) - W(n'|n)P(n,t|n_1,t_1)\} . \tag{6}$$

This is the <u>master equation</u>. One recognizes it as a gain-loss equation of probabilities. The initial value n_1 at t_1 enters only as a parameter and is often not indicated explicitly. However, it must be borne in mind that <u>the master equation is an equation for the transition probability of a Markov process</u>. Its role in the theory of stochastic processes is analogous to that of ordinary differential equations in fluctuationless physics.

The simplest Markov process has two possible values, say $n = \pm 1$, with transition probabilities per unit time

$$W(1|-1) = \alpha , \qquad W(-1|1) = \beta .$$

It is convenient to write its master equation in vector notation

$$\frac{\partial}{\partial t}\begin{pmatrix} P(1) \\ P(-1) \end{pmatrix} = \begin{pmatrix} -\beta & \alpha \\ \beta & -\alpha \end{pmatrix} \begin{pmatrix} P(1) \\ P(-1) \end{pmatrix} .$$

The solution is

$$\begin{pmatrix} P(1,t|1,t_1) & P(1,t|-1,t_1) \\ P(-1,t|1,t_1) & P(-1,t|-1,t_1) \end{pmatrix} = \frac{1}{\alpha+\beta}\begin{pmatrix} \alpha+\beta\varepsilon & \alpha(1-\varepsilon) \\ \beta(1-\varepsilon) & \beta+\alpha\varepsilon \end{pmatrix} ,$$

where $\varepsilon = \exp[-(\alpha+\beta)t]$. For $t \to \infty$ it tends to the stationary distribution $\{\alpha/(\alpha+\beta), \beta/(\alpha+\beta)\}$. Its autocorrelation function is $4\alpha\beta\varepsilon/(\alpha+\beta)^2$. This two-valued or dichotomous Markov process describes for instance a two-level atom in a radiation field whose temperature is given by $\alpha/\beta = \exp[-(E_2-E_1)/kT]$. It also often serves for construction of simple models.

For a continuous-valued Markov process the master equation takes the form

$$\frac{\partial P(y,t)}{\partial t} = \int \{W(y|y')P(y',t) - W(y'|y)P(y,t)\}dy' \quad . \tag{7}$$

To find the transition probability $P(y,t|y_1,t_1)$ of the process one has to solve this equation for $t > t_1$ with initial condition $P(y,t_1) = \delta(y-y_1)$. If the process is stationary its one-point distribution function $P_1(y)$ is the time-independent or stationary solution of the same master equation:

$$W(y|y')\ P_1(y')dy' = P_1(y) \int W(y'|y)dy' \quad .$$

As an example of a soluble master equation consider a quantized harmonic oscillator with states $n = 0,1,2,\ldots$ in a radiation field with temperature T. The interaction is determined by the electric dipole, whose matrix element between the states n and n+1 is proportional to (n+a). Hence

$$W(n|n+1) = \alpha(n+1) , \qquad W(n+1|n) = \beta(n+1) .$$

The master equation for the probability of the oscillator to be in state n obeys

$$\partial_t\ P(n) = \alpha(n+1)P(n+1) + \beta n\ P(n-1) - (\alpha n+\beta n+\beta)P(n) . \tag{8}$$

This equation can be solved by means of the characteristic function, but in this case the 'probability generating function' $F(z,t) = \sum_0^\infty z^n\ P(n,t)$ is simpler. If one multiplies (8) with z^n and sums over all n one obtains a first-order linear partial differential equation for F(t,z), which can be solved by the method of characteristics. We leave it as an exercise to the reader.

However, the time-independent solution $P^s(n)$ of (8) can be obtained directly. It obeys

$$\alpha(n+1)P^s(n+1) - \beta(n+1)P^s(n) = \alpha n\ P^s(n) - \beta n\ P^s(n-1) .$$

The two members are the same quantity with the value of n shifted by 1. Hence this quantity is independent of n. As it clearly vanishes for n = 0, it must vanish for all n. This reduces the two-step difference equation to a one-step equation, which is easily solved to give

$$P^s(n) = C(\beta/\alpha)^n = \frac{\alpha}{\alpha-\beta}\left(\frac{\beta}{\alpha}\right)^n$$

This is, indeed, the thermal distribution $P^e(n)$, provided that $\beta/\alpha = \exp[-h\nu/kT]$. Thus we have established a relation between the absorption rate β and the emission rate α, thanks to the fact that the stationary solution $P^s(n)$ must be equal to the thermodynamic equilibrium $P^e(n)$, which is known. This is the Einstein relation, a form of the fluctuation-dissipation theorem.

V. FOKKER-PLANCK AND LANGEVIN

The master equation has the general form

$$\partial_t P = W\,P \quad ,$$

where W is the matrix or, in the case of a variable y with continuous range, a linear integral operator, see (7). Suppose the jumps from y' to y are very small, i.e., $W(y|y')$ is sharply peaked around $y-y' = 0$. Then the operator W can be approximated by a differential operator of the form [9]

$$\frac{\partial P(y,t)}{\partial t} = -\frac{\partial}{\partial y}[F(y)P] + \tfrac{1}{2}\frac{\partial^2}{\partial y^2}[G(y)P] \quad , \tag{9}$$

where F and G are the first two 'jump moments'

$$F(y) = \int (y'-y)W(y'|y)dy' \quad , \qquad G(y) = \int (y'-y)^2\, W(y'|y)dy' \ .$$

(9) is called the <u>Fokker-Planck</u> equation.

A brownian particle is a heavy particle surrounded by a fluid of much lighter molecules. Einstein [10] derived for the <u>position</u> x of such a particle an equation of type (9),

$$\frac{\partial P(x,t)}{\partial t} = -\mu E\frac{\partial P}{\partial x} + D\frac{\partial^2 P}{\partial x^2} \ . \tag{10}$$

Here D is the diffusion constant, E a constant external force, and μ a reciprocal friction coefficient called mobility. Also by taking for E the gravitational force and confronting the resulting stationary solution with the known barometric formula he found

$$D = \mu kT \ . \tag{11}$$

Rayleigh [11] had already described the <u>velocity</u> distribution (without field) by

$$\frac{\partial P(v,t)}{\partial t} = \gamma\left(\frac{\partial}{\partial v} vP + \frac{kT}{M}\frac{\partial^2 P}{\partial v^2}\right) , \qquad (12)$$

where γ is the friction coefficient. Kramers wrote for the joint distribution a bivariate equation of type (9),

$$\frac{\partial P(x,v,t)}{\partial t} = -v\frac{\partial P}{\partial x} - \frac{E}{M}\frac{\partial P}{\partial v} + \gamma\left(\frac{\partial}{\partial v} vP + \frac{kT}{M}\frac{\partial^2 P}{\partial v^2}\right) .$$

Moreover he proved that for large γ it reduces to (10).[12]

The general FP equation (9) has been widely used to describe noisy systems, but some caution is needed. Following Planck we regarded it as an approximation to the actual master equation (7). Kolmogorov [13] has proved that the equation (9) as it stands does indeed correspond to a well-defined Markov process. That is no longer true if one tries to improve on Planck's approximation by adding more terms involving higher derivatives [14]. Nonetheless one must not conclude that (9) is anything but an approximation of the actual physical process. This erroneous conclusion is aided and abetted by the fact that mathematicians call those processes whose master equation is a FP equation 'continuous Markov processes' because their sample paths are almost all continuous. This has sometimes led to the wrong idea that all processes with a continuous range of y-values must obey a FP equation. On the contrary, the 'continuous' processes fluctuate through infinitely many infinitely small jumps, but infinitely small jumps do not occur in physical reality.

VI. THE LANGEVIN APPROACH

An alternative approach to Brownian motion is due to Langevin [15]. He wrote the equation of motion of the particle,

$$\dot{v} = -\gamma v + L(t) \qquad (M=1) .$$

The first term is the macroscopic damping while the Langevin force $L(t)$ describes the effect of the irregular kicks of the fluid molecules that is left after their average effect has been taken into account by the damping term. To express the irregularity of the kicks one assumes that $L(t)$ is a random process with the properties

$$\langle L(t)\rangle = 0 , \qquad \langle L(t_1)L(t_2)\rangle = C\Delta(t_1-t_2) , \qquad (13)$$

where $\Delta(\tau)$ is a sharply peaked function with a width of the order of the duration of an individual collision. It is normalized by $\int \Delta(\tau)d\tau = 1$, while the constant C measures the strength of the collisions.

This Langevin equation can be solved explicitly [5]

$$v(t) = v(0)e^{-\gamma t} + e^{-\gamma t}\int_0^t e^{\gamma t'} L(t')dt' .$$

It then follows that

$$\langle v(t)^2\rangle = v(0)^2\, e^{-2\gamma t} + e^{-2\gamma t} \int_0^t e^{\gamma(t'+t'')} \langle L(t')L(t'')\rangle dt'dt'' .$$

$$= v(0)^2\, e^{-2\gamma t} + (C/2\gamma)(1-e^{-2\gamma t}) .$$

When $t \gg 1/2\gamma$ the effect of the initial value has disappeared and $\langle v^2\rangle = C/2\gamma$. On the other hand, we know $\langle v^2\rangle = kT/M$, where T is the temperature of the fluid, so that we have found $C = 2\gamma kT/M$. Thus the strength C of the fluctuations is related to the damping constant γ, which is the fluctuation-dissipation theorem. It can be understood as follows. In equilibrium the random collisions tend to cause the mean square velocity to grow without bounds, but the damping keeps the velocity down; the net effect of these competing tendencies is the mean square velocity that we know from equilibrium statistical mechanics. This Langevin approach has been applied with great success to other linear systems, such as pendulums, galvanometers, and electrical circuits.

Of course, later our sharply peaked $\Delta(\tau)$ was replaced with the infinitely sharply peaked $\delta(\tau)$, which did not change the results but streamlined the calculations. Mathematically it has the effect that the stochastic process v(t), which previously hid a memory due to the finite width of $\Delta(\tau)$, becomes an actual Markov process. It can be proved that it is a 'continuous' Markov process, i.e., that its transition probability obeys the Fokker-Planck equation

$$\frac{\partial P(v,t)}{\partial t} = \gamma \frac{\partial}{\partial v} vP + \tfrac{1}{2}C \frac{\partial^2 P}{\partial v^2} .$$

This FP equation is therefore mathematically equivalent with the above Langevin equation.

That statement is not strictly true, because so far only the first and second moment of L(t) have been prescribed. One needs to add the stipulation that L(t) is Gaussian, so that its higher moments als determined, comp. (1). That makes L(t) Gaussian white noise, see (2). It was mentioned that this is not a proper stochastic process, but its integral is the well-defined Wiener process,

$$\int_0^t L(t')dt' = W(t) .$$

Mathematicians often write dW(t) rather than L(t)dt to soothe their consciences.

The trouble starts if one tries to apply the same procedure to nonlinear systems. Take an electrical circuit consisting of a condenser and a non-ohmic resistance, for instance a diode. The macroscopic equation for the potential difference x on the condenser has the form $\dot{x} = f(x)$. As the resistance is dissipative it is a source of fluctuations, so one adds a Langevin term,

$$\dot{x} = f(x) + L(t) . \tag{14}$$

However, if one now takes the average one does not retrieve the macroscopic equation that served as starting point:

$$\partial_t \langle x \rangle = \langle f(x) \rangle \neq f(\langle x \rangle) \ .$$

Hence there is no justification for using in (14) the same f as in the macroscopic equation. In fact, in the case of the diode it leads to a nonzero equilibrium potential, which violates the second law of thermodynamics (Brillouin paradox [16]). Conclusion: The Langevin approach does not work for nonlinear systems. One cannot describe the fluctuations simply by adding a Langevin term to the macroscopic equation.

The insistence on a Langevin formulation has given rise to some weird consequences. In many applications, e.g. the laser, there are physical reasons for expecting the strength of the noise to depend on the variable x; accordingly one writes

$$\dot{x} = f(x) + g(x)L(t) \ . \tag{15}$$

However, this equation has no meaning: the fluctuations in L(t) cause simultaneous fluctuations in x, so that one does not know which value of x is to be inserted in g(x). This is a consequence of the singular nature of the process L(t). The obvious physical answer is to restore the original Δ-correlation (13). That amounts to assigning a certain interpretation to the otherwise meaningless equation (15), called the Stratonovich interpretation. If one subsequently goes to the limit of infinitely narrow Δ one obtains in this way a Markov process that is alternatively described by the FP equation.

$$\frac{\partial P(x,t)}{\partial t} = -\frac{\partial}{\partial x} f(x)P + \tfrac{1}{2}\frac{\partial}{\partial x} g(x) \frac{\partial}{\partial x} g(x)P$$

$$= -\frac{\partial}{\partial x} (f+\tfrac{1}{2}g'g)P + \tfrac{1}{2}\frac{\partial^2}{\partial x^2} g^2P \ .$$

Unfortunately an alternative convention for assigning an interpretation to (15) was cooked up by Itô. It gives rise to another Markov process described by

$$\frac{\partial P(x,t)}{\partial t} = -\frac{\partial}{\partial x} fP + \tfrac{1}{2}\frac{\partial^2}{\partial x^2} g^2P \ .$$

His scheme, however, applies only to strictly delta-correlated L(t), which is unphysical. In spite of the enthusiasm of the mathematicians for the Itô interpretation it is safer for a physicist to ignore it [17].

These problems arose from postulating a Langevin equation to describe fluctuations without going into their physical origin. A master equation gives a more fundamental description, but is usually hard to solve. The following example demonstrates a method for obtaining sensible approximate solutions.

VII. EXPANSION OF THE MASTER EQUATION

The master equation (8) could be solved because its coefficients are linear in n. In contrast, consider the case of a semiconductor of volume Ω with N donors. We want to know the probability P(n,t) that there are n charge carriers in the conduction band. (It is assumed that they survive long enough to traverse the example, so that their location need not be taken into account.) The probability per unit time for an excitation is proportional to the number N-n of occupied donors. The probability for de-excitation is proportional to the number of carriers n, and to the density n/Ω of empty donors.

$$W(n+1|n) = \beta(N-n) \qquad W(n-1|n) = \alpha n^2/\Omega \ .$$

The master equation for $P(n,t)$ is therefore

$$\partial_t P(n,t) = \beta(N-n+1)P(n-1,t) - \beta(N-n)P(n,t) + \frac{\alpha}{\Omega}(n+1)^2 P(n+1,t) - \frac{\alpha}{\Omega} n^2 P(n,t) \ .$$

As it cannot be solved explicitly we shall develop an expansion method for large systems, that is, for

$$\Omega \to \infty \ , \qquad N \to \infty \ , \qquad N/\Omega = \nu = \text{constant} \ .$$

One expects that in this limit the values of n crowd around some average value of order Ω. In the time-dependent solutions this average will depend on time and we denote it by $\Omega\psi(t)$, where $\psi(t)$ is one of the things to be determined. The width of P around this average is expected to be of order $\Omega^{\frac{1}{2}}$. Hence we set

$$n = \Omega\psi(t) + \Omega^{\frac{1}{2}}\xi \ , \tag{17}$$

where ξ is the new variable replacing n. Accordingly

$$P(n,t) = P\left(\Omega\psi(t)+\Omega^{\frac{1}{2}}\xi,t\right) = \Pi(\xi,t) \ .$$

We transform the master equation (16) into an equation for Π. Note that

$$P(n\pm1,t) = \left\{1\pm\Omega^{-\frac{1}{2}}\frac{\partial}{\partial\xi} + \tfrac{1}{2}\Omega^{-1}\frac{\partial^2}{\partial\xi^2} \pm \ldots\right\}\Pi(\xi,t) \ .$$

Then (16) becomes

$$\frac{\partial\Pi}{\partial t} - \Omega^{\frac{1}{2}}\frac{d\psi}{dt}\frac{\partial\Pi}{\partial\xi} = \beta\Omega\{-\Omega^{-\frac{1}{2}}\frac{\partial}{\partial\xi} + \tfrac{1}{2}\Omega^{-1}\frac{\partial^2}{\partial\xi^2} - \ldots\}\left(\nu-\psi-\Omega^{-\frac{1}{2}}\xi\right)\Pi + \alpha\Omega\{\Omega^{-\frac{1}{2}}\frac{\partial}{\partial\xi} + \tfrac{1}{2}\Omega^{-1}\frac{\partial^2}{\partial\xi^2} + \ldots\}\left(\psi+\Omega^{-\frac{1}{2}}\xi\right)^2\Pi \ .$$

The terms of order $\Omega^{\frac{1}{2}}$ cancel if one now chooses for $\psi(t)$ a solution of

$$\partial_t\psi = \beta(\nu-\psi) - \alpha\psi^2 \ . \tag{18}$$

This determines the time dependence of the macroscopic part of n. In fact, it is precisely the macroscopic rate equation that one would expect.

In terms of order Ω^0 are

$$\frac{\partial\Pi}{\partial t} = (\beta+2\alpha\psi)\frac{\partial}{\partial\xi}\xi\Pi + \tfrac{1}{2}[\beta(\nu-\psi)+\alpha\psi^2]\frac{\partial^2\Pi}{\partial\xi^2} \ . \tag{19}$$

This is a Fokker-Planck equation, but the coefficients depend on time through $\psi(t)$. Yet it can be solved explicitly (for instance by means of the characteristic function of Π) thanks to the fact that the first coefficient is linear in ξ and the second does not depend on it. The result is a Gaussian [18], but we do not need its explicit form. It is sufficient to find equations for the first two moments by multiplying the equation by ξ and ξ^2 respectively and integrating.

$$\partial_t\langle\xi\rangle = -(\beta+2\alpha\psi)\langle\xi\rangle$$

$$\partial_t\langle\xi^2\rangle = -2(\beta+2\alpha\psi)\langle\xi^2\rangle + \beta(\nu-\psi) + \alpha\psi^2 . \qquad (20)$$

This solves our problem. To find the solution of (16) with initial condition $P(n,0) = \delta_{n,n_o}$ one first solves (18) with initial value $\Omega\psi(0) = n_o$. Then one solves (19) with initial distribution $\Pi(\xi,0) = \delta(\xi)$. This implies $\langle\xi\rangle_t = 0$ at all $t > 0$, while $\langle\xi^2\rangle_t$ is to be found from (20) with initial value $\langle\xi^2\rangle_o = 0$. The result is that $P(n,t)$ is a Gaussian with mean $\Omega\psi(t)$ and variance $\Omega\langle\xi^2\rangle_t$:

$$P(n,t) = [2\pi\Omega\langle\xi^2\rangle_t]^{-\frac{1}{2}} \exp\left[-\frac{\{n-\Omega\psi(t)\}^2}{2\Omega\langle\xi^2\rangle_t}\right] .$$

The stationary solution can be obtained from this by taking $t = \infty$.

<u>Conclusion</u>. The Ω-expansion of the master equation yields in lowest approximation the macroscopic rate equation. The next approximation gives the fluctuations around the macroscopic value in Gaussian approximation. Higher orders in Ω^{-1} add non-Gaussian corrections to the fluctuations.

This expansion can be applied to numerous cases, not only excitations in atoms and semiconductors, but also magnetic systems, electric devices, radio-active decay, chemical reactions, and populations and epidemics. There is one essential restriction, however; it must be true that $\langle\xi^2\rangle_t$ does not grow beyond bounds. If for instance $\langle\xi^2\rangle_t$ were to grow exponentially in time, then the second term of (17), although formally of order $\Omega^{-\frac{1}{2}}$ with respect to the macroscopic term, ceases to be small after a time of order $\log\Omega$. Hence one must require that in the right-hand member of (20) the coefficient of $\langle\xi^2\rangle$ is negative, in this case that $\beta+2\alpha\psi > 0$. This is precisely the condition for the solution $\psi(t)$ of (18) to be stable. <u>The stability of the macroscopic equation keeps the fluctuations in check</u>.

This restriction is violated at points of instability of the macroscopic equation, in particular at critical points [19]. However, we shall here take the case that the macroscopic equation turns out to be $\psi = 0$. Then the first term in (20) vanishes, and $\langle\xi^2\rangle_t$ in the present approximation grows linearly with t. Master equations for which this happens are called of diffusion type. It turns out that in this case the

Ω-expansion leads to a FP equation (9) as its lowest approximation. We shall again demonstrate this on an example.

VIII. HOPPING IN AN INHOMOGENEOUS MEDIUM

The following model is somewhat elaborate and its choice should therefore be justified. It is motivated by three considerations.

(i) It demonstrates a master equation of diffusive type and how it leads to a FP equation of the form (9).

(ii) It illustrates how one really has to start from a detailed physical model in order to establish the correct equations for the fluctuations.

(iii) It sheds some light on the controversy concerning the form of the diffusion equation in inhomogeneous media.

An electron hops in a one-dimensional medium in the presence of an externally applied potential $U(x)$. The temperature is constant. There are traps, placed at random points independent from each other, with density $\Omega\sigma(x)$. The factor Ω separates the microscopic distance between traps from the macroscopic scale on which $\sigma(x)$ varies. Starting at an arbitrary point x' the probability to meet a first trap at a point between x and $x+dx$ is

$$dx.\Omega\sigma(x) \exp[-\Omega|\int_{x'}^{x} \sigma(x'')dx''|] .$$

Each trap is a potential pit of depth Φ. The equilibrium distribution will be proportional to the density of traps, times the probability that a trap is occupied:

$$P^e(x) = c\ \sigma(x) \exp \beta[\Phi-U(x)] . \tag{21}$$

In hopping models the time spent traveling between traps is neglected.

The probability per unit time for a trapped electron to pick up the energy Φ needed to escape is $ce^{-\beta\Phi}$. This suffices to roll down the slope of $U(x)$ into the next trap, but not to move up that slope. Suppose $U'(x) > 0$; then the probability per unit time for a jump from a trap at x' to some $x < x'$ is, apart from a constant factor

$$\Omega\sigma(x) \exp[-\Omega \int_{x}^{x'} \sigma(x'')dx''] \exp[-\beta\Phi] . \qquad (x < x') .$$

However, in order to hop to a trap at $x > x'$ the electron has to surmount the additional energy $U(x) - U(x')$; the probability is

$$c\Omega\sigma(x) \exp[-\Omega \int_{x'}^{x} \sigma(x'')dx''] \exp[-\beta\{\Phi+U(x)-U(x')\}] \qquad (x > x') .$$

These equations establish the transition probability $W(x|x')$.

The master equation is

$$\partial_t P(x,t) = \int_x^{\infty} W(x|x')P(x',t)dx' - \int_{-\infty}^{x} W(x'|x)dx' \, . \, P(x,t)$$

$$+ \int_{-\infty}^{x} W(x|x')P(x',t)dx' - \int_x^{\infty} W(x'|x)dx' \, . \, P(x,t) \, . \qquad (22)$$

We consider its Ω-expansion. It is clear that $W(x|x')$ vanishes unless $|x-x'|$ is small of order Ω^{-1}. Hence we set $x' = x + r/\Omega$ and $x' = x - r/\Omega$ respectively. Moreover, for brevity I omit some constant factors. The first line of (22) becomes

$$\int_0^{\infty} dr\, \sigma(x) \exp[-\Omega \int_x^{x+r/\Omega} \sigma(x'')dx''] \, P\left(x + \frac{r}{\Omega}, t\right)$$

$$- \int_0^{\infty} dr\, \sigma\left(x - \frac{r}{\Omega}\right) \exp[-\Omega \int_{x-r/\Omega}^{x} \sigma(x'')dx''] \, P(x,t) \, .$$

These two lines differ only be a shift r/Ω of x. Hence to order Ω^{-2} they combine to give

$$\frac{1}{\Omega}\frac{\partial}{\partial x}\{ \int_0^{\infty} r\, dr\, \sigma\left(x - \frac{r}{\Omega}\right) \exp[-\Omega \int_{x-r/\Omega}^{x} \sigma(x'')dx''] \, P(x,t)\}$$

$$+ \frac{1}{2\Omega^2}\frac{\partial^2}{\partial x^2}\{ \int_0^{\infty} r^2\, dr\, \sigma(x)\, e^{-r\sigma(x)} \, P(x,t)\} \, .$$

The second line of (22) gives in the same way

$$- \frac{1}{\Omega}\frac{\partial}{\partial x}\{ \int_0^{\infty} r\, dr\, \sigma\left(x + \frac{r}{\Omega}\right) \exp[-\Omega \int_x^{x+r/\Omega} \sigma(x'')dx''] .$$

$$. \exp[-\beta\{U(x+r/\Omega) - U(x)\}] \, P(x,t)\}$$

$$+ \frac{1}{2\Omega^2}\frac{\partial^2}{\partial x^2}\{ \int_0^{\infty} r^2\, dr\, \sigma(x)\, e^{-r\sigma(x)} \, P(x,t)\} \, .$$

On adding all this up one sees that the terms of order Ω^{-1} cancel, as expected. To order Ω^{-2} one obtains

$$\frac{\partial P(x,t)}{\partial t} = \frac{2}{\Omega^2}\left[\frac{\partial}{\partial x}\left(\frac{\beta}{\sigma^2} U' + \frac{\sigma^1}{\sigma^3}\right)P + \frac{\partial^2}{\partial x^2} \frac{1}{\sigma^2} P\right].$$

Restoring the omitted factors and rearranging some terms finally gives

$$\frac{\partial P(x,t)}{\partial t} = \frac{2c}{\Omega^2}\left[\frac{\partial}{\partial x}\{\frac{\beta}{\sigma^2} U' e^{-\beta\Phi} P\} + \frac{\partial}{\partial x}\frac{1}{\sigma}\frac{\partial}{\partial x}\frac{1}{\sigma} e^{-\beta\Phi} P\right] . \qquad (23)$$

This result calls for a few comments.

(i) There is no term of order Ω^{-1} and therefore no macroscopic equation as in (18). It appears that in this case the terms of order Ω^2 have the form of the FP equation (9) rather than (19). Thus we have deduced (9) as the first term in an approximation scheme instead of postulating it. This conclusion holds for all master equations whose zeroth order term (18) vanishes.

(ii) One recognizes the mobility and the diffusion coefficient:

$$\mu = \frac{2c}{\Omega^2}\frac{\beta}{\sigma^2}\, e^{-\beta\Phi} , \qquad D = \frac{2c}{\Omega^2}\frac{1}{\sigma^2}\, e^{-\beta\Phi}$$

and sees that the Einstein relation $D = kT\mu$ is satisfied. However, in the last term $D(x)$ does not appear as such, but its two factors σ^{-1} are separated by a differentiation. This could not have been guessed without starting from a detailed model.

(iii) We performed the calculation for constant temperature, but it can easily be extended for β depending on x. It is natural to assume that the escape probability from a trap is determined by the temperature $\beta(x)$ at the position x of that trap. A similar derivation then leads again to (23), where β now depends on x.

(iv) The stationary solution is

$$P^s(x) = \text{const.}\ \sigma(x)\ \exp[\beta(x)\Phi - \int \beta(x)U'(x)dx] .$$

For constant temperature it reduces to the expected thermal equilibrium distribution (21). The present form for non-constant temperature gives rise to some interesting effects [20].

REFERENCES

The main source of these lectures is of course
N.G. van Kampen, Stochastic Processes in Physics and Chemistry (North-Holland, Amsterdam 1981).

For a more conventional approach see
C.W. Gardiner, Handbook of Stochastic Methods (Springer, Berlin 1983).

More mathematical, but still readable is
W. Feller, An Introduction to Probability Theory and its Applications, I and II (Wiley, New York 1957, 1966).

The classic papers of Uhlenbeck and Ornstein, of Chandrasekhar, Doob, and others are reprinted in
N. Wax ed., Selected Papers on Noise and Stochastic Processes (Dover, New York 1954).

Another classic, but less relevant for these lectures, is
R. L. Stratonovich, Topics in the Theory of Random Noise. I and II (Gordon and Breach, New York 1963, 1967).

Additional references quoted in the text.

1. E. W. Montroll and B. J. West, in Fluctuation Phenomena (E. W. Montroll and J. L. Lebowitz eds., North-Holland, Amsterdam 1979).
2. M. L. Mehta, Random Matrices and the Statistical Theory of Energy Levels (Acad. Press, New York 1967); R. J. Elliott, J. A. Krumhansl, and P. L. Leath, Rev. Mod. Phys. 46, 465 (1974).
3. S. Alexander, J. Bernasconi, W. R. Schneider, and R. Orbach, Rev. Mod. Phys. 53, 175 (1981).
4. M. B. Weissman, in Proc. 6th Intern. Conf. on Noise in Physical Systems (P. H. E. Meijer et al. eds., NBS Washington, D.C. 1981); F. N. Hooge, T. G. M. Kleinpenning, and L. K. J. Vandamme, Repts. Prog. Phys. 44, 479 (1981).
5. G. E. Uhlenbeck and L. S. Ornstein, Phys. Rev. 36, 823 (1930).
6. G. Czycholl, Phys. Repts. 143, 277 (1986).
7. J. L. Doob, Annals Math. 43, 351 (1942).
8. A. Ramakrishnan, in Encyclopedia of Physics 3/2 (S. Flügge ed., Springer. Berlin 1959), Sec. 33; Stratonovich, I. Ch. 6; Van Kampen, Ch. 2.
9. M. Planck, Sitzber. Preuss. Akad. Wissens. (1917), p. 324.
10. A. Einstein, Ann. Phys. [4] 17, 549 (1905); 19, 371 (1906).
11. Lord Rayleigh, Phil. Mag. 32, 424 (1891).
12. H. A. Kramers, Physica 7, 284 (1940); U. M. Titulaer, Physica 91A, 321 (1978); 100A, 234, 251 (1980).
13. A. Kolmogorov, Mathem. Annalen 104, 415 (1931).
14. R. F. Pawula, Phys. Rev. 162, 186 (1967).
15. P. Langevin, Comptes Rendus (Paris) 146, 530 (1908).
16. L. Brillouin, Phys. Rep. 78, 627 (1950).
17. R. E. Mortensen, J. Stat. Phys. 1, 271 (1969); N. G. van Kampen, J. Stat. Phys. 24, 175 (1981).
18. S. Chandrasekhar, Rev. Mod. Phys. 15, 1 (1943).
19. R. Kubo, K. Matsuo, and K. Kitahara, J. Stat. Phys. 9, 51 (1973).
20. R. Landauer, Phys. Rev. A12, 636 (1975); Helv. Phys. Acta 56, 847 (1983); N. G. van Kampen, IBM J. Res. Dev. 32, 107 (1988).

DYNAMICAL PROCESSES IN DISORDERED SYSTEMS

R. Silbey*

Department of Chemistry
Massachusetts Institute of Technology
Cambridge, MA 02139

ABSTRACT

We discuss the effect of disorder on dynamical processes in solids. In particular, we treat donor (to acceptor) decay, excitation diffusion, trapping and optical line shapes in disordered systems. The modelling of disorder with fractals is discussed.

I. INTRODUCTION

The presence of disorder in solids has a substantial effect on the energy levels (both electronic and vibrational) and on the dynamics of particles and excitations travelling through the solid. An enormous amount of effort has gone into studying these effects, both experimentally and theoretically. In this article, I will necessarily be forced to restrict my comments to a small subset of the possible interesting topics. I will in fact discuss issues which arise in the optical spectroscopy of such solids; however these will exhibit the interesting problems which occur when disorder is present in a solid. The plan of the article is to begin with the dynamics of the transfer of excitation from a single excited donor to a collection of randomly arranged acceptors. The disorder gives rise to a dramatic change in the time dependence which is highly dependent on density (of acceptors) and on the dimensionality of the space. Next, we will examine the transport of an excitation among similar molecules randomly arranged on a lattice. We will discuss the time dependence and the diffusion coefficient of this transport and their dependence on density and dimensionality. The trapping of excitations during this diffusion process will also be discussed. We will then go on to examine the effect of disorder on the optical line shapes of molecules in solids. Here we will discuss the ways of dealing with the weak disorder case and the strong disorder case.

After these detailed studies, we will go on to discuss the modelling of disorder fractals. After an introduction to this concept, we will discuss the effects of fractal structure on the dynamic problems we discussed above.

---

*Supported in part by the NSF

II. FÖRSTER TRANSFER [1-10]

The simplest process we can imagine which exhibits the effects of disorder is that of excitation transfer from an excited molecule to a large number of acceptor molecules arranged in space in some way. The time dependent probability of finding the excitation on the αth donor molecule is

$$\frac{dp_\alpha(t)}{dt} = -\sum_i W_{i\alpha} p_\alpha(t) - \frac{1}{\tau} p_\alpha(t) \tag{1}$$

where $W_{i\alpha}$ is the rate of energy transfer from the donor to the ith acceptor at position r_i and τ is the lifetime of the excitation in the absence of acceptors. We take the donor position to be at r_α and assume that the donors are very far apart, so they do not interact with one another. In addition, we assume that $W_{i\alpha}$ is a function of $\underset{\sim}{r}_i-\underset{\sim}{r}_\alpha$. In fact, we normally assume that it is a function of $|\underset{\sim}{r}_i-\underset{\sim}{r}_\alpha|$ only, although although some concern has been raised about this. The solution to Eq. (1) is immediate:

$$p_\alpha(t) = e^{-(\sum_i W_{i\alpha})t} e^{-t/\tau} = [\prod_i e^{-W_{i\alpha}t}]e^{-t/\tau} \tag{2}$$

The experimenter measures the fluorescence decay of the donor molecules and infers the "extra" decay due to the donor-acceptor transfer. If the acceptors were arranged in exactly the same manner around each donor, then the problem is done and the donor-acceptor transfer has changed the exponential decay rate from $1/\tau$ to $1/\tau + \sum W_{i\alpha}$. However, the assumption of order is incorrect: each donor sees a different arrangement of acceptors, so the experimentally determined variable is the average of $p_\alpha(t)$ over all possible arrangements:

$$\langle p(t)\rangle = \langle \prod_i e^{-W_{i\alpha}t}\rangle e^{-t/\tau} \equiv \Phi(t)\, e^{-t/\tau} \tag{3}$$

where we have assumed that the lifetime in the absence of acceptors is independent of donor position in the solid. In order to carry out this configurational average, we can assume that the acceptors are arranged on a lattice and that the probability of any lattice site being occupied by an acceptor is p and that it is unoccupied is (1-p). Then, we have for uncorrelated sites,

$$\Phi(t) = \prod_i [(1-p) + p\, e^{-W_{i\alpha}t}] \tag{4}$$

where the product is now over all lattice sites. We can rewrite this as

$$\Phi(t) = \exp[\sum_i \log\{1-p + pe^{-W_{i\alpha}t}\}] \tag{5a}$$

$$\simeq \exp -[p\sum_i \{1 - e^{-W_{i\alpha}t}\}] \tag{5b}$$

for small p. Finally we can convert the sum over lattice sites to an integral over the volume. Defining concentration of acceptors by

$$c = p(N/V) \tag{6}$$

we find

$$\Phi(t) = \exp[- c \int_{R_{min}}^{R_{max}} d\underset{\sim}{r} \{1 - e^{-tW(r)}\}] \qquad (7)$$

Here we have replaced $W_{i\alpha}$ by $W(|\underset{\sim}{r}_i - \underset{\sim}{r}| = W(r)$ and placed the donor molecule at the origin. If we now assume that W(r) has a multipolar form: $W(r) = \alpha/r^s$ (s = 6 is the Förster dipole-dipole transfer rate) and that the transfer takes place in a space of dimension d, we find (with spherical volume element $d\underset{\sim}{r}$ is replaced by $r^{d-1} A_d dr$ since the integrand is independent of angle)

$$\Phi(t) = \exp\{-cA_d \int_{R_{min}}^{R_{max}} dr\, r^{d-1}[1-e^{-\alpha t/r^s}]\} \qquad (8)$$

The upper limit, R_{max}, is the dimension of the region containing acceptors in which acceptors interact with the donor. The integral can be transformed into a Γ function for $\alpha t/R^s{}_{max} \ll 1$ (i.e. for times short compared to the inverse rate to the farthest donor) and $\alpha t/R^s{}_{min} \gg 1$ (times long compared to the inverse rate to the nearest neighbor molecule.)

$$\Phi(t) = \exp\{-A_d(\alpha t)^{d/s}\, \Gamma(1 - \frac{d}{s})\} \qquad (9)$$

So here we see our first example of a strong effect of disorder: the exponential decay law has been modified to a decay <u>nonlinear</u> in time. The usual case of d = 3, s = 6 yields a $t^{1/2}$ which has been known since Förster's pioneering work. Note that a crucial assumption in the above was to replace the actual distribution of acceptor molecules by an average distribution proportional to $c d\underset{\sim}{r}$.

The form of the decay of ⟨p(t)⟩ is an example of a"stretched exponential":

$$e^{-(t/t_o)^\beta} \quad . \qquad (10)$$

In the Förster case β = 1/2, but as we see from the above β = d/s is possible for s > d + 1.

If instead of a multipole transfer rate, an exponential $w(r) = W_o e^{-\gamma r}$ (exchange transfer) were chosen, then we find

$$\Phi(t) = \exp - \{A_d \frac{c}{\gamma^d} g_d(W_o t)\} \qquad (11)$$

where $g_d(Wt)$ are functions introduced by Inokuti and Hirayama [2] and discussed extensively by Blumen [7,8]. In the limit of long times (but still short compared to $W_o^{-1} e^{\gamma R_{max}}$) these authors find

$$g_d(x) \sim \ln^d(W_o t) \qquad (12)$$

This decay is slower than the stretched exponential form found for multipolar transfer. This is sometimes called an exponential log decay.

We should point out for completeness, that often the experimentally relevant parameter is the branching ratio, i.e. the ratio of the radiative rate to the total rate. From Eq. (3), we can write this as

$$\eta = \left\langle \frac{\tau^{-1}}{\tau^{-1} + \sum_i W_{i\alpha}} \right\rangle \tag{13}$$

which is just τ^{-1} multiplied by the Laplace transform of $\Phi(t)$ evaluated at τ^{-1}:

$$\eta = \tau^{-1}\tilde{\Phi}(u)\Big|_{u = \tau^{-1}} \tag{14}$$

where $\tilde{\Phi}(u)$ is

$$\tilde{\Phi}(u) = \int_0^\infty e^{-ut}\Phi(t)dt \tag{15}$$

As a final point in our present discussion of donor-acceptor transfer, we can ask what would happen if the acceptors were arranged in a non-isotropic manner. A simple example will suffice: suppose the acceptors are arranged in a cylinder of radius R_1 and length $R_2 << R_1$. Assume multipolar decay $W(r) = \alpha/r^s$, then for times such that $\alpha t/R_2^s << 1$, the decay is three-dimensional. However, for times such that $\alpha t/R_2^s >> 1$ (but $\alpha t/R_1^s << 1$) the decay will resemble the 2-dimensional rate. This is an example of the effect of restricted geometries.

III. EXCITATION OR PARTICLE TRANSPORT [11-23]

III.A. Formal Manipulations

Another important process which is strongly affected by disorder is the transport of excitations in a solid. We will only be concerned with incoherent transport, that transport which can be described by the master equation. The simplest physical situation is to assume that many donors are arranged at random throughout the solid. The excitation is placed on one donor initially and the probability of finding the excitation on any donor, say the jth, at time t is $p_j(t)$. The equation of motion for $p_n(t)$ is

$$\dot{p}_n(t) = \sum_i W_{nm}p_m(t) - \sum_i W_{mn}p_n(t) - p_n(t)\tau^{-1} \tag{16}$$

The first term on the right represents the rate of transfer to the nth donor from the other donors; the second term is the rate of transfer from the nth donor. (If we assume that the lifetime of the donors, τ, is independent of n, we can eliminate the last term by writing $p_n(t) = p_n(t)e^{-t/\tau}$. This is so straightforward we will merely drop the lifetime term altogether and, if needed, put it back later). For simplicity, we now take $W_{ji} = W_{ij}$.

For an ordered system, say a crystal, Eq. (8) can be solved by assuming periodic boundary conditions and writing

$$p(k,t) = \frac{1}{N}\sum_n e^{i\underset{\sim}{k}\cdot\underset{\sim}{R}_n} p_n(t) \quad . \tag{17}$$

Then

$$\dot{p}(k,t) = -[W(\underset{\sim}{0}) - W(\underset{\sim}{k})]p(k,t) \tag{18}$$

or

$$p(k,t) = \frac{1}{N} \exp\{[-W(\underset{\sim}{0}) + W(\underset{\sim}{k})]t\} \quad (19)$$

where we have assumed the excitation was originally at the origin and we have defined

$$W(\underset{\sim}{k}) \equiv \sum_j \varepsilon^{i\underset{\sim}{k}\cdot(\underset{\sim}{r}_j - r_i)} W_{ji} \quad (20)$$

Thus the probability of finding the excitation on donor n is

$$p_n(t) = \frac{1}{N} \sum_k \exp\{[-W(\underset{\sim}{0}) + W(\underset{\sim}{k})]t\}\, e^{ik\cdot R_n} \quad (21)$$

At very low short times,

$$p_o(t) = e^{-tW(\underset{\sim}{0})} \quad (22)$$

and at along times we find

$$p_o(t) \sim t^{-d/2} \quad (23)$$

(d again the dimension of the system). We can calculate the mean square displacement of the excitation (initially at the origin)

$$\langle R^2(t)\rangle = \sum_n |\underset{\sim}{R}_n - \underset{\sim}{R}_o|^2\, p_n(t) = 2d\, Dt \quad (24)$$

with

$$D = \frac{1}{2d} \sum_j W_{nm}\, |\underset{\sim}{R}_n - \underset{\sim}{R}_m|^2 \quad (25)$$

Thus for regular lattices, the transport of an excitation is diffusive at all times. Note however that the probability of being at the origin, $p_o(t)$, has a more complicated time dependence.

When disorder is present, the time dependence of the mean square displacement is more complex than Eq. (24). A number of different approaches have been taken for this problem. Here we will try to show the connections among some of these.

Consider a partially occupied lattice for which each site can be occupied by a molecule capable of carrying the excitation with probability p. Then the equation of motion for the probabilities of finding the excitation at one of the occupied sites is once again the master equation.

$$\dot{p}_n(t) = \sum_{m\neq n} W_{nm} p_m(t) - \sum_{m\neq n} W_{mn} p_n(t) \quad (26)$$

Note that probability is conserved so that $\Sigma \dot{p}_n(t) = 0$. An average over all possible starting points for the excitation for one configuration is the same as an average over all possible configurations of occupied sites on the lattice with the excitation starting at the origin. Eq. (26) can be formally solved by writing it in vector form

$$\dot{\underset{\sim}{p}}(t) = \underline{\underline{V}} p(t) \quad (27)$$

with

$$V_{nm} = (1-\delta_{nm})W_{nm} - \delta_{nm} \sum_{q \neq n} W_{qn} \tag{28}$$

Then

$$\underset{\sim}{p}(t) = (\exp t\underline{\underline{V}}) \cdot \underset{\sim}{p}(0) \tag{29}$$

and, after configurational averaging,

$$\langle p_n(t) \rangle = \sum_m \langle e^{t\underline{\underline{V}}} \rangle_{no} \tag{30}$$

where we have now assumed the excitation began on the n = 0 siute. By Laplace transforming we find

$$\langle \hat{p}_n(u) \rangle = \langle (u\underline{\underline{I}} - \underline{\underline{V}})^{-1} \rangle_{no} \tag{31}$$

Using a projection operator formalism [18], we can rewrite this formally as

$$\langle \hat{p}_n(u) \rangle = [u\underline{\underline{1}} - \underline{\underline{M}}(u)]^{-1}_{no} \tag{32}$$

where M(u) is a self-energy matrix. In the time domain, this can be rewritten

$$\langle \dot{p}_n(t) \rangle = \int_0^t d\tau \{ \sum_{\ell \neq n} M_{n\ell}(t-\tau) p_\ell(\tau) + M_{nn}(t-\tau) p_n(\tau) \} \tag{33}$$

Since probability is conserved

$$\sum_{\ell \neq n} M_{n\ell}(t-\tau) = - M_{nn}(t-\tau) \tag{34}$$

so that the final equation can be written in a form reminiscent of the original master equation, in which the time <u>independent</u> rates have been replaced by time <u>dependent</u> kernels: this is the generalized master equation:

$$\langle \dot{p}_n(t) \rangle = \int_0^t d\tau \{ \sum_{\ell \neq n} M_{n\ell}(t-\tau) \langle p_\ell(\tau) \rangle - \sum_{\ell \neq n} M_{\ell n}(t-\tau) \langle p_n(\tau) \rangle \} \tag{35}$$

or in Laplace transform

$$u\langle \hat{p}_n(t) \rangle - \langle p_n(0) \rangle = \sum_{\ell \neq n} \{ \hat{M}_{n\ell}(u) \langle \hat{p}_\ell(u) \rangle - \hat{M}_{\ell n}(u) \langle \hat{p}_n(u) \rangle \} \tag{36}$$

The effect of averaging over all possible configurations is to make the system translationally invariant, so that these equations can be solved by transforming to k space (just as we did for nondisordered crystals), and we find (taking $\langle p_n(0) \rangle = \delta_{no}$)

$$\langle p_n(u) \rangle = \frac{1}{N} \sum_k \frac{e^{i\underset{\sim}{k} \cdot \underset{\sim}{R}_n}}{u + [\hat{M}(0,u) - \hat{M}(k,u)]} \tag{37}$$

where

$$M(k,u) = \sum_n e^{ik\cdot(\underset{\sim}{R}_n-\underset{\sim}{R}_\ell)} M_{n\ell}(u) \tag{38}$$

We can define a generalized diffusion kernel D(k,u) by

$$\hat{D}(k,u) \equiv \hat{M}(0,u) - \hat{M}(k,u) \tag{39}$$

and we find that the long time diffusion constant is given by

$$D = \frac{1}{2d} \left[\frac{d^2}{dk^2} D(k,u)\right]_{k=0,\ u=0} \tag{40}$$

Thus the averaging procedure has produced an effective medium in which the excitation moves. In this medium, there is a memory term in the dynamical equation.

We can rearrange the GME into another equation, also often used to model disordered systems: the continuous time random walk equation [13] or CTRW. In this model, a probability density distribution for jumping from $\underset{\sim}{R}_n$ to $\underset{\sim}{R}_m$ at time t is given by $\Psi(\underset{\sim}{R}_n-\underset{\sim}{R}_m,t)$ and the probability for finding the excitation at $\underset{\sim}{R}_n$, given that it was at the origin initially is

$$\langle p_n(\underset{\sim}{R}_n,t)\rangle = \Phi(t)\delta_{n,o} + \sum_{m\neq n} \int_o^t d\tau\ \psi(\underset{\sim}{R}_m-\underset{\sim}{R}_n,\ t-\tau)\langle P(\underset{\sim}{R}_m,\tau)\rangle \tag{41}$$

where $\Phi(t)$ is the probability that the excitation has not left the initial site at time t. By Laplace and Fourier transforming this equation and comparing it to Eq. (37), we find that (in Laplace variable u)

$$\tilde{\psi}(R_m,u) = N^{-1}\sum_k \tilde{S}(k,u)\ e^{ik\cdot\underset{\sim}{R}_m} \tag{42}$$

where

$$\tilde{S}(k,u) = \frac{\tilde{M}(\underset{\sim}{k},u)}{u + \tilde{M}(\underset{\sim}{0},u)} \tag{43}$$

The probabilities are given by

$$\langle p_n(u)\rangle = \frac{1-S(\underset{\sim}{0},u)}{Nu} \sum_k \frac{e^{ik\cdot\underset{\sim}{R}_n}}{1-S(k,u)} \tag{44}$$

and so in the CTRW, the Laplace transform of the mean square displacement is given by

$$\langle R^2(u)\rangle = \frac{1}{u(1-S(\underset{\sim}{0},u))} \sum_n R_n^2\ \tilde{\psi}\ (R_n,u) \tag{45}$$

and the (time dependent) diffusion coefficient is given as

$$D(u) = \frac{(1/2d)}{1-S(\underset{\sim}{0},u)} \sum_n R_n^2\ \tilde{\psi}\ (\underset{\sim}{R}_n,u) \tag{46}$$

Thus in the CTRW approach, we must find $\tilde{\psi}(\underset{\sim}{R}_n,u)$ in order to obtain expressions for the diffusion coefficient and site probabilities. In the GME approach, we must find expressions for the kernels $\tilde{M}(\underset{\sim}{k},u)$. But, it should be emphasized that the information in one formal theory can be found from that in the other. Unfortunately, it is impossible in all but the simplest (and least interesting) models to find exact expressions for these quantities. We are then forced to approximate in order to go further. It turns out that for certain models one form of the theory is easier to use than the other form. We will illustrate this in the following examples.

III.B. Effective Medium Approximation [21,24,25]

As we said above, averaging over all configurations has produced an effective medium with memory. One direct approach to approximate this exact effective medium is to limit the range of the kernels $\tilde{M}_{n\ell}(u)$, see Eq. (36), to a small distance and solve for these in a self-consistent fashion. For example, suppose the original system is a partially filled lattice in which only nearest neighbor transfer rates are allowed in Eq. (26) (and for simplicity, we take a d-dimensional lattice with equal transfer rates to all z of the nearest neighbors):

$$W_{mn} = \begin{cases} W_o & m,n \text{ nearest neighbors} \\ 0 & \text{otherwise} \end{cases} \tag{47}$$

The transport in this system will exhibit a percolation transition: below some value p_c of fraction of sites occupied on the lattice, there will be a bound on the mean square displacement, so that the long time diffusion constant is zero. We can approximate this dynamics in the GME formalism by replacing the $\tilde{M}_{n\ell}(u)$ in the GME by a <u>nearest neighbor</u> effective rate, $\tilde{W}_E(u)$, so that

$$\tilde{M}_E(k,u) = \tilde{W}_E(u) \sum_{nn} e^{ik\cdot \underset{\sim}{R}_{nn}} \tag{48a}$$

$$\tilde{M}_E(\underset{\sim}{0},u) = z\tilde{W}_E(u) \tag{48b}$$

where the sum is over the nearest neighbors of any site. In order to determine $\tilde{W}_E(u)$, we must find a self-consistent equation for it. There are of course many possibilities for this. Following earlier workers [21,24,], we choose a method closely related to the coherent potential approximation (CPA) used to approximate the density of electronic states in disordered systems [37]. We take the effective medium (represented by Eq. (48)) as the zeroth order description and represent the exact equations (Eq. (29)) as a sum of this and "fluctuations" from this zeroth order description. It is easier to work in Laplace transforms for which the exact p(u) are given by (we have gone back to a vector description for succinctness)

$$\underset{\sim}{p}(u) = (u\,\underline{\underline{1}} + \underline{\underline{V}})^{-1}\ p(0) \equiv \underline{\underline{G}}(u)p(0) \tag{49a}$$

The matrix $\underline{\underline{G}}(u)$ is now written as

$$\underline{\underline{G}}(u) = [u\,\underline{\underline{1}} + \underline{\underline{M}}_E(u) + \delta\underline{\underline{V}}]^{-1} \tag{49b}$$

with $\delta\underline{\underline{V}} = \underline{\underline{V}} - \underline{\underline{M}}_E(u)$. We can expand $\underline{\underline{G}}(u)$ in terms of $\underline{\underline{G}}_M(u) = [u\,\underline{\underline{1}} + \underline{\underline{M}}_E(u)]^{-1}$ as

$$\underline{\underline{G}}(u) = \underline{\underline{G}}_M + \underline{\underline{G}}_M\,\underline{\underline{T}}\,G_M \tag{50}$$

where the T matrix is expressed in terms of bond t-matrices, t_{nm} [24]:

$$t_{nm} = \frac{W_E - W_{nm}}{1-2(W_E-W_{nm})((G_M)_{nn}-(G_M)_{nm})} \tag{51}$$

Since the effective medium is translationally invariant, we can easily find $(G_M)_{nm}$, etc. The self-consistency condition is to average Eq. (51) over the distribution of W_{nm}'s and set the average to zero. In this way, the "scattering" of the excitation due to fluctuations in the nearest neighbor transfer rates is minimized in the same way as the scattering of the wave function is in the CPA. Using the effective medium G_M, we find

$$\langle t_{nm}\rangle = \int \rho(W) \frac{(W_E-W)dW}{1-2(W_E-W)(1-uG_o)/zW_E} = 0 \tag{52}$$

Here $\rho(W)$ is the distribution function of nearest neighbor transfer rates W_{nm} (due to the occupation probabilities), $G_o = (G_M)_{nn}$ and z is the number of nearest neighbors in the lattice.

If we take a model in which the probability that the bond between n and m is present (i.e. $W = W_o$) is p and the probability that that bond is absent (i.e. $W = 0$) is $(1 - p)$, then we find from Eq. (52)

$$W_E(u) = W_o \frac{p - \frac{2}{z}(1-uG_o(u))}{1 - \frac{2}{z}(1-uG_o(u))} \tag{53}$$

This predicts a (long time) diffusion constant, D, which vanishes for $p < 2/z$ and is linear in $p - 2/z$ for $p > 2/z$. Thus the EMA predicts bond percolation with a critical concentration of bonds equal to $2/z$, independent of dimension. This result was found earlier in the percolation problem by Kirkpatrick [38]. This result is not exact for the bond problem, but is a reasonable approximation in almost all cases [21]. In addition to the diffusion constant, the approximate time dependence of $\langle R^2(t)\rangle$ and $D(t)$ can be found using the EMA [21]. The dependence of D and R^2 on time and $p - 2/z$ as p approaches $2/z$ (the percolating cluster) can also be found approximately by this method. Although the results are not exact, the capability of the EMA to get useful approximations in a simple way is excellent.

If instead of a percolation model, we take a specific form for $\rho(W)$ in Eq. (52), and if $uG_o(u)\to 0$ as $u\to 0$, then we can find the diffusion constant, D, from Eq. (52). In particular, in a <u>one-dimensional</u> system, we find $D \propto \langle W^{-1}\rangle^{-1}$ where the average is over the distribution $\rho(W)$.

III.C. CTRW Approximations

We return to the discussion of the CTRW equations and note that if we had an approximate form for the waiting time distribution functions $\psi(Rn,u)$, we could find a form for D from Eq. (46). A number of years ago, Scher and Lax [13] suggested a route to $\psi(R_n,t)$. They noted that the probability of remaining at the initial site, $\Phi(t)$ (see Eq. (41)) must be the form we have already discussed for the donor-acceptor problem. But in addition, by comparing eq. (41), (42), and (44), we can make the identification that

$$\dot{\Phi}(t) = - S(0,t) = -\sum_m \psi(R_m,t) \tag{54}$$

Since, from Eq. (4)

$$\Phi(t) = \prod_i [(1-p) + p\, e^{-W_{i\alpha} t}] \qquad (55)$$

we can identify

$$\psi(R_m,t) = pW(R_m)e^{-tW(R_m)}\{\prod_{n\neq m}[(1-p) + p\, e^{-W(R_n)t}]\} \qquad (56)$$

or in the limit of small concentrations

$$\psi(R,t) = pW(R)\, e^{-tW(R)}\Phi(t) \qquad (57)$$

and $\Phi(t)$ given by Eq. (7). We may now evaluate the mean square displacement using Eq. (45) and the diffusion coefficient using eq. (46). As an example, choose the multipolar transition rate given by $W(r) = \alpha/r^s$ the we saw above that (Eq. (9))

$$\Phi(t) = \exp - Bt^{d/s} \qquad (58)$$

The Laplace transform of the diffusion coefficient $\tilde{D}(u)$ is then given by Eq. (46) (and Eqs. (54) and (57)) as

$$\tilde{D}(u) = \frac{1/2d}{u\tilde{\Phi}(u)}\, L\{c\int d^dRR^2W(R)\, e^{-tW(R)}\Phi(t)\} \quad . \qquad (59)$$

Using the multipolar form for W(R),

$$\int d^dRR^2W(R)e^{-tW(R)} \sim t^{(d+2/s)-1}\, e^{-Bt^{d/s}} \qquad (60)$$

We can immediately find the short and long time dependences of D from these equations. At short times

$$D(t) \sim t^{(d+2)/s-1} \quad , \qquad (61)$$

and at long times D is a constant but depends on concentration of occupied sites as $c^{(s-2)/d}$.

These results agree with a density expansion by Haan and Zwanzig [11] of the Förster energy transfer problem: $W(R) = 1/\tau(R_0/R)^6$. By a clever scaling argument these authors showed that the mean square displacement of an excitation starting at the origin and moving through a randomly arranged array of donor molecules is given by

$$\langle R^2(t)\rangle = t^{1/3}\, F(ct^{1/2}) \qquad (62)$$

where F is a function of the variable $ct^{1/2}$ the first few terms of which are known. Thus at short times, $\langle R^2(t)\rangle$ varies as $ct^{5/6}$ and at long times, if $\langle R^2(t)\rangle$ varies as t then D varies with concentration as $c^{4/3}$. Their arguments are easily generalized to $W(R) = 1/\tau(R_0/R)^s$ in d-dimensional space leading to the results found using the CTRW. This shows explicitly that the dependence of the dynamics on concentration and time are inevitably linked.

A number of authors have applied the CTRW technique to problems of diffusion in random media, othere have used density expansions of the kernels appearing in the GME, and while these calculations agree with one another qualitatively, the correct value of D is still uncertain in any model with long range transfer. Burshtein [16] has recently reviewed the various theoretical models for energy transfer in disordered systems.

III.D. Anomalous Diffusion and Trapping

A standard approximation in the CTRW is to make the assumption that (see Eq. (42))

$$\psi(R_m,t) = \underline{P}(R_m)\psi(t) \tag{63}$$

Then

$$\tilde{S}(k,u) = \lambda(k)\,\tilde{\psi}(u) \tag{64}$$

where

$$\lambda(k) = \sum_m e^{ik\cdot R_m}\,\underline{P}(R_m) \quad . \tag{65}$$

Then (see Eq. (45))

$$\langle R^2(u)\rangle = \frac{1}{u[1-\tilde{\psi}(u)]}\,\tilde{\psi}(u)\left(\sum_n R_n^2\underline{P}_n\right) \tag{66}$$

since $\Sigma\underline{P}(R_m) = 1$ (The total probability of making any jump must be 1.) and

$$D(u) = \frac{(\Sigma R_n^2 \underline{P}_u)}{2d}\;\frac{\tilde{\psi}(u)}{1-\tilde{\psi}(u)} \tag{67}$$

Now the time dependence of D(u) is governed solely by $\tilde{\psi}(u)$. In particular, the long time dependence of $\tilde{D}(t)$ is determined the small u dependence of $\tilde{\psi}(u)$. $\psi(t)$ represents the probability distribution of waiting times; if this distribution has a long time tail, there may be anomalous diffusion. Consider first the non-anomalous case, then

$$\tilde{\psi}(u) = \int_o^\infty dt\; e^{-ut}\psi(t) = 1 - u\int_o^\infty dt\; t\;\psi(t) + \cdots \tag{68a}$$

$$= 1 - u\,\tau_1 + \cdots \tag{68b}$$

If τ_1 is finite, then

$$D(u) = \frac{(\Sigma R_n^2 \underline{P}_n)/2d}{u\tau_1} \qquad u\to o \tag{69}$$

so that for long times

$$D = (\Sigma R_n^2 \underline{P}_n)/2d\tau_1 \quad . \tag{70}$$

If, however, the first moment of $\psi(t)$ is infinite, say for example

$$\psi(t) \sim t^{-1-\beta}\;,\; t\to\infty \qquad 0 < \beta < 1 \tag{71}$$

then

$$\psi(u) = 1 - \Gamma(1-\beta)u^{\beta}/\beta + \cdots \tag{72}$$

and

$$D(t) \sim t^{\beta-1} \tag{73}$$

Thus $\langle R^2(t)\rangle \sim t^{\beta}$ and the diffusion is anomalous. By the way, this approximation for $\psi(t)$ yields

$$\Phi(t) \sim t^{-\beta} \tag{74}$$

for the decay law of an excitation due to donor-acceptor transfer.

Montroll and Scher chose the form of Eq. (71) for $\psi(t)$ in their study of dispersive transport in amorphous materials. It is the extremely long waiting times (or trapping times) which cause the dispersion in the transport. In our discussion, this yields a time dependent diffusion coefficient (Eq. (73)) which decays to zero at long time. The form for $\langle R^2(t)\rangle$ (i.e. proportional to t^{β}, $\beta < 1$) is unusual; it is interesting to ask in what circumstances such a form can come about.

Another problem of current interest is the trapping of an excitation as it hops through a disordered system. We think about the excitation as a random walker which is removed from the problem if it walks onto a trap site. The trap sites are randomly arranged on a lattice with probability c of being on any particular site. This problem has been extensively treated in the literature [27,28] in terms of R_n, the number of distinct sites visited by the walker after n steps. The survival probability, Φ_n, is then the average over all realizations of the walk of the survival probability for each realization:

$$\Phi_n = \langle(1-c)^{R_n-1}\rangle = \langle e^{(R_n-1)\ln(1-c)}\rangle \tag{75}$$

R_n is a stochastic variable, and the average in Eq. (75) cannot be done in general. The standard approximation for small c (low concentration of traps) is to write

$$X_n \approx e^{-cS_n} \tag{76a}$$

where $S_n = \langle R_n\rangle$, the mean number of distinct sites visited after n steps. This is a known function of dimensionality and lattice.

The above was derived for random walks with a constant jump time. If we allow for a jump-time distribution as in the CTRW, then the survival probability is

$$X(t) \simeq e^{-cS(t)} \tag{76b}$$

and we can define the decay rate of the excitation to be

$$k(t) = -\frac{\dot{X}(t)}{X(t)} = c\dot{S}(t) \tag{76c}$$

Here S(t) is related to the $\tilde{\psi}(u)$ we introduced above (Eq. (64)) by

$$\tilde{S}(u) = \sum_{n=0} S_n[\tilde{\psi}(u)]^n \frac{(1-\tilde{\psi}(u))}{u} \tag{77}$$

Other approximations to Eq. (75) are possible. In particular, at not too long times we can go to the second cumulant to find [32]

$$X(t) \cong e^{-cS(t) + c^2\sigma^2(t)/2} \tag{78a}$$

The properties of S_n are known for d-dimensional lattices [27,28] and thus we can find the long time behavior of X(t) for a variety of situations.

Blumen and Zumofen [43] have carried out extensive simulations of trapping rates and survival probabilities, and compared them with the approximate results mentioned above as well as other approximations. For example, these authors found that if $\psi(t)$ has a finite first moment (Eq. (68)) then, in <u>3 dimensions</u>, X(t) was exponential in time (for not too long times). <u>If however</u> $\psi(t)$ has a long time algebraic tail (Eq. (17)), then τ_1 does not exist and the survival probability $X(t) \sim t^{-\beta}$.

The asymptotic decay of X(t) for very long times has been discussed by a number of authors [40-42] and using rather general arguments, it is found that

$$X(t) \sim \exp - t^{(d/d+2)} \tag{78b}$$

This long time decay is dominated by the probability of the excitation being born in a large trap-free region. It turns out that although this is correct, it occurs for such long times that it is unlikely to be observable in practice. In fact, except for 1 dimension it is difficult to see even in simulations.

Other problems, related to trapping, have been studied in the past few years. These are the bimolecular reactions $A + A \to 0$ and $A + B \to 0$ with rate constant k. The kinetics of these under the assumption of a "well-stirred" reactor are easy to find so that in the first

$$A(t) = A_o(1+2A_okt)^{-1} \sim 1/t \tag{79a}$$

and in the second, for $A_o = B_o$,

$$A(t) = A_o(1+A_okt) \sim 1/t \tag{79b}$$

If the well-stirred condition is not correct, then there are strong dependences on disorder for the long time limit of these processes. For example, for A's walking on regular lattices, it is found (by simulation) that for the first case

$$A(t) \sim [cS(t)]^{-1} \tag{80}$$

describes the dynamics very well. However, for the second case, another problem arises. Spatial fluctuations in the concentrations of A and B are <u>enhanced</u> by the chemical reaction, and regions containing only A or only B molecules appear. In this case, it turns out that the behavior (for regular lattices) of A(t) starts out as Eq. (80) but crosses over to a slower decay at longer times.

IV. OPTICAL LINE SHAPES IN DISORDERED SOLIDS

In this section, we will deal with two topics: A) the optical line shape of molecules in a solid in the presence of weak disorder at low T and B) the homogeneous line shape of a guest molecule in a glass at low T.

Both of these have their genesis in trying to answer specific experimental questions, so we will present them in that manner.

IV.A. Optical Line Shape: Weak Disorder [44,45]

The optical absorption spectroscopy of molecular crystals at very low temperature has been of interest for many years [46], but until recently, the true shape of the optical lines at very low temperature have been obscured by various effects. Recently, Port [45] was able to determine the absorption line of a particular molecular crystal at temperature below 2 K with such clarity that, I believe, we are seeing the scattering of the excitation by isotopic impurities in the solid. This is a case in which we know so many of the parameters that a clear comparison between experiment and theory is possible.

The molecule is 1,4-dibromonaphthalene (or DBN). It crystallizes, as do many planar molecules, in a form which exhibits linear stacks of molecules. The optical spectroscopists have discovered that, in the pure material, the lowest energy electronic excitation moves down these one-dimensional stacks preferentially (the transfer matrix element for off the stack is $\sim 10^{-4}$ of that for down the stack), so that the excitons are very nearly one-dimensional. At very low T, the spectroscopists have found that the optical absorption spectrum is a highly asymmetric line (independent of T for the lowest temperatures). They guessed that the cause of the asymmetry was impurities of DBN containing C^{13}. Since the natural abundance of C^{13} is 1%, approximately 10% of DBN molecules will have one C^{13} in it. Because of zero point vibrational energy effects, the optical absorption of these molecules are slightly blue shifted from that of the "pure" DBN. This energy difference causes scattering of the exciton and changes the optical spectrum. Another way to think about it is to say that these impurities break down the translational symmetry and cause intensity redistribution in the spectrum.

We describe the Hamiltonian of the system as

$$H = \sum_n \varepsilon_n |n\rangle\langle n| + \sum_{n,m} J_{nm} |n\rangle\langle m| \tag{81}$$

where ε_n is the excitation energy at site n, $|n\rangle$ is the localized state with excitation on site n and J_{nm} is the excitation transfer matrix element. We assume (because the spectroscopists tell us) that $J_{nm}=0$ unless n and m are nearest neighbors. If all the ε_n are equal, the eigenstates of H can be written down immediately (assuming periodic boundary conditions) as

$$|k\rangle = \frac{1}{\sqrt{N}} \sum_n e^{ikn} |n\rangle \tag{82}$$

We have written this in one-dimensional notation for convenience. The eigenvalue corresponding to $|k\rangle$ is 2Jcosk in one dimension. For the impure crystal, we assume that J is unchanged and independent of isotopic species. The optical line shape as a function of energy, I(E) is given by an average over all possible configurations of impurities of the imaginary part of the k = 0 Green function:

$$I(E) = \mathrm{Im}\langle G(k=0,E)\rangle_c \tag{83}$$

and

$$G(k,E) \; \langle k|(E-H-i\delta)^{-1}|k\rangle \tag{84}$$

with δ a small positive number. For the pure system, this gives a delta function line shape centered at $E = \varepsilon+2J$ (or if δ is the inverse of the radiative lifetime, a sharp Lorentzian of width δ). In the specific case of DBN, the lifetime is quite long, so that τ^{-1} is quite small. The effect of the impurities is to spread intensity from the zeroth order k=0 level to the other energies in the band, by the mixing of this state into other states. One way of carrying this out is to write

$$\langle(E-H-i\delta)^{-1}\rangle_c = [E-H_{eff}(E)]^{-1} \quad , \tag{85}$$

where $H_{eff}(E)$ is some complex function of E. A formally exact expression for $H_{eff}(E)$ can be written, but since it cannot be evaluated exactly, we will not bother. Instead, we will find an approximate form for $H_{eff}(E)$ using the coherent potential approximation [37]. We write ($\bar{\varepsilon}$ is the configurational average of the excitation energy)

$$H_{eff} = [\bar{\varepsilon}+\sigma(E)] \sum_n |n\rangle\langle n| + \sum_{n,m} J_{nm}|n\rangle\langle m| \tag{86}$$

and $H = H_{eff} + V$ with

$$V = \sum_n [\varepsilon_n-\bar{\varepsilon}-\sigma(E)]|n\rangle\langle n| \equiv \sum v_n|n\rangle\langle n| \tag{87}$$

The exact Green function can now be expanded in terms of the zeroth order (effective) Green function and the t matrices, and the site diagonal t is

$$t_{nn} = v_n[1-v_nG_{nn}^{(0)}]^{-1} \tag{88}$$

where

$$G_{nn}^{(0)} = \langle n|(E-H_{eff})^{-1}|n\rangle \tag{89}$$

We now set the average of t equal to zero:

$$\sum_{\alpha=1}^{p} c_\alpha t^{(\alpha)} = 0 \tag{90}$$

where c_α is the concentration of each isotopic species. This gives, after some manipulation

$$\sum_{\alpha=1}^{p} \frac{c_\alpha}{1-v^{(\alpha)}G_{nn}^{(0)}(E)} = 1 \tag{91}$$

This is the CPA equation. Note that since both $v^{(\alpha)}$ and $G_{nn}^{(0)}$ contain $\sigma(E)$, this must be solved self-consistently. We have solved this for the parameters of DBN (all the relevant parameters are given in ref. [45]). For this system, the energy defects (i.e. $\varepsilon^{(\alpha)}-\varepsilon_{pure}$) are small compared to J, being at most 0.2 J, thus we expect the CPa to be a reasonable description. We have computed the line shape in this way and compared it to the experiment. There is almost perfect agreement in shape and overall width between the two. This suggests that, at least in certain cases, it is possible to find a zero temperature line shape dominated by weak disorder effects such as isotopic scattering rather than other defect scattering.

IV.B. Homogeneous Line Widths of Molecular Transitions in Glasses [47]

When a small concentration of guest molecules is placed in a host, the optical absorption spectrum of these is inhomogeneously broadened because each molecule is in a different environment. If the host is crystalline, it is possible to observe the homogeneous line shape of the molecular absorption using a variety of techniques such as flourescence line narrowing, photon echos, and hole burning. When the homogeneous linewidth is measured as a function of temperature, the mechanism of homogeneous broadening or dephasing can be elucidated in the more favorable cases. For example, local mode or librational mode dephasing varying as $e^{-\omega_o/kT}$ at low temperature has been observed, as well as acoustic phonon dephasing by Raman processes (varying as T^7) have been identified.

When the host is a glass, things are not so easy. First of all, the inhomogeneous broadening is quite large, often complicating the interpretation of the more complex experiments. Secondly, when the homogeneous linewidth is uncovered, unusual and totally unexpected temperature dependences have been found. This field is of sufficient current interest that an issue of Journal of Luminescence was recently devoted to the problem of hole burning in optical lines in glasses.

The hole burning experiment is the following: an inhomogeneously broadened line is irradiated at a particular frequency by a narrow band laser for some time, and then the optical absorption recorded with low intensity light. A hole is found in the inhomogeneous line at the frequency of the laser. When care is taken not to heat the sample, etc., the hole shape is assumed to be related to the homogeneous line shape of the molecules at the laser frequency by a simple convolution. We assume that the experiments have been done well, and look at the homogeneous linewidths $\gamma(T)$. We find that they vary with temperature as

$$\gamma(T) = \gamma_o + aT^{1+\gamma} \tag{92}$$

where γ_o is the inverse lifetime of the transition and α is a parameter which depends (slightly) on system but $0 \leqslant \alpha \leqslant 1$. For many systems (that is many chromophores in different glasses) $\alpha = 0.3$.

This unusual temperature dependence has been ascribed to the presence in glasses of low frequency excitations, called two level systems (TLS), which have been invoked to explain the low temperature heat capacity and thermal conductivity of glasses. It was discovered about 15 years ago that at very low $T(T<1K)$, glasses have heat capacities varying approximately linearly with T. An explanation was presented by Phillips [48] and by Anderson et al. [49] based on the presence of many barriers and disorder in glasses, which would give rise to many tunneling states. If the levels on either side of the barrier differed in energy by Δ and there was a tunneling matrix element K/2 between them, then the energy separation between the final two levels would be $(\Delta^2+K^2)^{1/2}$. If we treat these as a TLS, then the intense energy of the glass at temperature T $(=1/k\beta)$ would be

$$E(T) = \int_o^{\omega_c} d\omega \rho(\omega)\omega \, \frac{e^{-\beta\omega}}{1+e^{-\beta\omega}} \tag{93}$$

where $\rho(\omega)$ is the density of states of the TLS and ω_c is some cutoff energy. If we assume that $\rho(\omega)$ varies as ω^α at low ω, then at temperatures low compared to ω_c/k, $E(T) \sim T^{2+\alpha}$ and $C_V \sim T^{1+\alpha}$. the standard model is $\alpha = 0$, but no one really knows for certain.

Although the exact nature of the TLS is still not known, this conjecture has proven to be of great utility in explaining phonon scattering, thermal conductivity etc. in glasses at low T. When the hole burning experiments were done, it was naturally assumed that the TLS were implicated, even though these experiments were usually done at temperatures above 1K (and the specific heat experiments were done below 1K). The basic model is straightforward: consider an optical transition between two levels and assume that there is a TLS strongly interacting with that optical system. This yields a 4-level system. Assume further that the TLS is slightly changed (in energy and eigenstates) when the chromophore is excited from what it is when the chromophore is in the ground state. Now, if phonons can "flip" the TLS during an optical transition of the chromophore, the optical transition will experience some frequency modulation, or dephasing. This is the origin of the TLS induced homogeneous width for the optical trasition. The TLS is changed when the chromophore is excited because the TLS will have a dipole moment and when this interacts with the chromophore, the energy depends on the state of the chromophore.

This model can be treated using the theory of spectral diffusion [50] or thé standard weak coupling (Redfield) theory [47] for a single TLS. The result must then be averaged over a distribution of TLS parameters. When this is done, it is found that the homogeneous linewidth depends on the temperature at low T as $T^{1+\alpha}$, where α is the exponent in the density of states of the TLS. This suggested to many workers that there was a (more or less) universal exponent for the TLS distribution of $\alpha \approx 0.3$. In fact, experimentally it was found that the specific heat of glassy silica below 700 mK varies as $T^{1.2}$, as well. Recently, a number of authors [51] have suggested that more careful averaging even over the standard parameters of the Anderson-Phillips model will give the $T^{1.3}$ dependence at very lot T.

In addition, other authors [47] have suggested that the hole widths and the photon echo decays are really sums of two mechanisms - one for the TLS (and nearly linear in T) and one for a direct phonon mechanism (such as Raman scattering or local mode scattering).

Finally, there is a recent speculation that the true cause of the anomalous exponent is due to the fractal nature of the glass [52].

V. FRACTALS AS MODELS OF DISORDERED SYSTEMS [53-63]

In the last five years, there has been an explosion of interest in fractals, as can best be seen by the number of conferences and papers devoted to the subject. This recent interest has its genesis in the work of Mandelbrot [53], and can in part be understood by the simplicity and beauty of both the science and the figures which are produced to illustrate the concepts. Very recently, a number of articles [64] have suggested that it was time to take stock of what had been done and to make certain that physics was the principal concern.

In this section, I will selectively review the work of others on the use of fractals to model disordered systems. In particular, I will be concerned only with the use of these structures for the modelling of the effect of disorder on dynamical processes in disordered systems.

V.A. Description of Fractal Structure

In 1919, Hausdorff suggested a way to generalize dimension by use of what we now call scaling. Suppose a very complex curve, for example the

Sicilian coastline, were to be measured by sticks of different length, ℓ. To do the measurement, we lay the stick down end-to-end along the coast. By this method, irregularities of length less than ℓ are smoothed out. The number of sticks of length ℓ needed to cover the coast line is $N(\ell)$. For a straight line, $N(\ell)$ depends on ℓ in a simple linear way: if we change the size of the stick from ℓ to ℓ/a we find

$$N(\ell/a) = (\frac{\ell}{a})N(\ell) \tag{94}$$

We can measure area in this manner as well using squares of side length ℓ or a to find that the number of squares needed in the first case is $(\ell/a)^2$ times the number needed in the first. In general, if we are measuring in units of ℓ^d we find that the number of units needed scales as $(\ell/a)^d$. In our usual world view, d can be an integer. However, Hausdorff and more recently Mandelbrot have shown us how to generalize this. A simple example is the Koch snowflake, which is formed by start-with an equilateral triangle and adding, in the middle of each side, an equilateral triangle of 1/3 the side length of the original. This produces a star of David. The process is repeated at 1/3 the scale to all the sides of the star. And so on, indefinitely. Our measurement process is blind to detail smaller than the stick. Then every time we measure the perimeter of this snowflake with a stick 1/3 the length of the previous stick, we find the number of sticks needed is 4 times the previous number. That is

$$N(\ell/a) = (\ell/a)^{\bar{d}}N(\ell) \tag{95}$$

where $\bar{d} = \log[N(\ell/a)/N(\ell)]/\log(\ell/a) = \log4/\log3 = 1.26$ in our case! Mandelbrot introduced the word fractal for objects with noninteger dimension.

Another example is the Sierpinski gasket which can also be formed from equilateral triangles, but with a different method. Stack three equivalent triangles together to form a new equilateral triangle with a hole in its center. Now stack three of these to form a new triangle with a larger hole in the center. Continue until exhausted. If we measure the area of such an object, we note that every time we decrease our length scale by 2 we "see" an area which increases by 3 (not 4 as in normal structures) so that $\bar{d} = \log3/\log2 = 1.58$. The Sierpinski gasket is a favorite toy for modelling disorder because we may assume a site exists on every vertex and bonds between every pair of near neighbors.

Regular fractals of this type come in many varieties. The dimension $\bar{d}$ is called the fractal or Hausdorff dimension (the space in which the structure is embedded has an integer dimension, so this particular Sierpinski gasket has fractal dimension 1.58 and embedding dimension 2).

Based on the construction we have outlined, fractals have the property of <u>self-similarity</u>. That is they look the same on every length scale. Every time we increase the magnification, the structure looks the same. (This is just the condition necessary to introduce the fractal dimension.) Real disordered systems are clearly not self-similar on <u>all</u> length scales; however, it may be a good approximation to assume it to be so over a restricted set of scales.

We do not restrict the term fractal only to geometric properties. This allows us to introduce fractal dimensions for many properties of random systems. For example, random walks can be considered to be fractal in certain properties.

Consider a random walker in a d dimensional regular lattice. Assume that at each step of the walk a molecule is placed on the site visited, and forget about self-avoidance. Then after N steps (or time $t = N\tau_1$) we find N molecules on the path of walker. This path on the average has a length, L, proportional to $N^{1/2}$. Therefore the mass of the path M is proportional to L^2, i.e. it is a fractal of dimension 2.

If we do not allow intersections, i.e. we have a self-avoiding walk, then L is proportional to N^{ν}, $\nu = 0.6$ on 3-dimensional lattices and $\nu = 0.75$ on 2-dimensional lattices, so that the mass of the path $M \sim L^{1/\nu}$, i.e. a fractal dimension of 1.67 in 3d and 1.33 in 2d.

Another statistical fractal is the percolation cluster. At the percolation threshold, $p = p_c$, the network of occupied sites is self-similar, that is, it looks the same on all length scales large compared to the bond length and short compared to the correlation length ξ. Since ξ varies as $(p-p_c)^{-\nu}$ as p approaches p_c, the object is self-similar on all scales at the percolation edge. By scaling arguments, it is found that the number of sites on the percolation network within a radius R of a randomly chosen origin is given by $N(R) \propto R^{\bar{d}}$ where $\bar{d} = d-\beta/\nu$. The exponent β is defined by the probability of a site belonging to the infinite cluster, $P_\infty - (p-p_c)^\beta$.

A final example of a statistical fractal are structures grown by diffusion limited aggregation, which have been studied in great detail in recent years [54].

How do experimenters tell if a real system can be described as a fractal? In practice, one measures a quantity, say the density, of objects of various radius and plots the log vs. the log of the radius, hoping for a straight line.

For example, if a real object has a mass depending on radius as $R^{\bar{d}}$, then the density depends on $R^{\bar{d}-d}$; that is, the bigger it gets the more tenuous it becomes. Scattering measurements are often the method of choice to determine the quantities we are after. Clearly, if the density scales in an unusual manner with R, the scattering as a function of $\underset{\sim}{K}$ will also scale with $\underset{\sim}{K}$ in a complementary manner [55]. Since we do not restrict attention to mass or volume, one fractal dimension is insufficient to determine all the quantities which we measure, Stanley [56] has discussed in his Cargese talk, 10 fractal dimensions. Happily, not all are independent quantities.

In the study of dynamics on fractals, for example, there has been considerable discussion in the past 10 years about a random walker walking on a fractal [57], for example the percolation cluster. Since the walker goes into dead-ends a lot of the time and must retrace its steps, we expect anomalous diffusion, i.e. $\langle R^2(t)\rangle \sim t^{2\gamma}$, with $\gamma = \frac{1}{d_w} < \frac{1}{2}$. Here we have introduced the dimension d_w often written as $2 + \theta$. The number of distinct sites visited on the walk scales with the Hausdorff or fractal dimension, $\bar{d}$; thus the number of distinct sites scales as $t^{\bar{d}/dw} = t^{d_s/2}$. Alexander and Orbach [58] suggested that the dimension d_s be called the spectral (or fracton) dimension. By a scaling argument they conjectured that the density of modes on the fractal would scale as ω^{d_s-1} (instead of ω^{d-1} in d dimensional Euclidean space). Using known values of various exponents, they found $d_s \cong 4/3$ for percolation clusters embedded in d dimensional lattices. This led to a number of studies which have suggested that d_s is not simply given by 4/3 for all d, but is close to that value and attains it for d>6. Later studies also refined their argument (because it was based on an isotropic model inappropriate for elastic modes) and redefined the spectral dimension [55]. In any case, the basic idea that the mode density of states varies on a fractal as ω^{d_s-1}, i.e. that the modes, which are localized, have an unusual form suggested that relaxation processes in such systems should exhibit interesting time or temperature dependences. For example, Stapleton et al. [59] suggested that the unusual temperature dependence of the spin lattice relaxation found in proteins was caused by the fractal structure of the protein. These authors used the Hausdorff dimension to fit their data; Orbach suggested using the spectral dimension instead. Unfortunately, the agreement between this theory and experiment is not very good. Recently, Elber and Karplus [60] suggested a model for calculating the spectral density in a protein directly and saw little evidence fo fractal or fracton behavior in their results.

V.B. Transport on Fractals [61]

We will now discuss how fractal geometry changes the dynamics of the various processes we discussed earlier.

First, we discuss donor-acceptor transfer to acceptors arranged on a fractal. Recall Eq. (8) for the probability, $\Phi(t)$, of the excitation remaining on the donor. If the acceptors are arranged on a fractal with Hausdorff dimension $\bar{d}$, the density of sites varies as $r^{\bar{d}-1}dr$, so the result is immediate for multiple transfer, $w(r) \sim r^{-s}$:

$$\Phi(t) = \exp\{-B_s \cdot ct^{\bar{d}/s}\} \tag{96}$$

and

$$\Phi(t) = \exp\{-B_\gamma \, \ell n^{\bar{d}} (W_o t)\} \tag{97}$$

This has been tested by the simulations of Blumen and Zumofen. The agreement with (96) and (97) is very good. A few years ago, Even et al. [62] put the donors and acceptors in a porous glass and measured the decay time of the donors (the donor was rhodamine 6G, the acceptor malachite green). When this was compared to the same system in a solid solution, these authors suggested that the structure in which the molecules sat was a fractal of dimension $\bar{d}$ = 1.74.

In recent years, Klafter, Blumen and Zumofen [63] simulated a series of random walks on Sierpinski gaskets embedded in Euclidean dimension of d = 2-6. These all have spectral dimensions $d_s < 2$. The simulations show that the mean number of distinct sites visited, S_n, varies as

$$S_n = an^{d_s/2} \tag{98}$$

The decay law for trapping agrees fairly well with $\Phi_n = \exp(-S_n)$ in all the simulations. In most cases, the second approximation was found to be slightly better, but the convergence is slow.

These same authors [63] have simulated the CTRW on fractals. Recall that the waiting time distribution can be of two kinds: having finite mean jump time τ_1 and having infinite τ_1. The case of finite τ_1 is much like the above case of fixed waiting times. However, for infinte τ_1 distributions, such as

$$\psi(t) \sim 1/t^{1+\gamma} \qquad 0 < \gamma < 1 \tag{99}$$

This leads to [63] a form for S(t), see eq. (76b),

$$S(t) \sim t^{\gamma d_s/2} \qquad \text{for } d_s < 2 \tag{100a}$$

and

$$S(t) \sim t^{\gamma} \qquad \text{for } d_s > 2 \tag{100b}$$

(Note that the spectral dimension, d_s, comes into play in this dynamics, while in the donor-acceptor case only the fractal dimension, $\bar{d}$, enters.) Eq. (100) show that the two exponents γ and d_s combine in a multiplicative manner (subordination).

These same authors [63] have also simulated the problem of trapping on fractals under CTRW dynamics and find

$$\Phi(t) \sim t^{-\gamma}/(c^{2/d_s}) \sim S(t)^{-2/d_s} \qquad \text{for } d_s < 2 \tag{101a}$$

$$\sim t^{-\gamma}/c \sim S(t)^{-1} \qquad \text{for } d_s > 2 \tag{101b}$$

where c is the concentration of traps.

Finally, these authors [63] have simulated the bimolecular reactions mentioned above (Eq. (79)) on fractals (Sierpinski gaskets) and find the decay is well approximated by

$$\Phi_n^{AA}(t) \simeq (1+2c\ S_n)^{-1} \tag{102}$$

as in Eq. (80). The reaction A + B → 0 (with $A_o = B_o$) again is anomalous because of the spatial fluctuations, as mentioned above. Here the decay, Φ_n^{AB} varies as $n^{-d_s/4}$ for $d_s < 4$.

VI. CONCLUSIONS

We have discussed a number of dynamical processes such as energy transfer, trapping, and optical line shapes in molecular systems, all of which are profoundly affected by the presence of disorder. One of the results of this is a rich variety of time dependencies. For example, disorder can give rise to algebraic decays,

$$\phi(t) \sim \exp t^{-\alpha} \quad ,$$

stretched exponentials,

$$\phi(t) \sim \exp-(t/\tau)^{\alpha}$$

and exponential log decays,

$$\phi(t) \sim \exp-[\ln^{\alpha}(t/\tau)].$$

Although these time dependencies are associated with disorder, it is often very difficult to uncover the underlying mechanism from the experimental results.

Optical line shapes are also profoundly affected by disorder, becoming broad and featureless. In some cases, however, it is possible to uncover the mechanism of scattering by careful experiment. However, there is still ambiguity in many cases, in particular in the study of optical lines in glasses.

ACKNOWLEDGEMENTS

I want to thank my former co-workers J. Klafter and A. Blumen with whom I worked on disordered systems a number of years ago. Their more recent work and discussions have enlightened me about the more modern aspects of the subject.

REFERENCES

1. T. Forster, Z. Naturf. A4:321 (1949).
2. M. Inokuti and F. Hirayama, J. Chem. Phys. 43:1978 (1965).
3. D. Rehm and K.B. Eisenthal, Chem. Phys. Lett. 9:387 (1971).
4. D. Dexter, J. Chem. Phys. 21:836 (1953).
5. D. Huber, in: "Laser Spectroscopy in Solids," W. Yen and P. Selzer, eds., Springer, Berlin (1981).
6. D. Huber, D. Hamilton, B. Barnett, Phys. Rev. B 16:4642 (1977).
7. A. Blumen and J. Manz, J. Chem. Phys. 71:4694 (1979).
8. A. Blumen, Nuovo Cimento 63B:50 (1981).
9. J. Klafter and A. Blumen, Chem. Phys. Lett. 119:377 (1985).
10. C. Yang, P. Evesque and M. El-Sayed, J. Phys. Chem. 89:3442 (1985).
11. S. Haan and R. Zwanzig, J. Chem. Phys. 68:1879 (1978).
12. C. Gouchanour, H.C. Anderson and M. Fayer, J. Chem. Phys. 70:4254 (1979).
13. H. Scher and M. Lax, Phys. Rev. B 7:4491, 4502 (1973).
14. H. Scher and E. Montroll, Phys. Rev. B 12:2455 (1975).
15. G. Pfister and H. Scher, Adv. Phys. 27:747 (1978).
16. A. Burshtein, Usp. 27:579 (1984); J. Lum. 34:167 (1985); ibid. 34:201 (1985); Chem. Phys.
17. V. Kenkre, E. Montroll, M. Schlesinger, J. Stat. Phys. 9:45 (1973).
18. J. Klafter and R. Silbey, J. Chem. Phys. 72:843 (1980).
19. J. Bernasconi, S. Alexander and R. Orbach, Rev. Mod. Phys. 53:175 (1981).
20. B. Movaghar and W. Schirmacher, J. Phys. C 14:859 (1981).
21. M. Sahimi, B. Hughes, L. Scriven, H.T. Davis, J. Chem. Phys. 78:6849 (1983).
22. J. Machta, Phys. Rev. B 24:5260 (1981).
23. J. Knoester and J. van Himbergen, J. Chem. Phys. 80:4200 (1984); 81:4380 (1984).
24. T. Odagaki and M. Lax, Phys. Rev. B 24:5284 (1981).
25. I. Webman, Phys. Rev. Lett. 47:1496 (1981).
26. K. Kundu and P. Phillips (in press).
27. E. Montroll and M. Schlesinger in: "Non-Equilibrium Phenomena II," J. Lebowitz and E. Montrol, eds., North-Holland, Amsterdam (1984).
28. "Random Walks and their Applications in the Physical and Biological Sciences," M. Schlesinger and B. West, eds., American Institute of Physics (1984).
29. G. Weiss and R. Rubin, Adv. Chem. Phys. 52:363 (1983).
30. E. Montroll and G. Weiss, J. Math. Phys. 6:167 (1965).
31. A. Blumen, J. Klafter and R. Silbey, J. Chem. Phys. 72:5320 (1980).
32. A. Blumen and G. Zumofen, J. Chem. Phys. 77:5127 (1982).
33. G. Zumofen, J. Klafter and A. Blumen, J. Chem. Phys. 79:5131 (1983).

34. P. Argyrakis and R. Kopelman, J. Chem. Phys. 83:3099 (1985).
35. J. Morgan and M. El-Sayed, J. Phys. Chem. 87:2178 (1983).
36. P. Klymko and R. Kopelman, J. Phys. Chem. 87:4565 (1983).
37. P. Soven, Phys. Rev. 156:804 (1967).
38. S. Kirkpatrick, Rev. Mod. Phys. 45:574 (1973).
39. A. Blumen, J. Klafter and G. Zumofen, in: "Optical Spectroscopy of Glasses," ed. I. Zschokke, Reidel, Dordrecht (1986).
40. B. Balagurov and V. Vaks, JETP 38:968 (1974).
41. P. Grassberger and I. Proccacia, J. Chem. Phys. 77:628 (1982).
42. R. Kayser and J. Hubbard, Phys. Rev. Lett. 51:6281 (1982).
43. G. Zumofen and A. Blumen, Chem. Phys. Lett. 88:63 (1982).
44. D. Burland, U. Konzelmann, and R. Macfarlane, J. Chem. Phys. 67:1926 (1977).
45. H. Port, H. Nissler, R. Silbey, J. Chem. Phys. (in press).
46. See, e.g. A. S. Davydov, "Molecular Excitons" Plenum, New York (1962).
47. A recent review is: K. Kassner and R. Silbey, J. Lum. 36:283 (1987).
48. W. Phillips, J. Low Temp. Phys. 7:351 (1972).
49. P. W. Anderson, B. Halperin and C. Varma, Phil. Mag. 25:1 (1972).
50. D. Huber, J. Lum. 36:307 (1987). J. Klauder and P. W. Anderson, Phys. Rev. 125:912 (1962).
51. K. Kassner and P. Reineker, in: "Optical Spectroscopy of Glasses," ed. I. Zschokker, Reidel, Dordrecht (1986); R. Jankowicek and G. Small, Chem. Phys. Lett. 128:377 (1986).
52. K. Lyo and R. Orbach, Phys. Rev. B 29:12300 (1984). G. Dixon, R. Powell and X. Gang, Phys. Rev. 33:2713 (1986).
53. B. Mandelbrot, "The Fractal Geometry of Nature," Freeman, San Francisco (1982).
54. "Kinetics of Aggregation and Gelation," eds. F. Family and D. Landau, North-Holland, Amsterdam (1984).
55. See, e.g. S.H. Liu, Sol. State Phys. 39:207 (1986).
56. H.E. Stanley in: "On Growth and Form," eds. H.E. Stanley and N. Ostrowsky, Nijhoff, Dordrecht, (1985).
57. P.G. deGennes, Recherche 7:919 (1976); Comptes Rendus 296:881 (1983).
58. S. Alexander and R. Orbach, J. Phys. Lett. 43:L625 (1982).
59. H. Stapleton, J. Allen, C. Flynn, D. Stimson and S. Kurtz, Phys. Rev. Lett. 45:1456 (1980); H. Stapleton, Phys. Rev. Lett. 54:1734 (1985).
60. R. Elber and M. Karplus, Phys. Rev. Lett. 56:394 (1986).
61. R. Rammal and G. Toulouse, J. Phys. Lett. 44:13 (1983).
62. U. Even, K. Rademann, J. Jortner, N. Manor and R. Reisfeld, Phys. Rev. Lett. 52:2164 (1984); See, D. Schaefer et al., Phys. Rev. Lett. 58:284 (1987).
63. a) J. Klafter and A. Blumen, Chem. Phys. Lett. 119:377 (1985); b) A. Blumen, J. Klafter and G. Zumofen, J. Chem. Phys. 84:6679 (1986); c) A. Blumen, J. Klafter, B. White and G. Zumofen, Phys. Rev. Lett. 53:1301 (1984).; d) G. Zumofen, J. Klafter and A. Blumen, J. Chem. Phys. 79:5131 (1983).
64. J. Krumhansl, Phys. Rev. Lett. 56:2696 (1986). See also, S. Alexander in: "Statistical Physics," ed. H.E. Stanley, North-Holland, Amsterdam (1986), p. 397.

SHARP SPECTRAL LINES OF LUMINESCENT CENTERS IN SOLIDS

B. Di Bartolo

Department of Physics
Boston College
Chestnut Hill, Massachusetts 02167, U.S.A.

ABSTRACT

In the first part of this article we consider the interactions of luminescent centers in solids with radiation and review the basic mechanisms present in spectral lines, such as saturation, and homogeneous and inhomogeneous broadening, and relate them to the lineshapes. The phenomena that affect the width of sharp lines in crystals are due to the interaction of the optically active centers with the thermal vibrations of the solid and to the spatial randomness of these centers. This randomness is present even in nominally ordered solids because the sites occupied by the luminescent centers in such solids are not exactly equal, due to the slight changes in the local crystalline field. In such systems the effects of randomness become evident at very low temperatures where the thermal broadening is drastically reduced, and the spectral lines appear mainly affected by a residual inhomogeneous broadening due to the spatial randomness. In glasses this randomness may represent the prevalent cause of broadening even at room temperature and may produce lines as wide as 100 cm^{-1} for rare earth centers. Inhomogeneous broadening may have relevant effects on the spectral characteristics of a system of luminescent ions when monochromatic laser light is used to excite selectively the ions residing in a particular environment. The technique used in this case is called FLN (Fluorescence Line Narrowing); the conditions for its applications are examined.

In the second part of the article we present a simple theory that, starting from the vibrations of solids, leads to the formulation of the interaction between the optically active ion and the phonon system. The possible mechanisms that may produce the thermal broadening of a sharp line are examined; it is found that the predominant mechanism is the Raman scattering of phonons. The mechanisms by which the spectral lines change their positions with temperature are also examined; it is found that the main mechanism is due to the stationary effects of the ion-vibration interaction, namely the emission and absorption of virtual phonons. The similarities and differences between the thermal shift and the electrmagnetic (Lamb) shift are examined. Finally some experimental findings on sharp spectral lines are reported together with representative examples.

I. SPECTROSCOPY OF LUMINESCENT SOLIDS

I.A. Systems and Interactions

The systems that we shall consider are laser-type solids doped with optically active ions. These dopants consist generally of transition metal ions or rare earth ions.

The relevant interactions are the ion-photon interaction, i.e. the interaction of the optically active ions with the electromagnetic radiation and the ion-phonon interaction, i.e. the interaction of the optically active ions with the thermal vibrations of the solid. The former interaction leads to the phenomena of absorption and induced and spontaneous emissions; the latter interaction to radiationless processes, vibrational electronic (also called vibronic) transitions, thermal line broadening and line shift, etc.

We shall examine "sharp" spectral lines and the mechanisms that may affect their characteristics. We shall define a line sharp if its half width in units of energy is much smaller than $\hbar\omega_D$, where ω_D is the highest phonon frequency of the solid.

I.B. Basic Concepts of Spectroscopy

1. Classical Bound and Radiating Electron. Let us consider the equation of motion of a classical bound electron that we assume to be nonradiating:

$$F = -kx = m\frac{d^2x}{dt^2} \tag{1}$$

or

$$\frac{d^2x}{dt^2} + \omega_o^2 x = 0 \tag{2}$$

where

$$\omega_o = \sqrt{\frac{k}{m}} \tag{3}$$

The energy is given by

$$E = KE + PE = \frac{1}{2}mv^2 + \frac{1}{2}kx^2 = \text{const} \tag{4}$$

We take as solution of the equation of motion

$$x = x_o \cos \omega_o t \tag{5}$$

This means that at time t = 0, $x = x_o$ and $\dot{x} = v = 0$; the energy is then given by

$$E = \frac{1}{2}kx_o^2 \quad \frac{1}{2}m\omega_o^2 x_o^2 \tag{6}$$

A radiating electron will produce radiation due to the vibrating dipole

$$d = ex = ex_o \cos\omega_o t = \text{Re}\left[ex_o e^{i\omega_o t}\right] \tag{7}$$

The average energy radiated in the unit time is given by

$$\overline{S} = \frac{\overline{2(\ddot{d})^2}}{3c^3} = \frac{\omega_o{}^4 e^2 x_o{}^2}{3c^3} \tag{8}$$

The radiated energy is given out at the expense of the internal energy of the electron

$$\overline{S} = -\frac{\partial E}{\partial t} \qquad (E = \frac{1}{2}\, m\omega_o{}^2 x_o{}^2) \tag{9}$$

or

$$\frac{\omega_o{}^4 e^2 x^2}{3c^3} = -m\omega_o{}^2 x_o \frac{\partial x_o}{\partial t}$$

$$\dot{x}_o = -\frac{\omega_o{}^2 e^2}{3mc^3}\, x_o \tag{10}$$

We can write

$$\dot{x}_o = -\frac{1}{2}\,\gamma x_o \tag{11}$$

where

$$\gamma = \frac{2\omega_o{}^2 e^2}{3mc^3} = \frac{2\omega_o{}^2}{3c}\,\frac{e^2}{mc^2} = \frac{2\omega_o{}^2}{3c}\, r_o = \frac{2\omega_o{}^2}{3}\,\tau \tag{12}$$

and

$$r_o = \frac{e^2}{mc^2} = 2.8 \times 10^{-13}\text{cm}$$

$$\tau = \frac{r_o}{c} = 9.38 \times 10^{-24}\text{sec}$$

If $\lambda = 6000$ A, $\omega_o \simeq 3 \times 10^{15}$ and $\gamma = 0.6 \times 10^8 \text{sec}^{-1}$. The solution of (11) is

$$x_o = X_o e^{-(1/2)\gamma t} \qquad (X_o = \text{const}) \tag{13}$$

The energy is given by

$$E = \frac{1}{2}\, m\omega_o{}^2 x_o{}^2 = \frac{1}{2}\, m\omega_o{}^2 (X_o e^{-(1/2)\gamma t})^2$$

$$= \frac{1}{2}\, m\omega_o{}^2 X_o\; e^{-\gamma t} = E_o\; e^{-\gamma t} \tag{14}$$

where

$$E_o = \frac{1}{2}\, m\omega_o{}^2 X_o{}^2 \tag{15}$$

γ is the rate of radiative decay for the classical electron.

The equation of motion (1) needs a revision on account of the presence of radiation. We proceed as follows

$$x = X_o\, e^{-(1/2)\gamma t} \cos \omega_o t \tag{16}$$

$$\dot{x} = -\frac{\gamma}{2}\, x - \omega_o X_o e^{-(1/2)\gamma t} \sin \omega_o t \tag{17}$$

and

$$\frac{\gamma}{2}\, \dot{x} = -\left(\frac{\gamma}{2}\right)^2 x - \omega_o \frac{\gamma}{2} X_o e^{-(1/2)\gamma t} \sin \omega_o t \tag{18}$$

$$\ddot{x} = -\frac{\gamma}{2}\, \dot{x} + \omega_o \frac{\gamma}{2} X_o e^{-(1/2)\gamma t} \sin \omega_o t - \omega_o^2 x \tag{19}$$

Summing the above two relations we obtain

$$\ddot{x} + \gamma \dot{x} + \left[\omega_o^2 + \left(\frac{\gamma}{2}\right)^2\right] x = 0 \tag{20}$$

that represents the equation of motion of the radiating, bound electron.

2. Quantum Mechanical Radiative Decay. Consider a two-level quantum system, where the energy and the eigenfunction of the lower (upper) level are E_a (E_b) and $\psi_a(\psi_b)$, respectively. If Ψ is the wavefunction of the system

$$i\hbar \frac{\partial \Psi}{\partial t} = H\Psi \tag{21}$$

where

$$\begin{cases} H\psi_a = E_a \psi_a \\ H\psi_b = E_b \psi_b \end{cases} \tag{22}$$

In general a quantum state of the system is given by

$$\Psi = a\, e^{-\frac{i}{\hbar} E_a t} \psi_a + b\, e^{-\frac{i}{\hbar} E_b t} \psi_b \tag{23}$$

For a nonradiating system H is independent of time,

$$\begin{cases} \dot{a} = 0 \longrightarrow a = \text{const} \\ \dot{b} = 0 \longrightarrow b = \text{const} \end{cases} \tag{24}$$

and

$$\langle H \rangle = |a|^2 E_a + |b|^2 E_b = \text{const} \tag{25}$$

Let us now allow for dipolar radiation

$$\begin{aligned} \langle ex \rangle &= \langle \Psi | ex | \Psi \rangle \\ &= \int \left[a e^{-i\frac{E_a}{\hbar}t} \psi_a + b\, e^{-i\frac{E_b}{\hbar}t} \psi_b \right]^* ex \left[a e^{-i\frac{E_a}{\hbar}t} \psi_a + b e^{-i\frac{E_b}{\hbar}t} \psi_b \right] d\tau \\ &= a^* b\, e^{-i\omega_o t} \mu_{ab} + a b^*\, e^{i\omega_o t} \mu_{ba} \end{aligned}$$

$$= \mathrm{Re}\left[d_o e^{i\omega_o t}\right] \tag{26}$$

where

$$\omega_o = (E_b - E_a)/\hbar \tag{27}$$

$$\mu_{ab} = \langle \psi_a | ex | \psi_b \rangle \tag{28}$$

$$\mu_{ba} = \langle \psi_b | ex | \psi_a \rangle \tag{29}$$

$$d_o = 2a^*b\ \mu_{ab} \tag{30}$$

and where we have assumed

$$\mu_{aa} = \mu_{bb} = 0 \tag{31}$$

We have from (8)

$$\bar{S} = \frac{\omega_o^4 d_o d_o^*}{3c^3} = \frac{4\omega_o^4 |\mu_{ab}|^2}{3c^3} |a|^2\ |b|^2$$

$$= \frac{4\omega_o^4 |\mu_{ab}|^2}{3c^3} |b|^2 (1-|b|^2) \simeq \frac{4\omega_o^4 |\mu_{ab}|^2}{3c^3} |b|^2 \tag{32}$$

assuming $|b|^2 \ll 1$. We can choose the phase of ψ_b that makes b real:

$$\bar{S} = \frac{4\omega_o^4 |\mu_{ab}|^2}{3c^3} b^2 \tag{33}$$

The energy of the system is given by

$$E = \langle H \rangle = b^2 E_b + a^2 E_a = b^2 E_b + (1-b^2) E_a$$

$$= b^2 (E_b - E_a) + E_a \tag{34}$$

If we set $E_a = 0$, we obtain

$$E = b^2 \hbar\omega_o \tag{35}$$

to be contrasted with the classical formula (6). For nonradiating systems E remains constant. For radiating systems

$$\bar{S} = -\frac{\partial E}{\partial t} = -2b\frac{\partial b}{\partial t}\hbar\omega_o = \frac{4\omega_o^4 |\mu|^2}{3c^3} b^2 \tag{36}$$

where we have dropped the subscripts ab from μ. Then

$$\frac{\partial b}{\partial t} = -\frac{1}{2}\gamma b \tag{37}$$

where

$$\gamma = \frac{4\omega_o^3 |\mu|^2}{3c^3 \hbar} \tag{38}$$

The solution of (37) is

$$b = b_o \, e^{-\frac{\gamma}{2} t} \tag{39}$$

Then

$$b^2 = b_o^2 \, e^{-\gamma t} \tag{40}$$

b^2 represents the probability of occupancy of the excited state. γ is the quantum-mechanical rate of radiative decay which we may equate to the Einstein coefficient A:

$$\gamma = A = \frac{4\omega_o^3 |\mu|^2}{3\hbar c^3} = \frac{64\pi^4 \nu^3 |\mu|^2}{3hc^3} = \frac{64\pi^4 |\mu|^2}{3h\lambda^3} \tag{41}$$

In general the lower level a may not be the ground level and both level b and a may decay to lower levels. In this case we may write

$$\begin{cases} \dot{b} = -\frac{\gamma_b}{2} b \longrightarrow b = b_o e^{-\frac{\gamma_b}{2} t} \\ \dot{a} = -\frac{\gamma_a}{2} a \longrightarrow a = a_o \, e^{-\frac{\gamma_a}{2} t} \end{cases} \tag{42}$$

where

$$\begin{cases} \gamma_b = \sum_i \gamma_{bi} \\ \gamma_a = \sum_i \gamma_{ai} \end{cases} \tag{43}$$

and

$$\begin{cases} \gamma_{bi} = \dfrac{64\pi^4 \nu_{bi}^3 |\mu_{bi}|^2}{3hc^3} \\ \gamma_{ai} = \dfrac{64\pi^4 \nu_{ai}^3 |\mu_{ai}|^2}{3hc^3} \end{cases} \tag{44}$$

In a non-magnetic medium of index of refraction n an oscillating dipole emits the power

$$S = \frac{2(\ddot{d})^2}{3c^3} n \tag{45}$$

whose time average is

$$\overline{S} = \frac{\omega_o^4 d_o d_o^*}{3c^3} n \tag{46}$$

Therefore in this case

$$A = \frac{4\omega_o^3 |\mu|^2}{3\hbar c^3} n \tag{47}$$

3. Absorption and Induced Emission. Consider a cavity whose walls are at temperature T, containing radiation and an ensemble of atoms and let each atom be represented by a two-level quantum mechanical system with an energy level separation of $\hbar\omega_o$. In thermal equilibrium the energy density per unit angular frequency range at ω_o is given by [1]

$$\rho_{\omega_o} = \frac{\hbar\omega_o^3}{\pi^2 c^3} \frac{n^3}{e^{\hbar\omega_o/kT}-1} = \frac{\omega_o^2}{\pi^2 c^3} \frac{\hbar\omega_o n^3}{e^{\hbar\omega_o/kT}-1} \tag{48}$$

Then

$$\rho_{\omega_o}(e^{\hbar\omega_o/kT}-1) = \frac{\hbar\omega_o^3 n^3}{\pi^2 c^3} \tag{49}$$

In addition, because of detailed balance

$$AN_2^e + B_{21}\rho_{\omega_o} N_2^e = B_{12}\rho_{\omega_o} N_1^e \tag{50}$$

where N_1^e and N_2^e are the equilibrium populations of atoms in the lower and upper levels, respectively, and $B_{21}\rho_{\omega_o}$ and $B_{12}\rho_{\omega_o}$ are the probabilities per unit time of induced downward and upward transitions, respectively. We can write

$$A \frac{N_2^e}{N_1^e} + B_{21}\rho_{\omega_o} \frac{N_2^e}{N_1^e} = B_{12}\rho_{\omega_o} \tag{51}$$

and

$$A\, e^{-\frac{\hbar\omega_o}{kT}} + B_{21}\rho_{\omega_o} e^{-\frac{\hbar\omega_o}{kT}} = B_{12}\rho_{\omega_o} \tag{52}$$

or

$$\rho_{\omega_o}(B_{12}e^{\frac{\hbar\omega_o}{kT}} - B_{21}) = A \tag{53}$$

We set

$$B_{12} = B_{21} = B \tag{54}$$

Then

$$\rho_{\omega_o}\left(e^{\frac{\hbar\omega_o}{kT}}-1\right) B = A \tag{55}$$

$$\frac{A}{B} = \rho_{\omega_o}\left(e^{\frac{\hbar\omega_o}{kT}} - 1\right) = \frac{\hbar\omega_o n^3}{\pi^2 c^3} \tag{56}$$

and

$$B = \frac{\pi^2 c^3}{\hbar\omega_o^3 n^3} A = \frac{\pi^2 c^3}{n^3 \hbar\omega_o^3} \frac{4n\omega_o^3|\mu|^2}{3\hbar c^3} = \frac{4\pi^2}{3\hbar^2 n^2}|\mu|^2 \tag{57}$$

or

$$\begin{cases} A = \dfrac{4n\omega_o^3}{3\hbar c^3}|\mu|^2 & \left[\sec^{-1}\right] \\ B = \dfrac{4\pi^2}{3\hbar^2 n^2}|\mu|^2 & \left[\dfrac{cm^3}{erg\ sec^2}\right] \end{cases} \tag{58}$$

$$\frac{A}{B} = \frac{\hbar\omega_o^3 n^3}{\pi^2 c^3} \qquad \left[\frac{erg\ sec}{cm^3}\right] \tag{59}$$

Let us now consider the more realistic situation in which the two atomic levels are not sharply defined, but have a certain width $\Delta\omega$, such that $\Delta\omega \ll \omega_o$. We have the following:

$N_2^e A_\omega d\omega$ = number of atoms per unit time that decay by spontaneous emission, giving out a photon with angular frequency in $(\omega, \omega+d\omega)$,

$N_2^e B_\omega d\omega$ = number of atoms which in the unit time undergo a downward transition by induced emission, giving out a photon with angular frequency in $(\omega, w+d\omega)$, and

$N_1^e B\rho_\omega d\omega$ = number of atoms which in the unit time undergo an upward transition by absorbing a photon with angular frequency in $(\omega, \omega+d\omega)$.

If we put a filter between the atoms and the walls that lets only the radiation in the narrow band $d\omega$ to interact with the atoms we have

$$N_2^e A_\omega d\omega + N_2^e B_\omega \rho_\omega d\omega = N_1^e B_\omega \rho_\omega d\omega \tag{60}$$

or

$$A_\omega \frac{N_2^e}{N_1^e} + B_\omega \rho_\omega \frac{N_2^e}{N_1^e} = B_\omega \rho_\omega \tag{61}$$

Then

$$A_\omega e^{-\frac{\hbar\omega_o}{kT}} + B_\omega \rho_\omega e^{-\frac{\hbar\omega_o}{kT}} = B_\omega \rho_\omega \tag{62}$$

$$\rho_\omega (e^{\frac{\hbar\omega_o}{kT}} - 1) B_\omega = A_\omega \tag{63}$$

and

$$\frac{A_\omega}{B_\omega} = \rho_\omega (e^{\frac{\hbar\omega_o}{kT}} - 1) \simeq \rho_{\omega_o} (e^{\frac{\hbar\omega_o}{kT}} - 1) = \frac{\hbar\omega_o^3 n^3}{\pi^2 c^3} = \frac{A}{B} \tag{64}$$

If $g(\omega)$ indicates the spectral lineshape

$$\int g(\omega) d\omega = 1 \tag{65}$$

and we can write

$$\begin{cases} A_\omega = A g(\omega) & ; \quad \int A_\omega d\omega = A \\ B_\omega = B g(\omega) & ; \quad \int B_\omega d\omega = B \end{cases} \tag{66}$$

Let us call $w(\omega)d\omega$ the probability per unit time that an atom undergoes an <u>induced</u> transition by absorbing or emitting a photon with angular frequency in $(\omega, \omega+d\omega)$. Then

$$\begin{aligned} w(\omega)d\omega &= B g(\omega) \rho_\omega d\omega = \frac{4\pi^2}{3\hbar^2 n^2} |\mu|^2 \rho_\omega g(\omega) d\omega \\ &= \frac{4\pi^2}{3nc\hbar^2} |\mu|^2 I(\omega) g(\omega) d\omega \end{aligned} \tag{67}$$

where

$$I(\omega)d\omega = \rho_\omega \frac{c}{n} d\omega \quad \text{intensity of radiation with angular frequency in } (\omega, \omega+d\omega) \tag{68}$$

We can write

$$\int w(\omega) d\omega = \frac{4\pi^2}{3\hbar^2 n^2} |\mu|^2 \rho_{\omega_o} = \frac{4\pi^2}{3\hbar^2 nc} |\mu|^2 I(\omega_o) \tag{69}$$

4. Absorption Coefficient and Absorption Cross Section

Let us assume that a plane wave goes through a certain medium in the x-direction. Let the medium consist of atoms which have two possible energy levels and let $N_1 (N_2)$ be the concentration of atoms in the lower (higher) energy level.

The energy intensity per unit angular frequency range $I(\omega)$, when the wave travels through the medium a distance dx, undergoes a change given by

$$dI(\omega) = -w(\omega)(N_1-N_2)\hbar\omega\ dx\left[\frac{erg}{cm^2}\right] \tag{70}$$

But from (67)

$$w(\omega) = \frac{4\pi^2}{3nc\hbar^2}|\mu|^2 I(\omega)g(\omega) \tag{71}$$

Then

$$dI(\omega) = -\left[\frac{4\pi^2}{3nc\hbar^2}|\mu|^2 I(\omega)g(\omega)\right](N_1-N_2)\hbar\omega dx$$

$$= -\left[\frac{4\pi^2(N_1-N_2)}{3nc\hbar}|\mu|^2\omega g(\omega)\right]I(\omega)dx$$

$$= -\alpha(\omega)I(\omega)dx \tag{72}$$

where

$$\alpha(\omega) = \frac{4\pi^2(N_1-N_2)}{3nc\hbar}|\mu|^2\omega g(\omega) = \frac{n\hbar\omega}{c}B(\omega)(N_1-N_2)$$

$$= \underline{\text{absorption coefficient}}\ (cm^{-1}) \tag{73}$$

The solution of (72) is

$$I(\omega;x) = I(\omega;x\ 0)e^{-\alpha x} \tag{74}$$

We define the <u>absorption cross section</u> of the radiative transition as follows

$$\sigma(\omega) = \frac{\alpha(\omega)}{N_1-N_2} = \frac{4\pi^2}{3nc\hbar}|\mu|^2\omega g(\omega)\ (cm^2) \tag{75}$$

Note the following

$$\int\alpha(\omega)d\omega = \frac{4\pi^2(N_1-N_2)}{3nc\hbar}|\mu|^2\omega_o = \frac{n\hbar\omega_o}{c}B(N_1-N_2)\ (cm^{-1}sec^{-1}) \tag{76}$$

In Table I we have summmarized the results obtained so far.

I.C. Saturation

Let us consider a medium consisting of two-level atoms and transversed by radiation (see Fig. 1). Let, as usual, N_1 and N_2 be the

TABLE 1. SUMMARY OF FORMULAE USED IN SPECTROSCOPY

$$w(\omega) = \frac{4\pi^2}{3n^2\hbar^2} |\mu|^2 \rho_\omega\, g(\omega) = \frac{4\pi^2}{3nc\hbar^2} |\mu|^2 I(\omega) g(\omega) = \frac{\sigma(\omega) I(\omega)}{\hbar\omega}$$

$$w = \int w(\omega) d\omega = \frac{4\pi^2}{3n^2\hbar^2} |\mu|^2 \rho_{\omega_o} = \frac{4\pi^2}{3nc\hbar^2} |\mu|^2 I(\omega_o) = B\rho_{\omega_o} \ (\text{sec}^{-1})$$

$$\alpha(\omega) = \frac{4\pi^2 (N_1 - N_2)}{3nc\hbar} |\mu|^2 \omega g(\omega) = \frac{n\hbar\omega}{c} B_\omega (N_1 - N_2) \quad (\text{cm}^{-1})$$

$$\sigma(\omega) = \frac{\alpha(\omega)}{N_1 - N_2} \quad \frac{4\pi^2}{3nc\hbar} |\mu|^2 \omega g(\omega) \quad (\text{cm}^2)$$

$$\int \alpha(\omega) d\omega = \frac{4\pi^2 (N_1 - N_2)}{3nc\hbar} |\mu|^2 \omega_o = \frac{n\hbar\omega_o}{c} B(N_1 - N_2) \quad (\text{cm}^{-1}\ \text{sec}^{-1})$$

$$\int \sigma(\omega) d\omega = \frac{4\pi^2}{3nc\hbar} |\mu|^2 \omega_o = \frac{n\hbar\omega_o}{c} B = \frac{2\pi^2 e^2}{mcn} f \quad (\text{cm}^2\ \text{sec}^{-1})$$

where

$$f = \frac{2m\omega_o}{3\hbar e^2} |\mu|^2$$

$$A = \frac{1}{\tau_{sp}} = \frac{4n\omega_o^3}{3\hbar c^3} |\mu|^2 = \frac{2e^2\omega_o^2 n}{mc^3} f (\text{sec}^{-1})$$

$$A_\omega = Ag(\omega)$$

$$B = \frac{4\pi^2}{3n^2\hbar^2} |\mu|^2 = \frac{2\pi^2 e^2}{m\hbar\omega_o n^2} f \quad \left[\frac{\text{cm}^3}{\text{erg sec}^2}\right]$$

$$B_\omega = Bg(\omega) \quad \left[\frac{\text{cm}^3}{\text{erg sec}}\right]$$

$$\frac{A}{B} = \frac{\hbar\omega_o^3 n^3}{\pi^2 c^3} \quad \left[\frac{\text{cm}^3}{\text{erg sec}}\right]$$

$$\frac{A_\omega}{B_\omega} = \frac{A}{B}$$

$$f\tau_{sp} = 1.5 \frac{\lambda^2}{n}$$

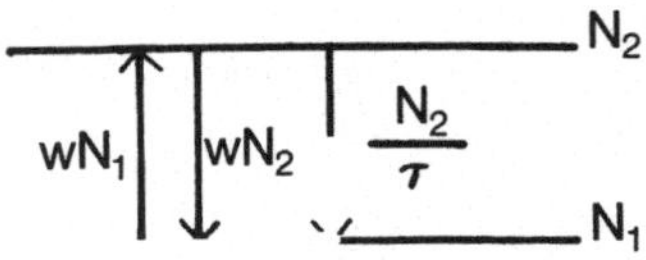

Fig. 1. Two-Level System Interacting with Radiation

populations per unit volume in the ground and excited levels, respectively. In addition, let us assume that w is restricted to a very small band within the linewidth of the transition. We can write

$$\begin{cases} N_1 + N_2 = N_t \\ N_2 = -w(N_2 - N_1) = -\dfrac{N_2}{\tau} \end{cases} \tag{77}$$

where $\tau^{-1} = A$. Let us set

$$\Delta N = N_1 - N_2 \tag{78}$$

Then

$$N = N_t - 2N_2 \tag{79}$$

and

$$\begin{cases} N_2 = \dfrac{N_t - \Delta N}{2} \\ N_1 = N_t - N_2 \end{cases} \tag{80}$$

Also

$$\dot{N}_1 = -\dot{N}_2 \tag{81}$$

and

$$\Delta\dot{N} = \dot{N}_1 - \dot{N}_2 = -2\dot{N}_2 = -2\left[-w(N_2 - N_1) - \frac{N_2}{\tau}\right]$$

$$= -2w\ \Delta N + \frac{2N_2}{\tau} = -2w\Delta N + \frac{N_t - \Delta N}{\tau} \tag{82}$$

or

$$\Delta\dot{N} = -\Delta N\left(\frac{1}{\tau} + 2w\right) + \frac{1}{\tau} N_t \tag{83}$$

In steady state

$$\Delta\dot{N} = 0 = -\Delta N(\frac{1}{\tau} + 2w) + \frac{1}{\tau} N_t \tag{84}$$

and

$$\Delta N = \frac{N_t}{1+2w\tau} \tag{85}$$

For $w \to \infty$, $\Delta N = 0$ and

$$N_1 = N_2 = \frac{N_t}{2} \tag{86}$$

In order to maintain the population difference ΔN, the material must absorb the following power per cm^3

$$\frac{\Delta P}{\Delta V} = \hbar\omega w \Delta N = \hbar\omega \frac{N_t w}{1+2w\tau} \tag{87}$$

At <u>saturation</u> ($w \gg \frac{1}{\tau}$)

$$\left(\frac{\Delta P}{\Delta V}\right)_{sat} = \frac{\hbar\omega N_t}{2\tau} \tag{88}$$

The above relation expresses the fact that the power per unit volume that is absorbed by the system during saturation is equal to the power per unit volume lost by the system because of the decay of the upper state. Now

$$\frac{\frac{\Delta P}{\Delta V}}{\left(\frac{\Delta P}{\Delta V}\right)_{sat}} = \frac{\hbar\omega \frac{N_t w}{1+2w\tau}}{\frac{\hbar\omega N_t}{2\tau}} = \frac{w2\tau}{1+2w\tau} = \frac{\frac{\sigma J}{\hbar\omega} 2\tau}{1+ \frac{\sigma J}{\hbar\omega} 2\tau} = \frac{J/\left(\frac{\hbar\omega}{2\sigma\tau}\right)}{1+J/\left(\frac{\hbar\omega}{2\sigma\tau}\right)} = \frac{J/J_s}{1+J/J_s} \tag{89}$$

where J = intensity of the radiation, and

$$J_s = \frac{\hbar\omega}{2\sigma\tau} \tag{90}$$

If $\omega = \omega_o$ J_s depends only on the parameters of the transition and is called <u>saturation intensity</u>. Also

$$\frac{\Delta N}{N_t} = \frac{1}{1+2w\tau} = \frac{1}{1 + \frac{\sigma J}{\hbar\omega} 2\tau} = \frac{1}{1 + J/\left(\frac{\hbar\omega}{2\sigma\tau}\right)} = \frac{1}{1 + J/J_s} \tag{91}$$

For $J = J_s = \frac{\hbar\omega}{2\sigma\tau}$

$$\frac{\Delta N}{N_t} = \frac{1}{2} \quad ; \quad N_1 = \frac{3}{4} N_t, \; N_2 = \frac{1}{4} N_t \tag{92}$$

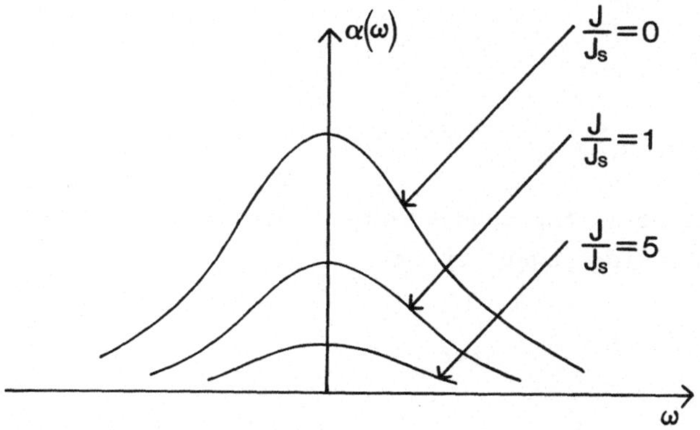

Figure 2. Behavior of a Homogeneously Broadened Line

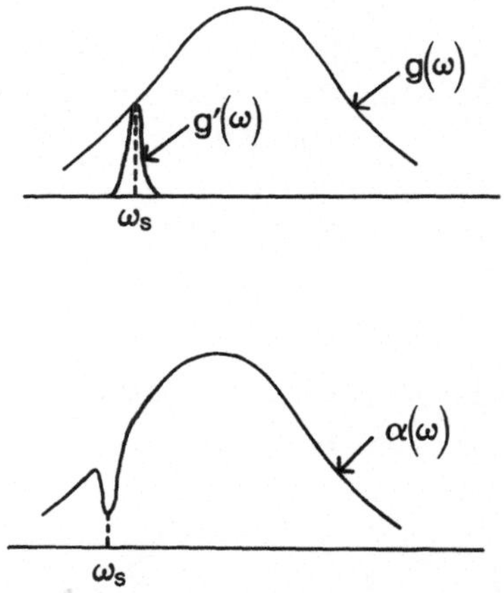

Figure 3. Behavior of an Inhomogeneously Broadened Line

I.D. Line Broadening

The expression for the absorption coefficient can be written

$$\alpha(\omega) = \frac{4\pi^2 \Delta N}{3nc\hbar} |\mu|^2 \omega g(\omega)$$

$$= \frac{4\pi^2}{3nc\hbar} |\mu|^2 \omega g(\omega) \frac{N_t}{1 + J/J_s} = \frac{\alpha_o(\omega)}{1 + J/J_s} \quad (93)$$

where use has been made of (91) and where

$$\alpha_o(\omega) = \frac{4\pi^2}{3nc\hbar} |\mu|^2 \omega \, N_t \, g(\omega) \quad (94)$$

For a homogeneously broadened line (see Fig. 2) when J is increased the absorption coefficient is reduced, but the line shape does not change.

For an inhomogeneously broadened line the situation is depicted in Figure 3. An incoming beam of intensity $J'(\omega_s)$ will interact only with the centers whose frequency of resonance is $\approx \omega_s$: only these centers' transitions will be saturated. The result is a hole in $\alpha(\omega)$ at $\omega \approx \omega_s$. The width of the hole is equal to the width of the homogeneously broadened line $g'(\omega)$.

Examples of homogeneously broadening mechanisms are: natural line broadening, collisional line broadening in gaseous system, broadening of spectral lines of ions in solids due to lattice-ion interaction. Examples of inhomogeneously broadening mechanisms are: Doppler broadening in gaseous systems, broadening of sharp spectral lines of ionic impurities in solids at very low temperatures, due to the inhomogeneities of the crystalline environment.

For neon atoms at a pressure of 0.5 Torr

pressure independent (Doppler broadening, natural broadening)

Doppler broadening >> natural broadening >> collision broadening

1.7 GH_z 20 MH_z 0.64 MH_z

pressure dependent (collision broadening)

(95)

At sufficient high pressures (i.e. CO_2 at atmospheric pressure) the collision broadening dominates over the Doppler broadening.

For "sharp" lines of transition metal ions or rare earth ions in solids:

at high T	T >> 77K	homogeneous broadening	(96)
at low T	T << 77K	inhomogeneous broadening (mostly)	

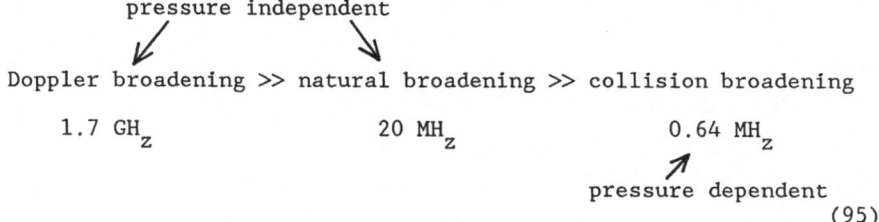

I.E. Lifetime-Broadening Mechanism. Lorentzian Lineshape

Let us assume that we have a system represented by a Hamiltonian H_o under the action of a time dependent perturbation H′ and let

$$H = H_o + H' \tag{97}$$

The wavefunction of the system is given by the equation

$$H\Psi(t) = i\hbar \frac{\partial\psi(t)}{\partial t} \tag{98}$$

and can be expanded in terms of the eigenfunctions $\psi_i(t)$ of H_o:

$$\Psi(t) = \sum_i c_i(t)\ \psi_i(t) \tag{99}$$

where

$$H_o\psi_i(t) = E_i\psi_i(t) \tag{100}$$

and

$$\psi_i(t) = \psi_i(0)e^{-\frac{i}{\hbar} E_i t} \tag{101}$$

Replacing (99) in (98) we obtain

$$\sum_i c_i(t)\ H'\ \psi_i(t) = i\hbar \sum_i \dot{c}_i(t)\ \psi_i(t) \tag{102}$$

Multiplying by $\psi_j^*(t)$ and integrating over space coordinates

$$i\hbar\ \dot{c}_j(t) = c_i(t)\ H'_{ji}\ e^{i\omega_{ji}t} \tag{103}$$

where

$$H'_{ij} = < \psi_j(0)|H'|\ \psi_i(0) > \tag{104}$$

$$\omega_{ji} = \frac{1}{\hbar}(E_j - E_i) \tag{105}$$

Assume now that the system may reside only in two states, say $\psi_i(t)$ or $\psi_j(t)$; in addition assume that $H'_{11} = H'_{22} = 0$. Then the equations (103) reduce to

$$\begin{cases} i\hbar\ \dot{c}_j(t) = c_i(t)\ H'_{ji}\ e^{i\omega_{ji}t} \\ i\hbar\ \dot{c}_i(t) = c_j(t)\ H'_{ij}\ e^{i\omega_{ij}t} \end{cases} \tag{106}$$

These two equations can be solved if we know the initial conditions and how H′ varies with time. We assume

$$\begin{cases} c_i(0) = 1 \\ c_j(0) = 0 \end{cases} \tag{107}$$

and, in addition, $H' = 0$ for $t < 0$, and $H' =$ const for $t \geq 0$. Then

$$\begin{cases} c_j(t) = \dfrac{H'_{ji}}{i\hbar a}\, e^{\frac{i\omega_{ji}t}{2}} \sin at \\ c_i(t) = e^{-\frac{i\omega_{ji}t}{2}} \left(\cos at + i\,\dfrac{\omega_{ji}}{2a} \sin at\right) \end{cases} \tag{108}$$

where

$$a = \frac{1}{2}\left[\omega_{ji}^2 + \frac{4\,|H'_{ji}|^2}{h^2}\right]^{1/2} \tag{109}$$

The probabilities of finding the system in the states j and i, respectively, can now be written

$$\begin{cases} |c_j(t)|^2 = \dfrac{|H'_{ji}|^2}{\hbar^2 a^2} \sin^2 at \\ |c_i(t)|^2 = \cos^2 at + \dfrac{\omega_{ji}{}^2}{4a^2} \sin^2 at \end{cases} \tag{110}$$

They are periodic functions of time, the period being π/a.

If we assume that the state j is in a continuum of states the equation (106) can be written

$$\begin{cases} i\hbar\dot{c}_j = c_i\, H'_{ji}\, e^{i\omega_{ji}t} & \text{(a)} \\ ih\dot{c}_i = \int c_j\, H'_{ij}\, e^{-i\omega_{ji}t}\, dj & \text{(b)} \end{cases} \tag{111}$$

If we set

$$c_i(t) = e^{-\frac{\gamma}{2}t} \tag{112}$$

(111a) gives

$$c_j(t) = \frac{1}{i\hbar} H'_{ji} \frac{e^{(i\omega_{ji} - \frac{\gamma}{2})t} - 1}{i\omega_{ji} - \frac{\gamma}{2}} \tag{113}$$

Replacing $c_j(t)$ with this expression in (111b) we obtain

$$\gamma = \frac{2i}{\hbar^2} \int |H'_{ij}|^2 \frac{1 - e^{(i\omega_{ij} + \frac{\gamma}{2})t}}{\omega_{ij} - \frac{i\gamma}{2}}\, dj$$

But

$$dj = \frac{dj}{d\omega_{ij}}\, d\omega_{ij} = \rho(\omega_{ij})\, d\omega_{ij} \tag{114}$$

where $\rho(\omega_{ij})$ is the density of j-states. Then

$$\gamma = \frac{2i}{\hbar^2} \int |H'_{ij}|^2 \frac{1 - e^{(i\omega_{ij} + \frac{\gamma}{2})t}}{\omega_{ij} - \frac{i\gamma}{2}} \rho(\omega_{ij}) d\omega_{ij} \tag{115}$$

In evaluating this integral we neglect γ in the integral and write

$$\gamma = \frac{2i}{\hbar^2} \int |H'_{ij}|^2 \frac{1 - e^{i\omega_{ij}t}}{\omega_{ij}} \rho(\omega_{ij}) d\omega_{ij} \tag{116}$$

Consider the function

$$\frac{1 - e^{i\omega t}}{\omega} = \frac{1 - \cos \omega t}{\omega} - i \frac{\sin \omega t}{\omega} \tag{117}$$

For very large t the first term may be replaced by $\frac{1}{\omega}$ for $\omega \neq 0$, because the rapidly oscillating cos ωt does not give any contribution to the integral. At ω 0, this term goes to zero. When it is multiplied by the rest of the integral and integrated over ω, the result is the principal value of the integrand

$$\lim \frac{1 - \cos \omega t}{\omega} \rightarrow \frac{\mathcal{P}}{\omega} \tag{118}$$

For very large t the second term in (117) can be expressed as $-i\pi\delta(\omega)$. Therefore

$$\lim_{t\to\infty} \frac{1 - e^{i\omega t}}{\omega} = \frac{\mathcal{P}}{\omega} - i\pi\delta(\omega) \tag{119}$$

Then

$$\begin{aligned} \gamma \xrightarrow[t\to\infty]{} & \frac{2i}{\hbar^2} \int |H'_{ji}|^2 \rho(\omega_{ij}) \left[\frac{\mathcal{P}}{\omega_{ij}} - i\pi\delta(\omega_{ij}) \right] d\omega_{ij} \\ &= \frac{2\pi}{\hbar^2} |H'_{ij}|^2 \rho(\omega_i = \omega_j) + \frac{i2}{\hbar^2} \mathcal{P} \int \frac{|H'_{ij}|^2 \rho(\omega_{ij})}{\omega_{ij}} d\omega_{ij} \\ &= w + i \mathcal{J} \end{aligned} \tag{120}$$

where

$$w = \mathrm{Re}(\gamma) = \frac{2\pi}{\hbar^2} |H'_{ij}|^2 \rho(\omega_i = \omega_j) \tag{121}$$

$$\mathcal{J} = \mathrm{Im}(\gamma) = \frac{2}{\hbar} \mathcal{P} \int \frac{|H'_{ij}|^2 \rho(\omega_{ij})}{\hbar\omega_{ij}} d\omega_{ij} \tag{122}$$

We note that w is the probability per unit time of the i $\rightarrow$ j transition. We can write

$$c_i(t) = e^{-\frac{\gamma}{2}t} = \exp\left(-\frac{w}{2} t - i \frac{\mathcal{J}}{2}\right)$$

$$= \exp\left[-\frac{w}{2}t - \frac{i}{\hbar}\delta E\, t\right] \tag{123}$$

where

$$\delta E = \mathcal{P}\int \frac{|H'_{ij}|^2\,\rho(\omega_{ij})}{\hbar\omega_{ij}}\,d\omega_{ij} \tag{124}$$

represents a correction to the energy to the second order in the perturbation.

Let us now go back to formula (113) for $c_j(t)$:

$$c_j(t) = \frac{1}{\hbar}H'_{ji}\,\frac{e^{(i\omega_{ji}-\frac{\gamma}{2})t}-1}{i\omega_{ji}-\frac{\gamma}{2}}$$

$$\xrightarrow[t\gg\frac{1}{w}]{} \frac{i}{\hbar}H'_{ji}\,\frac{1}{i(\omega_j-\omega_i)-\frac{1}{2}(w+i\mathcal{I})}$$

$$= \frac{i}{\hbar}H'_{ji}\,\frac{1}{i[\omega_j-(\omega_i+\frac{1}{2}\mathcal{I})]-\frac{1}{2}w} \tag{125}$$

and

$$|c_j(t)|^2 = \frac{1}{\hbar^2}|H'_{ji}|^2\,\frac{1}{\left[\left(\omega_i+\frac{1}{2}\mathcal{I}\right)-\omega_j\right]^2+\frac{w^2}{4}} \tag{126}$$

It is interesting to apply the above results to the case of an atom in an excited state interacting with the radiation field. In this case the unperturbed system consists of the atom and the radiation field and the perturbing Hamiltonian H′ is given by the interaction between them. The following observations can be made:

1. The real part of γ (equal to the transition probability per unit time) produces a finite width of the transition. In the case of an atom interacting with radiation this width is called the natural linewidth. Spectral lines with the shape (126) are called Lorentzian lines. The appearance of a line breadth derives from the fact that we have allowed the probability $|c_i|^2$ of the atom being in the initial state to decay for a time which is greater than the lifetime of the initial state. For times much shorter than this lifetime, the probability $|c_i|^2$ maintains practically the initial value and the line is very sharp.

2. The imaginary part of γ produces a shift in the energy of the initial level of the transition. This shift is in effect the Lamb shift.

3. By the use of the principal value $\mathcal{P}$, (122) takes care of the resonances that may occur in correspondence to the emission and reabsorption of virtual photons with energy equal to the energy difference between the initial level and a different level connected to it by H′.

From (126) we can see that <u>the width at half maximum intensity is equal to the total transition probability</u> per unit time

$$\Delta E = \hbar w \tag{127}$$

We may call w^{-1} the lifetime τ of the state i,

$$(\Delta E)\tau = \hbar \tag{128}$$

This is nothing but the uncertainty relation between energy and time which expresses the fact that we may know the energy of a system with an accuracy $\Delta E = \hbar/\tau$, if only the time τ is available to measure it.

Such effects as linewidth and energy shift are a consequence of the interaction between the atomic system and the radiation; similar effects are however found when an atom in a crystal interacts with the thermal vibrations of the lattice. The formalism developed in this section applies whenever an isolated atom interacts with a large number of other degrees of freedom whose density is practically continuous.

Several lifetime shortening mechanisms may be present at the same time; for example, thermal vibrations and radiation damping may be operating at the same time; also several transitions may originate from the same level. In this case the breadth of the level is given by the sum of the breadths due to all these processes,

$$\Delta E = \hbar\left(\sum_j w_{ij}\right) \tag{129}$$

We note also that when a transition connects two excited levels, one with breadth w_i and the other with breadth w_j, the linewidth of the transition is $w_i + w_j$.

A Lorentzian profile is due to interactions between the radiative or absorbing systems and some time dependent perturbations; these interactions are <u>the same</u> for each atom contributing to the emission or absorption.

I.F. <u>Time-Independent Random Perturbations. Gaussian Lineshape</u>

A Gaussian distribution of frequencies is in general due to a completely different type of mechanism.

An example of Gaussian distribution is presented by the so-called <u>Doppler broadened lines</u>. These lines are produced by the fact that atoms or molecules in a container have a Maxwellian distribution of velocities expressed by exp $(-mv^2/2kT)$. The light emitted by an atom in its excited state is seen in the x direction with the frequency shifted by

$$|\nu - \nu_o| = \frac{\nu_o |v_x|}{c} \tag{130}$$

for $v \ll c$. This fact produces a line with a profile given by

$$\text{const} \times \exp\left[-\frac{mc^2(\nu - \nu_0)^2}{2\nu_0^{\,2}kT}\right] \tag{131}$$

which describes a Gaussian line shape.

In this case the profile may be considered as the superposition of a great number of independent spectral lines, with each line corresponding to transitions that take place in a certain number of atoms. This fact turns out to be a general property of Gaussian lines.

Consider an ensemble of radiating ions in a crystal, in which the influence of the environment on each ion can be thought as due to the presence of a crystalline field. The ions in the crystal may actually see a slightly different crystalline field in dependence of their position; it is a fact that no crystal is perfect and that internal microscopic strains can be present. We may assume that these perturbations of the crystalline field are completely random in space; therefore each ion has an energy that is slightly different from the energy of another ion. The crystal will then produce a line which is the superposition of many lines of different frequencies.

It is known from the theory of probability that the probability distribution of the sum of a very great number of independent and random variables is a function of the type $Ce^{-\alpha^2 x^2}$ (C, α constant); namely, it is a Gaussian distribution function. This result is known as the central limit theorem [2]. The presence of microscopic random distortions in a crystal will then produce a Gaussian spectral line.

I.G. Superposition of Probability Densities: Lorentzian, Gaussian and Voigt Lineshapes

The probability dp that a continuous random variable x takes a value in $(x, x + dx)$ can be introduced as follows:

$$dp = p(x)dx \tag{132}$$

The function $p(x)$ is said to be a probability density. Probability densities respect the normalization condition,

$$\int_{-\infty}^{+\infty} p(x)dx = 1 \tag{133}$$

The nth moment of the variable x is given by

$$M_n = \overline{x^n} = \int_{-\infty}^{+\infty} x^n p(x)dx \tag{134}$$

If $p(x)$ is an even function $\overline{x^n} = 0$ for n odd.

Given a certain probability density $p(x)$, we define as characteristic function of $p(x)$ the Fourier transform:

$$s(t) = \int_{-\infty}^{+\infty} p(x)\, e^{-itx}\, dx \tag{135}$$

Given $s(t)$, the probability density is

$$p(x) = \frac{1}{2\pi} \int_{-\infty}^{+\infty} s(t)\, e^{itx}\, dt \tag{136}$$

We define also the variance of x from $\bar{x}$ as follows:

$$\sigma^2 = \int p(x)(x-\bar{x})^2 \, dx = \overline{x^2} - 2\bar{x} \int xp(x)dx + (\bar{x})^2 = \overline{x^2} - (\bar{x})^2 \quad (137)$$

The square root of the variance, σ, is known as <u>the standard deviation</u> of x from $\bar{x}$. The above relation expresses the fact that the variance is equal to the second moment minus the squared first moment. If $\bar{x} = 0$, $\sigma^2 = \overline{x^2}$.

Deriving s(t) once and twice we get

$$s'(t) = -i \int_{-\infty}^{+\infty} xp(x)e^{itx} \, dx \quad (138)$$

$$s''(t) = -\int_{-\infty}^{+\infty} x^2 p(x)e^{-itx} \, dx \quad (139)$$

Also

$$s'(0) = -i\bar{x} \quad (140)$$

$$s''(0) = -\overline{x^2} \quad (141)$$

Then, replacing these values in (137) we find

$$\sigma^2 = -s''(0) + [s'(0)]^2 \quad (142)$$

Given two independent probabilities density p(y) and q(w), the probability density of the variable x = y + w is given by the <u>convolution integral</u>

$$r(x) = \int_{-\infty}^{+\infty} p(y)q(x-y)dy \quad (143)$$

Let us call s(t), $s_p(t)$ and $s_q(t)$, the characteristic functions of r(x), p(y) and q(w), respectively. We have then

$$s_p(t)s_q(t) = \int_{-\infty}^{+\infty} p(y)e^{-ity} \, dy \int_{-\infty}^{+\infty} q(w)e^{itw} \, dw$$

$$= \int_{-\infty}^{+\infty} \int_{-\infty}^{+\infty} p(y)q(w)e^{-it(y+w)} \, dy \, dw$$

$$= \int_{-\infty}^{+\infty} \int_{-\infty}^{+\infty} p(y)q(x-y)e^{-itx} \, dx \, dy \quad s(t) \quad (144)$$

Therefore the characteristic function of the convolution of two probability densities is equal to the product of the characteristic functions of the two probability densities.

A probability density of the form

$$p_L(\nu) = \frac{\Delta\nu_L}{2\pi} \frac{1}{(\nu - \nu_0)^2 + (\Delta\nu_L/2)^2} \tag{145}$$

is called Lorentzian. The characteristic function of such a probability density is

$$s_L(t) = \int_{-\infty}^{+\infty} p_L(\nu)e^{-it\nu}\, d\nu$$

$$= \frac{\Delta\nu_L}{2\pi} \int_{-\infty}^{+\infty} \frac{e^{-i\nu t}}{(\nu - \nu_0)^2 + (\Delta\nu_L/2)^2}\, d\nu = e^{-(\Delta\nu_L/2)|t|}e^{-i\nu_0 t} \tag{146}$$

A probability density of the form

$$p_G(\nu) = \frac{1}{\sigma\sqrt{2\pi}} \exp\left[-\frac{(\nu - \nu_0)^2}{2\sigma^2}\right] \tag{147}$$

is called Gaussian. The characteristic function of such a probability density is given by

$$s_G(t) = \frac{1}{\sigma\sqrt{2\pi}} \int_{-\infty}^{+\infty} e^{-it\nu} \exp\left[-\frac{(\nu - \nu_0)^2}{2\sigma^2}\right] d\nu$$

$$= \frac{1}{\sigma\sqrt{2\pi}} \int_{-\infty}^{+\infty} e^{-i(\nu'+\nu_0)t} e^{-(\nu'^2/2\sigma^2)}\, d\nu'$$

$$= \frac{e^{-i\nu_0 t}\, e^{-(\sigma^2 t^2/2)}}{\sigma\sqrt{2\pi}} \int_{-\infty}^{+\infty} \exp\left[-\left(\frac{\nu'}{\sqrt{2\pi}} + \frac{i\sigma t}{\sqrt{2}}\right)^2\right] d\nu' . \tag{148}$$

Putting

$$\frac{\nu'}{\sqrt{2\pi}} + \frac{i\sigma t}{\sqrt{2}} = y,$$

we get

$$d\nu' = \sqrt{2}\,\sigma\, dy,$$

and

$$s_G(t) = \frac{e^{-i\nu_0 t} e^{-(\sigma^2 t^2/2)}}{\sigma\sqrt{2\pi}} \sqrt{2}\,\sigma \int_{-\infty}^{+\infty} e^{-\nu^2}\, dy = e^{-i\nu_0 t} e^{-(\sigma^2 t^2/2)} \tag{149}$$

The first and the second derivatives are given by

$$s'_G(t) = -(i\nu_0 + \sigma^2 t)e^{-i\nu_0 t-(\sigma^2 t^2/2)} \tag{150}$$

$$s_G''(t) = (i\nu_0 + \sigma^2 t)^2 e^{-i\nu_0 t-(\sigma^2 t^2/2)} - \sigma^2 e^{-i\nu_0 t-(\sigma^2 t^2/2)} \tag{151}$$

and

$$s_G'(0) = - i\nu_0 = - i\overline{\nu} \tag{152}$$

$$s_G''(0) = - \nu_0^2 - \sigma^2 = - \overline{\nu^2} \tag{153}$$

Therefore the average value of $p_G(\nu)$ is ν_0 and the variance is σ^2.

We may want to express the probability distribution (147) in terms of the width at half intensity $\Delta\nu_G$. The following relation between the standard deviation and the halfwidth $\Delta\nu_G$ can be easily derived:

$$\sigma = \frac{\Delta\nu_G}{2\sqrt{2 \ln 2}} \tag{154}$$

We can now express a Gaussian probability density in the following way:

$$p_G(\nu) = \frac{2\sqrt{\ln 2}}{\Delta\nu_G\sqrt{\pi}} \quad \exp\left[-\left(\frac{2(\nu - \nu_0)}{\Delta\nu_G}\sqrt{\ln 2}\right)^2\right] \tag{155}$$

Given now two probability densities $p_L(\ell)$ and $p_G(\delta)$, let us consider the probability density of the variable $\ell + \delta = \nu$. Setting $\nu_0 = 0$ we obtain

$$r_V(\nu) = \int_{-\infty}^{+\infty} p_G(\delta)p_L(\nu - \delta)\, d\delta$$

$$= \frac{2}{\pi\Delta\nu_L} \frac{2\sqrt{\ln 2}}{\Delta\nu_G\sqrt{\pi}} \int_{-\infty}^{+\infty} \frac{\exp\left[-\left(\frac{2\delta}{\Delta\nu_G}\sqrt{\ln 2}\right)^2\right]}{1 + \frac{2}{\Delta\nu_L}(\nu - \delta)^2}\, d\delta$$

$$= \frac{4\sqrt{\ln 2}}{\pi\sqrt{\pi}} \frac{1}{\Delta\nu_L\Delta\nu_G} \int_{-\infty}^{+\infty} \frac{\exp\left[-\left(\frac{2\delta}{\Delta\nu_G}\sqrt{\ln 2}\right)^2\right]}{1 + \frac{2}{\Delta\nu_L}(\nu - \delta)^2}\, d\delta \tag{156}$$

Let

$$y = \frac{2\delta}{\Delta\nu_G}\sqrt{\ln 2}; \qquad d\delta = \frac{\Delta\nu_G}{2\sqrt{\ln 2}}\, dy,$$

$$\omega = \frac{2\nu\sqrt{\ln 2}}{\Delta\nu_G} \tag{157}$$

$$a = \frac{\Delta\nu_L}{\Delta\nu_G}\sqrt{\ln 2}$$

With these, $r_V(\nu)$ becomes

$$r_V(\nu) = \frac{4\sqrt{\ln 2}}{\pi\sqrt{\pi}} \frac{1}{\Delta\nu_L \Delta\nu_G} \frac{\Delta\nu_L{}^2\sqrt{\ln 2}}{2\Delta\nu_G} \int_{-\infty}^{+\infty} \frac{e^{-y^2}}{a^2 + (\omega - y)^2} dy$$

$$= \frac{2 \ln 2}{\pi\sqrt{\pi}} \frac{\Delta\nu_L}{\Delta\nu_G{}^2} \int_{-\infty}^{+\infty} \frac{e^{-y^2}}{a^2 + (\omega - y)^2} dy \tag{158}$$

This probability density, already normalized, is called Voigt probability density.

In Figure 4 we report the three shapes of a Guassian, a Lorentzian and a Voigt profiles with the same half width. The Voigt probability density reduces to a Lorentzian if $\Delta\nu_G = 0$ and to a Gaussian if $\Delta\nu_L = 0$.

Given a Voigt profile, $\Delta\nu_L$ and $\Delta\nu_G$ may be derived in the following way. We may write

$$s_V(t) = s_G(t)s_L(t) = e^{-i\nu_0 t}e^{-(\sigma^2t^2/2)}e^{-i\nu_0 t}e^{-(\Delta\nu_L/2)|t|}$$

$$= e^{-2i\nu_0 t}e^{-(\sigma^2t^2/2)}e^{-(\Delta\nu_L/2)|t|} \tag{159}$$

Putting $\nu_0 = 0$, for $t > 0$, we have

$$\ln s_V(t) = -\frac{\sigma^2t^2}{2} \frac{\Delta\nu_L t}{2} = -\left(\frac{\sigma^2}{2} t^2 + \frac{\Delta\nu_L}{2} t\right) \tag{160}$$

where the first term predominates for large t's, and the second term for small t's. By plotting $\ln s_V(t)$ in a semilog paper versus t, and versus t^2 we may find σ and $\Delta\nu_L$ respectively for the Gaussian and the Lorentzian contribution [3]. $\Delta\nu_L$ and $\Delta\nu_G$ may also be obtained by using the Posener's tables [4].

I.H. Fluorescence Line Narrowing

Inhomogeneous broadening occurs when the sites occupied by the optically active ions are not exactly equal. In glasses the disorder produces an inhomogeneous line broadening of $\sim$ 100 cm^{-1} for rare earths.

Even in a "perfect" crystal there may be environments which are different for the various ions. The difference of the environments can be due to slight changes in the local crystalline field.

Consider now a system where the optically active ions are in different environments exemplified in Figure 5. Ion to ion energy transfer in this case takes place among nonresonant levels.

If $g_i(E)$ is the homogeneous line shape function of level i, the transfer rate between level i and level k is proportional to the overlap integral of the line shape functions corresponding to the two levels:

$$w_{ik} \propto \int g_i(E - E_i)\, g_k(E - E_k)\, dE \tag{161}$$

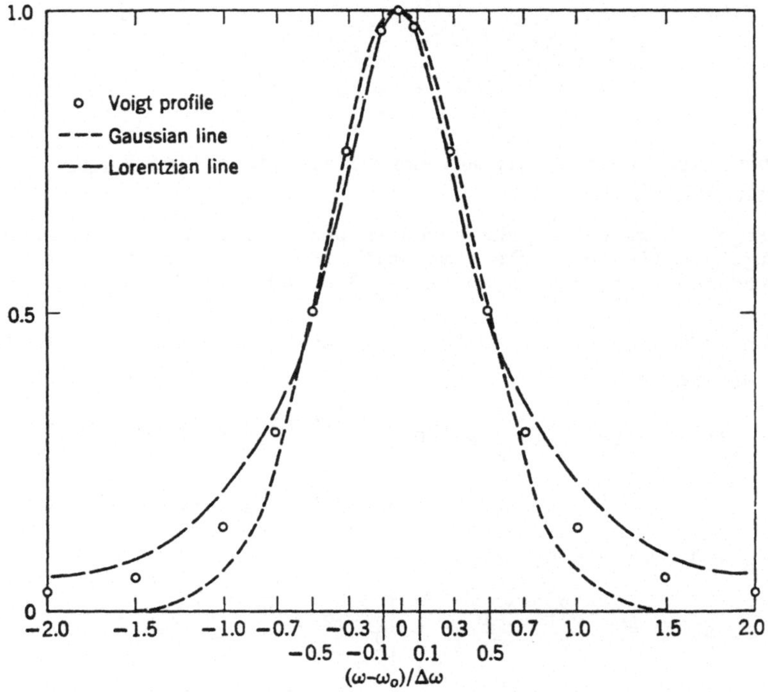

Figure 4. Lorentzian, Gaussian and Voigt Line Shapes

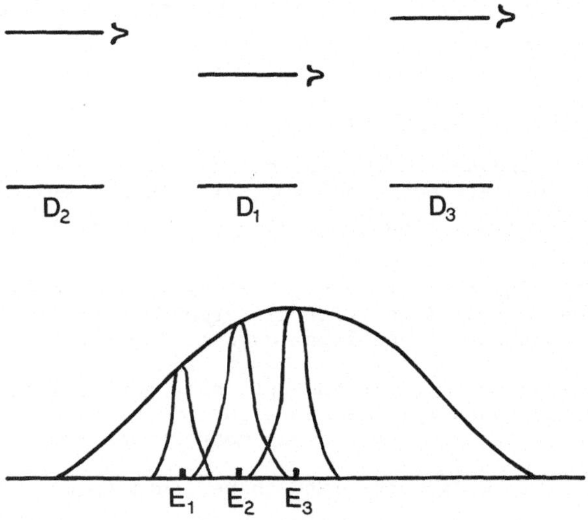

Figure 5. Inhomogeneous Broadening of Donor Levels

This means that the energy transfer rate between two ions i and k is a function not only of their relative distance $|\vec{R}_i - \vec{R}_k|$, but also of the difference of their excited state energies $(E_i - E_k)$:

$$w_{ik} = w_{ik} (|\vec{R}_i - \vec{R}_k|, E_1 - E_k) \qquad (162)$$

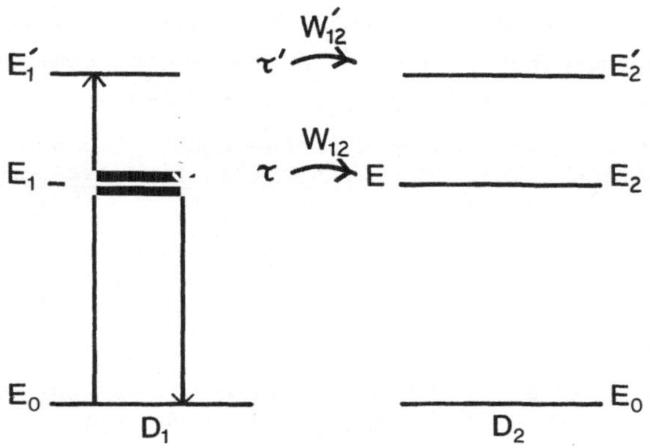

Figure 6. Energy Levels of Ions in Different Environments

Consider now the case in which $|\vec{R}_i - \vec{R}_k|$ is large. Because of the large distance energy transfer is improbable, even if $E_i = E_k$; the inhomogeneous broadening has then no effect in this case. If $|\vec{R}_i - \vec{R}_k|$ is small, the inhomogeneous broadening may reduce the overlap integral (161) and consequently prevent the $i \to k$ energy transfer.

If $\rho_i(t)$ is the probability that ion i is excited at time t:

$$\frac{d}{dt} \rho_i(t) = -\frac{1}{\tau} \rho_i(t) - \rho_i(t) \sum_k w_{ik} + \sum_k w_{ki} \rho_k(t) \qquad (163)$$

This relation gives us a system of equations which are coupled through the last term in the right member.

Inhomogeneous broadening may have relevant effects on the spectral characteristics of an emitting system, even if this system consists of ions of one type. Consider Figure 6, where for simplicity only two different environments are represented. In general the distribution of the ions in the available sites may be random, but, by the use of monochromatic (laser) light it is possible to excite selectively the ions residing in a particular environment, say D_1. Ions D_1 are then brougt to level E_1' from which they decay nonradiatively to level E_1. Assume the exciting light intensity to be constant in time and let τ and τ' be the intrinsic lifetimes of level E_1 and E_1', respectively. If $\tau' \gg (\sum_k w_{1k}')^{-1}$ <u>or</u> $\tau \gg (\sum_k w_{1k})^{-1}$, then it takes the excited D_1 ions much less time to transfer their energy to the other D_k ions than to relax. Under these circumstances the selective excitation has no effect, since the energy spreads rapidly among the ions in different sites; the luminescence spectrum will be similar to the one obtained with wideband excitation.

If, on the other hand, $\tau' \ll (\sum_k w_{1k}')^{-1}$ <u>and</u> $\tau \ll (\sum_k w_{1k})^{-1}$, then it takes the excited D_1 ions much less time to relax than to transfer their energy to the other D_k ions. Under these circumstances only the D_1 ions will luminesce with a definite photon energy E_1; the luminescence spectrum will be much narrower than the one obtained with wide band excitation. This effect is called <u>fluorescence line narrowing</u> (FLN) [5].

II. SPECTRAL LINES OF LUMINESCENT CENTERS IN ORDERED SOLIDS

II. A. <u>Systems</u>

We shall now concentrate our attention on the spectral characteristics of dopant ions in crystals. We shall assume the following:

a. The concentration of optically-active ions is very low: no ion-ion interaction.

b. The electronic energy levels of these ions do not overlap with the electronic energy bands of the solid.

II. B. <u>Vibrations of Solids</u>

In dealing with the radiative processes of ions in solids, we have to consider the fact that these ions are part of a structure which undergoes vibrational (thermal) motion. We can make the following observations:

1. The "electronic" energy eigenvalue $\epsilon_k(\vec{R})$ ($\vec{R}$ stand for the internuclear coordinates) plays the role of a potential in which the atoms of the solid perform their vibrational motions.

2. These motions can be thought of as a superposition of normal modes of vibration whcih are represented by normal coordinates q_i ($i = 1,2,\ldots\ldots,3N$), where N = number of atoms in the solid. (More precisely, since the solid has also three translational and three rotational degrees of freedom, the number of the normal modes of vibration is 3N-6).

3. A normal mode is a pattern of motion in which, in general, all the atoms of the solid participate. There may be, however, normal modes, called "localized", which are related to the motion of a relatively small number of atoms.

4. In treating the vibrations of solids, the "harmonic approximation" is used. According to this approximation, no

exchange of energy takes place between the normal modes and the potential energy and the kinetic energy are both sums of independent quadratic ($\sim q^2, \sim \dot{q}^2$) terms, one for each normal mode.

5. Each normal mode is equivalent to a harmonic oscillator. The vibrations of a solid are then equivalent to a collection of 3N harmonic oscillators. If an oscillator of frequency ω is in its nth excited state, this fact is also expressed by saying that n "phonons" of energy $\hbar\omega$ are present in the solid.

6. The different normal modes are not completely isolated, but are, rather, in speaking terms due to their anharmonicity: this fact provides the mechanism for reaching thermal equilibrium.

7. When considering the thermal vibrations of a solid, it may be interesting to have an idea of how many phonons may be present in a solid at, say, room temperature. The number of phonons in a frequency interval $(\omega, \omega + d\omega)$ is given by $\bar{n}(\omega)\rho(\omega)d\omega$ where: $\bar{n}(\omega) = (e^{\hbar\omega/kT}-1)^{-1}$ and $\rho(\omega)$ = density of phonon states $\propto \omega^2/c_s^3$. It is the value of c_s = velocity of sound in solids $\simeq 5 \times 10^5$ cm/sec (versus c = velocity of light = 3×10^{10} cm/sec) that makes the number of phonons extremely large. For the sake of comparison, the total number of photons/cm^3 in black body radiation at T = 300°K is $\sim 6.4 \times 10^8$, the number of phonons/cm^3 in a solid, with a Debye temperature T_D = 1000 K, at T = 300 K is 3.5×10^{23}! The sheer number of phonons may give us an idea of their importance in affecting the spectral characteristics of ions in solids.

8. The electronic energy $\epsilon_k(\vec{R})$ can be represented as a surface in a (3N+1)-fold space. If the system of normal coordinates q_i is used, a potential "curve" will correspond to each coordinate q_i. For a diatomic molecule only one vibrational mode is present (q_i = R).

9. The question now arises: is it legitimate, when considering the potential curves in correspondence to a certain normal coordinate q_i, to put the potential curves for the ground state and for an excited electronic state in the same diagram? Electronic excitation corresponds always to a rearrangement of the electron charges in the internuclear spaces: this may in turn produce changes in the molecular architecture, i.e., the actual positions of the nuclei. In a diatomic molecule, the electronic excitation produces a change in the relative distance of the two nuclei, and in the force constant that controls the vibrational motion; however, since we have only one vibrational coordinate, it is legitimate to represent all the potential curves in the same diagram. This is not the case for a complex molecule, where, in general, angular distortions and changes in the force constants may result from an electronic excitation. This means that, to be precise, a change of electronic state produces also a change in the system of normal coordinates: this change has been called pictorially by Rebane [6] the "scrambling of normal coordinates".

10. According to the adiabatic approximation, the quantum states of a solid are represented by the Born Oppenheimer products

$$\Psi_i(\vec{r},\vec{R}) = \psi_k(\vec{r},\vec{R})\phi_{k\ell}(\vec{R}) \tag{164}$$

where $\vec{r}$ stands for the electron coordinates. We note here that, due to the presence of nonadiabatic terms in the Hamiltonian, the states exemplified in (164) are not truly stationary. The non-adiabaticity may furnish the mechanism by which a system moves nonradiatively from a state Ψ_i, given by (164) above, to a state

$$\Psi_f(\vec{r},\vec{R}) = \psi_m(\vec{r},\vec{R})\phi_{mn}(\vec{R}) \tag{165}$$

11. Finally, we need to mention the so-called Jahn-Teller effect [7]: complexes with degenerate electronic states are unstable to symmetry - lowering distortions that take place in the way of vibrational modes. This fact was established by Jahn and Teller on the basis of symmetry considerations which do not furnish any quantitative information.

II.C. Eigenstates and Eigenfunctions of Vibrations

Let

u_s = displacement of sth atom $(s = 1,2,\ldots 3N)$

m_s = mass of sth atom $(m_1 = m_2 = m_3)$

The potential and kinetic energies are given by

$$\begin{cases} V = V_0 + \frac{1}{2}\sum_{s\,1}^{3N}\sum_{s'\,1}^{3N} A_{ss'}\, u_s\, u_{s'} \\ T = \frac{1}{2}\sum_{s\,1}^{3N} m_s\, \dot{u}_s^2 \end{cases} \tag{166}$$

respectively, where

$$A_{ss'} = \left.\frac{\partial^2 V}{\partial u_s\, \partial u_{s'}}\right|_o = A_{s's} \tag{167}$$

The equations of motions are

$$F_s = m\ddot{u}_s = -\frac{\partial V}{\partial u_s} = -\sum_{s'} A_{ss'}\, u_{s'} \tag{168}$$

$A_{ss'}$ is the sth component of a force due to the displacement $u_{s'}=1$.

We use the coordinate transformation

$$u_s = \frac{1}{\sqrt{m_s}}\sum_k y_{ks}\, q_k \tag{169}$$

where

$$q_k = \sum_s \sqrt{m_s}\; y_{ks}\, u_s \tag{170}$$

are the normal coordinates expressed in terms of the displacements. The following relations give us the properties of the y_{ks} coefficients:

$$q_k = \sum_s \sqrt{m_s}\, y_{ks}\, u_s = \sum_s \sqrt{m_s}\, y_{ks} = \frac{1}{\sqrt{m_s}} \sum_{k'} y_{k's}\, q_{k'}$$

$$= \sum_{k'} q_{k'} \sum_s y_{ks}\, y_{k's} \longrightarrow \sum_s y_{ks}\, y_{k's} = \delta_{kk'} \tag{171}$$

$$u_s = \frac{1}{\sqrt{m_s}} \sum_k y_{ks}\, q_k = \frac{1}{\sqrt{m_s}} \sum_t y_{ks} \sum_{s'} \sqrt{m_{s'}}\, y_{ks'}\, u_{s'}$$

$$= \sum_{s'} \sqrt{\frac{m_{s'}}{m_s}}\, u_{s'} \sum_k y_{ks}\, y_{ks'} \longrightarrow \sum_k y_{ks}\, y_{ks'} = \delta_{ss'} \tag{172}$$

We can then say that the matrix of the y_{ks} coefficients is unitary. The Lagrangian of the vibrations is given by

$$L = T - V = \frac{1}{2} \sum_k \dot{q}_k^2 - \frac{1}{2} \sum_k \omega_k^2\, q_k^2 \tag{173}$$

The frequencies ω_k are function of the $A_{ss'}$ coefficients. The Lagrange equations

$$\frac{d}{dt} \left(\frac{\partial L}{\partial \dot{q}_k} \right) - \frac{\partial L}{\partial q_k} = 0 \tag{174}$$

give

$$\ddot{q}_k + \omega_k^2\, q_k \quad 0 \tag{175}$$

The momentum conjugate to the normal coordinate q_k is given by

$$p_k = \frac{\partial L}{\partial \dot{q}_k} = \dot{q}_k \tag{176}$$

We can then write the expression for the Hamiltonian

$$H = \sum_k p_k\, \dot{q}_k - L = \sum_k p_k^2 - \frac{1}{2} \sum_k \dot{q}_k^2 + \frac{1}{2} \sum_k \omega_k^2\, q_k^2$$

$$= \frac{1}{2} \sum_k (p_k^2 + \omega_k^2\, q_k^2) \tag{177}$$

Also

$$m\dot{u}_s = p_s = \sqrt{m_s} \sum_k y_{ks}\, p_k \tag{178}$$

$$p_k = \dot{q}_k = \sum_k \sqrt{m_s}\; y_{ks}\, \dot{u}_s \tag{179}$$

q_k, p_k are real variables that satisfy Hamilton's equations:

$$\begin{cases} \dfrac{\partial H}{\partial p_k} = p_k = \dot{q}_k \\ \dfrac{\partial H}{\partial q_k} = \omega^2 q_k = -\dot{p}_k \end{cases} \tag{180}$$

At this point we move to a quantum-mechanical treatment. u_s, p_s, q_k, and $p_k = \dot{q}_k$ become operators with the following relations

$$\begin{cases} \left[u_s, p_t\right] = i\hbar\, \delta_{st} \\ \left[u_s, u_t\right] = \left[p_s, p_t\right] = 0 \end{cases} \tag{181}$$

and

$$\begin{cases} \left[q_k, p_{k'}\right] = i\hbar\, \delta_{kk'} \\ \left[q_k, q_{k'}\right] = \left[p_k, p_{k'}\right] = 0 \end{cases} \tag{182}$$

We also introduce the following non-Hermitian dimensionless operators

$$\begin{cases} a_k = \sqrt{\dfrac{\omega_k}{2\hbar}}\, (q_k + \dfrac{i}{\omega_k}\, p_k) \\ \\ a_k^+ = \sqrt{\dfrac{\omega_k}{2\hbar}}\, (q_k - \dfrac{i}{\omega_k}\, p_k) \end{cases} \tag{183}$$

The commutation relations of these operators are

$$\begin{aligned} &\left[a_k, a_{k'}^+\right] = \delta_{kk'} \\ &\left[a_k, a_{k'}\right] = \left[a_k^+, a_{k'}^+\right] = 0 \end{aligned} \tag{184}$$

The Hamiltonian of the vibrations is given by

$$H = \sum_k \hbar\omega_k\, (a_k^+ a_k + \frac{1}{2}) \tag{185}$$

Note that

$$\begin{cases} q_k = \sqrt{\dfrac{\hbar}{2\omega_k}}\, (a_k + a_k^+) \\ \\ p_k = \dfrac{1}{i} \sqrt{\dfrac{\hbar\omega_k}{2}}\, (a_k - a_k^+) \end{cases} \tag{186}$$

In summary we can write the following formulae for the relevant entities:

Hamiltonian:

$$H = \sum_k H_k = \sum_k \hbar\omega_k\, (a_k^+ a_k + \frac{1}{2}) \tag{187}$$

Eigenstates:

$$\Psi_{n_1 n_2 \cdots\cdots} = |n_1> |n_2> \ldots \tag{188}$$

Energy eigenvalues:

$$E_{n_1, n_2 \cdots\cdots} = \sum_k \hbar\omega_k (n_k + \frac{1}{2}) \tag{189}$$

The operators displacement and velocity are represented by

$$u_s = \frac{1}{\sqrt{m_s}} \sum_k y_{ks}\, q_k = \frac{1}{\sqrt{m_s}} \sum_k y_{ks} \sqrt{\frac{\hbar}{2\omega_k}}\ (a_k + a_k^+) \tag{190}$$

$$\dot{u}_s = \frac{1}{\sqrt{m_s}} \sum_k y_{ks}\, \dot{q}_k = \frac{1}{\sqrt{m_s}} \sum_k y_{ks}\, p_k$$

$$= \frac{1}{\sqrt{m_s}} \sum_k y_{ks} \frac{1}{i} \sqrt{\frac{\hbar\omega_k}{2}}\ (a_k - a_k^+) \tag{191}$$

II. D. <u>The Ion-Vibration Interaction</u>

The system we are considering is the optically active ion in the solid. It is represented by a wavefunction of the form

$$\Psi_{k\ell}\ (\vec{r},\vec{R}) = \psi_k\ (\vec{r},\vec{R})\ \phi_{k\ell}\ (\vec{R}) \tag{192}$$

where $\vec{r}$ stands for the coordinates of the electrons of the optically active ion, and $\vec{R}$ stands for the internuclear coordinates of the solid. The subscript k indicate the quantum state of the ion's electrons, and the subscript ℓ the quantum state of the solid's vibrations.

In order to deal with radiative transitions we use the Frank-Condon approximation. Accordingly the relevant matrix element for a radiative transition from an initial state (k,ℓ) to a final state (m,n) is given by

$$\vec{M}_{fi} = \vec{M}_{mn,k\ell} = < \phi_{mn}(\vec{R}) | \phi_{k\ell}(\vec{R}) > \vec{D}^o_{mk} \tag{193}$$

where

$$\vec{D}^o_{mk} = <\psi_m(\vec{r},\vec{R}_o) | e\ \vec{r} | \psi_k(\vec{r},\vec{R}_o) > \tag{194}$$

and where $\vec{R}_o$ corresponds to the nuclear coordinates when the ion is in the initial state k.

Another approximation that is commonly used in order to deal with laser (transition-metal or rare-earth) ions is that which uses a <u>crystalline field</u>. This approximation consists of the following:

a. Each electron of the optically active ion feels the influence of the electrons belonging to the ligand ions (a repulsion) and of the nuclei belonging to the ligand ions (an attraction).

b. This influence is taken into account by considering that the electrons of the active ion are subjected to the action of a crystalline field.

c. The crystalline field is completely external to the ion and has a definite symmetry set by the configuration of the ligands.

The charges of the ligand ions do not penetrate into the region occupied by the active ion and the perturbing potential satisfies Laplace equation

$$\nabla^2 V = 0 \tag{195}$$

Then

$$V(r,\theta,\varphi) = \sum_{\ell,m} A_{\ell m}\, r^{\ell}\, Y_{\ell}^{m}(\theta,\varphi) \tag{196}$$

An assumption that is implicit to the crystalline field approximation is that the wavefunction ψ_k of the optically active ion does not depend on the nuclear coordinates, so that the $\vec{R}$ dependence in the Born-Oppenheimer product drops. It is also generally assumed that the vibrations of the solid do not depend on the quantum state of the optically active ion, so that the ℓ subscript in the $\phi(\vec{R})$ function is dropped.

In order to deal with the problem of the ion-vibration interaction we shall consider the following case:

The solid consists of 1 atom/unit cell, and has N atoms; the optically active ion has n electrons and is located at $\vec{R}_1$.

The potential energy of the system is given by

$$V = V(\vec{r}_1, \vec{r}_2, \ldots \vec{r}_n, \vec{R}_2 + \vec{u}_2, \vec{R}_3 + \vec{u}_3, \ldots \vec{R}_n + \vec{u}_N) \tag{197}$$

where $\vec{r}_i$ = coordinate of the ith electron of the ion, $\vec{R}_i$, $\vec{u}_i$ = coordinate and displacement of the ith nucleus. Expanding in a Taylor series:

$$V = V_0 + \sum_{s\alpha}' \left.\frac{\partial V}{\partial u_{s\alpha}}\right|_0 u_{s\alpha} + \sum_{t=1}^{n}\sum_{\alpha=1}^{3} \left.\frac{\partial V}{\partial r_{t\alpha}}\right|_o u_{1\alpha}$$

$$+ \frac{1}{2}\sum_{s\alpha}'\sum_{s'\beta}' \left.\frac{\partial^2 V}{\partial u_{s\alpha}\,\partial u_{s'\beta}}\right|_o u_{s\alpha}\, u_{s'\beta} + \frac{1}{2}\sum_{t\alpha}\sum_{t'\beta} \left.\frac{\partial^2 V}{\partial r_{t\alpha}\,\partial r_{t'\beta}}\right|_o u_{1\alpha}\, u_{1\beta}$$

$$+ \sum_{s\alpha}'\sum_{t\beta} \left.\frac{\partial^2 V}{\partial u_{s\alpha}\,\partial r_{t\beta}}\right|_o u_{s\alpha}\, u_{1\beta}$$

$$= V_0 + \sum_{s} \vec{u}_s \cdot \vec{\nabla}_s V + \frac{1}{2}\Big(\sum_{s} \vec{u}_s \cdot \vec{\nabla}_s\Big)^2 V \tag{198}$$

In the above derivation the sum with a prime sign, such as $\sum'$, exclude the derivative with respect to $\vec{u}_1$. The operator $\vec{\nabla}_1$ includes the n derivatives with respect to the coordinates of the ion's electrons.

The interaction Hamiltonian is then given by

$$H_{ev} = V - V_0 = \sum_{s} \vec{u}_s \cdot \vec{\nabla}_s V + \frac{1}{2}\Big(\sum_{s} \vec{u}_s \cdot \vec{\nabla}_s\Big)^2 V$$

$$= \sum_{s\,1}^{3N} \left.\frac{\partial V}{\partial u_s}\right|_{u_s=0} + \frac{1}{2}\sum_{s=1}^{3N}\sum_{s'=1}^{3N} \left.\frac{\partial^2 V}{\partial u_s\,\partial u_{s'}}\right|_{u_s=u_{s'}=0} u_s\, u_{s'} \tag{199}$$

where the term V_0 includes the static value of the potential energy corresponding to the equilibrium value of the crystalline field.

But

$$u_s = \frac{1}{\sqrt{m_s}} \sum_{k\,1}^{3N} y_{ks}\, q_k \tag{200}$$

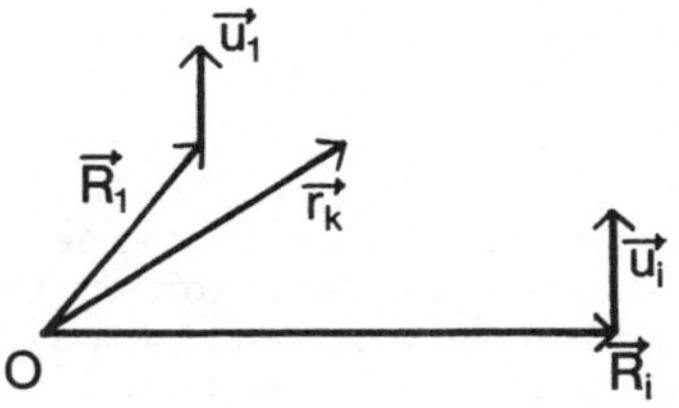

Figure 7. Relevant Coordinates

Then

$$H_{ev} = \sum_s \frac{\partial V}{\partial u_s}\Big|_o \frac{1}{\sqrt{m_s}} \sum_k y_{ks}\, q_k$$

$$+ \frac{1}{2} \sum_s \sum_{s'} \frac{\partial^2 V}{\partial u_s\, \partial u_{s'}}\Big|_o \frac{1}{\sqrt{m_s}} \sum_k y_{ks}\, q_k \frac{1}{\sqrt{m_{s'}}} \sum_{k'} y_{k's'}\, q_{k'}$$

$$= \sum_k \left[\sum_s \frac{\partial V}{\partial u_s}\Big|_o \frac{1}{\sqrt{m_s}} y_{ks} \right] q_k + \sum_k \sum_{k'} \left[\frac{1}{2} \sum_s \sum_{s'} \frac{\partial^2 V}{\partial u_s\, \partial u_{s'}}\Big|_o \frac{1}{\sqrt{m_s m_{s'}}} y_{ks} y_{k's'} \right] q_k q_{k'} + \ldots$$

$$= \sum_k V_k\, q_k + \sum_{kk'} V_{kk'}\, q_k\, q_{k'} + \ldots \tag{201}$$

V_k, $V_{kk'}$ are functions only of the coordinates of the electrons

But

$$q_k = \sqrt{\frac{\hbar}{2\omega_k}}\, (a_k + a_k^+)$$

Then

$$H_{ev} = \sum_k V_k\, q_k + \sum_{kk'} V_{kk'}\, q_k\, q_{k'}$$

$$= \sum_k V'_k\, (a_k + a_k^+) + \sum_{kk'} V'_{kk'}\, (a_k\, a_{k'} + a_k^+\, a_{k'} + a_k\, a_{k'}^+ + a_k^+\, a_{k'}^+)$$

$$= H'_{ev} + H'_{ev} \tag{203}$$

where

$$H'_{ev} = \sum_k V'_k (a_k + a_k^+) \quad (204)$$

$$H''_{ev} = \sum_{kk'} V'_{kk'} (a_k a_{k'} + a_k^+ a_{k'} + a_k a_k^+ + a_k^+ a_{k'}^+) \quad (205)$$

and

$$V'_k = V_k \sqrt{\frac{\hbar}{2\omega_k}} \quad (206)$$

$$V'_{kk'} = V_{kk'} \frac{\hbar}{2\sqrt{\omega_k \omega_{k'}}} \quad (207)$$

II.E. Thermal Line Broadening

We want to examine now the effect of temperature on the shape and width of sharp spectral lines. We restrict ourselves to the case where a purely radiative transition takes place, namely to the so-called no-phonon lines; other temperature-dependent lines which are produced by the simultaneously interaction of an ionic system with the thermal vibrations and the radiation field, namely the so-called vibronic lines, will not be considered.

The crystalline filed at the ion site, varying in time with the thermal vibrations of the neighboring ions, sets up an interaction between the ionic system and the normal modes of vibrations of the lattice. Such an interaction has been used in the past to explain the temperature dependence of the relaxation time of paramagnetic ions. Kiel [8] has treated the temperature dependent linewidth of excited states in crystals on the basis of a square law dependence from displacement. A model studied by Silsbee [9] introduces as possible source of broadening of narrow lines the dependence of the lattice vibrations upon the excitation state of the ion; this model, however, has not been applied to any quantitative analysis. Finally, McCumber and Sturge [10] explained the temperature dependence of the width and position of the R lines in ruby with a two-phonon Raman-process model, having a single charasteristic Debye temperature.

Let us first make the obvious point that the width of a line is the sum of the energy spreads of the two energy levels involved in the transition; the two energy levels may be broadened by the same mechanism or by different mechanisms. It is then proper to start considering the broadening of a single energy level. Consider the ith energy level of a multilevel system and the different mechanisms which may produce a broadening of this level. We may list the following processes:

1. Phonon radiationless decay by spontaneous or induced emission of one or more phonons; this process takes place when an ion in an excited state decays to a lower level by transforming its energy into vibrational energy.
2. Phonon excitation of an ion to a more energetic state; this process takes place by the absorption of one or more phonons.
3. Raman relaxation of an ion to a lower state; this process takes place in the following way. The ion in its excited state absorbs a phonon of a certain frequency and re-emits a phonon of higher frequency, the difference between the two frequencies being equal to the frequency difference between the initial and the final states. When the energy of the absorbed phonon is equal to the energy difference between the initial level and the upper level, and the energy of the emitted phonon is equal to the difference in energy between this upper level and the final level of the decay, the process taking place is an Orbach process.

4. Raman excitation of an ion to a higher state. This process takes place in the following way. The ion in a certain state absorbs a phonon of a certain frequency and re-emits a phonon of lower frequency, the difference between the two frequencies being equal to the energy difference between the initial and the final states. A two-phonon excitation may also take place by an inverse Orbach process [11].
5. Vibronic decay of an ion from an excited state. This process involves the emission of a photon and the absorption or emission of one or more phonons.
6. Raman scattering of phonons by an ion in an excited state; this process takes place while the ion remains in the same electronic state. The variations of the crystal field due to the lattice vibrations can be considered adiabatic or slow, since for an optical transition the ratio of the maximum lattice frequency to the optical frequency is much less than one. For this reason, this mechanism can cause relatively slow variations of the energy of the excited state, without affecting its lifetime.

We notice that all these processes, since they involve phonons, are temperature-dependent.

A very important distinction can be made between the first five processes that remove the ion from its electronic state and the last process that leaves it in the same electronic state. The first five processes may affect the linewidth by shortening the lifetime of the state, the last process may affect the width of the level but has no effects on its lifetime.

The vibronic decay contributes to the shortening of the lifetime of a metastable state. It has been found [12] that when the energy gap between a metastable state and the lower states is large, this type of decay may be responsible for the temperature dependence of the lifetime of the metastable state. However, the contribution of the vibronic decay to the linewidth may be in general considered small.

The line width of sharp spectral transitions when seen in emission is in general much greater than the inverse of the fluorescent lifetime. The relation between the lifetime of a level and the energy broadening of the same level can be derived from (128):

$$\Delta E(\mathrm{cm}^{-1}) = \frac{5.3 \times 10^{-12}}{\tau(\mathrm{sec})} \tag{208}$$

Take, for example, the R_1 line of ruby which corresponds to a radiative transition to the ground state from a metastable state which has the lifetime of about 3 msec at room temperature. The corresponding lifetime broadening is only 2×10^{-9} cm^{-1}, whereas in effect a linewidth of 10 cm^{-1} is observed. This shows how important are the mechanisms which broaden the energy level without removing the ion from it. We now examine these mechanisms.

A Raman phonon scattering process consists of the absorption of a phonon and the emission of another phonon. We can set up the problem as follows:

The initial state of the system is

$$| \, i > = | \, \psi_i^{el} > | \, n_n > | \, n_{k'} > \, . \; . \; . \; . \; . \tag{209}$$

The final state of the system is

$$| \, f > = | \, \psi^{el} > | \, n_k - 1 > | \, n_{k'} + 1 > \, . \; . \; . \; . \; . \tag{210}$$

where the ket containing ψ_i^{el} indicates the wavefunction of the optically active ion and the ket $| n_k >$ indicates the number of phonons of k-type (equal to the degree of excitation of the k-th vibrational oscillator).

The ion-vibration interaction Hamiltonian is given by

$$H'_{ev} + H''_{ev}$$

$$= \sum_k V'_k (a_k + a_k^+) + \sum_{kk'} V'_{kk'} (a_k a_{k'} + a_k^+ a_{k'} + a_k a_{k'}^+ + a_k^+ a_{k'}^+) \qquad (211)$$

The relevant matrix element is given by

$$<f|H_{Raman}|i> = <\psi_i^{el};\ n_k-1,\ n_{k'}+1|H_{Raman}|\psi_i^{el};\ n_k,\ n_{k'}>$$

$$= \sum_j \frac{<\psi_i^{el};\ n_k-1,\ n_{k'}+1|H'|\psi_j^{el};\ n_k-1,\ n_{k'}><\psi_j^{el};\ n_k-1,\ n_{k'}|H'|n_k,\ n_{k'}>}{E_i^{el} - (E_j^{el} - \hbar\omega_{k'})}$$

$$+ \sum_j \frac{<\psi_i^{el};\ n_k-1,\ n_{k'}+1|H'|\psi_j^{el};\ n_k,\ n_{k'}+1><\psi_j^{el};\ n_k,\ n_{k'}+1|H'|n_k,\ n_{k'}>}{E_i^{el} - (E_j^{el} + \hbar\omega_{k'})}$$

$$+ <\psi_i^{el};\ n_k-1,\ n_{k'}+1|H''|\psi_i^{el};\ n_k-1,\ n_{k'}+1>$$

$$= \sum_j \frac{<\psi_i^{el}|V'_{k'}|\psi_j^{el}><\psi_j^{el}|V'_k|\psi_i^{el}>}{E_i^{el} - (E_j^{el} - \hbar\omega_k)} <n_{k'}+1|a_{k'}^+|n_{k'}><n_k-1|a_k|n_k>$$

$$+ \sum_j \frac{<\psi_i^{el}|V'_{k'}|\psi_j^{el}><\psi_j^{el}|V'_k|\psi_i^{el}>}{E_i^{el} - (E_j^{el} + \hbar\omega_k)} <n_k-1|a_k|n_k><n_{k'}+1|a_{k'}^+|n_{k'}>$$

$$+ <\psi_i^{el}|V'_{kk'}|\psi_i^{el}><n_k-1,\ n_{k'}+1|2a_k\ a_{k'}^+|n_k\ n_{k'}>$$

$$\simeq \left[2\sum_j \frac{<\psi_i^{el}|V'_{k'}|\psi_j^{el}><\psi_j^{el}|V'_k|\psi_i^{el}>}{E_i^{el} - E_j^{el}} + 2<\psi_i^{el}|V'_{kk'}|\psi_i^{el}\right]\sqrt{n_k}\ \sqrt{n_{k'}+1}$$

$$= \alpha' \sqrt{\omega_k\omega_{k'}\ n_k(n_{k'}+1)} \qquad (212)$$

where

$$\alpha' = \frac{\hbar}{\omega_k\omega_{k'}} \left[\sum_j \frac{<\psi_i^{el}|V_{k'}|\psi_j^{el}><\psi_j^{el}|V_k|\psi_i^{el}>}{E_i^{el} - E_j^{el}} + <\psi_i^{el}|V_{kk'}|\psi_i^{el}>\right] \qquad (213)$$

If α' is independent of ω_k, then $V_k \propto \omega_k$, and $V_{kk'} \propto \omega_k\omega_{k'}$.

The result of our calculations is

$$<f|H_{Raman}|i> = \alpha' \sqrt{\omega_k\omega_{k'}\ n_k(n_{k'}+1)} \qquad (214)$$

The probability per unit time of a Raman scattering of phonon process is then given by

$$w = \frac{2\pi}{\hbar^2} |\langle f| H_{Raman}|i\rangle|^2 \rho(\omega_f) \tag{215}$$

For sharp lines

$$\rho(\omega_f) \simeq \rho(\omega_k) \rho(\omega_{k'}) \delta(\omega_k - \omega_{k'}) d\omega_k d\omega_{k'} \tag{216}$$

Then

$$w = \frac{2\pi}{\hbar^2} \iint |\langle f|H_{Raman}|i\rangle|^2 \rho(\omega_k) \rho(\omega_{k'}) \delta(\omega_k - \omega_{k'}) d\omega_k d\omega_{k'}$$

$$= \frac{2\pi}{\hbar^2} |\alpha'|^2 \int_0^{\omega_D} \left[\rho(\omega_k)\right]^2 \omega_k^2 n_k(n_k+1) d\omega_k \tag{217}$$

where ω_D = Debye angular frequency. But

$$\rho(\omega) = \frac{3V\omega^2}{2\pi^2 c_s^3} \tag{218}$$

Therefore

$$w = \frac{2\pi}{\hbar^2} |\alpha'|^2 \frac{9V^2}{4\pi^4 c_s^6} \int_0^{\omega_D} \frac{\omega^6 e^{\hbar\omega/kT}}{(e^{\hbar\omega/kT}-1)^2} d\omega$$

$$= \frac{9V^2}{2\pi^3 \hbar^2 c_s^6} |\alpha'|^2 \left(\frac{kT_D}{\hbar}\right)^7 \left(\frac{T}{T_D}\right)^7 \int_0^{T_D/T} \frac{x^6 e^x}{(e^x-1)^2} dx \tag{219}$$

where T_D = Debye temperature. We can then write

$$\Delta E \ (cm^{-1}) = \bar{\alpha}\left(\frac{T}{T_D}\right)^7 \int_0^{T_D/T} \frac{x^6 e^x}{(e^x-1)^2} dx \tag{220}$$

where

$$\bar{\alpha} = \frac{1}{c} \frac{9V^2}{2\pi^3 \hbar^2 c_s^6} (\alpha')^2 \left(\frac{kT_D}{\hbar}\right)^7 \tag{221}$$

We note that the coefficient $\bar{\alpha}$ is intrinsically positive.

The values of the expression $\Delta E/\bar{\alpha}$ are reported in Table 2.

We have from (220)

$\underline{T \ll T_D}$

$$\Delta E \simeq \bar{\alpha} \left(\frac{T}{T_D}\right)^7 \int_0^{\infty} \frac{x^6 e^x}{(e^x-1)^2} dx \propto T^7 \tag{222}$$

$\underline{T \gg T_D}$

$$\Delta E \simeq \bar{\alpha} \left(\frac{T}{T_D}\right)^7 \int_0^{T_D/T} \left(\frac{x^6}{x^2}\right) dx = \bar{\alpha} \left(\frac{T}{T_D}\right)^7 \left[\frac{x^5}{5}\right]_0^{T_D/T} \propto T^2 \qquad (223)$$

The formula (220) has been used to interpret the thermal broadening of sharp lines of transition metal and rare earth ions. [10,12-15]

II.F. Thermal Line Shift

The thermal shift of a purely electronic line is the algebraic sum of the shifts of the two levels involved in the transition. We shall then consider an optically active ion in a certain state and look for the mechanisms that can cause the shift of the energy of this state. We shall assume this energy to be a thermodynamic quantity depending on the two independent variables T, the temperature, and V, the volume of the crystal [16]:

$$E = E(T,V) \qquad (224)$$

Then

$$dE = \left(\frac{\partial E}{\partial T}\right)_V dT + \left(\frac{\partial E}{\partial V}\right)_T dV \qquad (225)$$

and

$$\left(\frac{\partial E}{\partial T}\right)_p = \left(\frac{\partial E}{\partial T}\right)_V + \left(\frac{\partial E}{\partial V}\right)_T \left(\frac{\partial V}{\partial T}\right)_p$$

$$= \left(\frac{\partial E}{\partial T}\right)_V + \left(\frac{\partial E}{\partial p}\right)_T \left(\frac{\partial p}{\partial V}\right)_T \left(\frac{\partial V}{\partial T}\right)_p \qquad (226)$$

But

$$K_T = \text{compressibility} = -\frac{1}{V}\left(\frac{\partial V}{\partial p}\right)_T \qquad (227)$$

and

$$\alpha = \text{coefficient of thermal expansion} = \frac{1}{V}\left(\frac{\partial V}{\partial T}\right)_p \qquad (228)$$

Therefore

$$\left(\frac{\partial p}{\partial V}\right)_T \left(\frac{\partial V}{\partial T}\right)_p = -\frac{1}{VK_T} V\alpha = -\frac{\alpha}{K_T} \qquad (229)$$

and

$$\left(\frac{\partial E}{\partial T}\right)_p = \left(\frac{\partial E}{\partial T}\right)_p + \left(\frac{\partial E}{\partial p}\right)_T \left(-\frac{\alpha}{K_T}\right) \qquad (230)$$

The first term in the right member of the equation above represents an intrinsic temperature dependence of the shift, the second term the part of the shift that is due to the thermal expansion of the solid. The results of hydrostatic pressure measurements, together with the knowledge of thermal expansion and compressibility can give us the value of the second term. This term cannot account for the observed shift and sometimes is even of sign opposite to that observed.

TABLE 2. THE VALUES OF

$$\left(\frac{T}{T_D}\right)^7 \int_0^{T_D/T} \frac{x^6 e^x}{(e^x - 1)^2}\, dx$$

T_D/T	0.0	0.1	0.2	0.3	0.4	0.5	0.6	0.7	0.8	0.9
0		19988.1×10^{-3}	4988.1×10^{-3}	2210.3×10^{-3}	1238.1×10^{-3}	788.2×10^{-3}	543.8×10^{-3}	396.4×10^{-3}	300.8×10^{-3}	235.3×10^{-3}
1	188.5×10^{-3}	153.9×10^{-3}	127.6×10^{-3}	107.1×10^{-3}	90.98×10^{-3}	77.95×10^{-3}	67.31×10^{-3}	58.52×10^{-3}	51.18×10^{-3}	44.99×10^{-3}
2	39.73×10^{-3}	35.23×10^{-3}	31.35×10^{-3}	27.99×10^{-3}	25.06×10^{-3}	22.49×10^{-3}	20.24×10^{-3}	18.25×10^{-3}	16.48×10^{-3}	14.92×10^{-3}
3	13.52×10^{-3}	12.28×10^{-3}	11.16×10^{-3}	10.16×10^{-3}	9.26×10^{-3}	8.44×10^{-3}	7.71×10^{-3}	7.05×10^{-3}	6.44×10^{-3}	5.90×10^{-3}
4	5.40×10^{-3}	4.95×10^{-3}	4.54×10^{-3}	4.17×10^{-3}	3.83×10^{-3}	3.52×10^{-3}	3.24×10^{-3}	2.98×10^{-3}	2.74×10^{-3}	2.53×10^{-3}
5	2.33×10^{-3}	2.15×10^{-3}	1.98×10^{-3}	1.83×10^{-3}	1.69×10^{-3}	1.56×10^{-3}	1.44×10^{-3}	1.33×10^{-3}	1.23×10^{-3}	1.14×10^{-3}
6	1.05×10^{-3}	9.77×10^{-4}	9.05×10^{-4}	8.38×10^{-4}	7.77×10^{-4}	7.20×10^{-4}	6.68×10^{-4}	6.19×10^{-4}	5.75×10^{-4}	5.34×10^{-4}
7	4.96×10^{-4}	4.60×10^{-4}	4.28×10^{-4}	3.98×10^{-4}	3.70×10^{-4}	3.45×10^{-4}	3.21×10^{-4}	2.98×10^{-4}	2.78×10^{-4}	2.59×10^{-4}
8	2.41×10^{-4}	2.25×10^{-4}	2.10×10^{-4}	1.96×10^{-4}	1.83×10^{-4}	1.71×10^{-4}	1.59×10^{-4}	1.49×10^{-4}	1.39×10^{-4}	1.30×10^{-4}
9	1.22×10^{-4}	1.14×10^{-4}	1.07×10^{-4}	1.00×10^{-4}	9.37×10^{-5}	8.79×10^{-5}	8.24×10^{-5}	7.73×10^{-5}	7.25×10^{-5}	6.80×10^{-5}
10	6.39×10^{-5}	6.00×10^{-5}	5.63×10^{-5}	5.30×10^{-5}	4.98×10^{-5}	4.68×10^{-5}	4.41×10^{-5}	4.15×10^{-5}	3.90×10^{-5}	3.68×10^{-5}
11	3.47×10^{-5}	3.27×10^{-5}	3.08×10^{-5}	2.91×10^{-5}	2.74×10^{-5}	2.59×10^{-5}	2.44×10^{-5}	2.31×10^{-5}	2.18×10^{-5}	2.06×10^{-5}
12	1.95×10^{-5}	1.84×10^{-5}	1.75×10^{-5}	1.65×10^{-5}	1.56×10^{-5}	1.48×10^{-5}	1.40×10^{-5}	1.33×10^{-5}	1.26×10^{-5}	1.19×10^{-5}
13	1.13×10^{-5}	1.08×10^{-5}	1.03×10^{-5}	9.73×10^{-6}	9.25×10^{-6}	8.79×10^{-6}	8.36×10^{-6}	7.95×10^{-6}	7.56×10^{-6}	7.20×10^{-6}
14	6.85×10^{-6}	6.52×10^{-6}	6.21×10^{-6}	5.92×10^{-6}	5.64×10^{-6}	5.37×10^{-6}	5.12×10^{-6}	4.89×10^{-6}	4.67×10^{-6}	4.46×10^{-6}
15	4.25×10^{-6}	4.06×10^{-6}	3.88×10^{-6}	3.71×10^{-6}	3.55×10^{-6}	3.39×10^{-6}	3.24×10^{-6}	3.10×10^{-6}	2.97×10^{-6}	2.84×10^{-6}
16	2.72×10^{-6}	2.60×10^{-6}	2.49×10^{-6}	2.39×10^{-6}	2.29×10^{-6}	2.19×10^{-6}	2.10×10^{-6}	2.02×10^{-6}	1.93×10^{-6}	1.86×10^{-6}
17	1.78×10^{-6}	1.71×10^{-6}	1.64×10^{-6}	1.58×10^{-6}	1.51×10^{-6}	1.46×10^{-6}	1.40×10^{-6}	1.34×10^{-6}	1.29×10^{-6}	1.24×10^{-6}
18	1.20×10^{-6}	1.15×10^{-6}	1.11×10^{-6}	1.06×10^{-6}	1.03×10^{-6}	9.87×10^{-7}	9.50×10^{-7}	9.15×10^{-7}	8.82×10^{-7}	8.50×10^{-7}
19	8.19×10^{-7}	7.90×10^{-7}	7.61×10^{-7}	7.34×10^{-7}	7.08×10^{-7}	6.83×10^{-7}	6.59×10^{-7}	6.36×10^{-7}	6.14×10^{-7}	5.93×10^{-7}
20	5.72×10^{-7}	5.52×10^{-7}	5.34×10^{-7}	5.16×10^{-7}	4.98×10^{-7}	4.81×10^{-7}	4.65×10^{-7}	4.50×10^{-7}	4.35×10^{-7}	4.20×10^{-7}

We shall then consider a possible mechanism related to the intrinsic temperature dependence of the shift. The ion-vibration interaction Hamiltonian, [see (211)] has a form that is in some way similar to that of the ion-radiation interaction Hamiltonian; this analogy becomes evident when the radiation field is quantized. [17] Both Hamiltonians have terms that are linear in the field and terms that are quadratic in the field. In the radiative case the interaction Hamiltonian is not diagonal in first order in the field, but in second order it accounts for the electromagnetic (Lamb) shift. In a similar way the ion-vibration interaction Hamiltonian is not diagonal in the first order in the normal vibrational coordinates q_k, but in the second order it may be diagonal and may contribute to the energy of the system. The result is a temperature dependent contribution to the energy of a certain state:

$$\delta E_i = \sum_j \frac{|H'_{ij}|^2}{E_i - E_j} + H''_{ij}$$

$$= \sum_j \frac{\langle i|H'|j\rangle\langle j|H'|i\rangle}{E_i - E_j} + \langle i|H''|i\rangle$$

$$= \sum_j \frac{\langle \psi_i^{el}, n_k|\sum_k V'_k(a_k+a_k^+)|\psi_j^{el}; n_k+1\rangle\langle\psi_j^{el}; n_k+1|\sum_k V'_k(a_k+a_k^+)|\psi_i^{el}; n_k\rangle}{E_i^{el} - (E_j^{el} + \hbar\omega_k)} +$$

$$+ \sum_j \frac{\langle \psi_i^{el}, n_k|\sum_k V'_k(a_k+a_k^+)|\psi_j^{el}; n_k-1\rangle\langle\psi_j^{el}; n_k-1|\sum_k V'_k(a_k+a_k^+)|\psi_i^{el}; n_k\rangle}{E_i^{el} - (E_j^{el} - \hbar\omega_k)} +$$

$$+ \langle \psi_i^{el}; n_k|\sum_{kk'} V'_{kk'}(a_k+a_k^+)(a_{k'}+a_{k'}^+)|\psi_i^{el}; n_k\rangle$$

$$= \sum_j\sum_k \left[\frac{\langle \psi_i^{el}|V'_k|\psi_j^{el}\rangle\langle\psi_j^{el}|V'_k|\psi_i^{el}\rangle\langle n_k|a_k a_k^+|n_k\rangle}{E_i^{el} - (E_j^{el} + \hbar\omega_k)} + \frac{\langle \psi_i^{el}|V'_k|\psi_j^{el}\rangle\langle\psi_j^{el}|V'_k|\psi_i^{el}\rangle\langle n_k|a_k^+ a_k|n_k\rangle}{E_i^{el} - (E_j^{el} - \hbar\omega_k)}\right] +$$

$$+ \sum_k \langle \psi_i^{el}|V'_{kk}|\psi_i^{el}\rangle\langle n_k|a_k a_k^+ + a_k^+ a_k|n_k\rangle$$

$$= \sum_j\sum_k |\langle\psi_i^{el}|V'_k|\psi_j^{el}\rangle|^2 \left[\frac{n_k+1}{E_i^{el} - (E_j^{el} + \hbar\omega_k)} + \frac{n_k}{E_i^{el} - (E_j^{el} - \hbar\omega_k)}\right] +$$

$$+ \sum_k \langle \psi_i^{el}|V'_{kk}|\psi_i^{el}\rangle (1+2n_k) \tag{231}$$

The "zero-field" part is given by

$$\sum_j \sum_k \left[\frac{|\langle \psi_i^{el} | V'_k | \psi_i^{el} \rangle|^2}{E_i^{el} - (E_j^{el} + \hbar\omega_k)} + \langle \psi_i^{el} | V'_{kk} | \psi_i^{el} \rangle \right] \tag{232}$$

This contribution is temperature-independent and is similar to the Lamb shift due to the interaction of an atomic system with the "zero" electromagnetic field.

Disregarding the zero-field contribution, and assuming $|E_i^{el} - E_j^{el}| \gg \hbar\omega_D$ we obtain:

$$\delta E_i = \sum_{j \neq i} \sum_k \left[|\langle \psi_i^{el} | V'_k | \psi_j^{el} \rangle|^2 \left(\frac{n_k}{E_i^{el} - E_j^{el} - \hbar\omega_k} + \frac{n_k}{E_i^{el} - E_j^{el} + \hbar\omega_k} \right) \right]$$

$$+ \sum_k 2n_k \langle \psi_i^{el} | V'_{kk} | \psi_i^{el} \rangle$$

$$\simeq \sum_k \left[\sum_{j \neq i} \frac{|\langle \psi_i^{el} | V'_k | \psi_j^{el} \rangle|^2}{E_i^{el} - E_j^{el}} + \langle \psi_i^{el} | V'_{kk} | \psi_i^{el} \rangle \right] 2n_k$$

$$= \alpha' \sum_k n_k \omega_k \tag{233}$$

where

$$\alpha' = \left[\sum_{j \neq i} \frac{|\langle \psi_i^{el} | V_k | \psi_j^{el} \rangle|^2}{E_i^{el} - E_j^{el}} + \langle \psi_i^{el} | V_{kk} | \psi_i^{el} \rangle \right] \frac{\hbar}{\omega_k^2} \tag{234}$$

But we can write

$$\sum_k n_k \omega_k \longrightarrow \int_0^{\omega_D} \frac{\rho(\omega_k)\omega_k}{e^{\hbar\omega_k/kT} - 1} d\omega_k$$

$$= \frac{3V}{2\pi^2 c_s^3} \int_0^{\omega_D} \frac{\omega_k^3}{e^{\hbar\omega/kT} - 1} d\omega_k = \frac{3V}{2\pi^2 c_s^3} \left(\frac{kT}{\hbar} \right)^4 \int_0^{T_D/T} \frac{x^3}{e^x - 1} dx \tag{235}$$

Then

$$\delta E_i = \alpha' \frac{3V}{2\pi^2 c_s^3} \left(\frac{kT}{\hbar} \right)^4 \int_0^{T_D/T} \frac{x^3}{e^x - 1} dx$$

$$= \alpha' \frac{3V}{2\pi^2 c_s^3} \left(\frac{kT_D}{\hbar} \right)^4 \left(\frac{T}{T_D} \right)^4 \int_0^{T_D/T} \frac{x^3}{e^x - 1} dx \tag{236}$$

or

$$\delta E(cm^{-1}) = \alpha \left(\frac{T}{T_D}\right)^4 \int_0^{T_D/T} \frac{x^3}{e^x-1}\,dx \tag{237}$$

where

$$\alpha = \frac{1}{hc}\,\alpha'\,\frac{3V}{2\pi^2 c_s^3}\left(\frac{kT_D}{\hbar}\right)^4$$

$$= \frac{1}{hc}\,\frac{3V}{2\pi^2 c_s^3}\left(\frac{kT_D}{\hbar}\right)^4\left[\sum_{j\neq i}\frac{|\langle\psi_i^{el}|V_k|\psi_j^{el}\rangle|^2}{E_i^{el}-E_j^{el}} + \langle\psi_i^{el}|V_{kk'}|\psi_i^{el}\rangle\right]\frac{\hbar}{\omega_k^2} \tag{238}$$

We note that the coefficient α can be positive or negative.

The values of $\delta E/\alpha$ are reported in Table 3.

In summary:

$T \ll T_D$

$$\delta E = \alpha\left(\frac{T}{T_D}\right)^4 \int_0^{\infty}\frac{x^3}{e^x-1}\,dx \;\propto\; T^4 \tag{239}$$

$T \gg T_D$

$$\delta E = \alpha\left(\frac{T}{T_D}\right)^4 \int_0^{T_D/T}\frac{x^3}{x}\,dx = \alpha\left(\frac{T}{T_D}\right)^4\left(\frac{x^3}{x}\right)_0^{T_D/T} \propto T \tag{240}$$

If there are intermediate states j such that $|E_i-E_j| \lesssim \hbar\omega_D$, then the approximation in (233) that consists in including the terms containing V_k' and V_{kk}' in the same square bracket will not be possible. In this case one would have to add a term to the thermal shift (237) with a much more complicate thermal dependence. [17]

Finally we notice that the total heat of the crystal is given by [17]

$$E(T) = 9NkT\left(\frac{T}{T_D}\right)^4\int_0^{T_D/T}\frac{x^3}{e^x-1}\,dx \tag{241}$$

Therefore we may expect the thermal line shift to be proportional to the total heat in several instances; this fact has been confirmed experimentally [19].

II.G. The Optical Analog of the Thermal Line Shift

The mechanism by which sharp spectral lines change their positions with temperature has been found to be due to the stationary effects of the ion-vibration interaction, namely the emission and reabsorption of virtual phonons. The thermal shift consists of two parts:

TABLE 3. THE VALUES OF

$$10^5 \times \left(\frac{T}{T_D}\right)^4 \int_0^{T_D/T} \frac{x^3}{e^x - 1}\,dx$$

T_D/T	0.0	0.1	0.2	0.3	0.4	0.5	0.6	0.7	0.8	0.9
0		321000	154499	99110.5	71498.7	54997.5	44051.3	36279.0	30489.9	26022.7
1	22480.5	19610.5	17244.3	15265.4	13590.3	12157.5	10922.3	9848.7	8909.3	8083.1
2	7352.1	6703.8	6124.1	56068.4	5140.7	4722.0	4343.5	4005.4	3690.4	3407.9
3	3150.8	2915.9	2701.9	2505.8	2326.0	2160.9	2009.5	1870.0	1741.8	1623.3
4	1514.4	1413.8	1320.6	1234.4	1154.7	1080.8	1012.3	948.6	889.7	835.6
5	783.9	736.6	692.4	651.1	612.8	577.2	543.8	512.6	483.5	456.0
6	431.0	407.2	385.0	364.2	344.6	326.1	308.9	292.8	277.6	263.4
7	250.0	237.4	225.6	211.4	204.0	194.0	184.6	175.9	167.6	159.7
8	152.3	145.4	139.0	132.7	126.5	120.9	115.6	110.6	105.8	101.3
9	97.0	93.0	89.1	85.3	81.9	78.6	75.4	72.5	69.6	66.9
10	64.3	61.8	59.4	57.1	55.0	53.2	51.1	49.2	47.4	45.7
11	44.2	42.6	41.1	39.7	38.3	37.0	35.7	34.5	33.4	32.2
12	31.3	30.3	29.4	28.4	27.5	26.6	25.7	24.8	24.1	23.4
13	22.6	21.9	21.2	20.6	20.0	19.5	18.9	18.4	17.8	17.3
14	16.7	16.3	15.8	15.4	15.0	14.6	14.2	13.8	13.5	13.1
15	12.8	12.4	12.0	11.7	11.3	11.1	10.8	10.6	10.3	10.1

i) the phonon zero-point contribution, which is given by (232) and is temperature independent, and

ii) the temperature dependent part which is given by (237).

The phonon zero-point contribution depends on the mass of the optically active ion and may cause the appearance of several peaks in correspondence to the same radiative transition if more than one isotope of the ion is present [18]. This contribution is also similar to the electromagnetic (Lamb) shift found in atomic systems. In the Lamb shift case we encounter the problem of an interaction between the atomic system and an infinite number of oscillators representing the electromagnetic field; this shift was calculated by Bethe by using the technique of mass renormalization [20]. In the present case the number of perturbing modes with which the ion interacts is finite and the evaluation of (237) would give directly the value of the shift.

In the Lamb shift case the assumption is made that all the oscillators are in their ground states, whereas here, when the temperature is different from zero, many phonon oscillators may be in excited states. The analog of the thermal shift in the radiative case is a "light shift" which depends on the intensity of the radiation present at the atomic site. This effect was predicted by the author of this article [17], who, however, pointed out that it was very small and "not observable" experimentally due to the fact that the density of photon oscillators is much smaller than the density of phonon oscillators; it has since been observed by Liverman, Pinard and Taleb [21]. These workers measured the light shift of a Rydberg level of a Rb atom, induced by strong nonresonant laser radiation. The agreement between the experimental results and the theory was satisfactory.

II.H. About Sharp Lines: Facts and Examples

Sharp lines are found in the absorption spectra of transition metal ions; some of these lines are also seen in emission. They present the following features:

1. The lines shift toward longer wavelengths as the temperature increases.

2. The lines broaden as the temperature increases.

3. The line widths tend to become independent of temperature at low temperatures, typically below $\sim$ 77K.

4. The line position changes even in the region in which the line width is constant.

A system that illustrates well these different points is that provided by the crystal $MgO:V^{2+}$. The ion V^{2+} enters MgO by substituting for the ion Mg^{2+} and resides in a field of perfectly cubic symmetry. The energy level diagram of this system is presented in Figure 8. The ^2E level is unsplit and a line, called R line, originates from it with a wavelength of $\sim$8707A. Two broad bands, one at $\sim$7000A and 1500A wide, and the other at $\sim$5000A and 1000A wide are used to excite the system.

The luminescence spectrum of this system is presented in Figure 9, that shows its emission at different temperatures and includes the R line and the vibrational satellites. Note the following:

1. Certain characteristic frequencies appear in the spectrum.

2. As the temperature goes down, the vibronic peaks become increasingly sharp and shift with the (no-phonon) R line. Also, the high-energy vibronic band decreases in intensity until it cannot be detected. This is due to the fact that at low temperatures no phonon is available to be absorbed; on the other hand, the low energy band is still present at low temperatures, for the spontaneous emission of phonons can take place even in the absence of phonons.

3. With increasing temperature the continuum of the spectrum expands more and more to wavelengths further removed from that of the R line and grows in intensity, while the peaks tend to smooth out and disappear. This is due to the increasing probability of multi-phonon emission.

The vibrational sidebands of the $MgO:V^{2+}$ spectrum reflect well the frequency distribution of lattice modes [22,23]; W.A. Wall [24] examined in great detail these sidebands and used group theoretical arguments to interpret them. It is a challenge to the researcher in this field to correlate such luminescence data with information obtained from infrared, Raman, and neutron scattering measurements.

In Figure 10 we report the temperature dependence of the lifetime of the $MgO:V^{2+}$ emission together with the same data for other Cr-based materials. We may note that, as usual, the widths of the sharp luminescence lines of these materials are much larger than the inverse of the lifetimes at all temperatures.

The changes of the width of the R line of $MgO:V^{2+}$ with temperature are reported in Figure 11. The temperature-dependent part was simply obtained by subtracting the strain width 0.45 cm^{-1} from the total width; the experimental points obtained in this way were fit by using the formula (220) with a T_D = 760K, and a coefficient $\bar{\alpha}$ = 377 cm^{-1}. The changes of the shape of the R line with temperature are presented in Figure 12: at high temperatures the line shape is Lorentzian, as expected, and at 4K it appears close to a Gaussian.

The changes of the position of the R line of $MgO:V^{2+}$ with temperature are reported in Figure 13. The experimental data were fit by using the formula (237) with a T_D = 760K, and a coefficient α = 400 cm^{-1}.

The spectral behavior of the Cr^{3+} ion in MgO and ruby is similar to that of V^{2+} in MgO.

The behavior of sharp lines of rare earth ions has been studied by several workers [14,15,25]. As for the thermal line broadening, there is evidence that the coupling coefficients $\bar{\alpha}$ [see (220] for spectral lines of rare earth ions are in most cases < 100 cm^{-1}, whereas for transition metal ions they are < 300 cm^{-1}.[10, 12-15, 25] In several instances the lines shift to the red with temperature, but in certain cases shifts to the blue have been observed [26]; however, the shifts are very small, most probably due to the fact that the presence of several levels which are at distance $\lesssim \hbar\omega_D$ from the level under consideration produces a more complicated and smaller temperature dependence.

In the $MgO:V^{2+}$ system the agreement between the Debye temperature used to fit the thermal width and the thermal shift data is in good

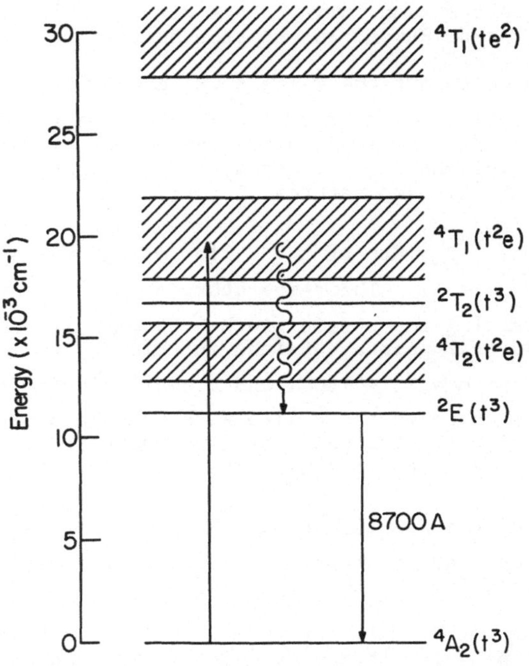

Figure 8. Energy Level Diagram of MgO:V^{2+}

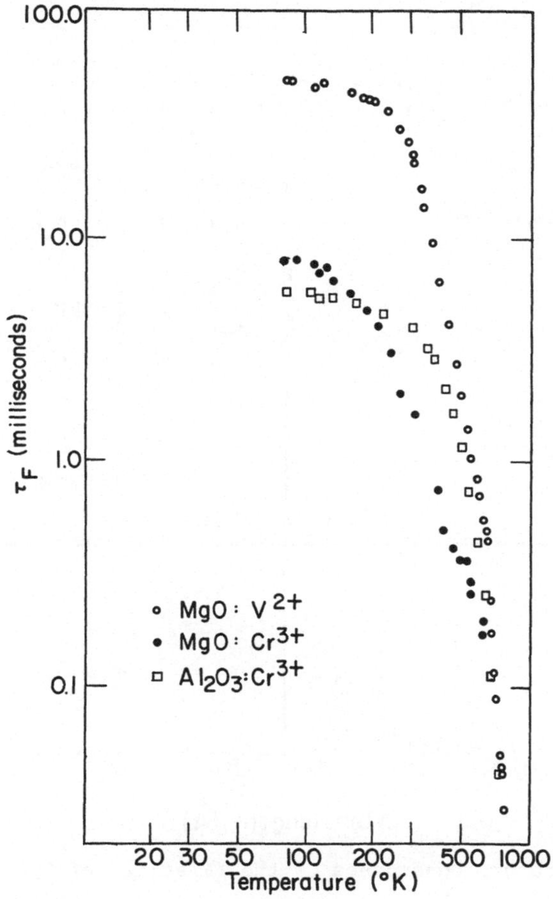

Figure 9. Luminescence Spectrum of MgO:V^{2+} at Various Temperatures Showing the R Line and the Vibrational Satellites

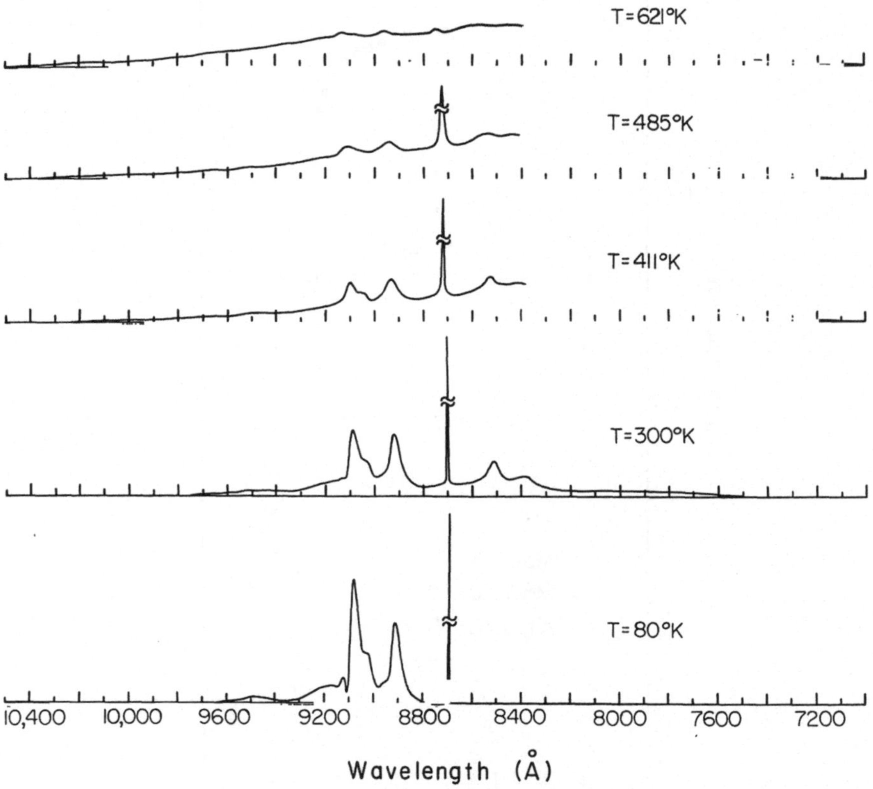

Figure 10. Temperature Dependence of the Lifetimes of $MgO:V^{2+}$, $MgO:Cr^{3+}$ and $Al_2O_3:Cr^{3+}$.

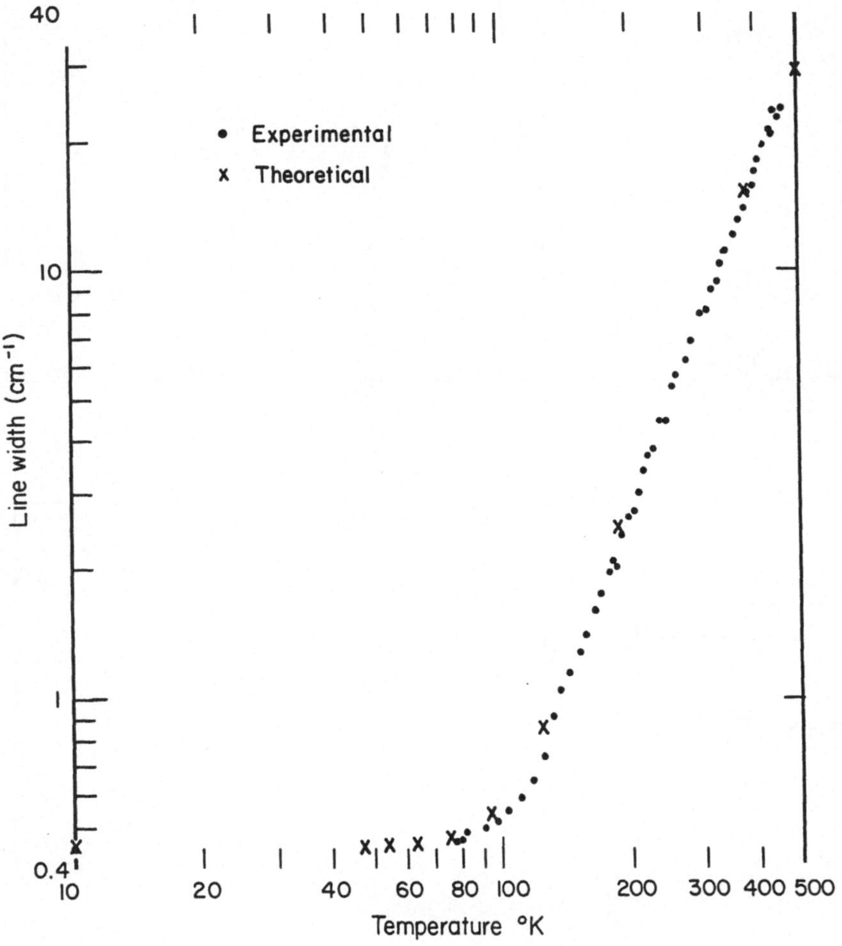

Figure 11. Thermal Broadening of the MgO:V^{2+} R Line.

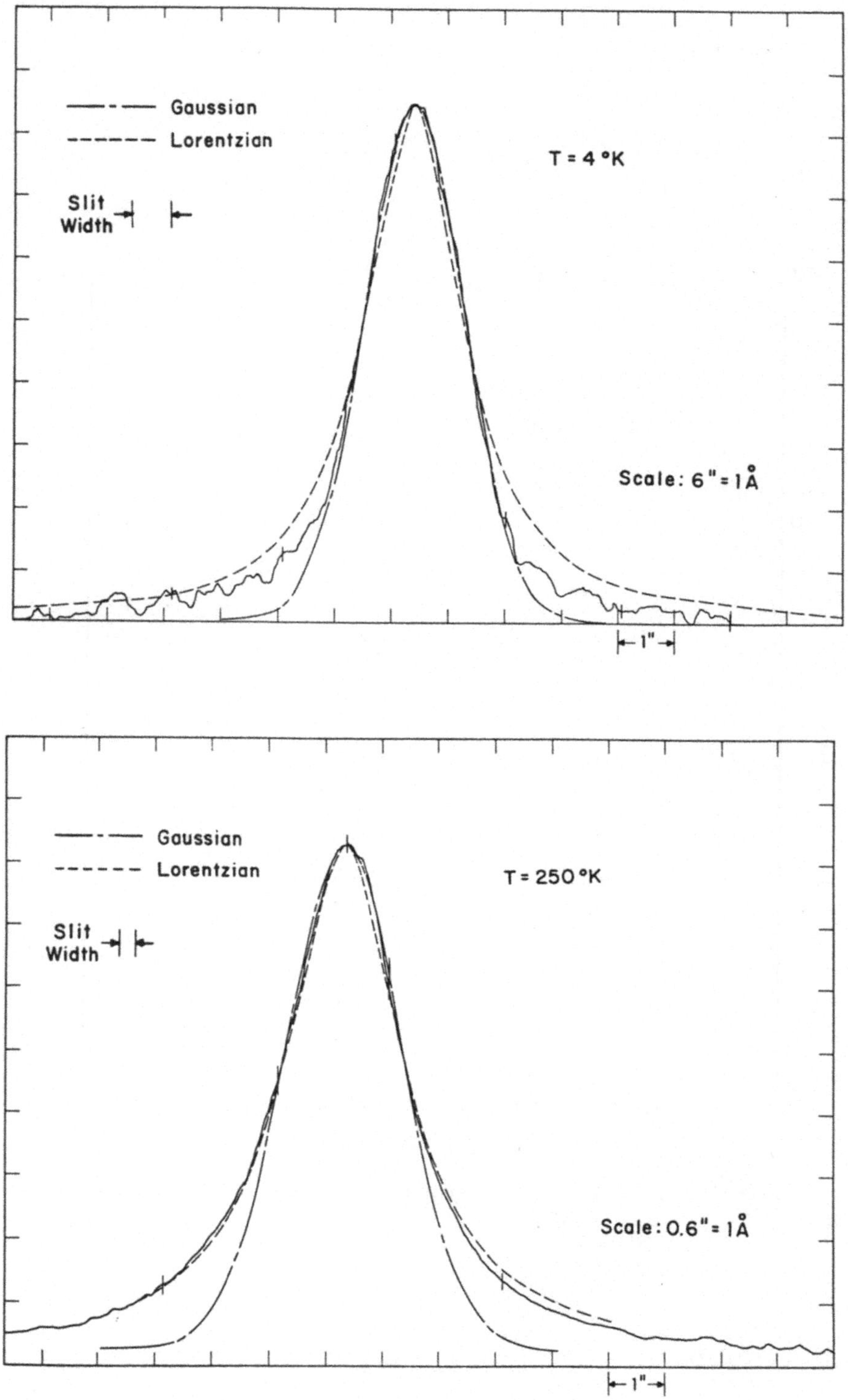

Figure 12. Changes of the Shape of the $MgO:V^{2+}$ R Line with Temperature.

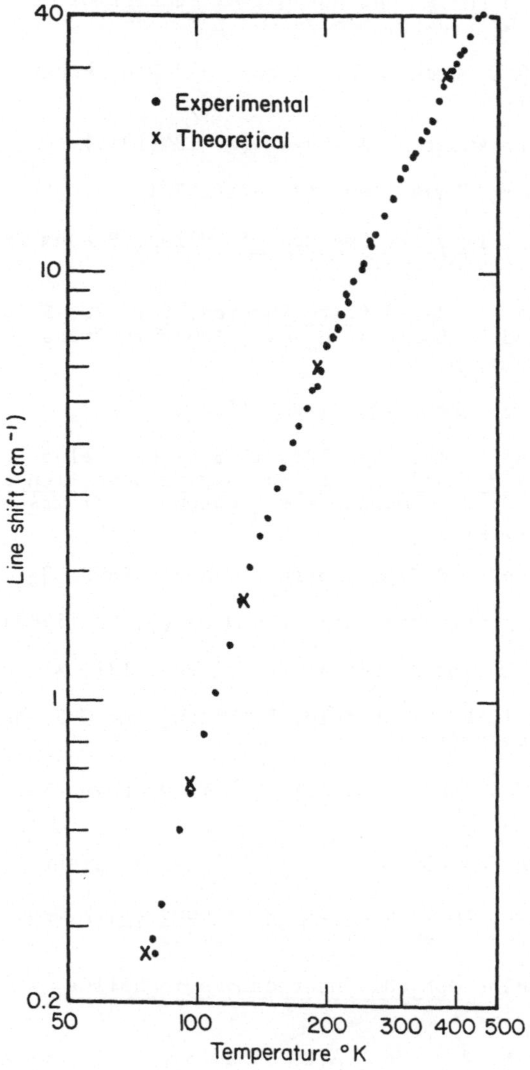

Figure 13. Thermal Shift of the $MgO:V^{2+}$ R Line

agreement with the Debye temperature derived from heat capacity measurements on MgO.[27] However, such an agreement is not always present.[13]

REFERENCES

1. F. Reif, Statistical Thermal Physics, McGraw New York (1965), p.375.

2. H. Cramer, Mathematical Methods of Statistics, Princeton, Princeton, NJ (1946), p.532.

3. W.M. Yen, R.L. Greens, W.C. Scott and D.L. Huber, Phys. Rev. 140, A1188(1965).

4. D.W. Posner, Austral. J. Phys. 12, 184(1959).

5. L.A. Riseberg, Phys. Rev. A7, 671(1973).

6. K.K. Rebane, Impurity Spectra of Solids, Plenum Press, New York and London (1970), p.13.

7. M.D. Sturge, in Solid State Physics, Vol. 20, F. Seitz, D. Turnbull and H. Ehrenreich eds., Academic Press, New York and London (1967), p.91.

8. A. Kiel, Phys, Rev. 126, 1292 (1962).

9. R.H. Silsbee, Phys. Rev. 128, 1726 (1962); also in Quantum Electronics Proceedings of the Third International Congress, P. Grivet and N. Bloembergen eds., Columbia University Press, New York (1964), p.774.

10. D.E. McCumber and M.D. Sturge, J. Appl. Phys. 34, 1682 (1963).

11. J.C. Gill, Proc. Phys. Soc. (London) 79, 58 (1962).

12. B. Di Bartolo and R. Peccei, Phys. Rev. 137, A1770 (1965).

13. R. C. Powell, B. Di Bartolo, B. Birang and C.S. Naiman, J. Appl. Phys. 37, 4973 (1966).

14. W.M. Yen, W.C. Scott and A.L. Schawlow, Phys. Rev. 136, A271 (1964).

15. J.T. Karpick and B. Di Bartolo, Il Nuovo Cimento, 7B, 62 (1972).

16. D.B. Fitchen, in Physics of Color Centers, W.Beall Fowler ed., Academic Press, New York (1964), p.293.

17. B. Di Bartolo, Optical Interactions in Solids, Wiley, New York (1968).

18. G.F. Imbusch, W.M. Yen, A.L. Schawlow, G.E. Devlin and J.P. Remeika, Phys. Rev. 136, A481 (1964).

19. G.F. Imbusch, W.M. Yen, A.L. Schawlow, D.E. McCumber and M.D. Struge, Phys. Rev. 133, A1029 (1967).

20. H.A. Bethe, Phys. Rev. 72, 339 (1947).

21. S. Liberman, J. Pinard and A. Taleb, Phys. Rev. Lett. 50, 888 (1983).

22. M.J. Sangster, G. Peckam and D.H. Saunderson, J. Phys. C: Solid St. Phys. 3, 1026 (1970).

23. H. Bilz and W. Kress, Phonon Dispersion Relations in Insulators, Springer-Verlag, Berlin, Heidelberg and New York (1979), p.50.

24. W.A. Wall, Doctoral Dissertation, Boston College 1973, (unpublished).

25. T. Kushida, Phys. Rev. 185, 500 (1969).

26. B. Di Bartolo, J.T. Karpick and B. Birang, Bull. Am. Phys. Soc. 15, 487 (1970).

27. T.H.K. Barron, W.T. Berg and J.A. Morrison, Proc. Roy. Soc. (London) A250, 70 (1959).

LOCALIZATION AND PERCOLATION IN ALLOY SEMICONDUCTORS

C. Klingshirn*

Physikalisches Institut der Universität
Robert Mayer Straße 2-4
D-6000 Frankfurt am Main
FR Germany

ABSTRACT

In this contribution we first try to shortly outline some concepts relevant to localization of electrons in semiconductors or metals. Then we want to verify the concept of localization in alloy semiconductors. The emphasis will be on optical properties and not on electric conductivity. In doing so, we first shortly review the optical properties of so-called pure semiconductors and then present a concept of localization together with some selected experimental results. The contribution is written from an experimentalist's point of view.

I. A PEDESTRIAN APPROACH TO LOCALIZATION AND PERCOLATION

Localization problems in disordered metals, semiconductors and insulators and the transition from metallic to insulating behaviour has attracted great interest and effort in the last few decades. The topic has been treated in numerous monographs, summer schools, and conferences. Without trying to be complete we mention e.g. [1-15,18] and the references therein. In this section we want to outline some of the relevant concepts. We restrict ourselves mainly to cite the reviews given above and refer the reader for original publications to the references given therein. We start with some simple classical aspects to introduce some of the basic notations. We want to apply these ideas to alloy semiconductors as an example of a disordered system while amorphous semiconductors are treated in the contribution of E.A. Shiff to this book. We shall investigate the problem of localization by means of optical spectroscopy. Consequently, there is a short outline of the optical properties of almost perfect semiconductors in section 2, while we apply in section 3 concepts of localization to alloy semiconductors and we present results from optical spectroscopy which confirm these results.

---

* New address: Universität Kaiserslautern, Fachbereich Physik, Erwin-Schrödinger-Straße, D-6750 Kaiserslautern, FR Germany

I.A. Simple Classical Aspects

In our first example we assume that we are flying in a balloon over a one-dimensional (1d) landscape. In Fig. 1 we show the potential relief V(x) of some mountains. We have chosen a balloon because it moves so slowly that its total energy is almost equal to its potential energy so that we can use the y-axis both for V and $E_{balloon}$.

In case A we are above the highest peak of the mountains and we can move as far as we want (i.e. from $-\infty$ to $+\infty$). This corresponds to a so-called extended or delocalized state. The situation is different in B, where our journey is limited to a finite region of space. States in which the motion is restricted to finite parts of space are called localized. The energy which separates extended from localized states is called mobility edge (ME). The situation C is somewhat peculiar. We start in the morning, we travel the whole day, and we land in the evening because we are tired or because we simply run out of champagne. If the air is somewhat hazy so that we cannot see the mountains in front or behind us, we might think that we were in an extended state because we were travelling as far as we wanted. By looking at the total of Fig. 1 it is however clear that we were in a localized state. What we learn from this is that various length scales are very important. The concept of localization applies to the real world only if the localization length is not much larger than the e.g. length of the sample in which an electron moves or the distance an exciton can travel during its lifetime. A similar statement is true concerning time scales.

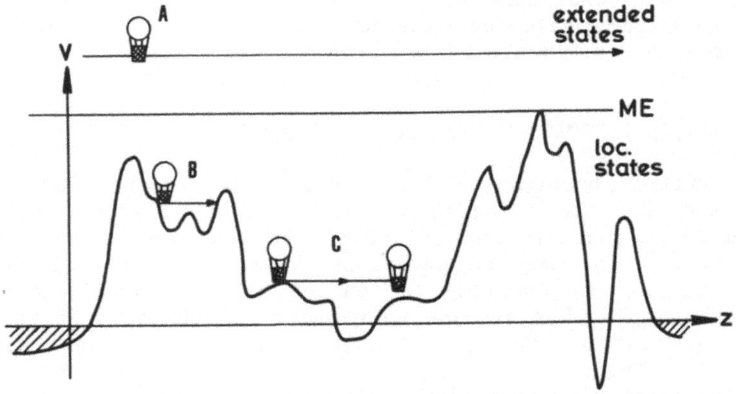

Fig. 1. Flying in a balloon over a one-dimensional landscape.

In our 1d world of Fig. 1 the ME is given by the height of the highest peak V_{max}. This is no longer true in 2d as shown schematically for a balloon-ride in a famous landscape (Fig. 2). If the wind is permitting we can travel as far as we want at an energy which is significantly below V_{max} as shown e.g. by the dashed line. I.e. the ME is situated in 2d below V_{max}. From this experience we deduce that the dimensionality of the system is very important.

Fig. 2. Flying in a balloon through a two-dimensional landscape.

I.B. <u>What is Different in Quantum Mechanics</u>?

As already mentioned, we want to apply the concepts of localization later on preferentially not to balloons or other classical particles, but to electrons and excitons in solids. These quasi-particles are properly described by quantum mechanics. In Fig. 3 we show a 1d potential barrier, for simplicity of rectangular shape. The classical ME_c from 1.1. is indicated. We discuss now what happens, if a plane wave with total energy E falls from the left hand side on this potential barrier.

For $E<ME_c$ the main part of the incident wave will be reflected (not shown in Fig. 3). However, due to the quantum mechanical tunnel effect there is also a certain probability for the particle to continue to propagate on the other side of the barrier in contrast to classical mechanics. The tunneling probability decreases exponentially both with the barrier height above E and its width. The quantum mechanical tunnel effect evidently favours delocalization.

For $E>ME_c$ a part of the incident wave is reflected from the potential discontinuities. Consequently, the transmission is smaller than one in contrast to the classical expectation. The interference of incident and backscattered waves will lead to the formation of a standing wave of a certain amplitude. As the name says, these waves do not propagate and

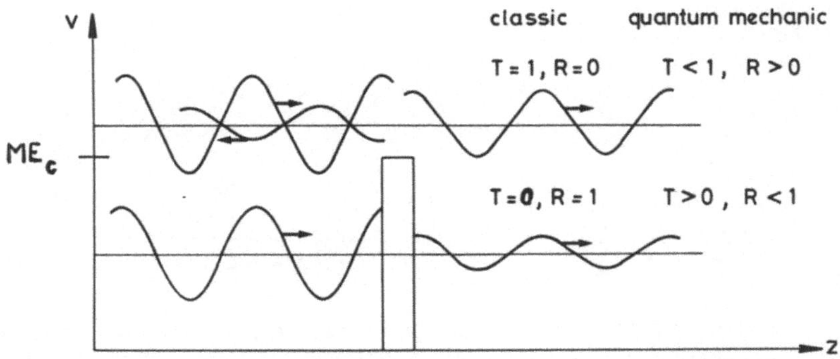

Fig. 3. A plane wave falling on a potential barrier (schematic).

we can state that scattering and the interference between incident and (back-)scattered waves favour localization. An example for delocalization by tunneling is the band structure of an ideal crystal. We show in Fig. 4a schematically the periodic potential of a semiconductor. The bands have a homogeneous width throughout the sample. The electrons in the valence band (VB) have a lower but finite probability to be found in the regions of the potential barriers as compared to the wells. Nevertheless, they, - or more precisely - the unoccupied states, the holes, can move freely through the lattice in contrast to classical results. The same holds for electrons, also on the bottom of the conduction band (CB). In Fig. 4b we show the eigenfunctions, which are of the Blochwave type

$$\varphi(\vec{k},\vec{r}) = u_{\vec{k}}(\vec{r})\, e^{i\vec{k}\vec{r}} = \sum e^{i\vec{k}\vec{R}_n} \varphi_{at}(\vec{r}-\vec{R}_n) \qquad (1)$$

$U_{\vec{k}}(\vec{r})$ is the lattice periodic part, which depends slightly on $\vec{k}$ and which has a plane wave factor as an envelope. In tight binding approximation the Bloch-waves can be generated by a superposition of atomic orbitals with suitable phase-factors (1). We have chosen here s-type atomic orbitals which are adequate for the CB of many of the more ionic bound semiconductors. In Fig. 4c we show finally again CB and VB of a real semiconductor which contains some point defects, here a donor and an acceptor. As shown schematically they can bind an electron (or a hole) with spatially localized wavefunction according to

$$D^o \leftrightarrow D^+ + e \quad \text{and} \quad A^o \leftrightarrow A^- + h \qquad (2)$$

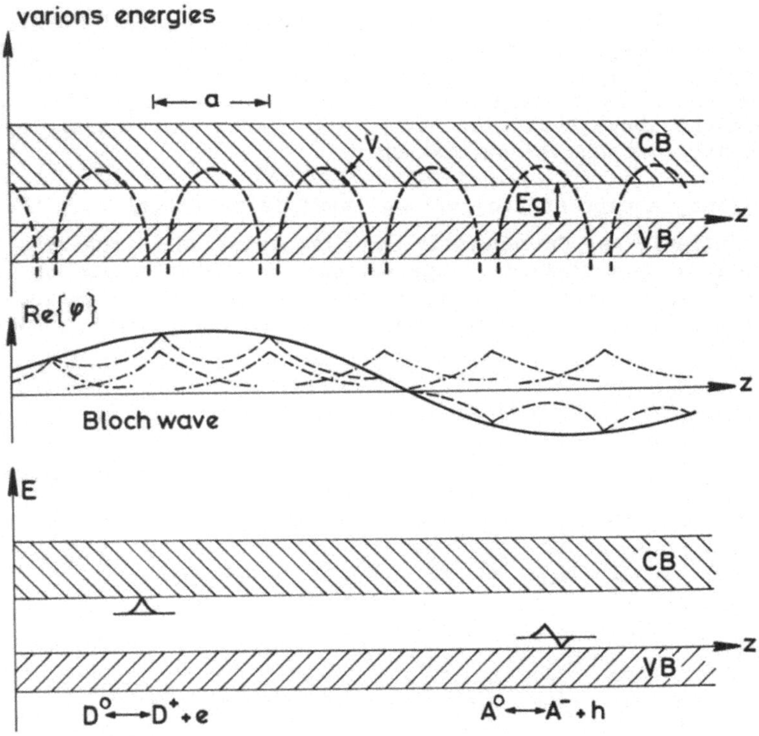

Fig. 4. Schematic drawing of the periodic potential V(r) and the band structure of an idealized semiconductor (a) an eigenfunction of the Bloch-wave type (b) and a real semiconductor containing donors and acceptors.

For shallow centers the wavefunctions in the bound states can be described by a superposition of Bloch-functions

$$\psi(\vec{r}) = \sum_{\vec{k}} a_{\vec{k}} e^{i\vec{k}\vec{r}} u_{\vec{k}}(\vec{r}) \tag{3}$$

We call states which are localized at a point defect in the following bound states to distinguish them from states localized by disorder, which will be described later. It should be mentioned that the meaning of donor and acceptor as used here in semiconductor physics is different from the use of these terms in the spectroscopy of ionic materials [16]. There a donor gives away energy which is taken over by an acceptor.

I.C. Concepts of Localization

Now we shall shortly outline some concepts of localization by disorder. We can come from extended states to localized ones by introducing some disorder as will be the case for weak localization. On the other hand we can also start with bound states and end up with delocalized ones as will be shown below.

1. Weak Localization. Weak localization is a phenomenon, which is essentially based on the interference of waves. It can be considered as a "precursor" of localization. The concept is strictly valid for T=0K. In our outline we follow the ideas developed in more detail e.g. in [10,14,17].

We consider a simple, isotropic metal at T=0. Then all states below the Fermi energy E_F and Fermi vector $|\vec{k}_F|$ are filled, all states above are empty. Further we assume that there are randomly distributed scattering centers with infinite mass and without spin or magnetic moment. The average distance of these centers d shall be large as compared to k_F^{-1}. The potential of the scattering centers can be repulsive or attractive. Under the above conditions only elastic scattering of electrons at the Fermi edge is possible, i.e. $|\vec{k}_i| = |\vec{k}_f| = |\vec{k}_F|$. We consider now the diffusive motion of a particle from A→B (Fig. 5a). There are of course many different ways on which the electron can travel from A to B. Two of them are shown in Fig. 5a. The transition probability $w_{A\to B}$ is given by summing over the amplitudes of all paths and taking the square of this sum.

$$w_{A\to B} = |\sum_i A_i|^2 = \sum_i |A_i|^2 + \underbrace{\sum_{i\neq j} A_i^* A_j}_{=0} = \sum_j |A_i|^2 \tag{4}$$

Since the different ways involve randomly distributed phase shifts, the sum over the mixed terms vanishes.

We consider now backscattering, i.e. $w_{A\to A}$ (Fig. 5b). We have again to sum over the amplitudes of the various possibilities. But what is now different from the case of Fig. 5a is, that we have to every path i the time-reversed path. This one has the same phase-shift and consequently (4) changes to (5)

$$w_{A\to A} = \sum_i A_i|^2 = \sum |A_i|^2 + \sum_{i\neq j} A_i^* A_j = 2 \sum |A_i|^2 \qquad (5)$$

This means that the probability of backscattering is twice as large as the probability for scattering from a point A to any other point B. This enhanced backscattering is the essence of weak localization.

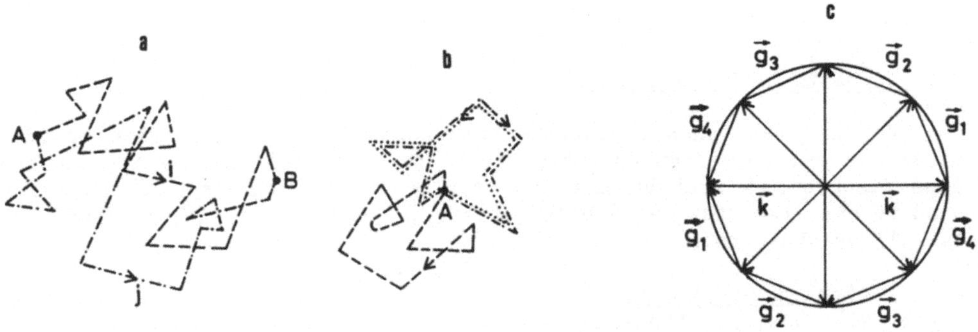

Fig. 5. Diffusive motion of a particle from A to B (a), backscattering showing for one process also the time reversed one (b), same as (b) but now representation in $\vec{k}$-space.

The same type of arguments holds in $\vec{k}$-space as schematically shown in Fig. 5c. The waves $\vec{k}$ and $-\vec{k}$ interfere to give a standing wave.

Weak localization is based on time-reversal symmetry. This gives a handle to verify the concept experimentally by destroying the time-reversal symmetry by applying a magnetic field. In Fig. 6a we show measured resistivity curves of quasi-2d Mg films as a function of an applied magnetic field for various temperatures [14]. The decrease of R for increasing B is due to the disappearance of weak localization. The fact that R changes by less than one percent justifies again the prefix "weak". It is also clear that increasing T destroys the weak localization since inelastic scattering processes (e.g. absorption and emission of phonons or electron-electron scattering) become increasingly important with temperature. Applying a small amount of Au on top of the Mg film also destroys the weak localization and may even lead to a weak "antilocalization" ($\partial R/\partial|B| > 0$). This is due to the significant spin-orbit scattering introduced by the high nuclear charge Z of the Au-atoms. For more details, the reader is referred e.g. to [14,17].

Bearing in mind all the assumptions made, one might think that weak localization is a highly specialized phenomenon, occuring only in metals with low Z. This is not the case [18]. We only need elastic scattering of waves with time reversal symmetry. The enhanced backscattering has e.g. been observed in the scattering of light from an aqueous solution of sub-μm size polyshyrene spheres [19]. Taking into account the finite angular resolution of the experiment, the data revealed the expected enhancement of a factor of two of the backscattered light intensity as compared to the scattering into any other direction. Weak localization of phonons is considered e.g. also in [18].

2. Anderson Localization. In this subsection we treat the basic ideas of Anderson-localization [20]. For more elaborate reviews the reader is referred to [6,8,9].

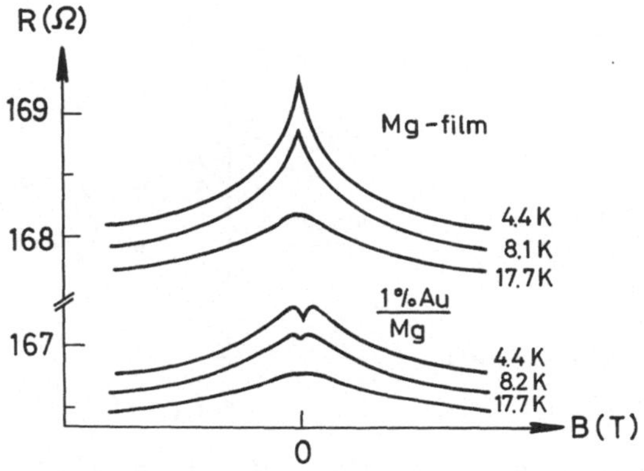

Fig. 6. The resistance of a Mg-film (a) and of a Mg-film covered with 1% of an atomic layer of Au (b) as a function of an appical magnetic field for various temperatures. From [14].

The basic idea is to introduce disorder into a periodic lattice. For simplicity we use a Kronig-Penney type potential (Fig. 7a). Due to the overlap (tunneling) of the wavefunctions centered in the individual wells we get a lowest band in the density of states with a width B given by

$$B = 2\,z\,I \tag{6}$$

where I is the overlap-integral between neighbouring "atomic" wavefunctions and z is the coordination number, i.e. the number of next neighbours (Fig. 7b). The wavefunctions are again of the type of eq. (1).

Disorder can now be introduced by either changing arbitrarily the distance between the wells and thus the coupling or overlap between the wavefunctions (off-diagonal disorder) or by changing the depth of the wells within a certain width V_o (diagonal disorder). The first case corresponds to the so-called Edwards - or Lifshitz - model [6], the second one represents the Anderson model [6,20] (Fig. 7c). The physically most realistic situation would be a combination of both but this involves the biggest theoretical problems. We outline here some important results of the Anderson model. The crucial parameter is the ratio of V_o and the

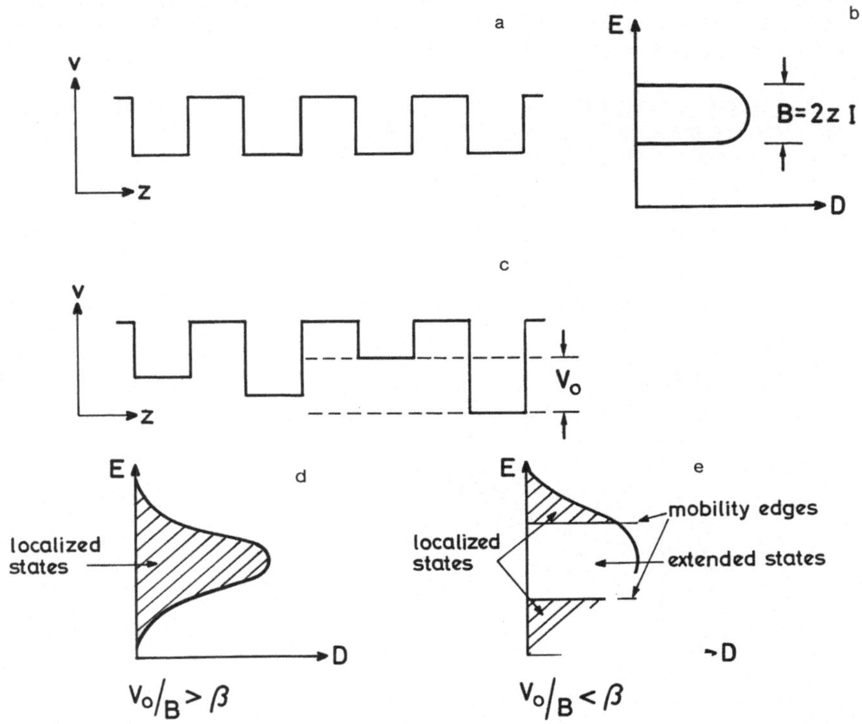

Fig. 7. A periodic potential of the Kronig-Penny type (a) the resulting density of states D of the lowest band (b), the same potential but with some fluctuation in depth within a range V_o (c) and the resulting densities of states for $V_o/B < \beta$ (d), and $V_o/B > \beta$ (e). According to e.g. [1,4].

width of the unperturbed band B. For $V_o/B < \beta$, where β is of the order of unity, we have still extended states in the center of the band and localized ones in the exponential tails of the density of states (Fig. 7e). If the fluctuations in the potential depth are even larger, i.e. $V_o/B > \beta$ all states are localized as shown schematically in Fig. 7d.

Let us now have a look on the wavefunctions in the various cases (Fig. 8). The upper row presents again a Bloch-type function in a strictly periodic lattice like in Fig. 7a,b. As already mentioned, the wavefunction is described by (7a)

$$\psi_{\vec{k}}(\vec{r}) = \sum_n e^{i\vec{k}\vec{R}_n} \phi_n(\vec{r}-\vec{R}_n) = e^{i\vec{k}\vec{r}} u_{\vec{k}}(\vec{r}) \quad \text{(a)}$$

$$\psi(\vec{r}) = \sum_n A_n \phi_n(\vec{r}-R_n) \quad \text{(b)} \qquad (7)$$

$$\psi(r) = \sum_n B_n \phi_n(r-R_n)^{-r/\xi} \quad \text{(c)}$$

where the sum runs over all lattice sites $\vec{R}_n$. An extended state in the disordered lattice is described by eq. (7b) and Fig. 8b. The main difference to the perfect lattice is, that the long-range phase-correlation is lost. The localized state is finally described by an envelope function decaying exponentially with a localization length ξ. A lower limit of ξ is given by the lattice constant.

The mobility edge in Fig. 7d, which separates localized from extended states (i.e. $\xi \to \infty$ in eq. (7c)) is in this model sharp: If a localized and a delocalized state would coexist at the same energy, the smallest perturbation will mix these states with the consequence that both states are extended. A necessary precondition that the above argument holds is an infinite lifetime of the particles under consideration. If the coupling-time between the localized and the extended states is long as compared to the lifetime of the particle (e.g. an exciton), localized and extended states may coexist at the same energy with the consequence that the mobility "edge" is smeared out.

In the following we want to discuss consequences of the model described above on the conductivity of semiconductors, especially the possible transition from an insulating to a metallic state, the so-called Mott-transition [1,6,9,11]. In Fig. 9 we show the square-root dependence of the density of states D(E) at the lower edge of the conduction band of an idealized semiconductor together with the D(E) introduced by donors at

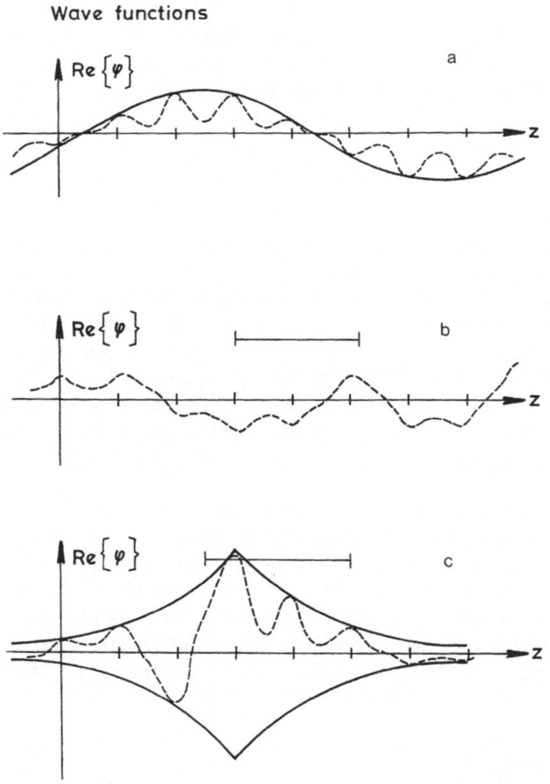

Fig. 8. Schematic representation of the wavefunction for a strictly periodic lattice (a), and extended (b), and a localized state (c) in a disordered lattice. According to e.g. [1,4].

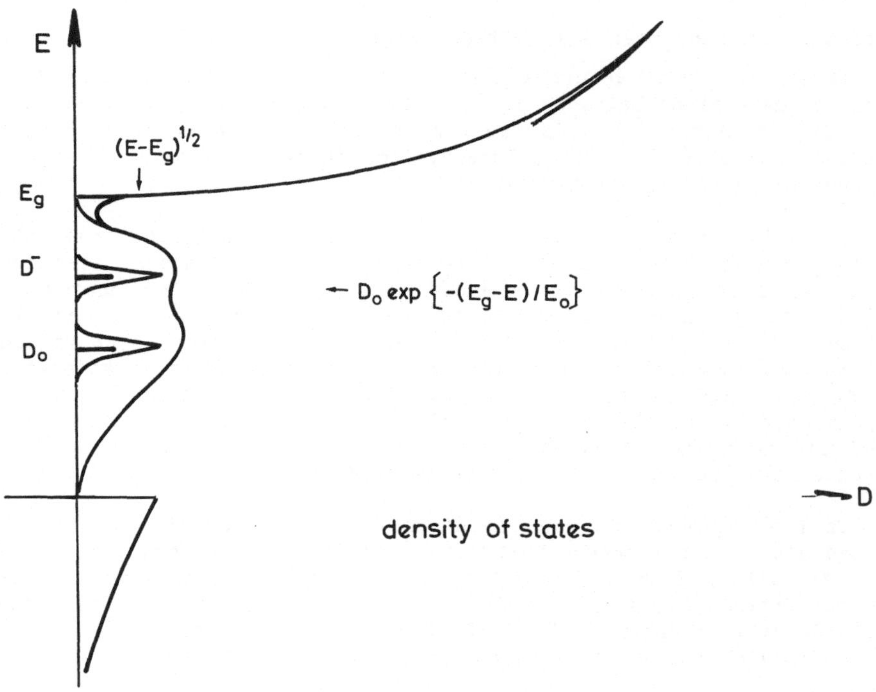

Fig. 9. Schematic drawing of the density of states at the bottom of the conduction band and the additional density of states introduced by donors of various concentrations.

various concentrations. We assume that there is only one type of donors each of which brings one electron. For a donor concentration, which is so low that there is virtually no interaction between them, D(E) looks like a δ-function. The electron is in its ground state in a 1s orbital as shown schematically in Fig. 5c. This state is in principle capable to accomodate two electrons with opposite spin. Due to the Coulomb-interaction between these electrons, they are usually situated at higher energy. (There are exceptions from this plausible argument, however [21]). So D_o denotes the energy level of the neutral donor with one electron, D^- the one of a donor with two electrons. If we increase the number of donors, they start to interact with each other and the eigenenergies will depend on the surrounding geometrical configuration of the other host and dopant atoms. Both effects will lead to a broadening of the donor levels into bands. As long as these two bands do not overlap, the material will still be an "insulator". Insulator means in this context that the conductivity goes to zero with decreasing temperature due to same activation energy as shown in eq. (8a)

$$\sigma(T) = \sigma_o\, e^{-E_a/k_B T} \qquad \lim_{T\Rightarrow 0} \sigma = 0 \tag{8a}$$

$$\lim_{T\Rightarrow 0} \sigma = \sigma_o \neq 0 \tag{8b}$$

The activation energy could be the energy between D_o and D^- levels or between D_o and the bottom of the CB.

If we further increase the donor concentration, the D_o and D^- bands will overlap to form one single band which will be half-occupied. According to the Anderson-model, extended states will be situated in the center of the band (if there are any) and the material will consequently be a "metal". Metal means in this context not that the electrical conductivity is very high as e.g. in Cu but only that the conductivity goes for $T \Rightarrow 0$ to a finite value (8b). What is common with usual metals is, that the Fermi-energy is situated (for $T \Rightarrow 0$) in a band in the region of extended states. The transition between insulating to metallic behaviour defined in the above sense is usually referred to as a Mott transition [6]. This Mott-transition can also be considered from another point of view: with increasing doping we introduce more and more carriers into the system. These carriers tend to screen the Coulomb-interaction between the positively charge donor atom and the electron bound to it. If the Coulomb potential is sufficiently screened, eq. (9)

$$V(r) = \frac{1}{4\pi\ \epsilon\epsilon_o}\frac{1}{r} \Rightarrow V(r) = \frac{1}{4\pi\ \epsilon\epsilon_o}\frac{e^{-\lambda r}}{r} \qquad (9)$$

it will be no longer capable to bind an electron. This situation occurs, if the screening length λ reaches a critical value λ_c.

A similar phenomenon occurs for excitons. Excitons are compound particles consisting of an electron and a hole (see below). They can be optically created. If they are excited in a sufficiently large number, there occurs a transition from an insulating gas of electrically neutral excitons into a metal-like electron-hole plasma. More details about this type of Mott-transitions are found in [22,23] and the references given therein.

If we further increase the density of donors, the band which they from will merge with the conduction band to give a usually exponential tail of states according to eq. (10)

$$D(E) = D_o \cdot \exp\{- E/\epsilon_o\} \text{ for } E < E_g \qquad (10)$$

By comparison with Fig. 7d we expect the mobility edge to be located somewhere in this exponential tail. The material will be an insulator or a metal if E_F falls below or above ME for $T \Rightarrow 0$, respectively.

As is well known from crystalline material, the position of E_F can be influenced by other types of dopants. Especially, if we put acceptors in the system, they are able to accomodate a part of the electrons and can thus influence the metal-insulator transition. This phenomenon is known as compensation and is treated in detail in [6]. However, it should be mentioned that there is in strongly disordered, i.e. amorphous, semiconductors not always the possibility to shift E_F by doping. This phenomenon of pinning of E_F is treated e.g. in the contribution to this book by E.A. Shiff [21].

3. Percolation. The concept of percolation has been introduced in various contributions to this School or in [4,6,7]. So we can be rather short. In Fig. 10 we show a 6 x 6 square lattice in which about 40% of the sites are randomly occupied by throwing a dice. Groups of neighbouring occupied sites form a cluster. One of the problems which can be considered, concerns the critical occupation probability p_c at which one gets for increasing p for the first time a cluster which percolates from $-\infty$ to $+\infty$ in an infinite system. The connection to the localization problem is evident. Finite clusters correspond to localized states, percolating ones to extended states. P_c can be found e.g. by computer simulations on finite systems by averaging over many realizations. The values of p_c depend on the dimensionality of the system and on the specific type of lattice considered. The values of p_c tend to be somewhat larger for the site-percolation problem considered here as compared to bond-percolation. While rather precise values of p_c are given in [6,7], we summarize in (11) onle crude numbers for various dimensions.

$$d = 1 \qquad p_c = 1 \tag{11a}$$

$$d = 2 \qquad p_c \approx 0.6 \pm 0.1 \tag{11b}$$

$$d = 3 \qquad p_c \approx 0.2 \pm 0.05 \tag{11c}$$

Obviously, there is a trend of p_c to decrease with increasing dimensionality. This can be considered as a hint that extended states are more easily realized in higher dimensions.

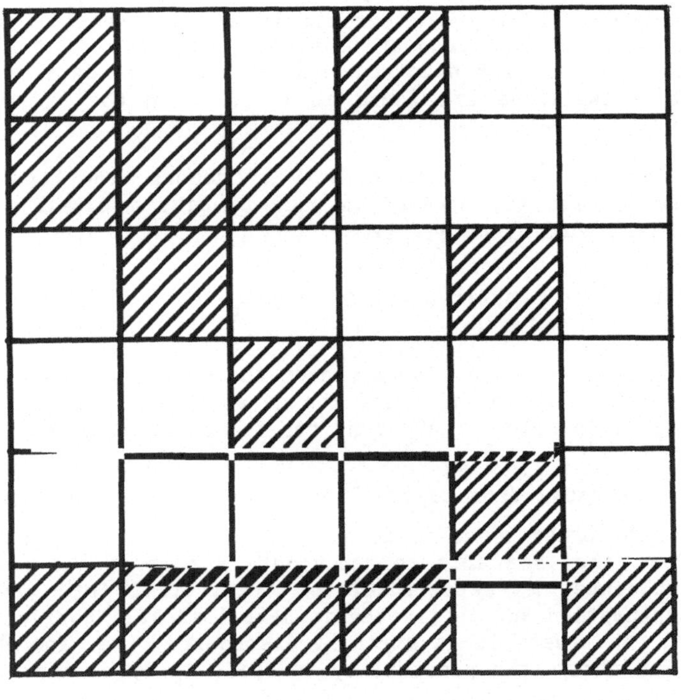

Fig. 10. Square lattice with sites occupied randomly with a probability $p \cong 0.4$. Two clusters are formed, which contain more than one square.

The value p_c for 1d systems is exact (and trivial) and leads to the simple statement that all states are localized in a 1d disordered system: for $p_c = 1$ all states are occupied and the system is no longer disordered.

To demonstrate how p influences diffusive motion, we consider the problem of an ant moving on a cluster of occupied sites in a 2d square lattice. The ant tries to hop at every tick of a clock randomly from one site to a neighbouring one and is allowed to do so only if this site belongs to the cluster on which the ant lives. This problem is treated in detail in [7] and we summarize here the main results. For p = 1 the distance R from the origin reached by the ant as a function of time, averaged over many realizations increases as typical for diffusive motion with $\alpha = \frac{1}{2}$ in eq. (12) and the prefactor β is the square root of the diffusion constant.

$$R = \beta\, t^{\alpha} \tag{12}$$

For p slightly smaller than 1, but well above p_c, the only difference is that the ant has to wait sometimes for one clock-time, if it wanted to hop on one of the few unoccupied sites. Consequently, α will still be close to 1/2 but the prefactor β decreases. For $p < p_c$ the ant moves on a finite cluster. This means that R(t) will start with a finite slope but will become constant for longer times (i.e. $\alpha = 0$) when R approaches a value limited by the largest geometrical distance between any two sites of the finite cluster. For $p = p_c$ one expects $\alpha \approx \frac{1}{3}$ [7], representing thus a case of anomalous diffusion.

Since physical objects can be represented only to a certain extent by square lattices in any dimensions, more general percolation problems have been considered in the literature. One can assume that there are randomly distributed centers of density N, occupied by species of radius r e.g. donor electrons. Assuming that those sites contribute to a cluster, which have more than one point in common, one has to consider the critical value of a parameter B which is given for d = 3 and d = 2 in (13)

$$d = 3:\ B = \frac{4}{3}\pi r^3 N;\ d = 2:\ B = \pi r^2 N \tag{13}$$

The next step of generalization is to allow r to vary randomly within a certain bandwidth, or to consider continuous percolation. In two dimensions this problem can be visualized by a continuous potential relief V(x,y) fluctuating randomly and symmetrically around $V_o=0$. It is gradually filled with water up to a constant level h. The transition from isolated lakes in a continent to islands in an ocean occurs for $h = V_o = 0$ [6]. Quantum percolation as another aspect of the transition from localized to extended state is considered e.g. in [24].

4. Scaling Theory. To conclude this general section we want to introduce the scaling theory which gives under certain assumptions information on the existence of localized and extended states in various dimensions [6,9,11,12,15,25].

The basic idea is the following: we consider the conductance G(L) of a d-dimensional cube of length L. Assume L has the length of the bar shown in Fig. 8. Then it is eventually not possible to decide from measuring G if the states in the sample are localized or not. However, if we make the sample-length L larger and larger (e.g. by a series of doubling of L [25b]) we should be finally able to get a decision concerning the nature of the states contributing to the electrical current. If we have extended states and the finite resistance is due to scattering events (without weak localization) then Ohm's law applies which reads in various dimensions:

$$\left.\begin{array}{lll} d = 3 & G(L) = R^{-1} & = \sigma \frac{L^2}{L} \\ d = 2 & G(L) & = \sigma \frac{L}{L} \\ d = 1 & G(L) & = \sigma \frac{1}{L} \end{array}\right\} = \sigma L^{d-2} \tag{14}$$

For current transport in localized states we expect G(L) to drop with some characteristic localization lengths ξ independent of d

$$G(L) = G_o \, e^{-L/\xi} \tag{15}$$

Equations (14) and (15) give the asymptotic behaviour of G(L) in the two limiting cases.

A one parameter scaling function β has been introduced in [25a] which is defined in (16)

$$\beta(G) \; := \frac{d \ln G}{d \ln L} = \frac{L}{G}\frac{dG}{dL} \tag{16}$$

It should be mentioned that this β-function concept only works if there is only one parameter (here L). It cannot be applied for a more than one-dimensional parameter space [25].

The limiting cases of eq. (14) and (15) lead with (16) to (17a) and (17b) for extended and localized states, respectively, to

$$\beta = d - 2 \tag{17a}$$

$$\beta = \ln G - \ln G_o \tag{17b}$$

This asymptotic behaviour is shown in Fig. 11 by dashed lines. If the concept of weak localization applies (i.e. $T \Rightarrow 0$; only elastic scattering, no spin-orbit scattering, no magnetic fields), then it can be shown that the actual curve $\beta(G)$ approaches the limiting case of the extended states asymptotically from below. If we further assume that $\beta(G)$ is not wildly oscillating for smaller ln G, we expect a behaviour as shown schematically in Fig. 11.

If we assume for the moment that the shape of $\beta(G)$ in Fig. 11 is correct, we come to some interesting conclusions:

Assume that we start with a 1d sample of some finite values of L and G (x in Fig. 11) and we increase L, then Fig. 11 tells us with (16) that

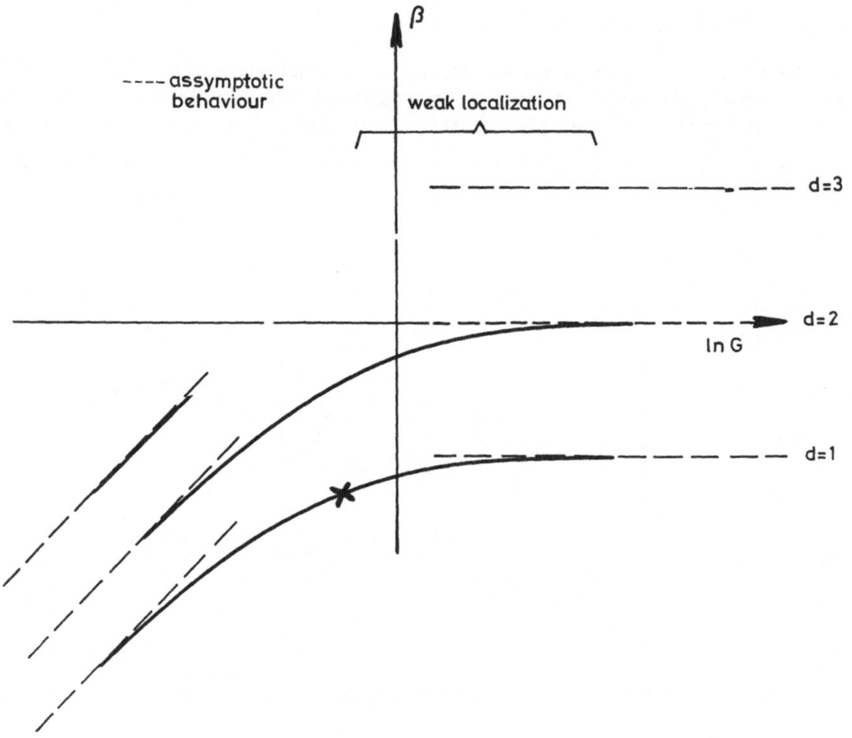

Fig. 11. The scaling function $\beta(G)$. According to e.g. [6,9,11,12,15,25].

$\frac{dG}{dL} < 0$, since L and G are positive and $\beta(G)$ is negative for all values of G. This means an increase of L leads to a decrease of G. We can continue the procedure and will end up for sufficiently large L with a behaviour characteristic for localized states, independent what our starting values were. The same holds for $d = 2$. Things are different for $d = 3$. Here we can find starting conditions which lead to a G(L) behaviour which is characteristic for extended states and Ohm's law and others which lead to localized states. So we can conclude that we have only localized states for $d \leq 2$ and both localized and extended states for $d > 2$ in disordered systems, provided that the assumptions made above are valid.

The concept of the β-function breaks down, however, if there are several parameters e.g. in addition electron-electron interaction and it does not give any quantitative information about the localization lengths which have to be expected.

To summarize, we may state that the various concepts to describe disordered systems indicate that localized states have to be expected. The detailed consequences depend on the model used.

II. OPTICAL PROPERTIES OF (ALMOST) PERFECT SEMICONDUCTORS

In this chapter we shall shortly outline the optical properties of semiconductors and give some experimental proofs for the free or delocalized nature of excitons.

II.A. Band Structure and Excitons

In Fig. 12 we show schematically the band structure of a semiconductor; on the left hand side the energies of the bands as a function of a space coordinate z, on the right hand side the so-called dipersion relation $E(\vec{k})$ where $\vec{k}$ is the wavevector and $\hbar\vec{k}$ the quasi-momentum (see also eq. (7a)).

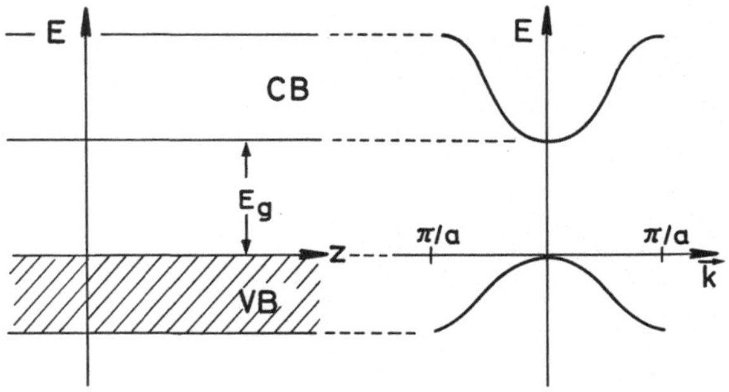

Fig. 12. Band structure of a semiconductor in E(z) and E(k) representation.

The conduction and valence bands describe the one-particle states of electrons and holes, which are used to describe electric transport phenomena like electrical conductivity, Hall effect, etc.. In contrast, the optical properties are always due to two particle transitions. In an absorption (emission) process we create (annihilate) always an electron and a hole. The hole which is created has wavevector and spin opposite to the ones of the electron which is excited into the conduction band.

Electrons and holes interact with each other due to their electric charges and form a series of hydrogen-like bound states below the band-to-band transition region. The quanta of these bound states are called excitons in the concept of quasiparticles [26]. Their dispersion relation is described in Fig. 13 and eq. (18) for the case of isotropic, parabolic, nondegenerate, bands with extrema at K=0, the so-called Γ-point

$$E_x(\vec{K},n_B) = E_g - E_x^b \frac{1}{n_B^2} + \frac{\hbar^2 K^2}{2M}$$

$$E_x^b = R_y \frac{\mu}{\epsilon^2} \; ; \quad \vec{K} = \vec{k}_e + \vec{k}_h \; ; \tag{18}$$

$$M = m_e + m_h \; ; \quad \mu = \frac{1}{m_o} \frac{m_e m_h}{m_e + m_h}$$

n_B is the main quantum number of the hydrogen problem: E_x^b is the exciton binding energy. This is the Rydberg energy (13.6 eV) reduced by the material parameters ϵ and μ to values from 5 to 200 meV in typical semiconductors, with a simultaneous increase of the Bohr radius to values 10 Å $\lesssim a_B \lesssim$ 200 Å for typical semiconductors, i.e. larger than the lattice constant (Wannier-limit). The good quantum number is the momentum $\vec{K}$ of the center of mass, where M is the translational mass. ϵ gives the dielectric constant and μ the dimensionless reduced mass. Excitons dominate the optical properties of semiconductors in the vicinity of the absorption edge. For dipole-allowed, direct band-to-band transitions excitons are coupling strongly to the radiation field, e.g. in GaAs, CdS, or CuCl. In this case the proper description is in terms of exciton-photon mixed states, the so-called exciton-polaritons. This concept is, however, beyond the scope of this contribution and the reader is referred e.g. to [26].

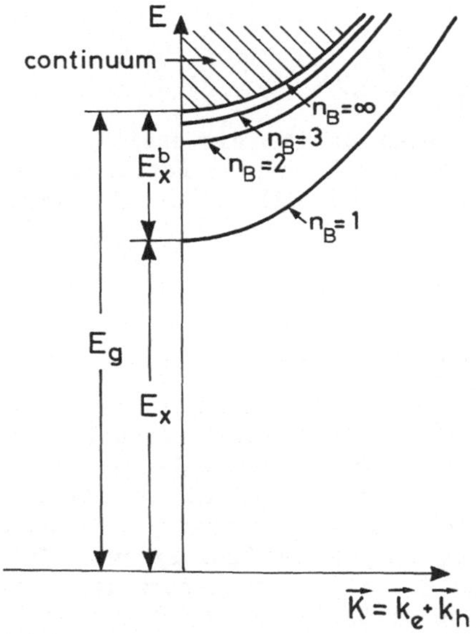

Fig. 13. Dispersion of excitons for a direct gap material with VB maximum and CB minimum at the Γ-point (K=0).

II.B. Some Optical Properties of Semiconductors

The strong coupling of excitons to the radiation field gives typical resonance structures in the spectra of the absorption coefficient $\alpha(\omega)$ and the reflectivity $R(\omega)$ as shown schematically in Fig. 14. In (a) one sees the hydrogen-like series of resonances with oscillator strength

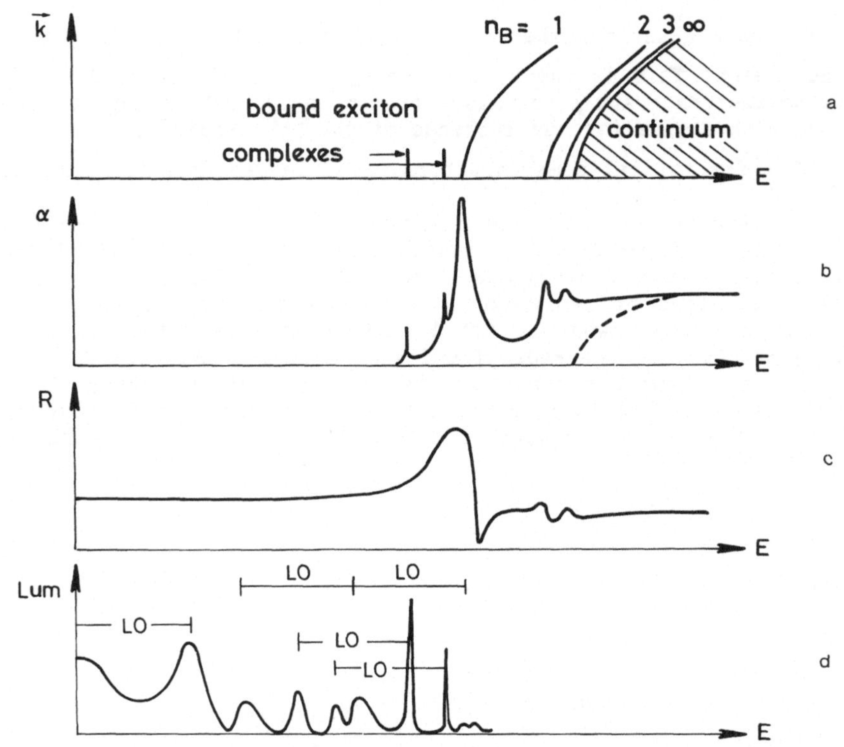

Fig. 14. Schematic drawing of the dispersion relation (a), of the absorption spectrum (b), of the reflection spectrum (c), and of a luminescence spectrum (d) of an idealized direct gap semiconductor.

decreasing like n_B^{-3} and the continuum transitions (compare to Fig. 13). The dashed line gives the hypotetic square-root dependence of the absorption coefficient expected without electron-hole interaction. The strong absorption bands show also up in the reflection spectrum (b). Figure 14c finally gives a schematic drawing of the luminescence spectrum. The direct emission at the free exciton resonance is generally rather weak, due to the exciton-like nature of the polaritons. The recombination under emission of LO phonons is usually much stronger. At low temperatures one often observes sharp emission lines, which are due to the recombination of excitons bound to various points defects, like neutral donors and acceptors, and their LO-phonon replica. At still lower photon energies one finds band-impurity or donor-acceptor pair recombination together with the LO-phonon replica. This emission is sometimes called "edge-emission". At even lower photon energies, there are recombination processes involving deep centers or intra-atomic recombination in some dopants, which are partly strongly coupling to the lattice. An example of luminescence spectra of the type shown in Fig. 14c is found in [27].

II.C. Some Proofs for Thermal Exciton Motion

Though most of the emission bands mentioned above are connected at low temperatures with impurities and defects, the rich variety of luminescence processes is generally considered as a proof for the mobility of excitons. If they would not move through the crystal, they would not find so many different recombination centers. The fact that the total luminescence yield in "pure" materials is often around 10^{-2} to 10^{-3} indicates that the excitons find during their motion through the lattice preferentially nonradiative recombination centers.

Another proof that excitons are mobile quasiparticles comes from the line shape analysis of the exciton-nLO emission bands [28,26d]. They can be described by a Boltzmann-distribution for the occupation probability and a square root density of states, which is characteristic for (quasi-) particles with finite effective mass.

Further evidence for the motion of free excitons came from spectroscopy with laser induced gratings [29]: Two pulsed coherent laser beams with photon energy in the band-to-band transition region and equal intensities I_{exc} of about 5 kW/cm² were brought to interference under a small angle on the surface of a CdS crystal. Due to the interference, a spatially periodic generation rate was produced, described by

$$G(z) \sim I_a + I_b + 2\sqrt{I_a I_b}\cos(2\pi z/\Lambda)$$

$$\cong 4\,I\,(1 - \sin(2\pi z/\Lambda)) \text{ for } I = I_a = I_b \tag{19}$$

with

$$\Lambda = \frac{\lambda}{2\sin\frac{\theta}{2}} \tag{20}$$

where Λ is the grating constant of the population grating, θ the angle between the two incident pump beams and λ their wavelength.

The population grating of excitons, which is formed according to eq. (19), decays by recombination and diffusion. While the recombination of excitons is independent on Λ, diffusion becomes more dominant for decreasing Λ. The population grating modulates e.g. via collision broadening of the exciton resonces the optical properties. In the experiment described here, the resulting phase grating was read by a synchroneous pulse in the transparent spectral region of the sample. The normalized diffraction efficiency η for the first diffrated order of the probe pulse is plotted in Fig. 15 as a function of Λ^{-1} for two different temperatures. The experimental data points drop for increasing Λ^{-1} (i.e. for decreasing Λ) due to the influence of diffusion. The theoretical curves are fits to deduce the values of the diffusion length l_D. The theory which includes two-dimensional diffusion and the optical nonlinearity is described in detail [29b]. As a rule of thumb η drops to one half for a grating constant connected with l_D by

$$l_D\, 2\pi \cong \Lambda_{1/2} \tag{21}$$

The fit reveals for excitons in CdS $l_D \cong 1\ \mu m$ with a tendency to decrease with increasing temperature due to scattering with phonons. The lifetime of the excitons is of the order of 1 ns.

To conclude we can state that excitons in reasonably pure semiconductors are mobile or delocalized quasi-particles. Due to their finite lifetime and consequently limited value of diffusion length, we are only able to detect localization effects by disorder, which have a localization length ξ below $l_D \cong 1$ μm and coupling time constants between free and localized states below 1 ns.

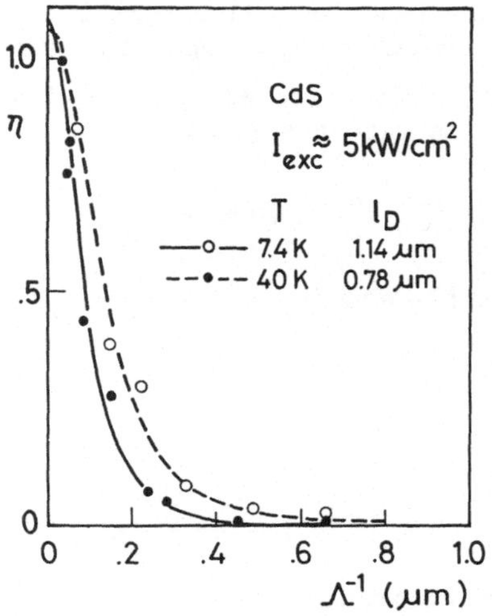

Fig. 15. The normalized diffraction efficiency η of a laser induced population grating in CdS as a function of the inverse grating constant Λ^{-1} for two different temperatures. Open and closed circles: experiment, solid lines: theoretical fit to deduce the diffusion length. From [29b].

III. WHAT CAN WE LEARN FROM OPTICAL SPECTROSCOPY ABOUT LOCALIZATION IN ALLOY SEMICONDUCTORS?

In this section we want to investigate a specific example of a disordered system, namely alloy semiconductors. We have chosen $CdS_{1-x}Se_x$ since this system is one of the most widely investigated alloy semiconductors. In 3.5 we will shortly review exciton localization in some other disordered systems. The experimental techniques, with which we want to investigate localization effects, come, as already mentioned, from optical spectroscopy.

III.A. Some Properties of Alloy Semiconductors

Alloy semiconductors are mixed crystals. There are anion substituted materials of the type AB → $AB_{1-x}C_x$ → AC. Examples are $CdS_{1-x}Se_x$, $ZnSe_{1-x}Te_x$ or $GaAs_{1-x}P_x$. Kation substituted materials of the type AB → $A_{1-y}D_yB$ → DB are e.g. $Al_{1-y}Ga_yAs$, $Zn_{1-y}Cd_yTe$ or $Cd_{1-y}Mn_yTe$. More examples and their properties are compiled e.g. in [30]. The systems form either mixed crystals for the whole range $0 \leq x,y \leq 1$ or for a fraction of it. The crystals have a well defined, crystalline lattice, usually of zinkblende (T_d) or wurtzite structure (C_{6v}). The disorder consists in fluctuations of the chemical composition on a microscopic scale. This compositional disorder is often called "weak disorder" to distinguish it from "strong" structural disorder of amorphous materials. Just for the sake of clarity (or confusion) we want to mention that the localization effects, which are introduced by this weak disorder are of the type of strong localization as defined in the first section.

The properties of the semiconductors depend on x. There may be a continuous variation of some property with x (e.g. the width of the forbidden gap $E_g(x)$ or the phonon frequency $\hbar\omega_{LO}(x)$) or a discontinuous one. The first case is referred to as amalgamation type, the second one as persistent mode type. $CdS_{1-x}Se_x$ is of the persistent mode type concerning the optical phonon energies. The Cd-S and Cd-Se LO modes are present simultaneously for $0 < x < 1$ with a weight function depending on x and with eigenenergies around 35 meV and 25 meV, respectively, which vary only weakly with x. The width of the forbidden gap varies on the other hand continuously with x from the value of pure CdS ($E_g \cong 2.58$ eV) to the one of CdSe ($\cong 1.85$ eV).

One often finds a relation

$$E_g(x) = E_g^{AB} + (E_g^{AC} - E_g^{AB} - b)x + bx^2 \tag{22}$$

where b is the "bending parameter" which describes the deviation of $E_g(x)$ from a linear relation $b_{CdS_{1-x}Se_x} \cong 0.31$ eV [31,32]. In Fig. 16 we show the eigenenergy of the exciton resonance as a function of x, deduced e.g. from the resonances in reflection.

III.B. A Review of the Optical Properties

In Fig. 17a we show the transmission spectra of various $CdS_{1-x}Se_x$ crystals. The continuous shift of the absorption edge is obvious. The position of the absorption edge, defined arbitrarily by T = 0.1 (the thickness of all samples is around a few tens of μm) decreases continuously with increasing x and is situated slightly below the exciton energies in Fig. 16.

Figure 17b shows luminescence spectra for various x obtained under weak band-to-band excitation. For the pure constituents x=0 and x=1 one sees narrow emission bands labelled I_1 and I_2 which are due to excitons bound to neutral acceptors and donors. Their spectral width is limited by the setup. For $0.03 \lesssim x \lesssim 0.8$ one observes a rather broad, unstructured zero-phonon emission band and rather strong LO-phonon replica. For x around 0.5 it consists of two components which correspond to the LO phonon modes of CdS and CdSe, respectively. The ratio of the intensities of the lower energy CdSe mode (which is closer to the zero phonon band) and the CdS mode increases with x, according to the persistent mode character. For $x > 0.8$ the luminescence spectra start to show again the sharp features of bound exciton complexes. Obviously, there is some asymmetry in the spectra if one replaces x by 1-x.

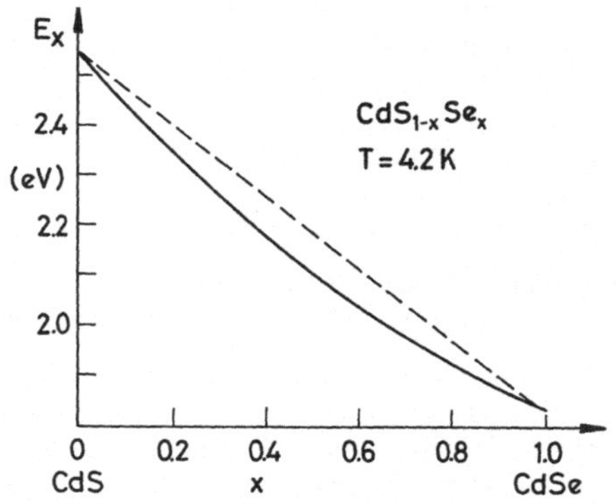

Fig. 16. The eigenenergy of the free exciton resonance as a function of x in $CdS_{1-x}Se_x$. According to [31].

By the pioneering work of [33,34] it has been found that the broad zero-phonon emission bands observed for $0.03 \lesssim x \lesssim 0.8$ are not simply broadened recombination lines of excitons bound to point defects as the I_1 and I_2 lines, but are due to excitons localized by compositional disorder.

Before we describe some of the experiments, which support this idea, we shall say a few words about compositional alloy disorder.

III.C. The Concept of Alloy Disorder

As already mentioned, the concept of alloy disorder is based on the fact that we have on a microscopic scale (lattice constant to exciton Bohr radius) random spatial fluctuations of x, which result via the relation $E_g(x)$ in fluctuations of the width of the forbidden gap. This is visualized in Fig. 18. CdS and CdSe are rather strongly ionic bound. Consequently, the lowest conduction band comes mainly from the empty 5s states of Cd^{2+} while the upper valence bands originate from the occupied 3p levels of S^{2-} or the 4p levels of Se^{2-} with some admixture of lower lying d states. Consequently, fluctuations in x affect mainly the VB. For

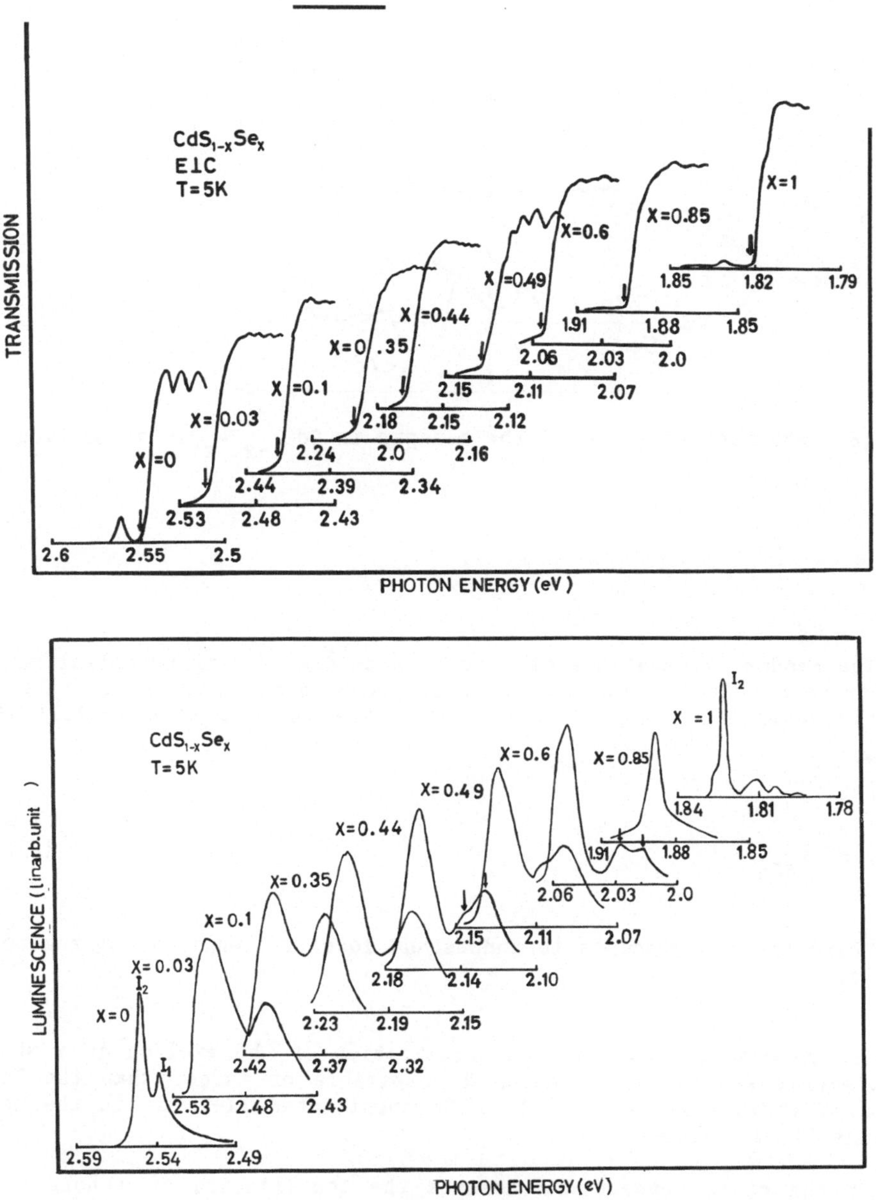

Fig. 17. Transmission (a) and luminescence spectra (b) of $CdS_{1-x}Se_x$ samples for various values of x. From [32].

more covalent binding, e.g. in $Ga_{1-x}Al_xAs$ fluctuations in x influence both the CB and the VB as is known e.g. from the study of the band-offsets in $Ga_{1-x}Al_xAs$ quantum well structures [35].

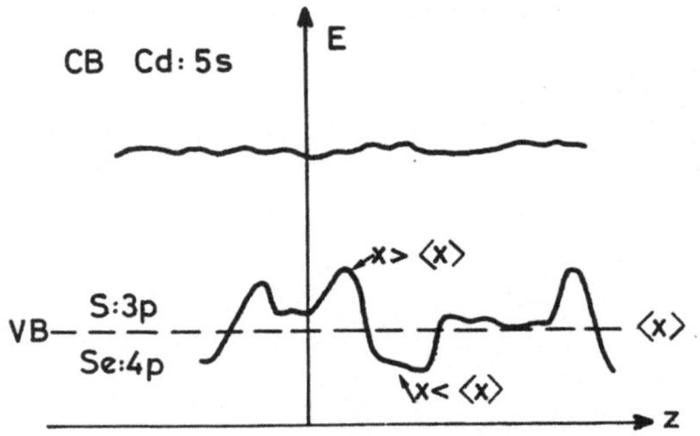

Fig. 18. Schematic drawing of the bandgap in $CdS_{1-x}Se_x$ as a function of space.

The random fluctuations of x form three-dimensional potential wells and barriers. These wells can bind in three dimensions a particle of effective mass m_{eff} only if the depth V and the radius R fulfill the inequality

$$V \cdot R^2 \cdot m_{eff} \gtrsim \hbar \qquad (23)$$

There are two concepts to understand localization in alloy semiconductors:

In one case one assumes the localization of the exciton as a whole in potential wells with a radius R comparable or larger than the Bohr radius. This idea seems to work in Mn substituted II-VI or in the more covalent III-V alloys.

In the other concept one assumes the localization of a hole in a deep and spatially narrow potential well which then binds an electron to form a localized exciton. Due to the fact that holes are usually considerably heavier, this concept explains via eq. (23) why it is easier to observe excitons localized by alloy disorder in some anion substituted materials as compared to kation substituted ones [36].

Most probably both concepts may work even in one and the same material with a continuous transition from the first to the second with increasing depth of localization below the mobility edge. Therefore, we shall not stress this difference in the following.

Both models yield a density of state as shown schematically in Fig. 19. There are extended states above a mobility edge with an almost square root dependence and there is an exponential tail of localized states below.

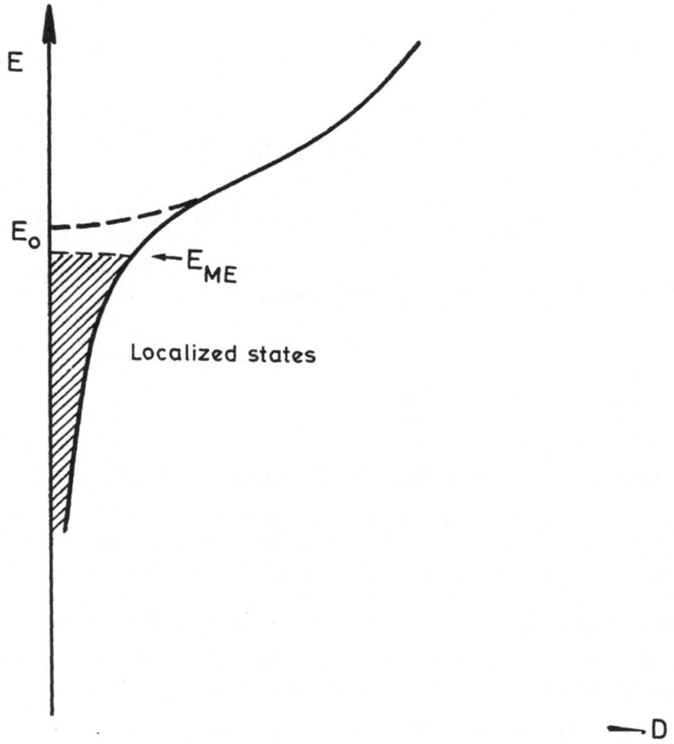

Fig. 19. Schematic density of states in an alloy semiconductor.

According to [37] it is possible to describe the density of states quantitatively in the following way.

We assume that the crystal potential $V(\vec{r})$ fluctuates spatially around its average value $\bar{V}$ by $v(\vec{r})$, i.e.

$$V(\vec{r}) = \bar{V} + v(\vec{r}) \tag{24}$$

The fluctuation of the width of the forbidden gap is connected with $v(\vec{r})$ via (25)

$$\frac{dE_g}{dx}\bigg|_{\bar{x}} \cdot \Delta x(\vec{r}) \cong \frac{dV}{dx}\bigg|_{\bar{x}} \Delta x(\vec{r}) = v(\vec{r}) \tag{25}$$

where $\Delta x(\vec{r})$ describes the random spatial fluctuation of the composition x. The exciton energy $E_x(\vec{r})$ fluctuates spatially in a similar way and one has to include also the dependence of E_x^b on x.

$$\epsilon(\vec{r}) = E_x(\vec{r}) - \bar{E}_x \tag{26}$$

For the density of states one finds

$$D(\epsilon) \sim \epsilon^{1/2} \quad \text{for} \quad \epsilon > 0 \tag{27}$$

$$D(\epsilon) = D_o \exp\{(-|\epsilon|\epsilon_o^{-1})^{\beta}\} \quad \text{for} \quad \epsilon < 0 \tag{28}$$

The parameters D_o and ϵ_o are given in (29). β is close to one:

$$D_o \cong (\epsilon_o a_B^3)^{-1} ; \epsilon_o \sim m_x a_B^2 \langle v^2\rangle/4\hbar^2, \ \beta \cong 1 \tag{29}$$

$$\langle v^2\rangle \sim \left[\frac{dE_x}{dx}\right]^2 x(x-1) \frac{\Omega}{\pi a_B^3} \tag{30}$$

D_o and ϵ_o decrease for excitons with small effective mass and large Bohr-radius. The first fact is a consequence of eq. (23). Heavy particles are easier to localize than light ones. The second one reflects the fact that large excitons average over more unit cells than small ones do. The relative fluctuations of the composition and via eq. (25,26) of the gap and the exciton energies are consequently smaller for large excitons. The meaning of the terms which describe $\langle v^2\rangle$ in eq. (30) is evident. Ω is the volume of the unit cell. Ω/a_B^3 is then proportional to the number of unit cells over which an exciton of radius a_B is averaging. The term x(x-1) describes that all fluctuation and localization phenomena disappear for the pure constituents x=1 and x=0. The term $(\frac{dE_x}{dx})^2$ tells us that there are no localized excitons in mixed crystals if the exciton energy E_x does not depend on the composition of x.

III.D. Experimental Proofs for the Existence of Localized Excitons in $CdS_{1-x}Se_x$

Now we want to present some experimental justifications for our hypothesis that the brood emission structures in the spectra of $CdS_{1-x}Se_x$ are due to the recombination of excitons localized by disorder and their LO phonon replica. Two hints are according to [36] the facts that at low temperature mixed crystals have a higher luminescence yield as compared to the pure constituents and that there are no other emission lines in contrast to the rich variety of lines (Fig. 14) in normally pure materials. In order to find the various radiative and nonradiative centers it is necessary that the excitons can travel through the lattice. If they relax after excitation rapidly into localized states, their chance to find one of the above mentioned centers is reduced and the main recombination channel is via emission of photons.

While these qualitative considerations are more a hint than a proof, stronger evidence came from the spectroscopic methods described in the following.

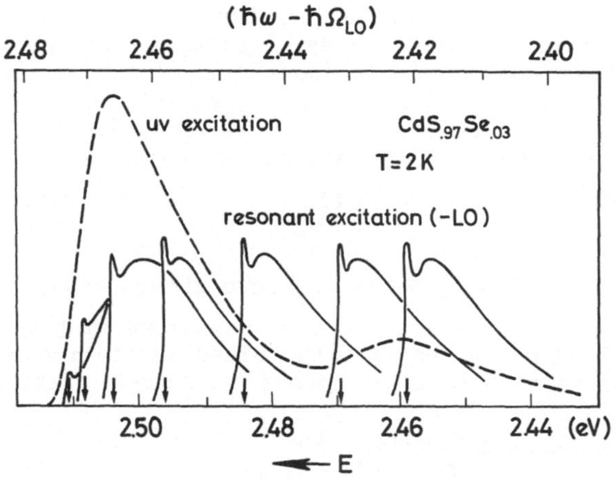

Fig. 20. Emission spectrum of $CdS_{1-x}Se_x$ under UV-excitation at low temperatures (dashed line) normalized spectra of the LO phonon replica under excitation at the photon energies indicated by an arrow (↓). These spectra are shifted by the LO phonon energy to higher energies to coincide with the zero phonon line. From [34,38,39].

1. <u>Site Selective Spectroscopy</u>. Site selective spectroscopy gave very valuable information about the nature of the emission bands in $CdS_{1-x}Se_x$ [33,34,36-39]. Figure 20 shows a typical emission spectrum under UV excitation. The spectra do not change in shape as long as the exciting photon energy $\hbar\omega_{exc}$ is above the high energy edge of the broad zero phonon line, but change drastically below.

Since the precise shape of the zero-phonon line is difficult to determine in the latter case due to resonant light scattering, one usually investigates the LO-phonon replica which reflect what happens in the zero phonon band. In Fig. 20 various spectra of the LO phonon satellite are shown, however shifted by the LO-phonon energy to higher energies to coincide with the zero phonon line. One observes a narrow spike which starts to appear when $\hbar\omega_{exc}$ reaches the high energy edge of the zero-phonon emission. In the representation of Fig. 20 this spike coincides with $\hbar\omega_{exc}$ (↓) and is interpreted as recombination from localized levels in the tail of the density of states (Fig. 19) which are excited resonantly and recombine without energy loss. The low energy tail which gets narrower with decreasing $\hbar\omega_{exc}$ is attributed to recombination of the resonantly excited, localized exciton levels under emission of acoustic phonons and/or recombination of lower lying localized states which are populated by the resonantly excited ones, e.g. by acoustic phonon assisted tunneling. Note that the sample temperature is so low that there is no thermal excitation to energies above $\hbar\omega_{exc}$. The fact that the line shape is independent on $\hbar\omega_{exc}$ for $\hbar\omega_{exc}$ above the zero phonon line can be understood as follows: in this case excitons are excited primarily in extended states. During the relaxation down to the mobility edge they can move and loose the memory of their excitation conditions. From the above discussion one expects that the mobility edge is situated at the high energy edge of the zero phonon line. Since both, this "edge" and the appearance of the narrow spikes with decreasing $\hbar\omega_{exc}$ are continuous functions, it is difficult to decide if this mobility edge is really sharp or not.

For $\hbar\omega_{exc}$ considerably below the zero phonon band, the narrow spike evolves into a usual Raman line without the broad wing on the low energy side. Time-resolved measurements with pulsed excitation should be useful to study this transition from luminescence to Raman scattering.

2. Polarization Spectroscopy. Our next example of optical spectroscopy concerns polarization spectroscopy. $CdS_{1-x}Se_x$ crystallizes in the uniaxial wurtzite structure (C_{6v}). Apart from a possible sixfold warping, the material is isotropic in the plane perpendicular to the $\vec{c}$-axis. In the experiments described in [38,39], samples have been excited on such a plane with the wavevectors of exciting and luminescence photons parallel to $\vec{c}$. Figure 21 shows the zero phonon band for UV excitation. The full circles give the maximum degree of polarization $\rho = (I_{\parallel}-I_{\perp})/(I_{\parallel}+I_{\perp})$ in the LO phonon band as a function of $\hbar\omega_{exc}$. The intensities $I_{\parallel}$ and $I_{\perp}$ are measured with respect to the linear polarization of the exciting light.

The open circles give the decay time of the luminescence at various photon energies after pulsed excitation. This aspect will be discussed in the following paragraph. The polarization ρ is zero for excitation above the high energy edge of the luminescence. At the edge one observes a steep increase of ρ to values around 0.35, followed by a gradual decrease. The interpretation is again as follows. For excitation above the mobility edge, the excitons loose their memory during relaxation. At the mobility edge and below, localized states are excited resonantly. One can expect a splitting of the otherwise degenerate Γ_x and Γ_y components of the excitons due to the random anisotropies of the wells which localize the excitons. The exciting linearly polarized photons excite preferentially excitons with the proper orientation. The maximum value of ρ can be .5 in this case and the maximum observed values come rather close to it. The rather sharp onset of ρ on the high energy edge of the luminescence indicates again that the mobility edge is at this energy and the position of this "edge" can be defined within 2 meV. The gradual decay of ρ for lower values of $\hbar\omega_{exc}$ is attributed in [37,38] to a decrease of the anisotropy of the localized excitons. Such a behaviour can be expected especially if only the hole is localized in a deep and narrow well with the electron orbiting around it in an isotropic 1s state.

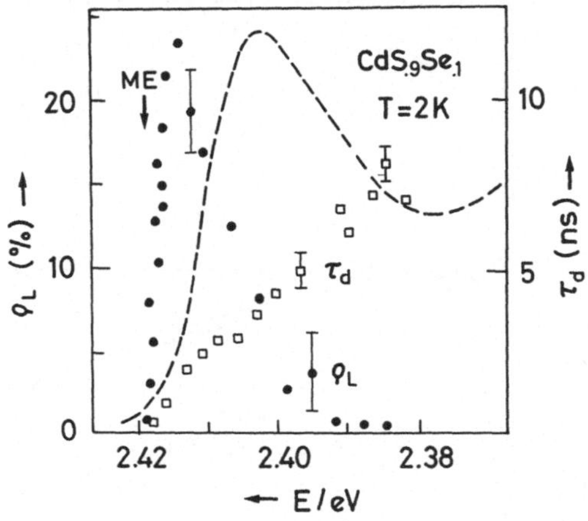

Fig. 21. The zero phonon line of a $CdS_{.9}Se_{.1}$ sample (dashed curve), the maximum degree of polarization in the LO phonon replica ρ as a function of the exciting photon energy (full circles), and the decay time as a function of the photon energy in the zero phonon line (squares). From [38,39].

3. Luminescence Dynamics. In the pure constituents one usually finds that the decay of the bound exciton emission after pulsed excitation is exponential and independent of the photon energy in a bound exciton band. Both statements are no longer true in mixed crystals. There is a significant non-exponential decay [37] and the decay time increases for decreasing photon energy in the zero phonon band [38-44]. An example is shown in Fig. 21. Due to the non-exponential decay the decay time "constants" have to be considered with some care.

In more recent experiments both rise and decay dynamics of the luminescence in $CdS_{1-x}Se_x$ ($0.5 \leq x \leq 1$) have been measured after band-toband excitation with pulses of 5 ps duration from a synchroneously pumped dye laser [43,44]. The temporal evolution of the luminescence has been recorded after a spectrometer by a streak camera. For pure CdSe the rise and decay kinetics are independent in the range of the bound excitons and the decay is exponential with a decay constant around 500 ps [43]. In the range of compositions where localized excitons are expected, the behaviour is quite different. In Fig. 22 we give an example. At the higher energy side of the zero phonon band the rise time of the luminescence is limited by the temporal resolution of the setup of about 20 ps and the decay is rather fast (solid line). If we go deeper into the band, i.e. to deeper localized states, both the rise and decay times of the luminescence increase (open cirlces, dash-dotted, and dashed curves). In the LO phonon satellite one finds the same behaviour. The dotted line gives e.g. the dynamics on the high energy edge of the LO replica. It coincides almost perfectly with the curve of the zero phonon line. In Fig. 23 we show a time-integrated luminescence spectrum at low temperature and the time τ_{max}, which elapses between the excitation pulse and the temporal maximum

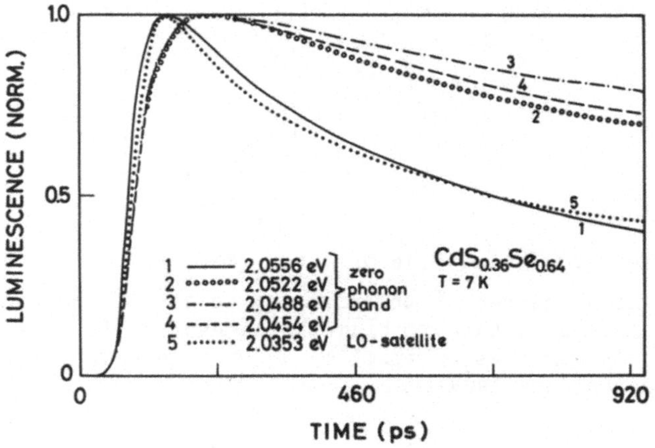

Fig. 22. Rise and decay curves of the luminescence in $CdS_{0.36}Se_{0.64}$ at low temperature for various photon energies. According to [43,44].

of the luminescence. One observes a drastic increase of τ_{max} from a resolution limited value of 20 ps to values around 350 ps over the zero phonon line and a similar behaviour for the LO phonon replica. This behaviour can be understood qualitatively if one assumes that the excitons relax after excitation fast through the extended states and then trickle down through the localized ones. The deeper the states are localized, the more time it takes the excitons to relax down to them e.g. by acoustic phonon assisted tunneling. The decay of the deeper states is slower because they are fed by relaxation from higher localized states and because their oscillator strength decreases. This latter phenomenon can be attributed to a decreasing overlap of electron and hole wave functions for the deeper localized states [36-44]. A simple phenomenological model which assumes a time constant τ_1 for the relaxation into a certain localized level and τ_2 for the decay out of this level yields for the normalized luminescence intensity the following expression:

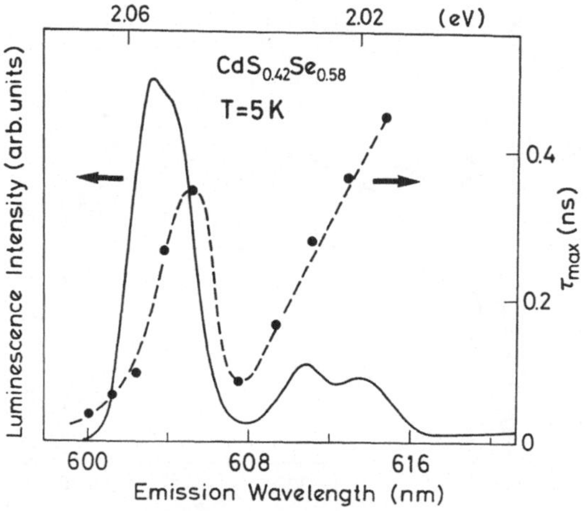

Fig. 23. A time integrated luminescence spectrum of $CdS_{.42}Se_{.58}$ and the time τ_{max} which elapses between the excitation pulse and the temporal maximum of the luminescence. From [44].

$$\frac{I(t)}{I_{max}} = \frac{e^{t/\tau_1} - e^{-t/\tau_2}}{\left[\frac{\tau_1}{\tau_2}\right]^{\tau_2/(\tau_2-\tau_1)} \left[1 - \frac{\tau_2}{\tau_1}\right]} \tag{31}$$

As seen in Fig. 24, this formula describes reasonably well the experimental findings of Fig. 22 with values of τ_1 and τ_2 ranging from 25 ps to 200 ps and from 400 ps to 700 ps, respectively.

A more elaborate hopping model to describe the reaction kinetics is given in [43], which uses the simplification of only one hop in the relaxation from the mobility edge to the final state from which the recombination takes place. This model fits the experimental data quite well and gives among others a parameter ϵ_0 in eq. (27) of 4.2 meV, which is a reasonable value.

4. <u>Thermal Effects</u>. The experiments on $CdS_{1-x}Se_x$ reported so far have been carried out at low temperatures (2K to 8K). This temperature is so low that thermal excitation of localized excitons to higher localized or extended states is negligible. This assumption breaks down, if the sample temperature is increased and indeed the luminescence dynamics change [44,45].

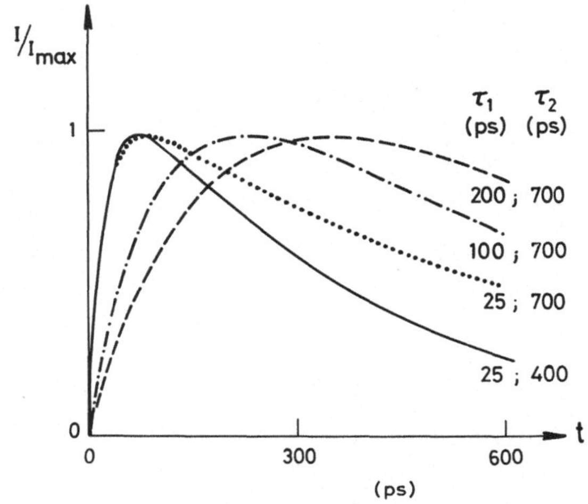

Fig. 24. Calculated rise and decay curves according to eq. (31) for various sets of parameters τ_1 and τ_2.

In Fig. 25 we show rise and decay curves for two different photon energies as in Fig. 22 but for 77K. Obviously, the luminescence kinetics are now independent of the photon energy. This phenomenon is further illustrated in Fig. 26 where we show τ_{max} as a function of the energetic distance from the emission maximum for various temperatures. At low temperatures one finds a similar behaviour as in Fig. 23. With increasing temperature the structures get flatter and disappear for $T > 50$ K.

A qualitative interpretation of this observation is as follows. At low temperatures there are only relaxation processes towards lower energies. Consequently, the "center of gravity" of the exciton population moves after pulsed excitation gradually to deeper localized states. At higher temperatures there is an equilibrium between relaxation and thermal excitation of excitons which may even be reexcited above the mobility edge after being trapped in localized states. Consequently, the "center of gravity" of the exciton population is stationary during the exciton lifetime after the thermalization time and the distribution function extends into the delocalized states. The temperature where the luminescence dynamics of the localized states becomes independent of photon energy $T \gtrsim 50$ and the energetic width of the exponential tail of localized states (ϵ_0 = 4.2 meV) nicely correspond to each other.

Putting all these data together we can be rather sure that there are localized exciton states due to composition fluctuations in $CdS_{1-x}Se_x$ for $0.03 \lesssim x \lesssim 0.8$. The transition from localized to delocalized states is situated on the high energy side of the zero phonon band. It is still an open question how sharp this transition might be. The upper limit of the width of the transition deduced from site-selective and polarization spectroscopy is between 5 meV and 2 meV. It can be given e.g. by the finite lifetime of the excitons or by macroscopic fluctuations of the composition on a length scale several orders of magnitude above a_B but in

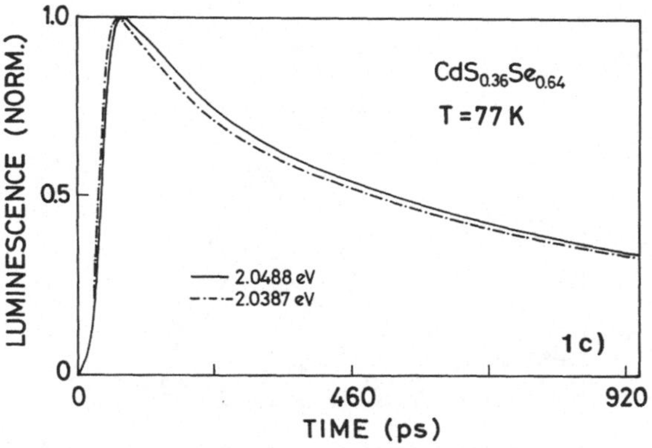

Fig. 25. Rise and decay curves as in Fig. 22 but for 77 K. From [45].

the order of the excitation spot diameters. Small excitation spots are therefore advisable (10 to 100 μm) but often values in the mm range are used. Finally, we should give a comment about the apparent asymmetry in the range of compositions where localized excitons occur. One is connected with the bending parameter b in eq. (22). As seen in Fig. 16 $\frac{dE_g}{dx}$ is much smaller on the Se rich side. This enters via eq. (30) in the width and density of localized states. The other argument is as follows. A single Se-atom in CdS produces already a deep (isoelectronic) trap. If their density is so large that the wavefunctions on the various Se sites overlap, a tail of localized states is easily formed. On the other hand, an S atom in CdSe produces a state in the valence band, which contributes litte to localization. Localization occurs only if $E_g(x)$ is larger than in CdSe and fluctuations in x result both in barriers and wells.

III.E. Exciton Localization in Other Materials

In this subsection we want to shortly outline exciton localization in some other semiconductors. Concerning the II-VI materials almost all ternary compounds of Zn and Cd as kations and S, Se, and Te as anions have been investigated. Examples are given in [36,46]. As already mentioned, there is a trend that exciton localization is stronger in anion substituted materials as compared to kation substituted ones. An exception from this rule are however the semimagnetic mixed crystals which contain Mn^{2+} like $Cd_{1-x}Mn_xSe$ and $Cd_{1-x}Mn_xTe$. In [47] the competition between exciton capture at impurities and exciton localization by alloy fluctuations is investigated in detail by time resolved measurements.

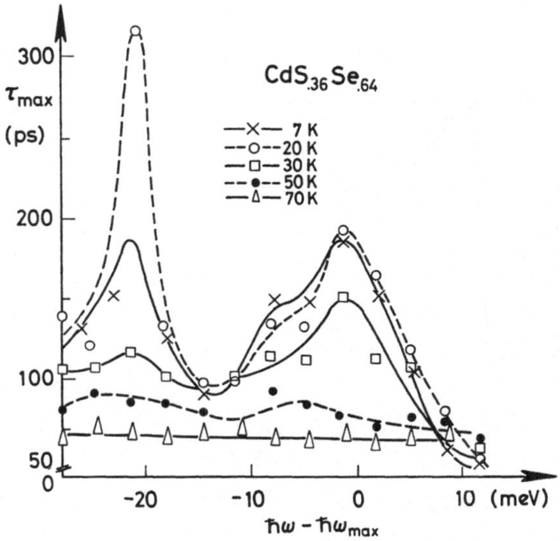

Fig. 26. The rise time τ_{max} as a function of the energetic distance from the emission maximum $\hbar\omega_{max} - \hbar\omega$ for various temperatures. From [45].

The Te substituted II-VI compounds like $CdS_{1-x}Te_x$ or $ZnSe_{1-x}Te_x$ with a small Te content show an unusual behaviour concerning the width of the emission bands attributed to the localized excitons [36,39,48,49]. Typical values are 200 meV or more for the Te materials as compared to about 20 meV for $CdS_{1-x}Se_x$ (see e.g. Figs. 17, 20 or 23). Since the material parameters of the various II-VI compounds are not so different [30], a mechanism different from the one described by eq. (26) to (30) should apply. In [39,48] small aggregates of Te are made responsible for the exciton localization. More recently, a model about exciton self-trapping [50] has been applied to the $ZnSe_{1-x}Te_x$ system [49]. It is assumed that self-trapping of excitons, which is usually absent in semiconductors due to the weak exciton-phonon coupling in the Wannier limit, occurs at single Te atoms or at aggregates of two Te. The strong lattice distortion (i.e. the large Huang-Rhys factor) explains then the width of the observed emission bands. The fact that free and localized excitons can be observed simultaneously in $ZnSe_{1-x}Te_x$ is taken into account in [49] by the assumption that the free and self-trapped excitons have not very different energies and are separated by potential wells which allow for tunneling.

The system $AgBr_{1-x}Cl_x$ has been investigated in some detail in [51]. Temperature variation, time-resolved resonant light scattering and the application of hydrostatic pressure are used to identify in this I-VII material the existence of excitons localized by alloy disorder.

In the III-V alloy compounds like $Ga_{1-x}Al_xAs$ [52], $GaAs_{1-x}P_x$ [53,54] or in heavily doped materials like GaAs·Si [55] localization effects have been found. These effects are partly less pronounced and need more careful investigation due to the larger exciton radius in connection with eqs. (29,30). Recently, time resolved luminescence excitation spectroscopy and resonant Raman scattering have been used to proof the existence and to determine the mobility edge in $GaAs_{1-x}P_x$ for small As content, i.e. for the indirect gap material [54].

Another exciting system and the last but not the least example, which we want to address here are (multiple) quantum well structures. By growing alternative layers of two materials with different bandwidth on top of each other it is possible to confine excitons in quasi-two dimensional layers of a thickness around 100 Å. The most widely investigated system is $GaAs/Ga_{1-x}Al_xAs$ where GaAs forms the wells and $Ga_{1-x}Al_xAs$ the barriers. Due to the finite barrier height, the exciton wavefunctions extend also into the barrier. The exciton energy depends sensitively on the width of the barrier. For details see the numerous reviews of this field like [35]. Fluctuations of the exciton energy arise in this system partly from random fluctuations of the width of the well. This fluctuation is in the best circumstance about one atomic layer. This leads to fluctuations in the confinement energy. For an idealized system with infinitely high barriers and a thickness L_z, the energy of the first confined level is given by

$$E_1 = \frac{\hbar^2 \pi^2}{2} \frac{1}{L_z^2 m_{eff}} \tag{32}$$

for single particles. For excitons m_{eff} has to be replaced by the reduced mass and the dependence of the exciton binding energy on L_z has to be considered.

For $L_z = 100$ Å and $\Delta L_z = 5$ Å one finds from (32)

$$\frac{\Delta E_1}{E_1} = \frac{2}{L} \Delta L_z \cong .1 \tag{33}$$

In addition, contributions to the total spatial fluctuation of the exciton eigenenergy have to be considered, which come from alloy disorder in the barriers (e.g. $GaAs/Ga_{1-x}Al_xAs$), in the wells (e.g. $Ga_{1-x}I_{nx}/InP$) or in both (e.g. $Ga_{1-x}In_xAs/Ga_{1-x}Al_xAs$). The localization in the $GaAs/Ga_{1-x}Al_xAs$ system has been investigated in detail in [56] by transient grating measurements, or measurements of the homogeneous line width by resonant Rayleigh scattering or optical hole burning and exciton localization has been clearly established. The mobility edge seems to be situated close to the center of the exciton absorption band in this quasi two-dimensional system. This finding has been connected in [56] with the fact that the percolation threshold in continuous two-dimensional systems occurs in the center of the potential relief, see 1.3.3.. It should be noted, however, that the width of the exciton band in the first, two-dimensional Brillouin-zone is much larger than the width of the absorption band, which represents only states close to $\vec{k} \cong 0$.

IV. CONCLUSION AND OUTLOOK

We have shown that optical spectroscopy is capable to reveal disorder induced localizaton in alloy semiconductors. In the II-VI materials which we emphasized here, actually the main part of our knowledge about localized and extended states of excitons came from optical spectroscopy. Transport measurements, which gave many valuable information about mobility edges e.g. in amorphous silicon [9,21] are almost completely missing for the disordered II-VI compounds. Another aspect which may receive some attention in the future concerns optical nonlinearities of semiconductor alloys.

The nonlinear optical properties of semiconductors, i.e. the variation of the optical properties under illumination, have been firstinvestigated mainly under the aspect of fundamental research, see e.g. [22,23,57]. Recently, however, a strong impetus came from applied research, since one has found that optically nonlinear materials may be used to construct optical memories and logic connections which open for the future the perspective of (digital) optical data handling. This technique may have some significant advantages over the presently used electronic data processing due to its inherent potential of parallel data handling which may help to overcome the van Neumann bottle neck which tends to limit the maximum speed of serial electronic computers. Some reviews and more details about this huge field, which we can address here only very shortly, are given e.g. in [58].

There is already a good understanding of the optical nonlinearities of pure semiconductors like Si, GaAs, ZnSe, CuCl or quantum wells (see e.g. [22,23,26e,35,37] and the references therein). The nonlinearities, which are created by optical excitation in the electronic system of the semiconductors are due to many particle effects like exciton-exciton collisions, formation of biexcitons or the transition to an electron-hole plasma (see also 1.3.2.). In the latter case one has simultaneously the effects of band-gap renormalization and screening of the Coulomb interaction due to many particle effects and state filling (or phase-space filling) due to the dominating Fermi character of electron-hole pair systems at high densities.

Optical nonlinearities tend to be maximal close to exciton resonances and the onset of band-to-band transitions. Consequently, every one of the above mentioned materials can be used efficiently only in a small spectral range.

As shown e.g. in Fig. 16, semiconductor alloys allow to shift the spectral position of the band edge at will by changing the composition. With the II-VI compounds it is possible to cover the whole range from the near UV to the far infrared with only a few systems. By a proper choice of the composition, the resonances of the electronic excitations, i.e. the excitons, can be shifted to every desired wavelength. The question arises now if alloy semiconductors will exhibit the same optical nonlinearities as the pure materials do. The attempts to answer this question are still at the very beginning. Obviously, there are optical nonlinearities e.g. in the system $CdS_{1-x}Se_x$ [32,59] or $Al_{1-x}Ga_xAs$ [60]. An electron-hole plasma is formed in the second example with the full variety of many-particle effects [60]. In contrast it has been found that in $CdS_{1-x}Se_x$ no plasma appears at low temperature under excitation conditions where a plasma is formed both in CdS and in CdSe [32]. The optical nonlinearities in $CdS_{1-x}Se_x$ at low temperature can be explained by filling of the localized states up to the onset of population inversion resulting in bleaching of absorptin, optical gain and laser emission. Many particle effects like band-gap renormalization seem to be much less pronounced. At temperatures around 60 K where the dynamics of the localized excitons indicate at thermal (re)excitation of a considerable fraction of the excitons above the mobility edge (see 3.4.4.), the optical nonlinearities of $CdS_{1-x}Se_x$ approach the behaviour of CdS and CdSe as found recently [45].

A lot of interesting physics and fascinating perspectives for possible applicatons can thus be expected from the investigation of the nonlinear optical properties of disordered semiconductors.

APPENDIX

In a simple mechanical experiment we demonstrated at the School the Anderson-localization in a one-dimensional system. Eight identical pendula are mounted on a solid bar (Fig. 27). For linearization purposes the pendula are suspended by a spring which consists of a thin plate made of bronze as shown in the inset of Fig. 27. The pendula have all the same oscillation period T_o of a little bit less than a second. As long as they are not coupled, they oscillate therefore all at the same frequency ω_o if we impose a wave-like excitation on them, independent of the wavelength and wavevector k of this excitation. The relation of the eigenfrequency ω as a function of k is thus simply a horizontal line at ω_o. The group velocity $\frac{d\omega}{dk}$ is consequently zero and a wavepaket shows neither propagation nor dispersion, as can be easily realized by exciting only one of the pendula. Now we couple the pendula by identical soft springs in a way that a pair of coupled pendula exhibitis a beat frequency which is more than a factor ten below its eigenfrequency. This coupling leads now to eight eigenfrequencies of the system of eight coupled pendula or to a band if the number of coupled oscillators would be larger, e.g. 10^{23} in a three dimensional solid, see eq. (6). Due to the finite dispersion of the new $\omega(k)$ relation we get now a finite group velocity. If we excite e.g. the pendulum at the right end, a wavepaket propagates through the chain, is reflected at the left end, returns, and so on. The wavepaket also clearly shows dispersion.

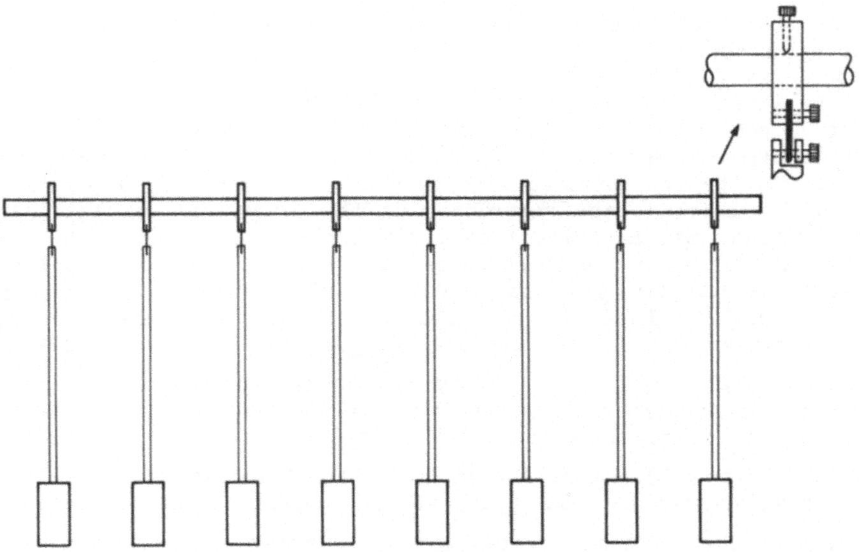

Fig. 27. A simple one-dimensional mechanical model to demonstrate Anderson localization.

It should be noted here that this model is significantly different from the linear chain which is usually treated to introduce lattice vibrations and phonons. There the restoring force only comes from the springs between adjacent masses. Here every mass has its eigenfrequency ω_o and the coupling between these oscillators serves only to split ω_o into a band. Our model is thus more similar to the Bloch-waves of Fig. 4 or to Fig. 7a than to phonons.

In the next step we introduce diagonal disorder. This means we do not change the coupling springs, but we randomly change the eigenfrequencies of the various oscillators. This corresponds to the fluctuations of the well depths in Fig. 7b. Technically we realize the change of ω_o by changing the length of the springs in the insert of Fig. 27. If we now excite a wavepaket in the center or at the edge of the linear chain of pendula, we find that the excitation, i.e. the wavepaket, spreads over about three neighbouring pendula, but does no longer propagate through the whole chain. This means that we have localized the excitation with a localization length ξ of about three lattice constants (compare Fig. 8) and thus demonstrated in this simple experiment the concept of Anderson localization.

Another mechanical model of this type of localization has been reported for a more phonon-like system in [61].

ACKNOWLEDGEMENTS

The author, who started only recently to work on disordered systems, namely on $CdS_{1-x}Se_x$, expresses his deep thanks for many fruitful, valuable, and stimulating discussions with Prof. Dr. S. Permogorov and Dr. A. Reznitsky (USSR), Prof. Dr. M. Sturge (USA), Prof. Dr. E. Cohen (Israel), Dr. S. Shevel (USSR), Prof. Dr. von der Osten, and Dr. Stolz (Germany), Prof. Dr. E. Göbel (Germany), Prof. Haidu (Germany), Prof. Dr. Bergmann (USA), and many other colleagues. Several of the experiments on

$CdS_{1-x}Se_x$ have been carried out by my coworkers F.A. Majumder and H.E. Swoboda in the frame of their PhD thesis and by visiting scientists Dr. S. Shevel and Dr. V.G. Lyssenko (USSR). The $CdS_{1-x}Se_x$ work is part of a project of the Sonderforschungsbereich 185 "Nichtlineare Dynamik" supported by the Deutsche Forschungsgemeinschaft.

REFERENCES

1. N.F. Mott and E.A. Davis, Electronic Processes of Non-Crystalline Solids, Clarendon Press, Oxford (1971).
2. J.M. Ziman, Models of Disorder, Cambridge University Press (1979).
3. Amorphous Semiconductors, in: Topics in Applied Physics, M.H. Brodsky, ed., Springer (1979).
4. R. Zallen, Physics of Amorphous Solids, Wiley, New York (1983).
5. G. Careri, Order and Disorder in Matter, Benjamin/Cummings Publishing Company, Menlo Park, Ca (1984).
6. B.I. Shklovskii and A.L. Efros, Electronic Properties of Doped Semiconductors, Springer Series in Solid State Sciences (1984).
7. D. Stauffer, Introduction to Percolation Theory, Taylor & Francis, London (1985).
8. Anderson Localization, Y. Nagaoka and H. Fukuyama eds., Springer Series in Solid State Sciences 39, (1982).
9. Physical Properties of Amorphous Materials, D. Adler, B.B. Schwartz, and M.C. Steele, eds.
   Physics of Disordered Materials, D. Adler, H, Fritzsche, and S.R. Ovshinsky, eds.
   Tetrahedrally-Bounded Amorphous Semiconductors, D. Adler and H. Fritzsche, eds.
   Localization and Metal-Insulator Transitions, H. Fritzsche and D. Adler, eds., Plenum Press, New York (1985).
10. Localization, Interaction, and Transport Phenomena, B. Kramer, G. Bergmann, and Y. Bruynseraede, eds., Springer Series in Solid State Sciences 61 (1985).
11. Localization and Interaction, D.M. Finlayson, eds., Scottish Universities Summer School in Physics (1986).
12. D.J. Thouless, Physics Reports 67, 5 (1980),
    E. Abrahams, ibid. p.9
    F. Wegener, ibid, p. 15.
13. T. Ando, A.B. Fowler, and F. Stern, Review of Modern Physics 54, 437 (1982).
14. G. Bergmann, Physics Reports 107, 1 (1984).
15. P.A. Lee and T.V. Ramakrishnan, Review of Modern Physics 57, 287 (1985).
16. see e.g. the contributions by B. Di Bartolo, W.M. Yen or R. Reisfeld to this volume and the references therein.
17. G. Bergmann, in Ref. [11], p. 149.
18. D. Vollhardt, Advances in Solid State Physics XXVII, Vieweg (1987) in press.
19. M.P. Albada and A. Lagendijk, Phys. Rev. Lett. 55, 2692 (1985),
    P.E. Wolf and G. Maret, ibid, p. 2696,
    E. Akkermans and R. Maynard, J. Physique Lett. 46L, 1045 (1985),
    E. Akkermans, P.W. Wolf and R. Maynard, Phys. Rev. Lett. 56, 1471 (1986),
    M.J. Stephen, ibid. 1809.
20. P.W. Anderson, Phys. Rev. 109, 1492 (1958).
21. E.A. Schiff, this volume.
22. C. Klingshirn and H. Haug, Physics Reports 70, 316 (1981).
23. H. Haug and S. Schmitt-Rink, Progress in Quantum Electronics 9, 3 (1984).

24. P.G. de Gennes, P. Lafore, and J.P. Millot, J. Phys. Chem. Solids 11, 105 (1959).
S. Kirkpatrick and T.P. Eggarter, Phys. Rev. B 6, 3598 (1972).
R. Raghavan and D.C. Mattis, Phys. Rev. B 23, 4791 (1981).
Y. Meir, A. Aharony, and A.B. Harris, Phys. Rev. Lett. 56, 976 (1986).
25. a. E. Abrahams, P.W. Anderson, D.C. Licciardello, and T.V. Ramakrishnan, Phys. Rev. Lett. 42, 673 (1979).
b. A. MacKinnon, in Ref. [11], p. 1
26. a. Excitons, K. Cho, ed. Topics in Current Physics 14, Springer (1979).
b. Excitons, E.I, Rashba and M.D. Sturge, eds. Modern Problems in Condensed Matter, Vol. 2, North Holland, Amsterdam (1982).
c. Collective Excitations in Solids, B. Di Bartolo, ed., NATO ASI Series 88, Plenum (1983).
d. C. Klingshirn in Energy Transfer Processes in Condensed Matter, B. Di Bartolo, ed., NATO ASI Series 144, p. 285, Plenum (1984).
e. B. Hönerlage, R. Levy, J.B. Grun, C. Klingshirn, and K. Bohnert, Physics Reports 12, 161 (1985).
27. E. Tomzig and R. Helbig, J. Luminesc. 14, 403 (1976).
D. Zwingel, J. Luminesc. 5, 385 (1972).
28. S. Peromogorov in Ref. [26b], p. 177.
29 a. R. Renner, Ch. Weber, U. Becker, and C. Klingshirn, Proc. Intern. Conf. on II-VI Compounds Montrery (1987) to be published in J. Crystal Growth.
b. Ch. Weber, Diploma thesis, Frankfurt (1987) to be published.
30. Landolt-Börnstein, New Series, Group III, Vol. 17, O. Madelung, M. Schulz, and H. Weiss, eds., Springer.
31. O. Goede, D. Henning, and L. John, phys. stat. sol. b 96, 671 (1979).
K.A. Dimitrenko, S.G. Shevel, L.V. Taranenko, and A.V. Marintchenko, phys. stat. sol. b 133, 1 (1986).
32. F.A. Majumder, S. Shevel, V.G. Lyssenko, H.E. Swoboda, and C. Klingshirn, Z. Physik B 60, 409 (1987).
33. E. Cohen and M.D. Sturge, Phys. Rev. B 25, 3828 (1982).
34. S.A. Permogorov, A. Reznitsky, P. Flögel, S. Verbin, G.O. Müller, and M. Nikiforova, Phys. Stat. Sol. b 113, 589 (1982).
35. IEEE J. Quantum Electronics 22, number 9, special issue on semiconductor quantum wells and superlattices, D.S. Chemal and A. Pinczuk, eds.
36. S. Peromogorov and A. Reznitsky, in Proc. Intern. Conf. "Excitons 84" p. 194, Gustrow (1984).
37. E. Cohen, Proc. 17th Intern. Conf. on the Physics of Semiconductors, J.D. Chadi and W.A. Harrison, eds., p. 1221, Springer (1984).
38. S.A. Permogorov, A.N. Reznitsky, S.Yu. Verbin, and V.G. Lysenko, JETP Lett. 37, 463 (1983) and Sol. State. Commun. 47, 5 (1983).
39. S. Permogorov, A. Reznitsky, S. Verbin, A. Naumov, W. von der Osten, and H. Stolz, J. Physique C7, 173 (1985).
40. J.A. Kash, A. Ron, and E. Cohen, Phys. Rev. B 28, 6147 (1983).
41. S.A. Permogorov, A.N. Reznitskii, S.Yu. Verbin, and V.A. Bonch-Bruevich, JETP Lett. 38, 25 (1983).
42. J. Aaviksoo, J. Lippmaa, S. Permogorov, A. Reznitsky, P. Lavallard, and C. Gourdon, JETP Lett. 45, 391 (1987).
43. S. Shevel, R. Fischer, E.O. Göbel, G. Noll, P. Thomas, and C. Klingshirn, J. Lumin. 37, 45 (1987).
44. H.E. Swoboda, F.A. Majumder, S. Shevel, R. Fischer, E.O. Göbel, G. Noll, P. Thomas, S. Reznitsky, and S. Permogorov, Proc. Intern. Conf. on Dynamic Processes in Condensed Matter, Tsukuba (1987) to be published in J. Lumin..

45. G. Noll, E.O. Göbel, F.A. Majumder, H.E. Swoboda, and C. Klingshirn, to be published.
F.A. Majumder, PhD Thesis, Frankfurt (1988).
46. H. Mariette, Y. Marfaing, and J. Camassel, Proc. 18th Intern. Conf. on the Physics of Semiconductors, Stockholm 1986, O. Engstrom, ed., p. 1405, World Scientific (1986).
47. a. A.V. Nurmikko, J. Luminesc. 30, 355 (1985).
b. J. Nakahora, S. Minomura, H. Kukimoto, F. Minami, and K. Era, J. Phys. Soc. Japan 56, 2252 (1987).
48. A. Reznitsky, S. Permogorov, S. Verbin, A. Naumov, Y. Korostelin, V. Novozhilov, and S. Prokovev, Solid State Commun. 52, 13 (1984).
49. D. Lee, A. Mysyrowicz, A.V. Nurmikko, and B.J. Fitzpatrick, Phys. Rev. Lett. 58, 1475 (1987).
50. Y. Toyozawa, Physica 117/118B, 23 (1983).
51. W. von der Osten, Physica B, to be published (1987).
52. M.V. Klein, M.D. Sturge, and E. Cohen, Phys. Rev. B 25, 4331 (1982).
M.D. Sturge, E. Cohen, and R.A. Logon, Phys. Rev. B 27, 2362 (1983).
53. Sh. Lai and M.V. Klein, Phys. Rev. B 29, 3217 (1984).
M. Oueslati, M. Zouaghi, M.E. Pistol, L. Samuelson, H.G. Grimmeiss, and M. Balkanski, Phys. Rev. B 32, 8220 (1985).
54. D. Gershoni, E. Cohen, and A. Ron, Phys. Rev. Lett. 56, 2211 (1986).
55. E.O. Göbel and W. Graudszus, Phys. Rev. Lett. 48, 1277 (1982).
56. J. Hegarty and M.D. Sturge, J. Opt. Soc. Am. B 2, 1143 (1985).
57. Optical Nonlinearities and Instabilities in Semiconductors, H. Haug ed., Academic Press, Boston (1987).
58. Optical Bistability, Dynamical Nonlinearity and Photonic Logic, S.D. Smith, B.S. Wherrett, and A. Miller, eds., Phil. Trans. R. Soc. Lond. A313 (1984).
Digital Optical Circuit Technology, B.L. Dove, ed., AGARD CP 362 (1985).
From Optical Bistability towards Optical Computing, P. Mandel, S.D. Smith and B.S. Wherrett, eds., North Holland (1987).
59. J. Puls and F. Henneberger, phys. stat. sol. b 121, K187 (1984).
60. H. Kalt, K. Bohnert, D.P. Norwood, T.F. Boggess, A.L. Smirl, I. D'Haenens, J. Appl. Phys. (1987) in press, and
SPIE Intern. Conf. on "Ultrafast Laser Probe Phenomena in Bulk- and Microstructured Semiconductors" 793, 37 (1987) and
Proc. IQEC TUDD3 (1987).
61. S. He and J.D. Maynard, Phys. Rev. Lett. 57, 3171 (1986).

ELECTRONIC PROPERTIES OF AMORPHOUS SEMICONDUCTORS: AN INTRODUCTION

E. A. Schiff

Department of Physics
Syracuse University
Syracuse, NY 13244-1130 U. S. A.

ABSTRACT

A self-contained, introductory review of several aspects of the electronic properties of amorphous semiconductors (primarily tetrahedrally bonded thin films and chalcogenide glasses) is given. The general features of the electronic bands in simple covalently bonded semiconductors are discussed. Principal emphasis is given to the relationship of these features to elementary viewpoints of covalent bonding; the cluster Bethe-lattice technique for calculating the density-of-states is described as an example of the possible theoretical refinements. Next the experimental knowledge of the bandedge density-of-states, and of electrical transport phenomena associated with bandedge states, is described. Optical spectroscopy, conductivity, thermopower, Hall effect, and transient drift mobility results are summarized. Finally an introduction to electronic defects in amorphous semiconductors is presented. The concepts of atomic relaxation and electronic correlation energies are introduced, and three special topics are treated: negative correlation energy defects in chalcogenide glass semiconductors, chemical doping of amorphous hydrogenated silicon, and the hydrogen-glass model for defect stability effects in amorphous hydrogenated silicon.

I. INTRODUCTION

It is the purpose of this article to provide an elementary introduction to several important aspects of the physics of electronic states in amorphous semiconductors. The term amorphous is used for materials without recognizable crystallinity or *long-range order*; for nearly 25 years amorphous semiconductors have attracted an intense international research effort. This effort is partly explained by the fact that amorphous semiconductors are a class of materials for which technological applications have greatly outpaced the fundamental understanding of their properties. Thus both Xerography and amorphous silicon solar cells depend on electrical transport processes which, while increasingly well characterized, remain poorly understood. However the primary reason for the sustained scientific scrutiny of these materials is surely the allure of a problem of fundamental importance: How are we to understand electronic states in materials in which we are deprived of the long-range order of crystals?

In this article I survey three aspects of this problem of primary interest in amorphous semiconductors:

Electronic Bands

Bandtails and Electrical Transport

Defects and Doping

Two other articles in this book provide introductions to closely related topics: the article by Klingshirn discusses the relationship of *localization* of electron wavefunctions and disorder, and the article by Zallen describes the characterization of non-crystalline structures. In addition there have been many excellent comprehensive review articles and monographs on the topics covered here; references 1-7 are examples of these, and other references will be given in the text.

I.A. Amorphous Semiconductor Terminology

Fig. 1 presents the fundamental viewpoint used in interpreting most electronic information about semiconductors: a one-electron *density-of-states* $g(E)$. The figure represents the present understanding of this function for amorphous silicon, denoted a-Si (the prefix *a*-denotes an amorphous material, and the prefix *c*- denotes a crystal). The upper portion of the figure illustrates the overall picture as obtained from *electron photoemission*; filled states below the *Fermi energy* E_f have been shaded. In photoemission experiments the sample is exposed to monochromatic light (X-rays or ultraviolet) and the energy spectrum of the photoemitted electrons is obtained; with sufficient care and the application of some cross-section corrections the spectrum may be identified with the density-of-states in the material. These experiments show three principle bands of states: (i) deeply bound *core levels* (in this case the Si 2p orbitals) originating from atomic levels which are only slightly perturbed by the structure of the material; (ii) the *valence band* just below the Fermi energy. As will be in Section II of this article, this band may be viewed as the occupied covalent *bonding* orbitals; and (iii) the unfilled *conduction band*, which may be viewed to some extent either as *antibonding* orbitals or as nearly free electron states.

The semiconductor properties of a material originate primarily in the very small *bandgap* region of the density-of-states illustrated in the lower portion of Fig. 1; note the use of a logarithmic scale for the vertical axis. The *bandtails* (in this case shown as exponentially distributed) at the edges of the conduction and valence bands make the identification of the *bandedge* energies E_v and E_c somewhat ambiguous. A reasonable theoretical definition for E_v and E_c is as the energy separating exponentially localized electronic states in the bandtails from "extended" states in the bands; E_v and E_c are then *mobility edges* [3-5]. This dividing energy between the two types of wavefunctions is in principle sharp; localized and extended states cannot have the same one-electron energy in the same specimen. The bandtail region of the density of states is extremely important for optical and electrical properties; it is, however, only roughly understood, as will be discussed further in Section III of this article (Bandtails and Transport). Finally the density of states between the bandtails (schematically represented as two humps in the density of states) is associated with *defect* states due either to chemical or structural deviations of the material; these *gap-states* control the performance of many semiconductor devices, and an introduction to their origins will be given in Section IV of this article (Defects and Doping).

I.B. The Independent Electron Approximation

> It surprises me, looking back, how easily we were satisfied with a model of a metal in which electrons occupy states up to a limiting energy, or in *k* space (wavevector space) up to a limiting surface, without taking into account the very large interactions between them.
>
> --Sir Nevill Mott [10]

The use of a density-of-states function $g(E)$ assumes the validity of an independent electron approximation in which electrons - or more precisely electron *quasiparticles* - occupy well-defined eigenstates of some effective one-electron potential. For "normal Fermi liquids" this

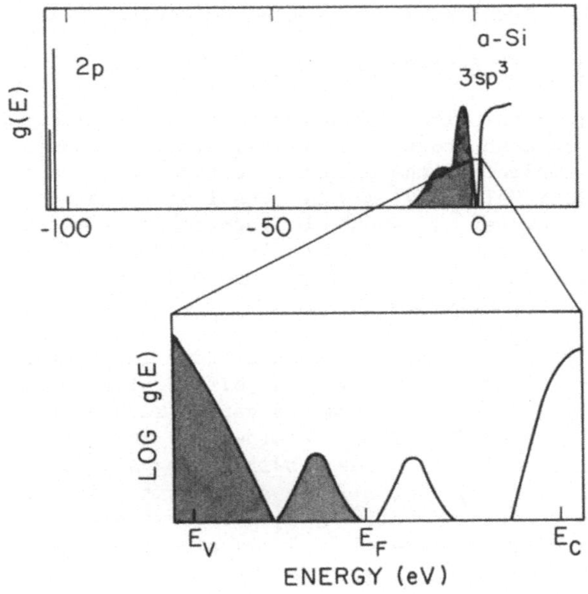

Fig. 1. The electronic density-of-states $g(E)$ for amorphous silicon (a-Si). *Upper section*: $g(E)$ as estimated from X-ray photoemission and Bremstrahlung isochromat spectroscopy (after Jackson, *et al* [8] and Ley [9]. *Lower section*: Schematic form of $g(E)$ near the Fermi energy showing valence and conduction *bandtails* and two *defect states*. Note the use of the logarithmic scale in the lower section; density-of-states features of electronic significance in semiconductors are unseen on the linear scale of the upper section.

approximation can be theoretically justified for quasiparticles occupying orbitals with energies near the Fermi energy [11]. However the approximation has also enjoyed remarkable success in describing the properties of most crystalline materials well outside this domain, and perhaps the best justification for its use is its empirical success. One convincing test of the approximation in crystals is *angle-resolved photoemission* [12]. Since in crystals the one-electron effective potential is periodic, the one-electron wavefunctions can be classified by their Bloch wavevectors $\mathbf{k}$. Angle-resolved photoemission experiments can explore the electron energies for a particular Bloch wavevector; the linewidth of the energy spectrum can in some cases be interpreted as an energy uncertainty due to the finite lifetime of an electron in a particular orbital of the periodic effective potential.

This energy uncertainty can exceed one eV, corresponding to a lifetime for an electron in a particular eigenstate of femtoseconds prior to a scattering event. Thus most photoemission spectra should be adequately explained by the independent electron approximation (and a one-electron density of states), although in some cases finite lifetime effects may indeed be required to interpret the spectra[13].

In amorphous semiconductors the effective potential is certainly not periodic; there is thus no experimental estimate for the energy uncertainty of states in the valence and conduction band beyond the crude limit of about 2 eV imposed by the width of the narrowest valence band feature. The use of the independent electron approximation beyond this limit should be viewed as a reasonable but untested assumption.

I.C. Importance of Short-Range Order

This article is primarily concerned with the electronic properties of amorphous semiconductors; relatively little emphasis will be given to the important subject of the atomic structure of these materials. The lucid textbook by Zallen [4] may be consulted for a more complete introduction. However, it should be recognized at the outset that most amorphous semiconductors are covalently bonded materials for which the bonding interactions create a *short-range order* which is little different in crystals and non-crystals.

This similarity of the short-range order in crystals and non-crystals is clearly shown by X-ray diffraction measurements. For an elemental material these measurements can be reduced to a *radial distribution function* $J(r)$ [3]. Consider the total electronic density in a thin spherical shell of radius r surrounding a particular atom. The average of this density over all the atomic sites in the material is the the radial distribution function $J(r)$. For example, in the physically unrealistic case of a completely uniform electronic density $\rho(\mathbf{r})=\rho_0$ in a material, $J(r) = 4\pi r^2\rho_0$.

In Fig. 2 the radial distribution functions (RDF) for amorphous and crystalline silicon are shown on the left. One sees immediately that the nearest neighbor peaks are essentially the same for both structures, implying that the number of nearest neighbors (4 for Si) and the separation (bond length) are essentially unchanged. The similarity of the second nearest neighbor peaks for the two structures further indicates that the bond angles are also not grossly perturbed. Past the second nearest neighbor the similarity between the structures is largely lost for these materials.

This information suggests the most common structural model for covalently bonded non-crystalline materials: the *continuous random network* (the *CRN*). In such a network nearly all atoms retain the coordination number required by covalent bonding, but sufficient deviation is permitted in the bond lengths and bond angles *vis a vis* their crystalline values to modify the *topology* of the network from that of a crystal. A simple example of a crystalline structure and a related CRN is presented in Fig. 3. Since most atoms retain the same coordination in various networks, the next simplest characterization of the network topology is *ring statistics*. The concept is easily described in two dimensions. In Fig. 3 the crystalline structure is characterized by the four four-membered rings of bonds which run through each atom. It is apparent that a CRN must have some rings with different orders, and in the diagram of the CRN a five-membered ring is prominent near the center of the structure.

The fact that *short-range* order is similar in the amorphous and crystalline forms of most semiconductors is only reasonable if the electronic interactions between the atoms of the material are also primarily short-ranged, and thus we should expect that photoemission and other measurements sensitive to the electronic structure of the material should be relatively unaffected by presence or absence of long-range order. This point is illustrated at the right of Fig. 2, where the X-ray photoemission spectra (XPS) for the filled valence band and the *Brehmstrahlung isochromat* spectra of the conduction band are shown for c-Si and a-Si [8]; although there are obvious differences in detail, the overall electronic structures are indeed similar.

The importance of short-range order thus leads to the following viewpoint regarding the electronic processes in covalently bonded semiconductors. The overall, large-scale electronic structure should be interpretable using covalent bonding arguments which are largely the same in crystalline and amorphous semiconductors; this viewpoint will be

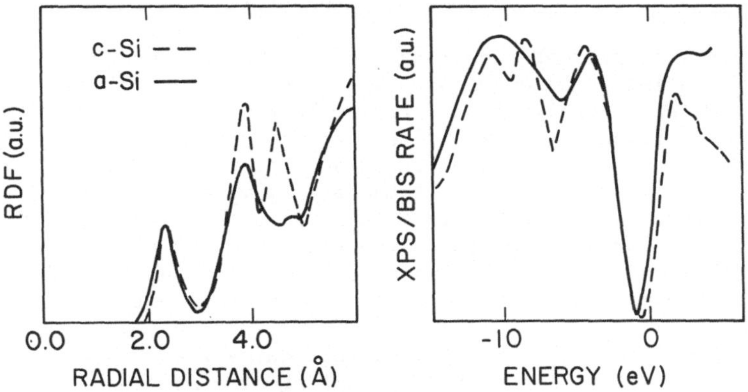

Fig. 2. (left) The radial distribution function (RDF) for amorphous and crystalline silicon (a-Si and c-Si; after Moss and Graczyk [14]). The figure illustrates the close correspondence of nearest neighbor positions in the two materials. (right) X-ray photoemission (XPS) and Bremstrahlung isochromat spectra (BIS) (after [8]) for a-Si and c-Si, illustrating the broad similarities of the valence and conduction band states in the two materials. a.u. denotes "arbitrary units."

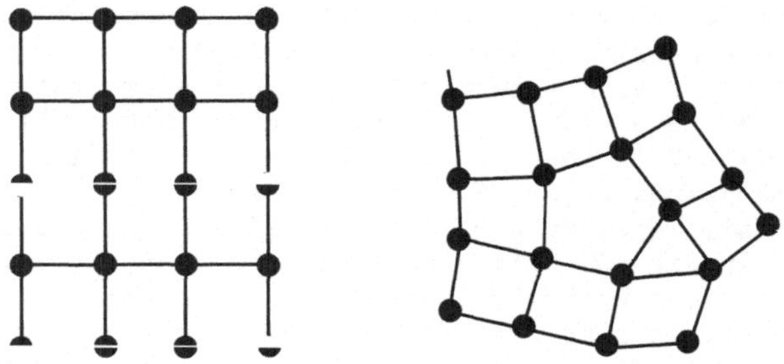

Fig. 3. Two-dimensional illustrations of networks requiring each atom to have four neighbors (fourfold coordination). In the crystalline network each atom lies on four distinct fourfold rings of bonds. In the continuous random network (CRN) threefold and fivefold rings may be encountered.

amplified in Section II of this article. However, many of the important optical and electrical properties of semiconductors are determined by subtleties of the electronic structure which are dramatically affected by intermediate and long-range order; these aspects will be discussed in Section III. Finally defects in semiconductors (the subject of Section IV of this article) may be viewed as disruptions of ideal covalent bonding in both crystalline and amorphous semiconductors; however detailed defect configurations are intimately associated with the overall structure of the material and are quite different in amorphous and crystalline materials.

I.D. Classification of Amorphous Semiconductors

There are at present two classes of amorphous semiconductors which are widely studied: *chalcogenide glasses* such as Se and As_2Se_3, and *hydrogenated tetrahedrally-bonded* semiconductors, primarily amorphous hydrogenated silicon (a-Si:H) and related alloys such as a-$Si_{1-x}Ge_x$:H. To these inorganic amorphous semiconductors might be added certain organic materials or polymers; however, these are conventionally considered as a distinct subject of research. In this article the discussion will be drawn entirely from experiments on the two inorganic classes of materials; in this section I present a very brief summary of their preparation, atomic structure, and electronic applications.

1. Chalcogenide Glasses. By definition, a glass is a non-crystalline material prepared by cooling of a melt of its constituents. Only certain non-crystalline materials may be prepared as glasses. The remainder are prepared using the much wider class of thin film preparation techniques including physical vapor deposition techniques (evaporation and sputtering), chemical vapor deposition (cf. thermal or plasma decomposition of source gases), and hybrid methods. Glass-forming materials can also be prepared with thin-film techniques; the properties and structure of the resulting thin-films do not appear to vary qualitatively from materials prepared from the melt.

Materials containing chalcogenide (Group VI) elements such as Se and Te often form glasses fairly readily, both as elemental materials and as alloys such as As_2Se_3. Oxide glasses such as SiO_2 will not be included in this discussion, although oxygen is technically a chalcogen. The alloys are reasonably-well *chemically ordered* (As is bound to Se) and preserve the coordination of the atoms expected from elementary chemistry [15]. The electrical properties of these glasses vary considerably, but certain generalizations may be offered [3,5,16]. Many of these glasses are quite insulating in the dark, but possess significant *photoconductivity* when they absorb light; indeed the phenomenon of photoconductivity was discovered in Se in 1826. This photoconductivity is due to the motion of holes in the filled valence bands of the material; electron transport has not been observed, although its absence does not appear to be understood. However, unipolar photoconductivity is a very useful property, and in particular the invention by Carlson of *Xerography* (or *electrophotography*) was based on this property of chalcogenide glasses [17]. As with most electronic applications of amorphous semiconductors, Xerography exploited the fact that large areas of the material could be inexpensively deposited. In addition a *switching* mechanism [18] was discovered by Ovshinsky in chalcogenide materials which has been exploited in the fabrication of electronic memories, although these have not found wide application. However, applications which require bipolar photoconductivity (such as solar cells) or the ability to modify the material's Fermi level (such as field-effect transistors) are not possible in these materials. The latter difficulty is often referred to by saying that chalcogenide glasses have *pinned Fermi levels*; the reasons for this pinning will be described in Section IV (Defects).

2. Hydrogenated Tetrahedrally-Bonded Amorphous Semiconductors. The prototypical semiconductor in the last several decades is of course crystalline Si (c-Si), and it is natural to inquire whether elemental amorphous Si might also be an important technological material. It is not. Si is not glass-forming, and the amorphous films prepared by evaporation or sputtering prove to be distressingly rich in microstructures, in particular voids [19]. The bonding defect density as measured by the electron spin resonance technique is very large (~10^{19} cm^{-3}) [20]. In common with the chalcogenides the Fermi energy is apparently pinned, and electronic applications of elemental a-Si appear impossible.

In the early 1970's Spear and LeComber demonstrated that amorphous silicon films which were *plasma deposited* by the decomposition of silane (SiH_4) gas had Fermi energies which could be moved, either by doping (with phosphorus or boron) [21] or by space charge (the field effect) [22]. Both electrons and holes were mobile, which led to the fabrication of a solar cell by Carlson and Wronski [23]. This new material proved to be an alloy of silicon with approximately 10 atomic percent of atomic hydrogen bonded to Si atoms, and it is known as hydrogenated amorphous silicon (a-Si:H). a-Si:H is widely used for the fabrication of solar cells and thin-film transistors [24].

X-ray diffraction studies of the structure of a-Si:H are primarily sensitive to the Si in the material; the radial distribution function is not profoundly different than that measured in elemental a-Si [25]. Imaging of microstructure with electron microscopy is difficult in a-Si:H, and there is no widely accepted view as to whether microstructures are present in material which is optimal for applications. However, proton nuclear magnetic resonance studies of the bonded hydrogen reveal two distinct types of bonded H [26]. Approximately half the hydrogen is well isolated from other hydrogen atoms; the remainder are clustered together in groups of about six hydrogens. They are presumably attached to a microstructure, possibly as small as a *divacancy defect* (two missing Si neighbors) in the network. Although there is not yet a quantitative explanation for the extraordinary differences between a-Si and a-Si:H, it is widely accepted that these materials differ in their *intermediate range order*. In particular a-Si:H should not be viewed simply as a hydrogenated form of elemental a-Si; there is far more hydrogen in the alloy than required to occupy the dangling bonds in elemental a-Si, and it is not possible to obtain the desirable electronic properties of a-Si:H simply by post-hydrogenation of the elemental material.

II. ELECTRONIC BANDS IN AMORPHOUS SEMICONDUCTORS

II.A. A Primer on Covalent Bonding and Molecular Orbitals

> The molecular orbital approximation is a useful empirical scheme which rationalizes much chemical experience...
>
> S. P. McGlynn [27]

The molecular orbital approximation is an independent electron approximation; the molecular orbitals are usually calculated as linear combinations of the one-electron atomic wavefunctions of the constituent atoms. As most readers will recall from textbooks on elementary quantum mechanics, the molecular orbital approximation for an H_2 molecule is constructed in two steps:

(i) Molecular orbitals which might be suitable for an H_2^+ ion are computed using the electrostatic Hamiltonian of two protons separated by a distance d. This Hamiltonian is solved in the approximation using only two hydrogen 1s atomic orbitals separated by an interatomic distance d (*linear combination of atomic orbitals* or *LCAO*). Two distinct molecular orbitals result: a symmetric *bonding orbital* with lower energy than the atomic 1s, and an antisymmetric, *antibonding orbital* with greater energy than the 1s.

(ii) It is assumed that the two electrons in the neutral molecule can occupy the lower-lying molecular orbital without considering exchange effects (because the electrons are identical particles) or correlation effects (due to Coulombic repulsion).

The approximation is a poor one for the hydrogen molecule, and textbooks on quantum mechanics usually describe the *Heitler-London* approach to computing the binding energy for the H_2 molecule as a superior alternative. However, the simpler molecular orbital approach offers a great deal of insight into the structures and electronic levels of more complex molecules; by using an empirically-determined, effective one-electron Hamiltonian it is often possible to obtain quantitative or semi-quantitative agreement between computation and experiment.

The value of the approximation can be assessed by considering the molecular photoemission data [28] summarized in Fig. 4. Photoemission bands are shown for a series of diatomic hydride molecules isoelectronic

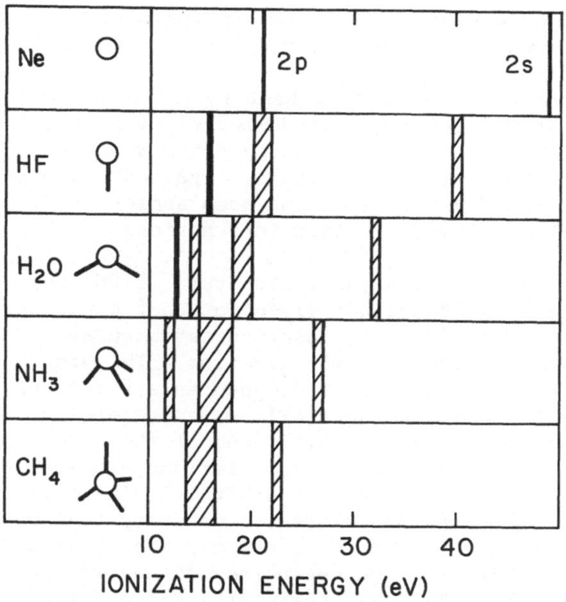

Fig. 4. Structures and molecular photoemission bands for Ne and four isoelectronic hydrides. Most of the features can be rationalized using elementary molecular orbital arguments (after Fig. 17.2 of ref. 28; photoemission spectra obtained with He light by Potts and Price [29]).

with Ne; the structures and identities of the molecules are illustrated at the left. The Ne 2s and 2p orbitals are clearly seen as bands in the photoemission data. A crude version of the molecular orbital picture accounts for many of the features shown by the photoemission data and the molecular structures [29]. In this version we shall assume that covalent bonds can be derived solely from a single atomic 2p orbital aligned with the hydrogen bond and an associated H 1s orbital. We neglect the role of charge transfer between the hydrogen and the central atom; inclusion of this effect would be required for calculations. For HF the deepest band then derives from the F 2s atomic orbital. Defining the molecular axis as the z-direction, the shallowest photoemission band may be identified with four electrons occupying F $2p_x$ and $2p_y$ atomic orbitals only slightly perturbed by molecular bonding. The remaining $2p_z$ atomic orbital is viewed as having bonded with the H 1s atomic orbital, creating the middle bonding orbital observed in the photoemission spectrum and an unoccupied antibonding orbital (lying above the F 2p) which is not observed.

For H_2O this simple molecular orbital picture applies with only slight modification. The atomic O 2s orbital is again only slightly perturbed by molecular bonding; in addition a *lone pair* of electrons occupies a slightly modified version of the O 2p atomic orbital perpendicular to the bonding plane. Two bands of bonding states are observed. In the crudest molecular orbital picture one might expect only a single bonding band, and that the bond angle of the molecule would be 90°. A complete molecular orbital calculation incorporating interactions (matrix elements) between all of the original atomic orbitals will of course split the degeneracy of the two bonding orbitals, and would also be expected to account for the observed bond angle.

In NH_3 vestiges of the most elementary picture still persist. A fairly distinct band originating predominantly from the N 2s exists, and the three H atoms in the molecule occupy sites similar to those expected from the three orthogonal N 2p orbitals. Again the complexity of the bonding band and the deviations of the bond angles from 90° indicate the need for refinements including other interactions.

Lastly the perfect tetrahedral structure of CH_4 signals the need for a qualitative change in the crude viewpoint that bonds are derived from individual atomic orbitals. The observed tetrahedral structure is rationalized using the idea of *hybridization*. The distinction between the atomic C 2s and 2p orbitals is dropped as a first approximation in favor of four orthogonal, tetrahedrally oriented atomic hybrid sp^3 orbitals; the viewpoint is a sensible one if the bonding matrix elements required for the molecular orbital calculations exceed the energy difference of the atomic C 2s and 2p orbitals. The splitting of the two bonding bands shown by the photoemission data may then be understood as due to refinements in which the original s-p energy difference is incorporated, thus breaking the degeneracy of the four bonding orbitals constructed using the hybrid Si atomic orbitals and the H 1s. One expects that the molecular orbitals for the lower bonding band is richer in C 2s than the orbitals yielding the upper band.

This article is primarily concerned with electronic structure in solids; Figs. 1 and 2 illustrate the structure of the valence band for solid Si. In the molecular orbital picture the band is viewed as due to bonding combinations of hybrid atomic Si sp^3 orbitals; this structure will be discussed again in section II.B. A somewhat more intricate example is given in Fig. 5, where the photoemission spectrum for trigonal (and crystalline) Se is reproduced. Se is a chalcogenide with 6 electrons in its outer shell; using the crude molecular orbital picture for bonding we expect bonding to be similar to that of H_2O: well-defined vestiges of the atomic Se 4s and lone pair 4p orbitals, two nearest neighbors and a bond angle of about 90°, and a photoemission bonding band lying below the lone pair. An inspection of Fig. 5 will find the required features.

II.B. Quantitative Bandstructure Theories

> We need to throw out ***k***-space, even when we have lattice periodicity ... because ***k***-space relates to individual eigenstates ψ_n and these are inappropriate to the systems and properties we wish to discuss.
>
> --V. Heine [31]

In this section a thumbnail sketch will be given of molecular orbital bandstructure calculations; the molecular orbitals $|j\rangle$ are treated as linear combinations of atomic orbitals (LCAO):

$$|j\rangle = \sum_{\alpha,l} c_{j,\alpha l}|\alpha l\rangle \tag{1}$$

where $|\alpha l\rangle$ denotes a particular atomic orbital of type α (s, p_x, etc.) at a particular atomic site l.

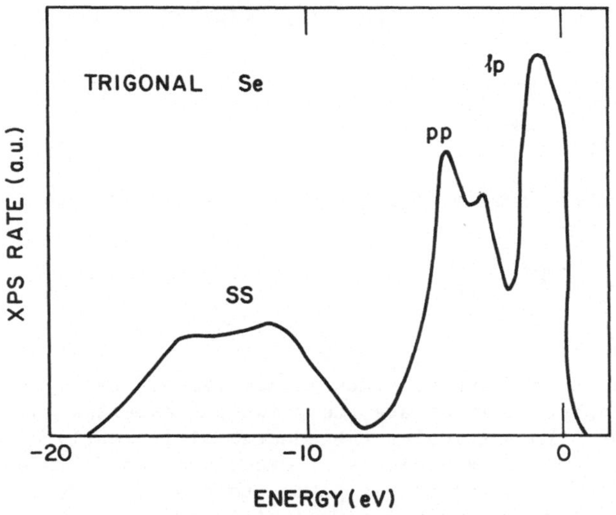

Fig. 5. X-ray photoemission spectra for trigonal Se as obtained by Shevchik, *et al* [30]. The three broad bands are labelled *ss* (broadened atomic *s* levels), *pp* (bonding levels constructed from atomic *p* orbitals) and *lp* (non-bonding or lone pair orbitals).

Given this viewpoint two additional inputs are required to make a computation:

(i) Specification of the structure of the material.

(ii) Specification of the one-electron effective Hamiltonian appropriate to the assumed structure.

It may be recognized that this procedure corresponds to the *adiabatic* or *Born-Oppenheimer* approximation that electronic properties may be calculated for a static atomic configuration. Finally it is rarely necessary to actually solve for the molecular orbital solutions $|j\rangle$; for non-crystalline materials such an approach would be extremely cumbersome because there is no simple, *a priori* classification of the solutions analagous to the Bloch wavevector $\boldsymbol{k}$ in crystals. In practice it is often sufficient to solve directly for densities of states using one-particle Green's function methods without further consideration of the actual wavefunctions which solve the effective Hamiltonian.

1. Objects for an LCAO/Molecular Orbital Theory. If an exact solution in terms of molecular orbital wavefunctions is to be avoided, which objects are appropriate for bandstructure calculations? The following are commonly computed:

(i) The global *density-of-states* $g(E)$.

$$g(E) = \sum_j \delta(E - E_j) \tag{2}$$

The sum runs over all eigenstates j of the effective Hamiltonian.

(ii) The *local density-of-states* for particular atomic site and orbital type:

$$N_{\alpha l}(E) = \sum_{j} \left| \langle \alpha l | j \rangle \right|^2 \delta(E - E_j) \tag{3}$$

(iii) The *partial density-of-states* for a particular atomic orbital type:

$$N_{\alpha}(E) = \sum_{l} N_{\alpha l}(E) \tag{4}$$

It may not be obvious how these densities-of-states are related to photoemission spectra, which are the principal experimental tool for exploring bands. In particular the spectra will clearly be affected by optical transition matrix elements or cross-sections. In at least some cases partial densities of states permit this difficulty to be resolved. The simplest procedure - known as *Gelius' Rule* [32] - utilizes the partial densities of states defined above: the angle-integrated photoemission current $J(E,\nu)$ corresponding to ionization energy E and measured with photon frequency ν is written:

$$J(E,\omega) = \sum_{\alpha} N_{\alpha}(E)\sigma_{\alpha}(\nu) \tag{5}$$

where $N_{\alpha}(E)$ is the partial density-of-states and $\sigma_{\alpha}(\nu)$ is the atomic photoemission cross-section for atomic orbital α.

As an example of the application of the rule consider the valence band photoemission spectra for a-Si and a-Ge in Fig. 6 [33]. The deeper feature may be considered to originate with molecular orbitals primarily constructed from atomic Si 3s or Ge 4s, and the shallower feature to originate primarily in atomic Si 3p or Ge 4p. This possibility was not described in the discussion of hybridization for the CH_4 molecule. It can easily be understood using the *sp^3 hybrid orbital* viewpoint described in Section II.A.. Hybrids neglect the energy difference in atomic s and p orbitals as a starting point; four degenerate bonding orbitals for each atom result when the bonding interaction between neighbors is considered. Finally the underlying distinction between s and p atomic orbitals can be viewed as a perturbation of the hybrid orbital calculation. The bonding orbitals split into two sub-bands which are relatively rich or poor in the weight of the atomic s orbital. In agreement with Gelius' rule, the different appearances of the spectra for the two materials are accounted for by the known atomic cross sections at 1486 eV: for Ge $\sigma_{4s}/\sigma_{4p} = 1$, whereas in Si $\sigma_{3s}/\sigma_{3p} = 3.4$!

Although Gelius' rule is certainly useful in interpreting photoemission spectra, a fully satisfactory theoretical explanation for the success of this rule is not obvious. Even within a molecular orbital context the empirical success is astonishing. Equation (5) amounts to an assertion that interference terms between different sites and different atomic orbitals of the underlying LCAO molecular orbital contribute negligibly to the photoemission matrix element! The key to understanding the success of Gelius' rule appears to be the large electron and photon energies used in the photoemission experiments from which density of states information is obtained. For Fig. 6 the incident X-ray photons

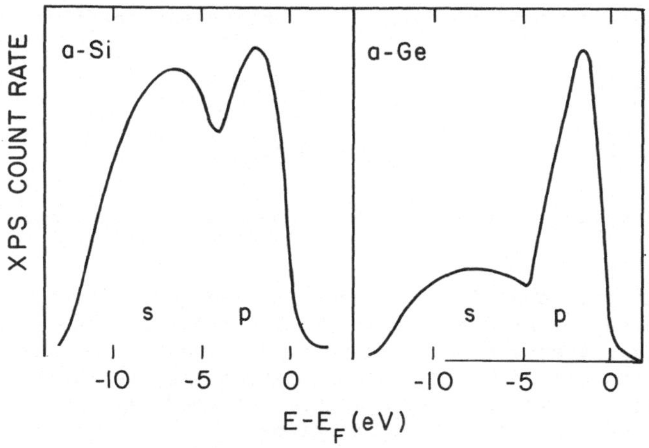

Fig. 6. 1486 eV X-ray photoemission spectra for a-Ge and a-Si (after Ley, *et al*, [33]). The decomposition of the valence band into atomic s-like and p-like molecular orbitals is illustrated. The decomposition accounts for the prominence of the p-like section in a-Ge as an atomic photoemission cross-section effect; the underlying densities of states are similar in the two materials.

had an energy of 1486 eV, and the outgoing electron only slightly less energy. The use of such high energies is required because lower energy excitation yields surface sensitive photoemission spectra. Thus the outgoing electrons have extremely short wavelengths - certainly much less than an interatomic length - and resultingly even relatively modest thermal or static disorder between neighboring atoms destroys the coherence of contributions to the photoemission cross-section corresponding to different atoms for the molecular orbital. A discussion of the absence of interference between the differing atomic orbitals on the same atom has been given elsewhere [34].

2. Empirical Tight Binding Hamiltonians. In this section one approach - the empirical tight binding (ETB) procedure - to obtaining a Hamiltonian for density-of-states calculations in non-crystalline materials will be described. For more detailed descriptions both of ETB and other procedures the reader should consult [31, 35, 36].

To give the flavor of this technique consider the extremely simple approach described in the textbook of Harrison [36]. For an elemental material such as Si each atom contributes four atomic orbitals (an s and three orthogonal p orbitals). If only nearest neighbor bonding interactions are considered, the Hamiltonian matrix is completely specified by the atomic energies ϵ_s and ϵ_p and by the four bonding interaction energies $V_{ss\sigma}$, $V_{sp\sigma}$, $V_{pp\sigma}$, and $V_{pp\pi}$ illustrated in Fig. 7; these energies of course depend upon the nearest neighbor distance d. Harrison's text proposes the following extraordinarily simple scheme for evaluating these matrix elements:

$$V_{ijn} = \eta_{ijn}(\hbar^2/md^2) \tag{6}$$

where m is the electron mass and the dimensionless coefficients η_{ijn} are:

$\eta_{ss\sigma} = -1.40$ $\quad\quad$ $\eta_{sp\sigma} = 1.84$

$\eta_{pp\sigma} = 3.24$ $\quad\quad$ $\eta_{pp\pi} = -0.81$

Note that these matrix elements are *independent of the atomic species involved*!

Obviously this approach represents the extreme of simplicity, and other approaches may give more quantitative results. In particular in some materials the crystal's electron bands have been measured the necessary matrix elements for ETB calculations may be obtained by curve-fitting; the matrix elements thus obtained may then be applied to calculating electronic densities of states for various structural models of non-crystalline materials. For Si Pandey found seven matrix elements including second nearest neighbor interactions which give a good account of the crystalline bandstructure [37]; Pandey's matrix elements are often used for computations for non-crystalline Si structures.

3. A Technical Note: Green's Functions. Heine's quotation given at the heading of this section alludes to the possibility of obtaining useful information about electronic densities of states without explicitly solving the Hamiltonian for wavefunctions. This possibility depends upon a fairly simple property of *one-particle Green's functions*. This note will simply acquaint the reader with the nomenclature of the Green's function approach; the textbook of Economou [38] or the volume edited by Heine [31] may be consulted for a thorough introduction to the technique and its applications.

For LCAO problems the Green's function is simply a matrix whose elements depend upon an energy variable E. For the atomic basis functions $\langle \alpha l|$ the matrix elements may be written in the notation:

$$G_{\alpha l, \alpha' l'}(E) = \langle \alpha l| \{E + i0 - \mathcal{H}\}^{-1} |\alpha' l'\rangle \quad (7)$$

0 is used to denote an infinitesimal real number, and $\mathcal{H}$ is the effective Hamiltonian for the problem (including the model atomic structure). One

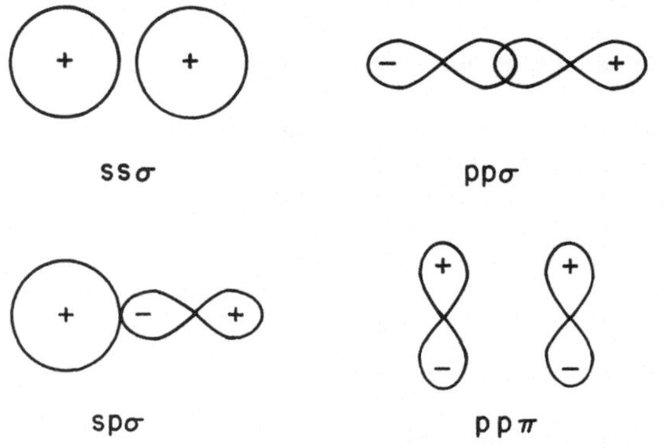

Fig. 7. The four independent interactions of atomic s and p orbitals required to compute molecular orbitals including only nearest neighbors.

use of the Green's function is to obtain the local density of states via a theorem:

$$N_{\alpha l}(E) = -\pi^{-1} \operatorname{Im}\{G_{\alpha l,\alpha l}(E)\} \tag{8}$$

Partial and global densities of states may be obtained if local densities of states for a sufficiently large number of orbitals are calculated.

4. Structure Models for Bandstructure Calculations. The most straightforward approach to generating a structural model for a non-crystalline material is simply to build - either by hand or on a computer - a model cluster of reasonably bonded atoms for which it can be argued that the cluster is representative of the bulk structure of the material. The Hamiltonian for the structure can then be generated using the empirical tight-binding or some other procedure. However, both the specific cluster as well as the techniques used to handle the atoms at the edges of the finite cluster can always be criticized. Here I shall describe another, subtler approach which bypasses these details and yet appears capable of accounting for many experimental features of bandstructures; [35] contains a comprehensive survey of calculations for a-Si:H, and includes most of the available computational techniques.

The approach is the *cluster Bethe lattice method* (the *CBLM*). The method is based on the relative ease with which the bandstructure of a special lattice - the Bethe lattice - can be computed [35]. The Bethe lattice is a mathematical entity in which atomic coordination (number of neighbors), the bond lengths and bond angles are precisely the same as in the crystal, but no closed rings of bonds exist. The lattice is illustrated in Fig. 8; of course such a lattice is not physically possible in ordinary three-dimensional space. However, as illustrated in

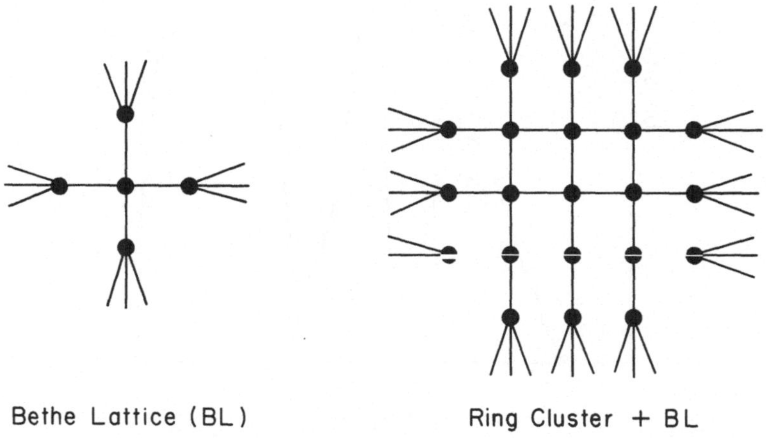

Fig. 8. The Bethe lattice is a convenient, although not physically realizable, structure which preserves all bond angles and bond lengths, but includes no closed rings of bonds. Bethe lattices can be used in bandstructure calculations to terminate clusters of atoms with structures of interest; a small "ring-cluster" (ie. a microcrystal) terminated by Bethe lattices is illustrated at the right.

the upper part of Fig. 9, the density of states computed for this lattice based on an empirical tight-binding Hamiltonian suitable for Si is not an unreasonable description for the experimental density of states in a-Si [35]! It is remarkable that the quantitatively perfect Bethe lattice should be similar to a-Si, and relatively dissimilar from the much more structured density of states in c-Si illustrated in the lower part of the figure.

One speculation for the structure in the c-Si $g(E)$ is the fact that in the crystal each atom lies on several sixfold rings of bonds. These rings may be sufficiently short to create structure in the smooth density of states which might be expected purely from nearest-neighbor, covalent bonding arguments. The Bethe lattice by definition has no rings of bonds, and a-Si has an unknown distribution of ring sizes which may also erode any overall structure in the density-of-states.

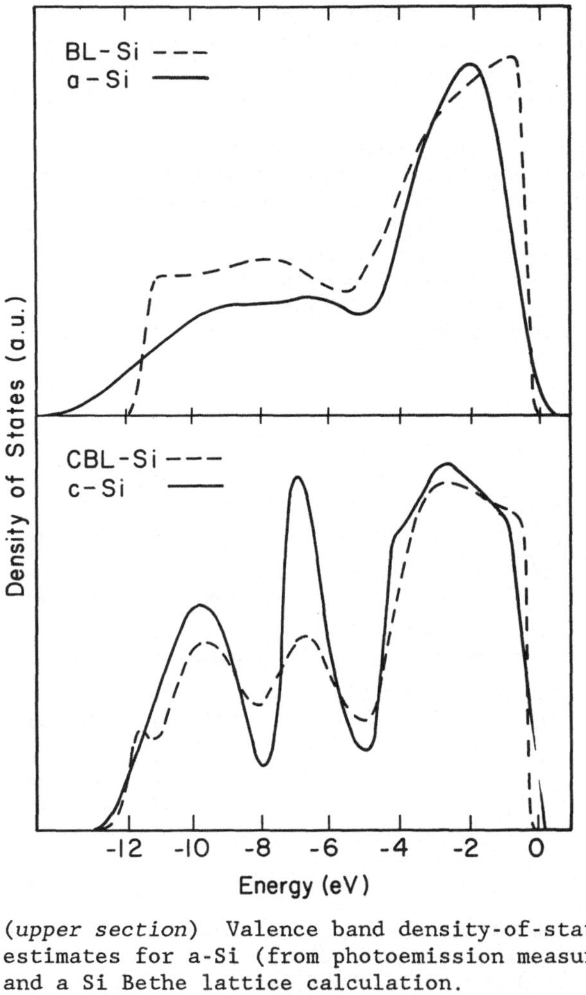

Fig. 9. (*upper section*) Valence band density-of-states estimates for a-Si (from photoemission measurements) and a Si Bethe lattice calculation.
(*lower section*) Valence band density of states for c-Si (from dispersion relations; broadened) and from a cluster Bethe lattice calculation for a "ring-cluster" microcrystallite terminated using Bethe lattices (CBL-Si). After [35].

This speculation was tested by a calculation in which a cluster containing six sixfold rings through a central Si atom was solved for the local density of states of the central atom. Bethe lattices were attached to any unsatisfied bonds of the cluster, thus embedding the cluster into an "effective medium." A two-dimensional version of this "Ring Cluster + Bethe Lattice" structure is illustrated in Fig. 8, and the density-of states is illustrated in the lower section of Fig. 9.

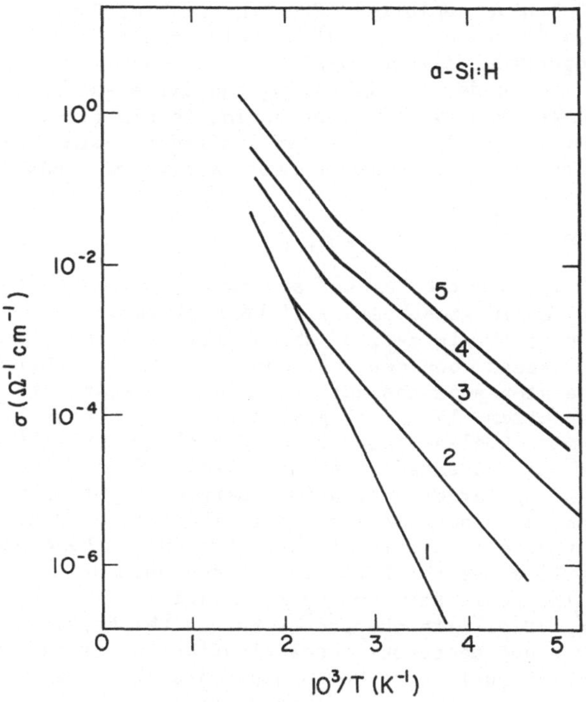

Fig. 10. The conductivity $\sigma(T)$ as a function of temperature T in hydrogenated amorphous silicon (a-Si:H) for specimens with different nominal phosphorus concentrations (in parts per million: 1-1, 2-3, 3-250, 4-1000, 5-10000). After Beyer [40].

Most of the structure in the density-of-states of crystalline silicon can in fact be attributed to the well-defined rings of bonds. The technique of embedding a relatively small cluster incorporating a structural feature of interest into a Bethe lattice, and then solving for the local density of states of the cluster atoms of interest, is known as the cluster Bethe-lattice method and has been widely applied by Joannopoulos and his collaborators to bandstructure problems in non-crystalline materials [35].

III. BANDEDGES AND ELECTRICAL TRANSPORT

Electrical transport in non-crystalline materials is a vast subject too vast to be even roughly summarized here. In this section I survey only those aspects of electrical transport which may reasonably be identified with electrons or holes occupying energy levels at the conduction or valence bandedges. For this reason I have also included in this section the closely related subject of the bandedge density of states in non-crystalline semiconductors. A more comprehensive introduction to transport in disordered materials may be found in Mott's recent book [5] or in the monograph of Shklovskii and Efros [39].

In Fig. 10 I have replotted (from the work of Beyer and his collaborators [40]) measurements of the electrical conductivity $\sigma(T)$ in amorphous hydrogenated silicon (a-Si:H) as a function of the specimen temperature T; the codes 1-5 labelling the different curves indicate different specimens having different *doping* levels (in this case phosphorus incorporation). The use of semilogarithmic scales (ie. $\ln(\sigma)$ vs. $1/T$) tests for an *exponentially activated* conductivity:

$$\sigma(T) = \sigma_0 \, e^{-(E_a/kT)} \tag{9}$$

For the present purpose the "kinks" separating segments of differing activation energies in these data will be neglected; these are related to defect metastability in a-Si:H, which will be briefly mentioned in Section IV.D. Instead consider the magnitude of the activation energy E_a derived from the slopes of the data at lower temperatures. For the most lightly doped specimen (1) the largest activation energy is obtained (of order 0.6 eV), with a clear trend towards smaller activation energies as the doping level is increased (specimens 2-5). Undoped specimens typically have still larger activation energies (typically 0.7 - 0.8 eV) in a-Si:H. These data suggest a crystal-like interpretation of $\sigma(T)$: only electrons thermally excited to an energy near the conduction bandedge (and well above the Fermi energy E_f) contribute to the conductivity. The activation energy E_a should be roughly E_c-E_f, and doping of the material with phosphorus raises the Fermi energy closer to the mobility-edge and decreases the activation energy accordingly. A similar description applies to boron doped a-Si:H, in which the charge carriers are holes near the valence band edge at E_v.

A more specific viewpoint regarding transport in non-crystalline materials is illustrated in Fig. 11. We assume that the contribution to transport of an electron in an energy level at E is given by a mobility function $\mu(E)$, and that consequently the conductivity of a specimen in thermal equilibrium may be obtained from a simple integral over states in the bandedge:

$$\sigma(T) = e\int f_{FD}(E)g(E)\mu(E)dE \tag{10}$$

where $g(E)$ is the electronic density-of-states, e is the electronic charge, and $f_{FD}(E)$ is the Fermi-Dirac distribution function. The use of a mobility function is appropriate only for spatially extended wavefunctions. In the vicinity of a mobility-edge (the energy separating exponentially localized wavefunctions and extended wavefunctions) one might expect a very rapidly falling mobility, as illustrated in the upper portion of Fig. 11; note the use of a logarithmic vertical axis. The Fermi-Dirac distribution function is also drawn in this portion of the figure for two values of the Fermi energy (corresponding perhaps to two

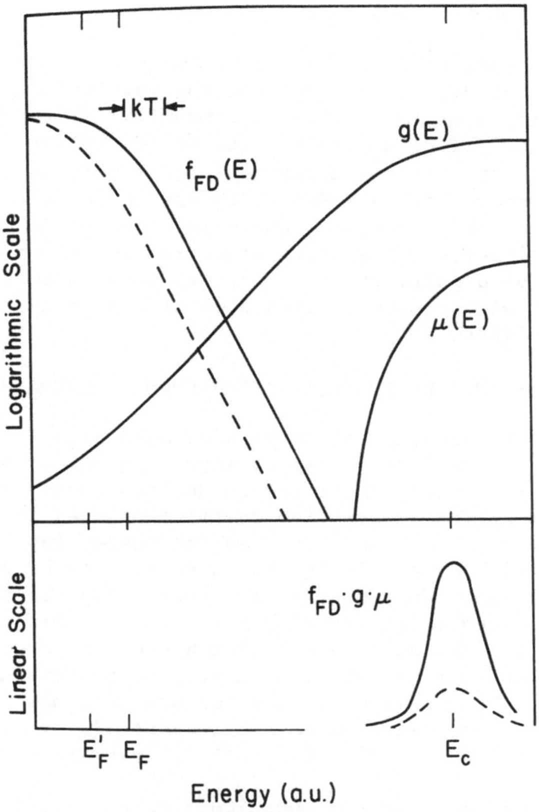

Fig. 11. Three functions important in the theory of transport in semiconductors: f_{FD} - the Fermi-Dirac distribution function. $g(E)$ - the density of states. $\mu(E)$ - a mobility-function indicating the relative contribution to transport of carriers with differing E. In the lower section the product of the three functions is shown, indicating a sharp peak. Dotted and dashed curves refer to two different Fermi energies.

doping levels of the material). Finally the bandedge density-of-states $g(E)$ has been illustrated as roughly exponential, as will be described in the next section.

With these choices the integrand of eq. (10) exhibits a sharp peak near E_c due to the competition between a sharp rise in $g\cdot\mu$ and the exponential decay of the Fermi-Dirac distribution function. $\mu(E)$ is not expected to be strongly temperature-dependent, since it describes what is primarily extended state transport; thus the activation energy of the conductivity is roughly the difference E_c-E_f. Doping the material raises the Fermi energy, increasing the conductivity and reducing its thermal activation energy.

This picture for transport in non-crystalline materials is largely due to Mott [3,5], who furthermore predicted that the quantity:

$$\sigma(E_C) \equiv (kT)(e)\mu(E_C)\cdot g(E_C)$$

should have a magnitude of order 10 - 100 $\Omega^{-1}cm^{-1}$ for amorphous semiconductors essentially independent of the particular material; this value is essentially a *minimum metallic conductivity*. As will be discussed in section III.C, an unequivocal determination of this number from steady-state transport measurements is extremely difficult, and at present it is not known how well Mott's theory agrees with experiment in amorphous semiconductors. However, there is clear evidence for the existence of a well defined transport edge from transient drift-mobility measurements. These experiments will be described in section III.B, after a discussion of the density-of-states at the conduction and valence bandedges has been given.

III.A. The Optical Absorption Edge in Amorphous Semiconductors

The most direct experimental studies of bandtails in amorphous semiconductors are optical measurements involving states near the top of the valence band and the bottom of the conduction band. As with all semiconductors the optical absorption coefficient $\alpha(\nu)$ is small for photon energies $h\nu$ which are smaller than the energy gap between the valence and conduction band, and increases rapidly as $h\nu$ becomes comparable to this gap. This *optical absorption edge* is thus very sensitive to the bandedge density-of-states, and in this section the present understanding of the interrelationship of the bandedge density-of-states and optical absorption will be described. In particular for amorphous semiconductors the bandedge has a tantalizingly universal form which ends with *exponential bandtails*, and which reflects the extent of thermal and static "disorder" in a particular specimen.

1. The Tauc Regime. In the upper portion of Fig. 12 I have plotted (after Dunstan [41]) the square-root of the optical absorption coefficient α as a function of photon energy $h\nu$ for several amorphous semiconductors. The optical absorption process involved is the "interband" transition of an electron from a filled valence band state to an empty conduction band state. The square-root functional dependence which this figure tests is expected from a simple argument due to Tauc [42].

First recall some related facts about interband optical transitions in crystals. In crystals the bandedge density-of-states $g(E)$ is of course proportional to $|E-E_C'|^{1/2}$ near the band extrema, where E_C' is the bandedge energy. However in crystals this information is inadequate for estimating the form of interband optical absorption because of the strong symmetry-based selection rules governing the rates of interband transitions. In particular for crystals pairs of valence and conduction band states separated by the same energy have greatly different optical transition (dipole) matrix elements depending upon whether or not they have the same Bloch $\boldsymbol{k}$-vectors (and thus whether or not the optical transition can conserve crystal momentum without involving phonons).

For non-crystalline materials the edifice of Bloch $\boldsymbol{k}$-vectors is largely irrelevant, and thus as a starting point one might consider simply neglecting altogether any variation of the dipole matrix element which governs the optical transition rate between various pairs of valence and conduction band states separated by a given photon energy $h\nu$. This "random-phase" approximation for the matrix elements governing the interband transitions has been experimentally verified by Jackson, *et al*, over quite large ranges of photon energy [8] by comparison of photoemission with conventional optical absorption experiments.

Using the random-phase approximation, the optical absorption coefficient in the one-electron approximation (neglecting electron-hole interactions) is proportional to a convolution integral:

$$\alpha(h\nu) \propto (h\nu)^{-1} \int_{E_C - h\nu}^{E_V} N_V(E) N_C(E + h\nu)\, dE \tag{11}$$

N_V and N_C are the valence and conduction band densities of states. The convolution integral simply counts the density of pairs of conduction and valence band states separated by $h\nu$; the factor of $h\nu$ appearing at the right of this proportionality originates from the detailed relationship

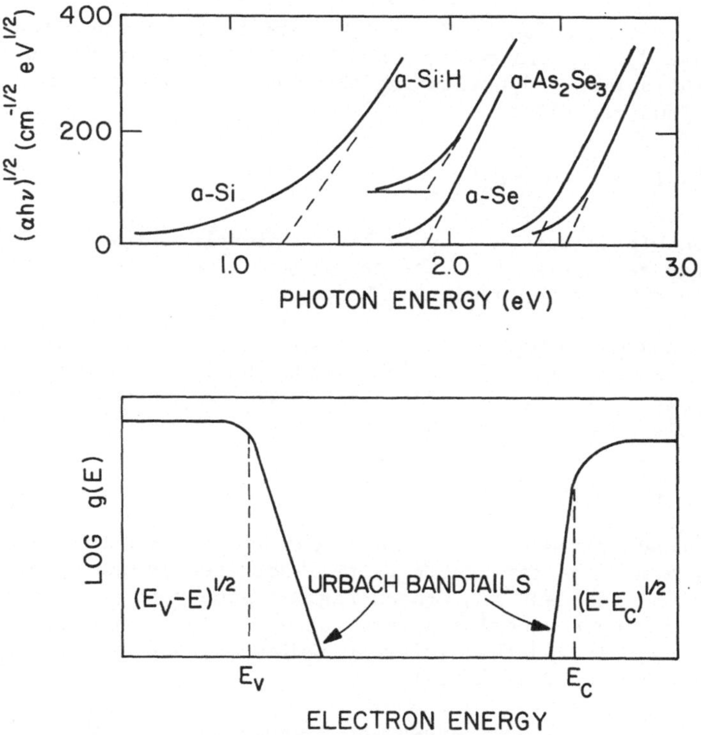

Fig. 12. (*upper*) The optical absorption coefficient α for the indicated amorphous semiconductors as a function of the photon energy $h\nu$. The curve for a-Si:H is shifted upward as shown; the two curves for a-As_2Se_3 indicate two measurement temperatures. After Dunstan [41]. (*lower*) Forms for the valence and conduction bandedge densities of states in amorphous semiconductors consistent with optical properties in amorphous semiconductors: exponential or Urbach bandtails well removed from the mobility edges E_V and E_C, and approximately square root forms near the edges.

between dipole transitions and optical properties [8,9]. For convenience only extended electronic states are included in the integral by the choice of limits indicated; as we shall see subsequently optical transitions between extended and localized states yield different limiting behavior. We neglect transitions between pairs of localized states altogether, although these are important for understanding *luminescence* in amorphous semiconductors [43].

In non-crystals we have no theoretical expectation for any particular form for the valence and conduction band densities-of-states; the reasonable success of the square-root dependence for the higher photon energies of Fig. 11 supports the assumption that N_V and N_C have the same square-root functional dependence upon energy as in crystals, at least over a limited region of energy:

$$N_C \propto (E - E_C')^{1/2} \tag{12a}$$

$$N_V \propto (E_V' - E)^{1/2} \tag{12b}$$

We expect that the bandedges E_C' and E_V' are close to the mobility edges E_C and E_V, but have indicated the possibility that they are not identical by the use of primes. Substituting these expressions into eq. (11) we obtain the Tauc behavior:

$$\alpha(h\nu) \propto (h\nu)^{-1}(h\nu - E_g)^2 \tag{13}$$

The relationship is tested graphically by plotting $\{\alpha(h\nu)\cdot h\nu\}^{1/2}$ versus photon energy as in Fig. 12; the energy gap $E_g \equiv E_C' - E_V'$ determined using this graph may be termed the *Tauc gap*.

2. <u>The Urbach Regime.</u> It is evident from Fig. 12 that the simple behavior for $\alpha(h\nu)$ (eq. (13)) which yields the Tauc gap fails at lower photon energies. In this *Urbach* regime [44] the optical absorption coefficient is found to obey the expression:

$$\alpha(h\nu) \propto e^{(h\nu - E_1)/E_0} \tag{14}$$

where the Urbach parameter E_0 depends upon both temperature and the state of the specimen. The remarkable fact illustrated by the upper portion Fig. 12 is that the Tauc and Urbach regimes appear to be different sections of a single underlying form for $\alpha(h\nu)$ in amorphous semiconductors; indeed the different semiconductors exhibit strikingly similar forms in absolute units!

There are several approaches to interpreting the optical absorption edge in amorphous semiconductors, but the simplest invokes a bandedge density-of-states having a surprisingly universal form: exponential behavior deep in the bandtail, and approximately square-root form at higher energies. An indication of the required form for the densities of states is given in the lower portion of Fig. 12. In this viewpoint the Tauc regime is dominated by transitions between extended valence band states (below E_V) and extended conduction band states (above E_C). The Urbach regime is largely due to transitions between localized states in one bandtail and extended states near the opposing bandedge; it is dominated by the broader exponential bandtail, which in Fig. 12 is the valence bandtail.

The close relationship of the exponential (Urbach) regime of interband absorption and the Tauc regime is further illustrated by the measurements of Cody and his collaborators [45] on a series of amorphous hydrogenated silicon (a-Si:H) specimens in differing states of *thermal* and *static disorder*. The results are plotted as Fig. 13; note the use of a logarithmic scale for the absorption coefficient, thus explicitly illustrating the exponential Urbach regime for interband absorption. These data show for a-Si:H that the exponential factor E_0 (cf. eq. (14)) in the Urbach regime shrinks as the bandgap (characterizing the Tauc regime) increases - either by lowering the measurement temperature or by direct modification of the material using annealing procedures. Remarkably the "focal-energy" E_1 of the exponential sections illustrated

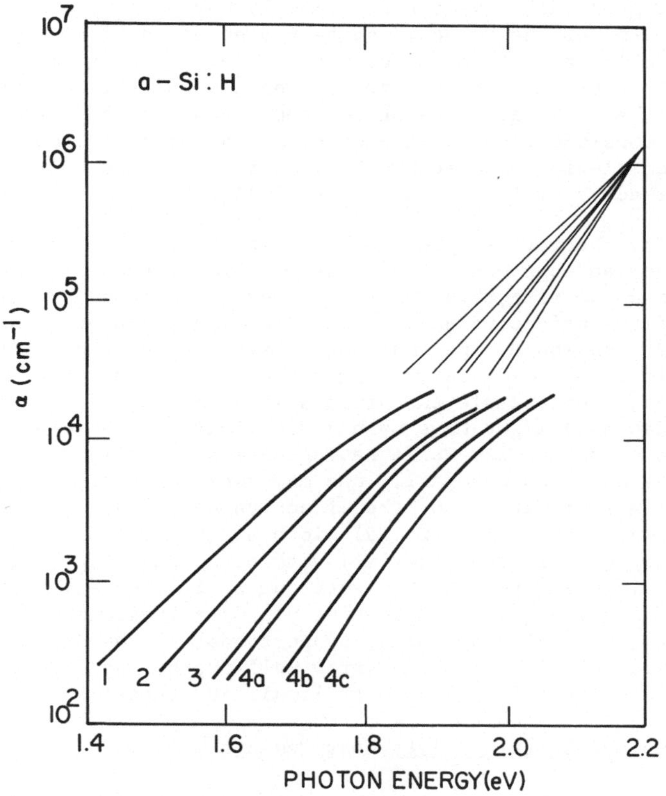

Fig. 13. (*bold curves near bottom*) The optical absorption coefficient α for an a-Si:H specimen at several temperatures (4a - 293 K, 4b - 151 K, 4c - 12.7 K) and at 300 K following an annealing procedure: (1 - 848 K, 2 - 798 K, 3 - 748 K). (*straight lines at upper right*) Extrapolations of α from the low-energy exponential domain indicating convergence to a single "focal energy." (After Cody [45]).

in Fig. 12 is not affected by temperature or the state of the specimen, suggesting that it is a more universal parameter of a-Si:H than either the apparent Tauc gap or the width of the Urbach tail.

3. Origins of the Bandedge Density-of-States. I have adopted here the simple viewpoint that optical absorption is the result of a well-defined bandedge density-of-states. Other models have been suggested, in particular that of Dow and Redfield which invokes excitonic or electron-hole interaction to account for the Urbach domain in crystals [46]. The present viewpoint is consistent with the photoemission experiments of Griep and Ley [47]; although the optical transition involved in photoemission is quite distinct from that of interband optical absorption, the same valence band density-of-states appears to apply. As we shall see in the next section describing transient drift-mobility experiments, a similar valence band density-of-states also describes certain electrical transport experiments. There is thus little reason in amorphous semiconductors to distrust the simple density-of-states viewpoint, at least for the valence bandedge.

Accepting the density-of-states model, how does disorder create the particular form for the bandedge suggested by the optical experiments? This question is not yet answered, but has received the sustained attention of theorists for over twenty years. For recent work the reader may consult the reviews of Sa-Yakanit and Glyde [48] and of Economou, *et al* [49]. I conclude this section by paraphrasing a particularly accessible conclusion reached by the latter group, and trust that interested readers will consult the original works for more information.

One major theoretical problem has been to explain the exponential region of the bandedge density-of-states. Simply stated, what are the requirements on an effective potential energy function $U(\mathbf{r})$ so that the solutions of the effective Hamiltonian yield an exponential distribution of energies? One would hope that some distribution of parameters in the problem would show the Gaussian form expected from the mean-value theorem of statistics; for example the various local minima of the potential (ie. "well-depths") might have such a distribution. However, a Gaussian distribution of depths for wells having several localized states in each well yields a Gaussian bandtail; one may consider these to be the slowly-varying fluctuations of the effective potential. On the other hand narrow wells (having a spatial width of the same order as the interatomic spacing in the material) with a Gaussian depth distribution can yield a different result. In particular, if the well-depths are not too deep, elementary quantum mechanics yields a single bound-state for each such well with binding energy proportional to the depth *squared*; thus a Gaussian distribution of narrow well-depths may indeed lead to the observed exponential distribution of localized states!

III.B. Transient Drift-Mobility Measurements

The simple qualitative picture for bandedges and bandedge transport that I have sketched so far has achieved perhaps its most convincing success as a description of the electrical transport of excess carriers photogenerated by a short laser pulse. This *photocarrier drift-mobility* experiment is illustrated in Fig. 14; photocarriers are drawn between the non-injecting electrodes by an external electric field, and the photocurrent $i(t)$ is monitored. If the excess carriers are generated near one electrode (by using strongly absorbed light) the current will be dominated by the motion of one type of photocarrier (electron or hole) selected by the polarity of the external field. Since the charge Q of photocarriers may be measured by sweeping them to the electrodes, the

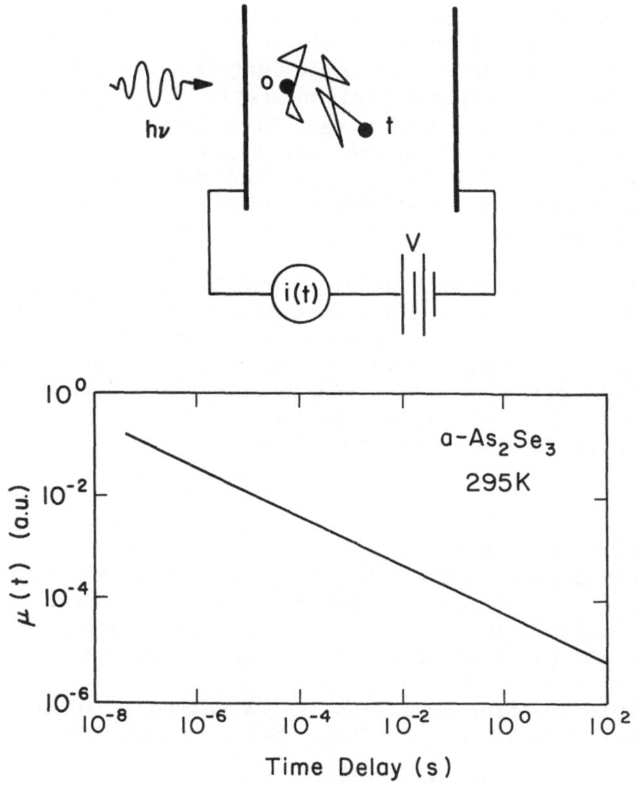

Fig. 14. (*upper*) The transient photocurrent measurement following an illumination impulse; motion of the photocarriers dissipates energy, which is equal to the power $i(t)\cdot V$ furnished by the external battery.
(*lower*) Transient drift mobility $\mu(t)$ for holes in a-As_2Se_3 obtained from the transient photocurrent (after Monroe and Kastner [51]).

photocurrent prior to sweepout simply measures the mean drift-velocity $\langle v(t)\rangle$ of the photocarriers at a delay t past the impulse [50]:

$$i(t)\cdot V = Q\cdot\langle v(t)\rangle\cdot E \tag{15}$$

This expression simply equates the power furnished by the external circuit with the dissipation of the drifting photocarriers in the uniform external field E. The *transient drift-mobility* $\mu(t)$ then relates $\langle v(t)\rangle$ and E:

$$\mu(t) = \langle v(t)\rangle/E \tag{16}$$

where of course one must check experimentally that $\mu(t)$ is independent of E.

In the lower half of Fig. 14 $\mu(t)$ is shown for holes in a-As$_2$Se$_3$ [51]; $\mu(t)$ exhibits a simple power-law decay over an extraordinary time-range - nearly ten orders of magnitude! Power-law decays for $\mu(t)$ are fairly typical in amorphous semiconductors, although they do not necessarily extend over the full range illustrated in Fig. 13. Such decays are quite different than the time-independent drift mobility expected for crystalline semiconductors, and the terms *Gaussian transport* (for time-independent $\mu(t)$) and *dispersive transport* (for power law decays of $\mu(t)$) are commonly used to distinguish the two cases [52].

Dispersive transport such as displayed in Fig. 14 is a consequence of mobility-edge transport and exponential bandtails of localized states. The particular description of transient drift mobilities involved is known as *multiple-trapping*, in which a fraction $\theta(t)$ of the photocarriers occupy relatively extended states and contribute to transport with an effective mobility μ_0, but the majority occupy traps and are immobile. The transient drift mobility is then simply:

$$\mu(t) = \mu_0 \cdot \theta(t) \tag{17}$$

The decay in $\theta(t)$ results from the competition between *trapping* (ie. capture) of mobile photocarriers by localized states and thermal emission of the trapped carriers back into the more extended states. For further details of the calculations the reader is referred to the review of Tiedje [53] or the paper of Orenstein, Vaninov, and Kastner [54].

The multiple-trapping equations may be solved given an exponential valence bandtail with N_{bt} traps. Using the form for $g(E)$:

$$g(E) = (N_{bt}/E_0) e^{-(E-E_v)/E_0} \tag{18}$$

the following expression for the drift-mobility obtains:

$$\mu(t) \cong [\mu_0(N_v/N_{bt})] \; (\nu t)^{-(1-\alpha)} \tag{19}$$

where N_v is the effective density of transport states at the mobility edge and $\alpha \equiv kT/E_0$ is the *dispersion parameter*. ν is the attempt-to-escape frequency describing the rate $\nu e^{-|E-E_v|/kT}$ of thermal emission of a photocarrier trapped in a state at energy E.

Inspection of this equation indicates that the exponential bandtail model determines the temperature dependence of $\mu(t)$; in particular the dispersion parameter $\alpha(T)$ must be proportional to temperature. The required temperature dependence has been observed both in a-As$_2$Se$_3$ and a-Si:H. For a-Si:H the valence bandtail width E_0 estimated from the drift-mobility agrees with the Urbach parameter from optical absorption [55]. The parameter $\mu_0(N_v/N_{bt})$ is nearly temperature-independent, as might be expected from identification of the transport-channel with extended-state transport or a mobility-edge. A value in the range 1 - 10 cm^{2}V^{-1}s^{-1} is often obtained. It is possible to extrapolate these data to estimate the mobility-edge parameter $\sigma(E_c)$, although these extrapolations will not be described here [5].

III.C. **Steady-State Transport Measurements in a-Si:H**

There appears to be little reason to doubt that electrical transport in amorphous semiconductors can in several important cases be associated with a well-defined energy in the bandedge density-of-states. The simplest viewpoint regarding this energy is that it is in fact the mobility-edge separating localized from extended wavefunctions; however, a detailed quantitative understanding of transport continues to elude scientists studying amorphous semiconductors. In this section I shall comment on three aspects of steady-state transport in amorphous semiconductors which present difficulties:

(1) Estimation of the mobility-edge conductivity prefactor
(2) The discrepancy between thermopower and conductivity measurements.
(3) The reversals in the sign of the Hall effect

1. The Conductivity Prefactor. Because of the wide range of its electronic properties, amorphous hydrogenated silicon (a-Si:H) would appear to be an ideal materials system in which to test for the conductivity prefactor $\sigma(E_c)$ expected from mobility-edge theories:

$$\sigma(T) = \sigma(E_c)e^{-(E_c-E_f)/kT}$$
$$= \sigma_0 e^{-E_a/kT} \tag{20}$$

At first glance the measured prefactor σ_0 for an exponentially activated conductivity might be identified with the prefactor $\sigma(E_c)$ from theory. Unfortunately this expectation is incorrect; both the Fermi energy E_f and the location of the mobility edge E_c may be temperature dependent Retaining only first-order terms for the temperature dependence of their difference over the range of the experiment:

$$E_c(T) - E_f(T) = (E_c - E_f)_0 - \gamma T \tag{21}$$

we obtain:

$$\sigma_0 = \sigma(E_c)e^{\gamma/k} \tag{22}$$

The temperature dependence $E_c(T)$ is due both to the shift of the bandedge density-of-states with temperature as well as to any temperature dependence of the mobility edge relative to this density-of-states. The temperature dependence $E_f(T)$ is known as the *statistical shift* and depends on the functional form for $g(E)$ near E_f. $(E_c-E_f)_0$ is simply a fitting parameter and can *not* usually be identified with the value $E_c(0) - E_f(0)$.

Regrettably the factor $e^{\gamma/k}$ is not small. The bandedge temperature-shift found in optical studies [9,45] yields $\gamma \cong -4\cdot10^{-4}$ eV/K, yielding a correction to σ_0 of order 10^2. Statistical shifts can be even larger, and in fact the measured values of σ_0 for a-Si:H vary over a range of nearly 10^5 [5]. It is of course possible to attempt to correct for these effects [56], but estimates for $\sigma(E_c)$ based on such large corrections are obviously suspect. In principle thermopower measurements offer another approach to estimating $\sigma(E_c)$, and this approach will be discussed in the next section.

2. Thermopower. A very useful auxiliary transport measurement in semiconductors is thermopower [40]. If a small temperature gradient is applied across a semiconductor along the x axis, a potential difference across the material also results (corresponding to a uniform, thermally-induced electric field E_x inside the specimen). The *Seebeck coefficient S* is simply the ratio of this field to the temperature gradient:

$$S \equiv E_x/(\partial T/\partial x) \tag{23}$$

For bandedge transport the approach used previously (cf. eq. (10)) for the conductivity of a specimen yields a closely related expression for the thermopower [40]:

$$-eST = \frac{1}{\sigma(T)} \int e(E - E_f) g(E) \mu(E) e^{-(E-E_f)/kT} dE \tag{24}$$

Inspection of this equation reveals a physical interpretation for the thermopower: $-eST$ is the mean energy with respect to the Fermi energy of charge transport. The sign of the thermopower directly indicates whether electrons or holes are responsible for transport.

If the integrand of eq. (24) is sharply peaked at a mobility-edge E_c, then $-eST$ is simply $E_c - E_f$ (cf. Fig. 11). Thus a comparison of the temperature-dependence of the thermopower and electrical conductivity in a single specimen provides a test for the model of a sharp mobility edge and a possible technique for determining $\sigma(E_c)$:

$$\sigma(T) = \sigma(E_c) e^{-(E_c-E_f)/kT} \tag{25}$$

$$-eST = E_c - E_f \tag{26}$$

$$Q \equiv \ln(\sigma) + (q/k)S \tag{27a}$$

$$\cong \ln(\sigma(E_c)) \tag{27b}$$

In Fig. 15 data in a-Si:H for several specimens of differing phosphorus incorporation are shown for the same specimens used in Fig. 10. The Q function is evidently *not* the simple constant expected from a sharp mobility edge at E_c. This thermopower discrepancy clearly challenges the simple interpretation of the transport edge in a-Si:H as a sharp mobility-edge; I shall defer discussing the transport models which have been proposed to account for this difficulty until after presentation of the Hall effect data for a-Si:H.

3. The Hall Mobility

> The Hall effect is perhaps the most striking of all the properties which distinguish non-crystalline from crystalline materials.
>
> --E. A. Davis [57]

The Hall effect is the generation by an external magnetic field H of a Hall electric field E_h transverse to a current flow; the magnetic field is transverse to the current and also to the Hall electric field. In semiconductors and metals for which an effective-mass theory of transport applies, the Hall field simply compensates for the Lorentz force experienced by the drifting charge carriers. The Hall mobility μ_h is defined from the three fields involved:

$$\mu_h \equiv E_h/(E \cdot H) \tag{28}$$

where E is the external electric field inducing the current flow. For a crystalline material with effective mass transport the Hall mobility should be comparable to the drift mobility of the dominant photocarrier, and the sign of the Hall field indicates the sign of the charge carrier dominating the current flow.

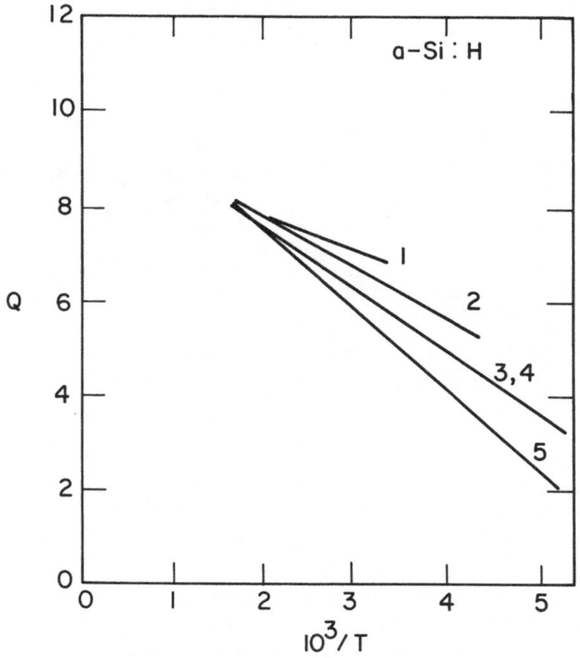

Fig. 15. The *Q*-function determined from conductivity (σ) and thermopower (*S*) ($Q = \ln(\sigma) - (e/k)S$) for the same phosphorus-doped a-Si:H specimens as in Fig. 10; the linear temperature dependence suggests a role for specimen inhomogeneity. (After Beyer [40]).

For non-crystalline materials the effective-mass model based on crystallinity does not apply; indeed even for polycrystalline materials the Hall mobility differs greatly from that of the single crystal. In Fig. 16 the data of Spear, *et al* [58] are illustrated both for microcrystalline silicon (μc-Si) and for amorphous silicon. All the specimens were fabricated in a plasma deposition reactor, and the differences in structure were obtained by varying the reactor conditions. Despite the considerable scatter in the data, it is evident that the Hall mobility declines rapidly as the average crystallite size is reduced. Remarkably the sign of the Hall effect *reverses* for the amorphous specimen; although electrons were still the charge carriers dominating the current, the Hall voltage has the sign expected for effective mass holes. A sign reversal is also found when holes are the dominant charge carriers, both in a-Si:H and also in chalcogenide glasses; the Hall voltage for hole currents has the same sign as that expected for effective mass electrons.

4. Transport Models. Both the Hall effect and thermopower measurements indicate the need for improvements in models for the transport edge in amorphous semiconductors in general and a-Si:H in particular. Here I simply note briefly three possibilities which have been studied theoretically. First, amorphous semiconductors may be *spatially inhomogeneous*, with long-range (compared to atomic dimensions) variations in the position of the mobility-edge or other transport parameters. Such inhomogeneities have been modeled using a semiclassical approach by Overhof and Beyer [59], who found a linearly decreasing form

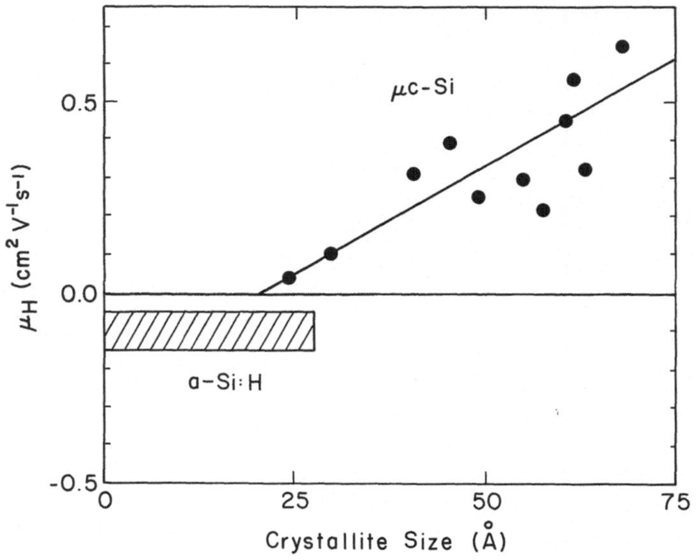

Fig. 16. The electron Hall mobility μ_h measured at room-temperature for a series of phosphorus-doped, plasma-deposited microcrystalline specimens (μc-Si) with the indicated characteristic crystallite size. The band at the bottom indicates the Hall mobility of amorphous silicon (a-Si:H); note the Hall mobility sign-reversal in the amorphous material. (After Spear, *et al* [58]).

for the Q-function ($Q = \ln(\sigma) + (q/k)S$) similar to that observed experimentally (Fig. 15). This approach suggests that a range for Δ (the magnitude of the potential variations) of 0.05 eV (undoped a-Si:H) to 0.25 eV (heavily doped a-Si:H). Such an approach may also be used to account for the decline in the Hall mobility in microcrystalline specimens, but it offers no real insight into the sign reversals for μ_h found in amorphous semiconductors.

Sign reversals are obtained in transport models based upon motion of localized charge carriers between sites - ie. *hopping* conduction[60-62]. If the sites have the same energy the tunneling between them need not be thermally activated. The sign reversal is possible because the net rate at which a carrier hops between two sites must be calculated including paths involving all other nearby sites, and interference between the amplitudes affecting the rate can cause the sign reversal with respect to the effective-mass expectation. If the charge carriers are considered to be localized on particular bonds or atoms, this approach predicts that transport in structures with odd-membered rings of atoms or bonds will have the Hall mobility sign reversals found in amorphous semiconductors, while transport in structures with even-membered rings will yield normal Hall mobilities [60,61].

It is certainly fair to say that the experimental results and theoretical work on transport in amorphous semiconductors have not yet been synthesized into a coherent whole, even for the particular type of bandedge transport discussed here. However, it also appears that the most essential pieces of this puzzle have been found. Bandedge transport occurs at a reasonably well-defined energy or transport edge. This transport edge may plausibly be identified with the *mobility-edge* (the energy at which the bandedge electron wavefunctions pass from localized to extended). However, electron wavefunctions at the transport-edge retain the character of localization sufficiently to allow for the sign reversals observed in the Hall mobility. Finally thermpower measurements suggest that long-range inhomogeneities of the material modulate transport; the case is especially strong for a-Si:H, where the *Q*-function's dependence upon doping (which increases the density of charged defects enormously) supports the inhomogeneity model.

IV. DEFECTS AND DOPING

Defects and impurities are rather rare even in the most miserable semiconductors. In elemental a-Si, for example, no more than 1 in 1000 Si atoms have unsatisfied chemical bonds resulting in the detection of an electron spin resonance signal [20]. But these defects are crucial to the electronic and photoelectronic properties of the materials. For example, consider the field-effect experiment. In this experiment the the amorphous semiconductor film is separated from a conducting plate (the *gate* electrode) by a thin insulator. The region of the amorphous semiconductor near the interface with the insulator can be charged with perhaps 10^{12} cm^{-2} electrons (or holes) when a voltage is applied to the gate; the device is essentially a parallel-plate capacitor. Of course the charge might lead to a conducting channel near the surface, increasing the conductance of the specimen and permitting the fabrication of a field-effect transistor.

In elemental a-Si the field-effect yields perhaps a factor of two increase in conductance [63]; the defect density detected by electron spin resonance is in the range 10^{18} - 10^{19} cm^{-3} [20]. In hydrogenated amorphous silicon, which has an electron spin resonance signal 1000 times or more smaller than a-Si [43], the field-effect is enormous, with an increase in conductance of nearly 10^6 in some cases [24]. Obviously semiconductor phenomena are enormously affected by defects and also by their close relatives *dopants*, which are impurities intentionally added to the material to modify the electronic properties.

In this section I shall briefly introduce defect phenomena in general, addressing in particular the definition of a defect in amorphous semiconductors and illustrating the phenomena of atomic relaxation and electronic correlation energies. I then survey three topics of importance: negative correlation energy defects in chalcogenide glasses, doping in a-Si:H, and defect metastability in a-Si:H.

IV.A. General Remarks

1. What is a Defect? As we have seen in earlier sections, amorphous semiconductors preserve fairly well the bonding arrangements expected from elementary chemistry. For Group IV, V, VI, and VII elements this arrangement is given by the "8-*N*" rule: the coordination of a Group *N* atom in the network is ideally 8-*N*. Thus a simple definition for an atomic defect in a non-crystalline semiconductor is as a deviation from the majority bonding arrangement - or a disruption of the short-range order of the material. This viewpoint is reflected in a conventional notation for the bonding arrangment of a particular atom [64]:

A_z^q

A: Type of atom (cf. C - chalcogen)
z: Bonding coordination of site
q: Local charge (units of e)

A tetrahedrally bonded Si atom is thus denoted T_4^0 (here T denotes a "tathogen"); T_4^0 is the arrangement of the vast majority of Si atoms. A *dangling bond* defect is a Si atom with only three bonds to three neighbors; in its neutral charge state the local arrangement is denoted T_3^0. In a non-crystalline semiconductor a dangling bond might in principle exist in isolation (as a point defect in a continuous random network) or it might be attached to a larger microstructure. This approach to identifying defects should be contrasted to that for crystals, where defects are viewed as disruptions of the long-range order of the crystal. Thus an isolated dangling bond is not possible in a bulk crystalline solid; simple point defects are instead the vacancy (an atom missing from the lattice) or the interstitial (an atom inserted inside a primitive cell of the lattice in violation of chemical bonding rules).

One other important distinction between defects should be noted. Simple bonding defects such as the T_3^0 have an unbonded, and *unpaired*, electron. Such electrons are intrinsically paramagnetic, and in principle should be detectable using an *electron spin resonance* (or ESR) spectrometer[43,65]. Covalent bonds involve pairs of electrons, and hence there is usually no net paramagnetism associated with the fully satisfied covalent bonds. In covalent solids ESR is thus primarily a defect spectroscopy; it should be contrasted to nuclear magnetic resonance, which is sensitive to *all* magnetic nuclei in a material. Of course an ionized - or a doubly occupied - bonding defect is not expected to be paramagnetic; thus it is possible for a defect to be unobservable by ESR (because the charge state of the defect is not paramagnetic), in addition to the possibility that the defect system is unobserved by ESR (due to poor sensitivity).

2. Relaxation and Correlation. Even if the chemical bonding configuration of a point defect is established, the total energy of the defect will depend upon the detailed nuclear configuration of the atoms near it and also upon its charge state. These dependences are illustrated in Fig. 17, which is based on the theoretical calculation of Bar-Yam and Joannopoulos [66] for the Si dangling bond. The particular structure used was an incompletely hydrogenated divacancy in c-Si. A dangling bond is illustrated in the the upper right of the figure, along with the definition of one of the coordinates (u) describing the atomic positions near the defect; one might expect that in the lowest energy configuration of the defect the T_3 atom (at the top of the tetrahedron) might be closer to the plane of its neighbors than if the atom were in the normal T_4 bonding arrangement. Of course other coordinates of atomic relaxation may be considered depending upon the elaborateness of the calculation.

The defect energy is illustrated as a function of a more general "configuration coordinate" Q at the left of Fig. 17; the definition of Q was chosen to illustrate the minimum energy of the defect in each of three distinct charge states. Each of the curves corresponds to an excitation of the neutral ground state of the defect; for example the curve labelled $(+)+e$ at the right is the sum of the energy of the positively charged (ionized) defect and an electron at the conduction-band edge. Similarly the curve labelled $(-)+h$ is the energy of the negatively-charged defect (binding an excess electron) and a hole at the valence-band edge. Lastly the curve labelled $(0)+e+h$ is simply the energy of the neutral defect plus the bandgap of a-Si:H. An important possibility is illustrated by this figure: the energy minima

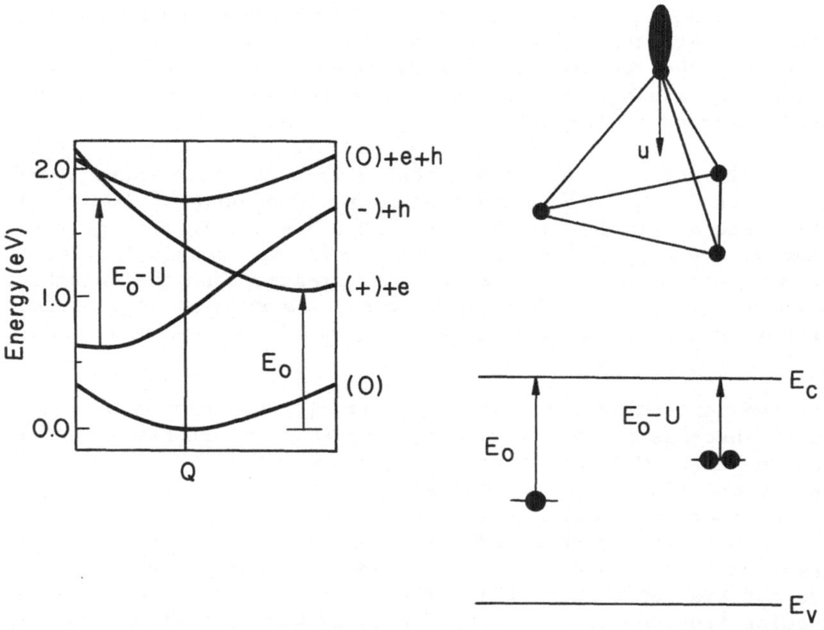

Fig. 17. (*upper right*) Atomic configuration near a dangling bond defect, illustrating the unbonded electron in a hybrid orbital. The variable u indicates a possible relaxation of the defect atom towards the plane of its three neighbors.
(*left*) Energy of the defect system as a function of a configuration coordinate Q indicating the atomic positions near the defect in four electronically excited states: (0) : the neutral defect in its electronic ground state. (+)+e : the positively charged defect plus the energy of an electron at the conduction band edge (E_c). (-)+h : the negatively charged defect plus the energy of a hole at the valence band edge (E_v). (0)+e+h : the neutral defect plus the energy of an electron and a hole (E_g). After Bar-Yam and Joannopoulos [66].
(*lower right*) Energy level diagram used in calculating relative populations of the defect's electronic states in thermal equilibrium. The neutral defect's thermal ionization energy E_0 and the negative ion's detachment energy E_0-U may be obtained from the configuration coordinate diagram as shown.

for the various charge-states of the defect do *not* correspond to the same atomic configuration coordinate! The phenomenon is similar to the *Franck-Condon* principle invoked to rationalize the optical properties of defects by considering differing atomic relaxation in the ground and excited states of a luminescence center.

Now consider the minimum energy required to ionize the neutral defect T_3^0. This energy is indicated as E_0 in the figure, and corresponds to the difference between the minima of the (0) and the (+)+e states. Since the minima do not correspond to the same configuration coordinate it is apparent that atomic relaxation affects this difference in addition to the purely electronic energies which might be calculated assuming fixed atomic positions. E_0 is the energy which would be required for statistical mechanics calculations of the relative populations of the two states. Similarly E_0-U is the energy required to detach the excess electron from T_3^-; U is defined as the *effective correlation energy* of the defect.

One would normally anticipate that U is a positive number, and that an electron on T_3^- would be easier to remove than on T_3^0. For example the ionization energy of ordinary hydrogen H^0 (13.6 eV) is much greater than the detachment energy for H^- (less than 1 eV), and the ionization energy for a hydrogenic donor D^0 in c-Si is much greater than the detachment energy for D^- [67]. A positive value for U was assumed in constructing the conventional semiconductor level diagram at the bottom right of Fig. 17.

For the particular calculation of Fig. 17 this expectation is incorrect; because of the large atomic relaxation effects the effective correlation energy U is *negative* for this model for T_3^-. Negative correlation energies lead to remarkable effects in chalcogenide glasses, as will be described in the next section. However, in a-Si:H these effects are not clearly observed. The defect observed by electron spin resonance, which may be denoted the D center, clearly does not have a negative correlation energy [43]. The implication drawn by Bar-Yam and Joannopoulos from the calculation of Fig. 17 was that T_3^0 may not be the correct microscopic configuration of the observed D defect. The issue of the correct identification of the D-center in a-Si:H is in fact still open [68].

IV.B. Negative Correlation Energy Defects in Chalcogenide Glasses

Chalcogenide glasses have several properties expected from amorphous semiconductors with large defect densities. In particular the materials have a small field-effect; from this result and from other electronic experiments an estimate for the electronically active defects of order 10^{18} cm^{-3} has been inferred. Unlike the situation in elemental a-Si, where a defect density adequate to explain the poor semiconductor quality is readily detected by electron spin resonance, no defects are observed by ESR in thermal equilibrium, and indeed careful studies have shown that the glasses can have a net *diamagnetism* [69].

An elegant explanation for the absence of an ESR-detectable defect in chalcogenides is that the defect has negative correlation energy behavior [64,,70,71]. If we denote the paramagnetic charge state of this defect as D^0, and the non-paramagnetic charge states as D^- and D^+, a negative correlation energy system has the following *exothermic* reaction:

$$2D^0 \longrightarrow D^- + D^+ - U \qquad (29)$$

A negative correlation energy defect thus neatly accounts for the absence of paramagnetism in chalcogenide glasses in thermal equilibrium. It can also accomodate the *light-induced electron spin resonance* (LESR) observed at lower temperatures: illumination below 77 K of As_2Se_3 induces a metastable density of more than 10^{17} cm^{-3} ESR-detectable defects which persists even when illumination is stopped [72]. In the negative correlation energy model illumination reverses (29) metastably:

$$D^+ + D^- + e^- + h^+ \longrightarrow 2D^0 \qquad (30)$$

The ESR signal disappears if the specimen's temperature is raised or if the specimen is illuminated with infrared (sub-bandgap) light.

A negative-U defect system also explains certain of the electrical properties of chalcogenide glasses such as the small field-effect. In particular *pinning* of the Fermi level E_f is a consequence of the statistical mechanics of negative-U defects. Consider the position of E_f at $T = 0$ K in the energy-level diagram at the bottom right of Fig. 17 as charge is added to or removed from the system of D defects. For a positive-U defect E_f lies at E_c-E_0 (in the lower energy band) as long as the defect system has a net positive electrical charge; some of the defects are neutral, and the remainder are positively charged. When the net charge becomes negative E_f rises to the energy E_c-E_0+U; again some defects are neutral, and the remainder negatively charged.

For a negative-U defect system half the defects are D^- and half D^+; there is never a substantial population of D^0 because as soon as two D^0 centers form the system can lower its energy via reaction (29). Hence E_f remains at E_c-E_0+$U/2$ and is independent of the net electrical charge of the system until *all* the defects are D^+ or all are D^-. In addition to accounting for Fermi level pinning, the negative correlation energy model also accounts for hole drift-mobility measurements above and below the *glass transition* temperature of As_2Se_3. An estimate $U = -0.7$ eV was obtained for this material, along with an estimate of the formation energy for the defect of +0.9 eV [73].

Regrettably it has proven very difficult to unequivocally identify the negative-U defect in a-As_2Se_3 and other chalcogenides. However, one particularly plausible model developed for chalcogenides has had a great deal of influence and will be described here. A *valence alternation pair* (VAP) defect [64] is a pair of oppositely charged point defects which crudely preserves ordinary covalent bonding. For example, C_1^0 (a singly bonded chalcogenide) has a dangling bond. A dangling bond is a rather gross disruption of covalent bonding, which for a Group VI chalcogen calls for twofold coordination. However, a negative chalcogen ion might be expected to bond as a Group VII element (ie. a halogen), and thus C_1^- crudely obeys the 8-N rule. The valence alternation pair of defects is $C_3^+ + C_1^-$. Each charged chalcogen atom is actually bonded as might be expected from crude covalent bonding arguments, with the C_3^+ acting similar to a neutral pnictide (Group V) atom with threefold coordination. Because the number of covalent bonds for a valence alternation pair is the same as for the ideally bonded network, the formation energy of such defect pairs is plausibly lower than that of uncharged defects.

A negative correlation energy property is not a necessary consequence of valence alternation pairs. For the chalcogenide glasses a negative correlation energy results from VAP's only if there is sufficient flexibility in the position of chalcogenide atoms to permit the coordination of an atom to change by bond breaking. In particular the following defect reaction must be exothermic and also reasonably unhindered by potential energy barriers:

$$2\ C_3^0 \longrightarrow C_1^- + C_3^+ \qquad (31)$$

Note that this reaction requires the bond between a C_3 and a C_2 to break, creating a C_2 and a C_1. The model seems violent, but it predicted the *transient field-effect* in chalcogenides observed by Frye and Adler [74]. The essence of this experiment was to show that, although a valence alternation pair has negative U behavior in thermal equilibrium, reaction (30) may proceed sufficiently slowly that the defect has positive U behavior for a short time following a change in its charge state.

IV.C. The Doping Paradox in Hydrogenated Amorphous Silicon

Doping of semiconductors is the intentional incorporation of impurity atoms in order to modify the material's Fermi level or other electronic properties. The phenomenon is readily understood in the crystalline form of Si. A small number of phosphorus (Group V) atoms in c-Si would be expected to substitute for Si atoms in the lattice. The resulting P_4 configuration of the phosphorus atom is *overcoordinated* relative to the ideal threefold coordination for phosphorus (P_3), but other configurations would require such a gross distortion of the diamond lattice of silicon that they become too costly in terms of formation energy. Covalent bonding of the P_4 utilizes only four of the available five electrons. The fifth electron is only loosely bound to the P_4, and the corresponding energy level is just slightly below the conduction band (about 25 meV). Consequently phosphorus-doped (n-type) c-Si has its $T = 0$ K Fermi energy extremely close to the conduction band.

In a-Si the P_4 configuration loses the energetic advantage it has in c-Si; it is more reasonable to expect that the network itself will be modified to accomodate the atom in its ideal 8-N rule P_3 configuration. This assertion has been experimentally tested in As doped a-Si:H [75,15]. Using the EXAFS (extended X-ray absorption fine structure) technique, the average coordination of the As was measured, and was found to agree quite well with the 8-N rule.

8-N rule bonding of phosphorus atoms does not lead to an energy level in the gap of a-Si:H; technically P_3 incorporation of phosphorus is an alloying effect and not doping at all. However, Fig. 10 illustrates conductivity data clearly showing that phosphorus incorporation does indeed lead to recognizable doping effects; in particular the Fermi level rises steadily towards the conduction band as phosphorus incorporation is increased. The simplest viewpoint which resolves this paradox is that doping effects are due to a minority of phosphorus atoms which have P_4 coordination in violation of the 8-N rule; doping effects are thus a consequence of bonding defects, and it is sensible to define a *doping efficiency* as the ratio of the densities of phosphorus atoms in the P_4 and P_3 configurations:

$$e \equiv \frac{[P_4]}{[P_3] + [P_4]} \tag{32}$$

A detailed understanding of doping in a-Si:H did not emerge for several years after its discovery. Street [76] has argued that a valence alternation pair model accounts in detail for several doping-related phenomena in a-Si:H, and I shall describe his argument here. Street's model has been criticized by Adler and his collaborators, and the reader should consult [77,78] for further discussion.

The application of the valence alternation pair idea to doping of a-Si:H is rather different than its application to chalcogenides, in particular in that the atoms of the pair are different. The defect pair $T_3^- + P_4^+$ is a valence alternation pair; the relative energy levels proposed for the two defects are indicated at the left of Fig. 18, where the notation D^- (associated with the ESR-detected defect D^0) has been used instead of T_3^-. If the energy gained by transfer of an electron from P_4^0 to D^0 exceeds the formation energy of the D defect, valence alternation pair formation will be favored over formation of a simple P_4.

A more detailed examination of the thermodynamics of valence alternation pair formation [79] shows that the density $[P_4]$ is proportional to the square-root of the available phosphorus density $[P]$. Some of the data supporting the valence-alternation pair model are summarized to the right of Fig. 18. Sub-bandgap optical absorption measurements in doped a-Si:H can be used to estimate the density of D defects; since the densities $[D]$ and $[P_4]$ are equal in the valence alternation pair model, the square root-dependence of sub-bandgap optical absorption upon the phosphorus concentration in the source gas used to grow the a-Si:H:P films agrees with the model. Finally the valence alternation pair predicts, as it did for chalcogenides, that no signal should be found in electron spin resonance. The curve labelled ESR in Fig. 18 indicates an upper bound to the densities of D and of P_4^0 in the ESR-detectable neutral charge state D^0, again in agreement with the prediction of the valence alternation pair that the densities of D and P_4 are approximately equal.

IV.D. Metastable Defects in Hydrogenated Amorphous Silicon

An important aspect of the defect systems in a-Si:H is their *stability* - or more correctly, their lack of stability. For example, the D^0 density detected by ESR is very strongly affected by the conditions

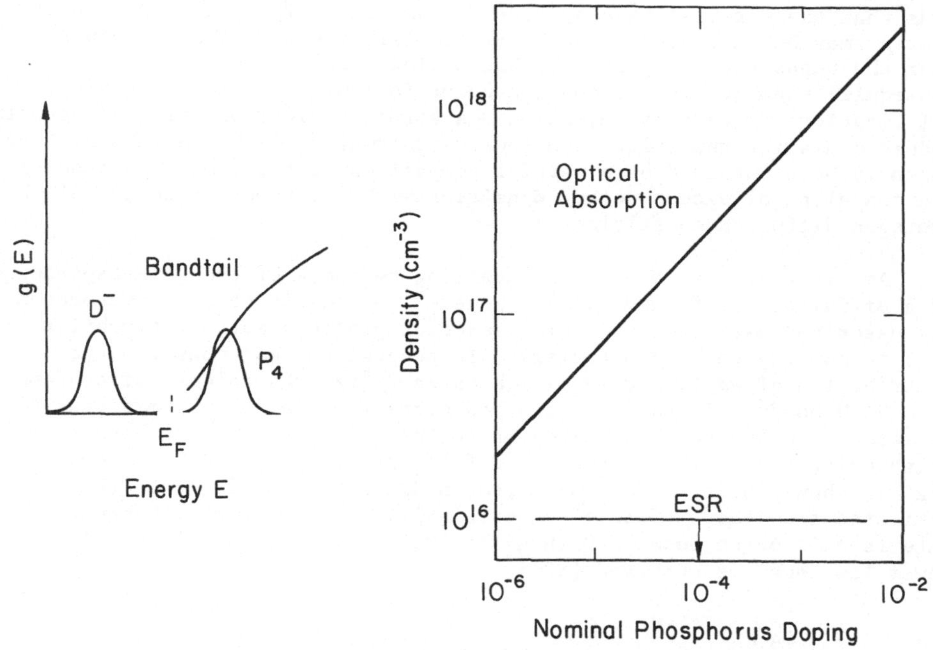

Fig. 18. The valence alternation pair model for doping in a-Si:H. At left the density of states $g(E)$ for this model is illustrated, showing nearly equal populations of negatively charged D^- and unpopulated P_4^+; the latter state may be masked by the conduction bandtail except at very high densities. To the right are two experimental results in agreement with the model: the absence of any significant D^0 or P_4^0 signal in electron spin resonance, but a clear indication that the defect density probed with infrared illumination rises essentially as the square-root of the nominal (gas phase) phosphorus doping. After Street [76].

during preparation of the film, in particular the growth rate and the substrate temperature; the electronic properties exhibit equally pronounced sensitivity to deposition conditions. A concomitant property of a-Si:H is that the defect densities are easily modified after the film is grown either by absorption of light (the *Staebler-Wronski* effect), by thermal annealing and quenching cycles, or by injection of currents from electrical contacts [80]. For example in the Staebler-Wronski effect the density of D^0 defects grows during illumination; the excess D^0 density is quite stable at 300 K in the dark, but can be removed by annealing the specimen at 475 K. Because of the importance of these defect stability phenomena for the performance of devices there has been a great deal of research upon them, but a microscopic understanding of defects in a-Si:H and of the reasons for these stability effects has proven elusive.

To conclude Section IV I shall discuss the recent proposal that the defect stability phenomena characteristic of a-Si:H reflect the relatively unhindered motion of protons in solids [81-83]. Given the presence of about 10 atomic percent of protons in a-Si:H this proposal is certainly a sensible starting point. Measurements of proton diffusion in a-Si:H yield a characteristic activation energy of about 1.4 eV for the diffusion coefficient, which is presumably the approximate energy required to break a Si-H bond and obtain a mobile proton.

Protons are vastly more mobile near the deposition temperature (about 500 - 600 K) than near 300 K. One viewpoint regarding a-Si:H is thus that the material is a *hydrogen glass* [81]. The configuration of protons may be in approximate thermal equilibrium with the Si network near the deposition temperature, but at lower temperatures non-equilibrium configurations can occur following thermal quenching, illumination, or carrier injection. A strong indication that hydrogen is indeed mediating the relaxation towards thermal equilibrium in a-Si:H has recently been obtained by a careful comparison of the time-dependence for the annealing of excess defect densities with the time-dependence of the hydrogen diffusion coefficient.

At first glance it may seem puzzling to speak of the time-dependence of a diffusion coefficient $D(t)$; however, a closely related phenomenon was described earlier in the discussion of photocarrier drift-mobility $\mu(t)$ in non-crystalline materials. In general one may consider the distribution of *waiting-times* which characterize the release of protons from Si-H bonds; if there is a broad *dispersion* in these times (due to variations in the release energy or in the distance for tunneling to a neighboring site), the diffusion coefficient will be time-dependent. Fig. 19 shows the time-dependent proton diffusion coefficient $D(t)$ estimated from the profile of an originally sharp deuterium-hydrogen interface in boron-doped a-Si:H as it evolved in time at 200 C. A power-law decay is observed [82]:

$$D(t) = D_0(\nu t)^{-(1-\beta)} \tag{33}$$

Power-law decays are reminiscent of the form of the photocarrier drift-mobility decay in non-crystalline materials (cf. Section III.B), where dispersion was discussed for electronic (as opposed to the present atomic) transport processes.

Also shown in Fig. 19 is the form for the decay of a density $\Delta N(t)$ of light-induced D^0 defects generated by prolonged illumination at room-temperature of an undoped a-Si:H specimen. The density was measured using electron spin resonance. The residual (not light-induced) density of D^0 centers remaining after prolonged annealing has been subtracted, and the measurements normalized by the excess density at the beginning of annealing $\Delta N(0)$. The data fit a form for decay known as a *stretched-exponential*:

$$\Delta N(t)/\Delta N(0) = e^{-(t/\tau)^{\beta}} \tag{34}$$

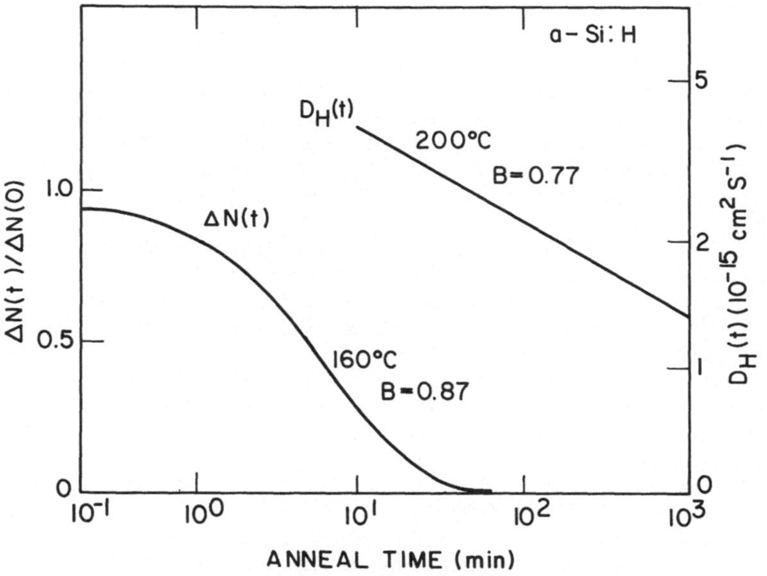

Fig. 19. Dispersive diffusion of hydrogen in doped a-Si:H, and stretched exponential decay of light-induced defects in undoped a-Si:H. The diffusion coefficient D_H for hydrogen measured using deuterium concentration depth profiling decreases with time as a weak power law (after Kakalios, *et al* [82]). The density of light-induced D^0 defects detected using ESR ΔN decays after the onset of annealing as a stretched exponential (after Jackson and Kakalios [83]). The dispersion parameters β for the two processes are similar, suggesting a possible causal relationship.

This stretched-exponential form for the decay of a metastable defect configuration can be argued [82] to result from proton mediated annealing events if the proton $D(t)$ has a power-law decay; the rough agreement of the *dispersion* parameters β obtained from the two independent experiments shown in Fig. 19 does indeed suggest a relationship of the two processes. Similar arguments invoking proton diffusion have been made to explain the annealing kinetics of other metastable electronic properties in a-Si:H, in particular the difference between the densities of P_4 and D defects in doped a-Si:H, and the reader is referred to references [81-83] for further discussion.

ACKNOWLEDGMENTS

I wish to take this occasion to remember David Adler's inspiring insight into physics and into amorphous semiconductors, which he coupled with an admirable talent for lucid expression. The writing of these lectures was greatly facilitated by conversations with P. A. Dowben and M. Silver, by the hospitality of J. Tauc and of Brown University, and by the discussions with and assistance of H. Antoniadis, K. A. Conrad, M. A. Parker, and S. Zafar. This work was supported by the U. S. Solar Energy Research Institute through grant XB-6-06005-2.

REFERENCES

1. J. I. Pankove, volume editor for *Semiconductors and Semimetals* **21A-D**: *Hydrogenated Amorphous Silicon* (Academic Press, Orlando, 1984; Series Editors R. K. Willardson and A. C. Beer).
2. J. D. Joannopoulos and G. Lucovsky, editors of *The Physics of Hydrogenated Amorphous Silicon, Volume II* (Springer-Verlag, Berlin, 1983).
3. N. F. Mott and E. A. Davis, *Electronic Processes in Non-Crystalline Materials, Second Edition* (Oxford University Press, Oxford, 1979).
4. R. Zallen, *The Physics of Amorphous Solids* (Wiley, New York, 1983).
5. N. F. Mott, *Conduction in Non-Crystalline Materials* (Oxford University Press, Oxford, 1987).
6. M. A. Kastner, G. A. Thomas, and S. R. Ovshinsky, editors, *Disordered Semiconductors* (Plenum Press, New York, 1987).
7. M. H. Brodsky, editor, *Amorphous Semiconductors, Second Edition* (Springer-Verlag, Berlin, 1985).
8. W. B. Jackson, S. M. Kelso, C. C. Tsai, J. W. Allen, and S. J. Oh, *Phys. Rev. B* **31**, 5187 (1985).
9. L. Ley, collected in ref. 1, p. 61.
10. N. F. Mott, *Reports on Progress in Physics* **46**, 909 (1984).
11. N. W. Ashcroft and N. D. Mermin, *Solid State Physics* (Holt, Rinehart and Winston, New York, 1976).
12. C. R. Brundle and A. D. Baker, editors, *Electron Spectroscopy - Theory, Techniques, and Applications* (Academic Press, New York, series 1977 -).
13. J. B. Pendry and J. F. L. Hopkinson, *J. Phys. F: Metal Phys.*, **8**, 1009 (1978).
14. S. C. Moss and J. F. Graczyk, *Proc. 10th Int. Conf. on the Physics of Semiconductors, Cambridge, Mass.*, edited by S. P. Keller, J. C. Hensel, and F. Stern (United States Atomic Energy Commission, Washington, D. C., 1970), p. 658.
15. G. Lucovsky and T. M Hayes, collected in ref. 7, p. 215.
16. M. A. Kastner, *J. Non-Cryst. Solids* **77&78**, 1173 (1985).
17. D. M. Burland and L. B. Schein, *Physics Today* **39**, No. 5, 46 (1986).
18. D. Adler, H. K. Henisch, and N. F. Mott, *Rev. Mod Phys.* **50**, 203 (1978).
19. T. M. Donovan and K. Heinemann, *Phys. Rev. Lett.* **27**, 1794 (1971).
20. M. H. Brodsky and R. S. Title, *Phys. Rev. Lett.* **23**, 581 (1969).
21. W. E. Spear and P. G. LeComber, *Solid State Comm.* **17**, 1193 (1975).
22. W. E. Spear and P. G. LeComber, *J. Non-Cryst. Solids* **8-10**, 727 (1972).
23. D. E. Carlson and C. R. Wronski, *Appl Phys. Lett.* **28**, 671 (1975).
24. Y. Hamakawa, P. G. LeComber, A. Madan, P. C. Taylor, and M. J. Thompson, editors, *Amorphous Silicon Technology* (Materials Research Society, Pittsburg, 1988).
25. J. Fortner, J. S. Lannin, and R. Fainchtein, *Bull. Am. Phys. Soc.* **33**, 474 (Abstract Only).

26. K. K. Gleason, J. Baum, A. N. Garroway, A. Pines, and J. Reimer, collected in *Materials Issues in Amorphous-Semiconductor Technology*, edited by D. Adler, Y. Hamakawa and A. Madan (Materials Research Society, Symposia Proceedings Vol. 70, Pittsburg, 1986), p. 83.
27. S. P. McGlynn, *Introduction to Applied Quantum Chemistry* (Holt, Rinehart, and Winston, New York, 1972).
28. R. E. Ballard, *Photoelectron Spectroscopy and Molecular Orbital Theory* (Adam Hilger, Bristol, 1978).
29. A. W. Potts and W. C. Price, *Proc. R. Soc.* **A326**, 181 (1972).
30. N. J. Shevchik, J. Tejeda, M. Cardona, and D. W Langer, *Solid State Comm.* **12**, 1285 (1973).
31 V. Heine, collected in *Solid State Physics, Volume 35*, edited by H. Ehrenreich, F. Seitz, and D. Turnbull (Academic Press, New York, 1980), p. 1.
32. U. Gelius, collected in *Electron Spectroscopy*, ed. by D. A. Shirley (North-Holland, Amsterdam, 1972), p. 311.
33. L. Ley, S. P. Kowalczyk, R. Pollak, D. A. Shirley, *Phys. Rev. Lett.* **29**, 1088 (1972).
34. J.-T. J. Huang and F. O. Ellison, *J. Elect. Spectroscopy and Rel. Phen.* **4**, 233 (1974).
35. D. Allan and J. D. Joannopoulos, collected in ref. 2, p. 5.
36. W. A. Harrison, *Electronic Structure and the Properties of Solids* (W. H. Freeman, San Francisco, 1980).
37. K. C. Pandey, *Phys. Rev. B* **14**, 1557 (1986).
38. E. N. Economou, *Green's Functions in Quantum Physics, Second Edition* (Springer-Verlag, Berlin, 1983).
39. B. I. Shklovskii and A. L. Efros, *Electronic Properties of Doped Semiconductors* (Springer-Verlag, Berlin, 1984)..
40. W. Beyer and H. Overhof, collected in ref. 1, Vol. **21C**, p. 258.
41. D. J. Dunstan, *J. Phys. C* **30**, L419 (1982).
42. J. Tauc, R. Grigorovici, A. Vancu, *Phys. Stat. Solidi* **15**, 627 (1966).
43. R. A. Street and D. K. Biegelsen, collected in ref. 2, p. 195.
44. F. Urbach, *Phys. Rev.* **92**, 1324 (1953).
45. G. Cody, collected in ref. 1, Vol. **21B**, p. 11.
46. J. D. Dow and D. Redfield, *Phys. Rev. B* **5**, 594 (1972).
47. S. Griep and L. Ley, *J. Non-Cryst. Solids* **59/60**, 253 (1983).
48. V. Sa-Yakanit and H. R. Glyde, *Comments Cond. Mat. Phys.* **13**, 35 (1987).
49. E. N. Economou, C. M Soukoulis, M. H. Cohen, and S. John, collected in ref. 6, p. 681 (1987).
50. E. A. Schiff, collected in ref. 24, *in press* (1988).
51. Don Monroe and M. Kastner, *Phys. Rev. B* **33**, 8881 (1986).
52. G. Pfister and H. Scher, *Adv. in Phys.* **27**, 747 (1978).
53. T. Tiedje, collected in ref. 2, p. 261.
54. J. Orenstein, M. Kastner, and V. Vaninov, *Phil. Mag.* **B46**, 23 (1982).
55. T. Tiedje, B. Abeles, and J. M. Cebulka, *Solid State Comm.* **47**, 493 (1983).
56. J. Stuke, *J. Non-Cryst. Solids* **97&98**, 1 91987).
57. E. A. Davis, *Phil. Mag. B* **38**, 463 (1978).
58. W. E. Spear, G. Willeke, P. G. LeComber, and A. G. Fitzgerald, *Journale de Phys.* **C4**, 257 (1981).
59. H. Overhof and W. Beyer, *Phil. Mag. B* **47**, 377 (1983).
60. D. Emin, collected in ref. 6, p. 751 (1987).
61. M. Grunewald, P. Thomas, and D. Wurtz, *J. Phys. C* **14**, 4083 (1981).
62. D. Monroe, *Phys. Rev. Lett.* **54**, 146 (1985).
63. A. Madan, P. G. LeComber, and W. E. Spear, *J. Non-Cryst. Solids* **20**, 239 (1976).
64. M. Kastner, D. Adler, and H. Fritzsche, *Phys. Rev. Lett.* **37**, 1504 (1976).

65. M. Lannoo and J. Bourgoin, *Point Defects in Semiconductors Volumes I & II* (Springer-Verlag, Berlin, 1984).
66. Y. Bar-Yam and J. D. Joannopoulos, *Phys. Rev. Lett.* **56**, 2203 (1986).
67. R. A. Stradling, collected in ref. 6, p. 57.
68. S. Pantelides, collected in ref. 24, *in press*.
69. H. Fritzsche, collected in *Electronic and Structural Properties of Amorphous Semiconductors: Proceedings of the 13th Session of the Scottish Universities Summer School in Physics*, edited by P. G. LeComber and J. Mort (Academic Press, New York, 1977), p. 55.
70. P. W. Anderson, *Phys. Rev. Lett.* **34**, 953 (1975).
71. R. A. Street and N. F. Mott, *Phys. Rev. Lett.* **35**, 1293 (1975).
72. S. G. Bishop and P. C. Taylor, *Phys. Rev. B* **15**, 2278.
73. M. A. Kastner, *J. Non-Cryst. Solids* **77&78**, 1173 (1985).
74. R. C. Frye and D. Adler, *Phys. Rev. Lett.* **46**, 1027 (1981); see also D. Adler, *Jour. de Phys.* **C-4**, No. 10, 3 (1981).
75. J. C. Knights, T. M. Hayes, J. C. Mikkelsen, Jr., *Phys. Rev. Lett.* **39**, 712 (1977).
76. R. A. Street, *J. Non-Cryst. Solids* **77&78**, 1 (1985).
77. D. Adler, M. Silver, M. P. Shaw, and V. Canella, collected in *Materials Issues in Amorphous-Semiconductor Technology* (Materials Research Society, Symposia Proceedings Vol. 70, Pittsburg, 1986), p. 113.
78. D. Adler, *Physics of Amorphous Semiconductor Devices*, edited by D. Adler (SPIE - The International Society for Optical Engineering, Proceedings Vol. 763, Bellingham, Washington, U. S. A.), p. 2.
79. R. A. Street, *Phys. Rev. B* **37**, 4209 (1988).
80. B. L. Stafford and E. Sabisky, editors, *Stability of Amorphous Silicon Alloy Materials and Devices* (American Institute of Physics, Conference Proceedings Vol. 157, New York, 1987).
81. R. A. Street, J. Kakalios, C. C. Tsai, and T. M. Hayes, *Phys. Rev. B* **35**, 1316 (1987).
82. J. Kakalios, R. A. Street, and W. B. Jackson, *Phys. Rev. Lett.* **59**, 1037 (1987).
83. W. B. Jackson and R. A. Street, *Phys. Rev. B* **37**, 1020 (1988).

EXCITATIONS AND PHASE TRANSITIONS OF DISORDERED MAGNETIC SYSTEMS

R.A. Cowley

Department of Physics
University of Edinburgh
Mayfield Road
Edinburgh EH9 3JZ
Scotland, United Kingdom

ABSTRACT

Randomly mixed transition metal fluorides are ideal systems with which to study the excitations and phase transitions of disordered systems. The crystal structures and magnetic interactions are well known and relatively simple, while different materials are magnetically two-or three-dimensional and have effectively Ising or Heisenberg interactions. The excitations of several of these mixed systems have been studied and the results show the importance of the Ising cluster modes. The results are consistent with computer simulations, but the coherent potential approximation does not describe the results in detail. The usefulness of fractons as a description of the results is still not established. The phase transitions of systems with well defined magnetic ground states are now well understood both experimentally and theoretically. The results for d = 2 and 3 Ising models are described. Close to the percolation point, a multicritical model based on geometric and thermal disorder and the importance of one-dimensional weak links describes the results for concentrations less than the percolation concentration. There are a number of unexplained results for concentrations above percolation. Random fields have a dramatic effect on the structure and phase transitions of magnetic systems. Experiments to investigate the resulting metastability and possible phase transitions are described, but the results are still not understood in detail.

I. INTRODUCTION

Transition metal fluorides and chlorides are ideal materials with which to study the properties of disordered systems. The transition metal ions Mn^{++}, Co^{++}, and Zn^{++} are chemically very similar and of very similar ionic radius. Consequently, single crystals can be grown in which these ions are distributed randomly over the transition metal sites. Furthermore, since these ions are magnetically very different, the study of the magnetic excitations and the magnetic phase transitions enables us to study in detail the excitations and phase transitions in randomly disordered systems. The comparison of experimental results and theory is further facilitated by the simple crystal structures, and the fact that the magnetic interactions are largely limited to nearest neighbours and are of known strength. There are then few, if any, unknown parameters which are needed before theory is confronted by the experimental results.

These 3d transition metal salts crystallise in a variety of different crystal structures. Many of the measurements described below have however been performed using materials having either the rutile or K_2NiF_4 structure. MnF_2 and CoF_2 crystallise in the rutile structure in which the transition metal ions are arranged on a body-centred tetragonal lattice. The largest magnetic interaction is between the ions at the corner sites and those at the body-centre sites and this is antiferromagnetic. There is a weaker ferromagnetic interaction between the nearest neighbours along the tetragonal c-axis. The low temperature magnetic structure of both materials is ferromagnetic with the magnetic moments aligned along the c-axis. The magnetic interactions are known from measurements of the spin waves in both MnF_2 [1] and CoF_2 [2]. In MnF_2 the exchange interactions are of Heisenberg character and antiferromagnetic between the ions at the corner and body centred neighbours, Δ;

$$H_{EX} = \frac{1}{2} J \sum_i \sum_\Delta \underline{S}_i \cdot \underline{S}_{i+\Delta} \quad (1)$$

The weaker dipolar interactions can be represented by a weak anisotropy field;

$$H_D = -H_A \sum_i (S_i^z)^2 \quad (2)$$

The magnetic properties of CoF_2 are more complicated because the Co^{++} ion has orbital and spin angular momentum [2]. The ground state of the ions is however a Kramers doublet which can be treated as an effective spin, $S = \frac{1}{2}$, but the exchange interactions are then anisotropic;

$$H_{EX} = \frac{1}{2} J \sum_i \sum_\Delta S_i^z S_{i+\Delta}^z + \alpha\left(S_i^x S_{i+\Delta}^x + S_i^y S_{i+\Delta}^y\right), \quad (3)$$

where the constant $\alpha = 0.81$. Due to the weakness of H_D compared with H_{EX}, the properties of MnF_2 are close to those of a $d = 3$ isotropic system, while the anisotropy of CoF_2 makes its properties closer to those of a $d = 3$ Ising system. Both materials can be grown mixed with ZnF_2 and since Zn^{++} is non-magnetic this dilutes the magnetic lattice.

The other structure is the K_2NiF_4 structure in which the transition metal ions are arranged on a square two-dimensional arrangement which is well separated, by the KF, from the other transition metal planes. The result is that many of the magnetic properties are then largely of two-dimensional character, although, in nearly all of the pure materials, the magnetic planes are ordered with respect to one another at low temperatures. There are a variety of different properties depending on the transition metal ions and the magnetic interactions are known from measurements of the spin wave spectrum. K_2MnF_4 [3] and K_2CoF_4 [4] are antiferromagnets with the spins aligned along the c-axis perpendicular to the magnetic planes. The magnetic interactions are similar to those of MnF_2 and CoF_2: in K_2MnF_4 the nearest neighbour spins interact through isotropic antiferromagnetic exchange interactions, eqn 1, together with a weak single ion anisotropy, eqn 2, while in K_2CoF_4 the exchange interactions are anisotropic, eqn 3. In K_2FeF_4 [5] the magnetic structure is different: the structure is antiferromagnetic but the spin direction lies in the magnetic sheets. Finally Rb_2CrCl_4 [6] has a similar crystallographic structure, but the magnetic structure is of ferromagnetic sheets with the spins aligned within the sheets. Clearly changing and mixing the different transition metal ions enables the behaviour of a wide variety of different systems to be studied with dimensionalities of 2 or 3, and with almost isotropic or very strongly anisotropic exchange interactions.

Single crystals of these mixed systems are grown by carefully purifying the starting materials and mixing them in the desired proportions. The crystals are then grown by either the Stockbarger or Bridgman techniques as for the pure materials. One of the very important advances in recent years has been the growth of large single crystals in which the concentration of the constituents is known, and more importantly uniform to about 0.1% over the whole of a macroscopic specimen, while the crystallographic quality is also excellent with a mosaic width of less than 0.02°. This development in the crystal growth has been essential for the detailed experiments on the phase transitions, described in sections III-V.

The properties of these disordered systems have been studied by a variety of techniques; specific heat measurements, capacitance, birefringence, Raman scattering, and neutron scattering. Neutron scattering has provided the most detailed information, and so most of the measurements described below were performed with this technique. Inelastic neutron scattering enables the spin correlation functions to be measured [7] as a function of the wavevector and frequency. These spin correlation functions then provide information about the effect of the disorder on the magnetic excitations. This work is reviewed in the next section. The elastic neutron scattering enables the magnetic structure, and the details of the phase transitions to be investigated. In section III, the effect of disorder on the phase transition is described when the disorder does not introduce frustration into the ground state - the strength of the magnetic interactions is disordered but not the sign.

When the magnetic lattice is diluted there is a critical concentration of magnetic sites below which long range magnetic order cannot be formed. This is known as the percolation concentration and the behaviour close to that concentration is described in section IV. In section V we describe the results obtained for systems in which the disorder introduces some frustration into the system - the results are then much more controversial both experimentally and theoretically. Throughout the article emphasis is placed on the experimental results, and the theory is described only in so far as it is necessary to understand the experiments. References are given to more extended descriptions of the theory.

II. EXCITATIONS OF DISORDERED SYSTEMS

II.A. Measurements on Unfrustrated Systems

Systems in which there is randomness but no frustration in the magnetic interactions have simple ground states. For example, both MnF_2 and CoF_2 have antiferromagnetic structures with the spins aligned along the c-axis, and in the mixed system the low temperature structure is a nearly fully aligned similar antiferromagnetic structure. The behaviour of the excitations in these systems is now fairly well understood. When there are two quite different types of magnetic ions, the excitation spectrum consists of two distinct branches each of which corresponds to the excitations propagating largely on one or other of the two different types of ions. We illustrate this behaviour in Fig 1 by showing measurements of the excitations of the $Mn_cCo_{1-c}F_2$ system [8], and similar results have been obtained for $KMn_cCo_{1-c}F_3$ [9], $Rb_2Mn_cNi_{1-c}F_4$ [10] and $K_2Fe_cCo_{1-c}F_4$ [11].

The existence of two bands can be understood on the basis of a simple Ising cluster model for the excitations. If the exchange constant between spins S_λ and $S_{\lambda'}$ is $J_{\lambda\lambda'}$, then the Ising energy needed to create a spin deviation on a site of type λ surrounded by r neighbours of type λ and (Z - r) neighbours of type λ' is:

$$E_\lambda(r) = r\, J_{\lambda\lambda}\, S_\lambda + (Z - r)\, J_{\lambda\lambda'}\, S_{\lambda'} \tag{4}$$

The two energies, $E_\lambda(r)$ and $E_{\lambda'}(r)$, then give a good approximation to the observed zone boundary energies as listed in table 1, where the $J_{\lambda\lambda}$ have been taken from spin wave measurements on the pure materials and $J_{\lambda\lambda'} = (J_{\lambda\lambda}\, J_{\lambda'\lambda'})^{\frac{1}{2}}$, while r is the mean number of neighbours if the concentration of type λ is c; $r = cZ$.

Extensive measurements have also been made of the excitations of the diluted systems; $Mn_cZn_{1-c}F_2$ [13, 14, 15], $Co_cZn_{1-c}F_2$ [16], $Rb_2Co_cMg_{1-c}F_4$ [17] and $Rb_2Mn_cMg_{1-c}F_4$ [18], for which the observed neutron

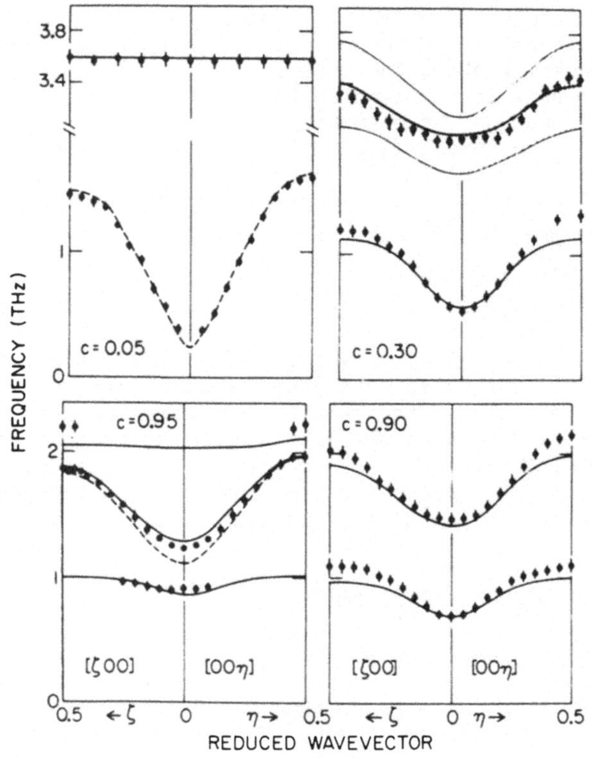

Fig. 1. Spin waves excitations in $Co_cMn_{1-c}F_2$ compared with coherent potential aproximation calculations [8]. The shaded area for c = 0.30 gives the full width of the excitations.

Table 1. Ising Cluster Energies of Mixed Antiferromagnets

System	c	Calculated (THz)	Observed (THz)
$Mn_cCo_{1-c}F_2$ [8]	0.95	3.46	3.57±0.05
		1.51	1.49±0.02
	0.30	2.42	2.32±0.10
		1.20	1.20±0.06
	0.10	2.05	2.02±0.08
		1.05	1.09±0.08
$KMn_cCo_{1-c}F_3$ [9]	0.80	6.96	6.55±0.15
		2.27	2.26±0.02
	0.29	7.04	6.80±0.10
		2.30	2.30±0.10
$KMn_cNi_{1-c}F_3$ [12]	0.25	11.70	11.80±0.50
		3.12	3.10±0.15
$Rb_2Mn_cNi_{1-c}F_4$ [10]	0.5	6.81	6.48±0.20
		1.92	1.79±0.05

scattering is shown for several wavevectors in fig 2. In all of these systems the widths of the excitations tend to be broader than in the mixed magnetic systems, and furthermore high resolution results reveal fine structure as shown in fig 2, where for the zone boundary wavevector, Q = (0.5,0,2.9), there are 4 peaks as a function of energy transfer. The origin of these peaks can again be understood, at least qualitatively, in terms of the Ising cluster modes. In $Rb_2Mn_cMg_{1-c}F_4$ there are 4 nearest neighbour ions and so there will be Mn^{++} ions surrounded by r = 0,1,2,3 or 4 other Mn^{++} ions. Each of the different environments will have a different energy, $E_\lambda(r)$, and the 4 peaks correspond to the 4 different energies with r = 1,2,3 or 4. Furthermore the probabilities of these different

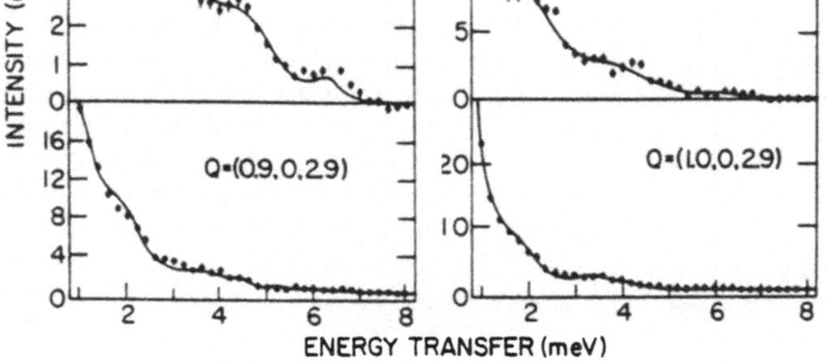

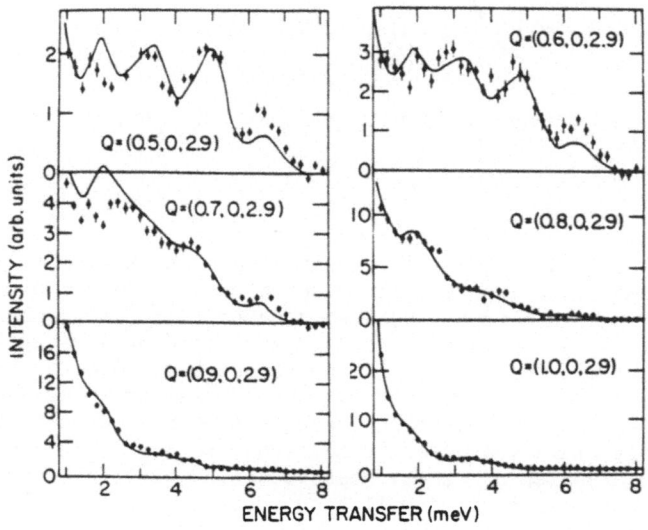

Fig. 2. Scattered neutron distributions [18] from $Rb_2Mn_{0.5}Mg_{0.5}F_4$ at 4.0 K. Q = (0.5, 0, 2.9) corresponds to the zone boundary of the magnetic Brillouin zone and Q = (1.0, 0, 2.9) the zone centre. The solid lines are the results of computer simulations.

environments gives a very reasonable account of the different intensities of the peaks. As the wavevector tends to the centre of the zone, the weight of the scattering moves to lower frequency transfers, but there are still vestiges of the Ising cluster modes. This behaviour cannot be understood by the simple Ising model. Similar results have been obtained [17] for the $Rb_2Co_cMg_{1-c}F_4$ system except that this system is a good Ising model and so the Ising cluster modes are more distinct and the results are largely independent of the wavevector.

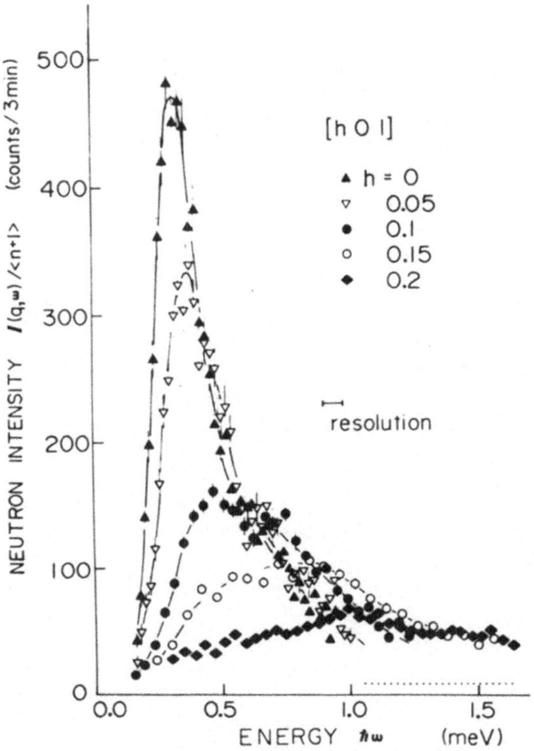

Fig. 3. Scattered neutron distributions from $Mn_{0.5}Zn_{0.5}F_2$ at wavevectors Q = [h01] [15].

There have been a succession of experiments on the $Mn_cZn_{1-c}F_2$ system with steadily improving experimental resolution. Since there are 8 ferromagnetic neighbours in the rutile structure, the Ising cluster effects are less well defined. Nevertheless they have been observed [14,15] close to the zone boundary. Close to the zone centre there are fairly well defined spin waves for c = 0.5 [15], which increase in energy as the wavevector increases. The width of these spin waves apparently, fig 3, increases rapidly mid-way across the Brillouin zone when the results become more similar to a series of barely resolved Ising cluster modes. A possible interpretation of these results will be discussed in section D.

II.B. Molecular Dynamics

The excitations of disordered systems can be calculated by using molecular dynamics or the equation of motion technique, as developed by

Thorpe and Alben [19], Alben et al [20], and by Buyers [21]. The equation of motion of the spin correlation functions are evaluated within the spin wave approximation, when S_λ^z is replaced by S_λ. The spin correlation functions are known from their definition at time t = 0, so they can be evaluated at later times by numerical integration of the equations of motion. The time dependence of the spin correlation functions can then be Fourier transformed to give the neutron scattering cross-section which is then compared directly with the experimental results. There are a number of variants of the method depending on exactly which quantities are required, and some experience is needed to choose an appropriate lattice size, time step, and apodizing function, to obtain reliable results in a reasonable computer time.

The agreement obtained between the molecular dynamics and the experimental results is in every case excellent, as illustrated in fig 2 for $Rb_2Mn_cMg_{1-c}F_4$. Similar agreement has also been obtained for the systems with two different types of magnetic ions, and more recently [22] with the results shown in fig 3 for $Mn_cZn_{1-c}F_2$.

These results show that the spin wave approximation, and the model for the exchange interactions in these disordered systems, are both satisfactory. This is despite the work of Halley and Holcomb [23] who showed that the zero point spin deviations are considerably larger in the d = 2 disordered system than in the analogous pure systems. These zero point spin deviations furthermore depended on the environment of the spin. These complications do not presumably significantly alter the observed spectra at least in part because the increased spin deviation can be compensated by increasing the effective exchange constant.

II.C. Coherent Potential Approximation

Despite the success of the molecular dynamics technique in giving agreement with the experimental results, there is still a need for an analytic description. In the 1970's the coherent potential approximation was developed to describe the electronic excitations in random alloys. This technique has been taken over to apply [24] to the magnetic systems but there are a variety of difficulties and problems.

The simplest form of the coherent potential approximation introduces a site diagonal self-energy for the spin waves which has both real and imaginary parts. This self-energy is then determined self-consistently by the requirement that on average there is no scattering from the disorder away from this average potential. In practice there are complications in implementing this procedure. Firstly the success of the cluster Ising model suggests that these Ising frequencies should be used [24] for the site-diagonal part of the self-energies and so there are correlations between the neighbouring sites. Secondly the exchange interactions are not site diagonal due to the $S_i^x S_j^x$ and $S_i^y S_j^y$ terms, and the disorder in these terms cannot be readily included. In the $Mn_cCo_{1-c}F_2$ system, fig 1, these transverse parts of the exchange interactions are, by chance, the same for Mn-Mn and Co-Co interactions. Consequently in this case the site diagonal coherent potential approximation can be evaluated and the results, fig 1 , give a very reasonable description of the measurements.

The approach is less satisfactory for other systems especially for the dilute systems with nearly Heisenberg or isotropic interactions. This is because the site diagonal coherent potential approximation erroneously breaks the isotropic symmetry [24], and various schemes have been used to fix up the resulting bad behaviour of the results. A further difficulty arises [25] in the dilute systems because the deficiencies of the coherent potential approximation allow some of the magnetic excitation to propagate on the non-magnetic sites. Again various ways of compensating for this in an ad hoc way have been developed. In view of these difficulties, the fact that the approximation converged slowly - it took more computer time than the equation of motion method - and that it gave unsatisfactory results [18], led to the coherent potential approximations being essentially abandoned since 1980 for describing magnetic excitations in disordered systems. It is, however, still used for electronic excitations in disordered alloys for which experiments provide a much less stringest test of the theory!

II.D. Fractons

An alternative approach to the theory of excitations in disordered systems was developed by Alexander and Orbach [26]. The theory arose from considering systems, such as percolation clusters, which for length scales, L, less than some correlation range, ζ, are fractal which means that the mass of the system scales not as L^d, where d is the Euclidean dimensionality of the system, but as L^D, where D is the fractal dimensionality. The dynamics of spins situated on these clusters is then known to be anomalous on length scales $L < \zeta$. In particular the elastic properties were predicted to be normal phonons for wavelengths $\lambda > \zeta$, but localised states called fractons for $\lambda < \zeta$. A similar theory is expected to be appropriate for magnetic excitations. The theory then predicts that there will be a crossover in behaviour, between conventional long wavelength behaviour in which the propagating excitations with a relatively small damping have a density of states $N(\omega) \sim \omega^{d-1}$, and the localised functions with a density of states $N(\omega) \sim \omega^{1/3}$ and with a damping comparable with their frequency [27]. Furthermore, thermal conductivity measurements on amorphous materials and glasses have been interpreted as giving evidence for the excitations being fractons even though these disordered systems are not fractal in the sense described above.

This theory then provides at least a qualitative account of the results shown in fig 3: the rapidly increasing width of the excitations mid-way across the zone is interpreted as the crossover between propagating and fracton-like excitations. This is of necessity only a qualitative conclusion at present, because further experiments at concentrations closer to the percolation concentration $c_p \sim 0.25$ are needed to find whether the cross-over wavevector does scale with the percolation correlation length. Unfortunately neutron scattering measurements cannot tell whether or not the excitations are localised or propagating. The validity of the fracton model might equally well be tested by numerical calculations, and evaluating the eigenvectors of the spin wave normal modes. If the excitations are then found to be localised above a certain critical frequency, and to follow the predicted density of states, this will provide striking evidence for the usefulness of the fracton model.

II.E. Systems with Frustrated Ground States

Much less work has been performed on the excitations of systems with frustrated or highly disordered ground states. Experimentally there are two different types of system available. Firstly there are the systems with competing anisotropies in which the anisotropies favour alignment of one of the magnetic ions along the c-axis and the other in the perpendicular plane. At suitable concentrations the moments are then aligned in an intermediate direction but there is considerable randomness in the directions of the different spins [28] as shown in Fig 4. Examples of these systems are $Fe_cCo_{1-c}C\ell_2$ and $K_2Fe_cCo_{1-c}F_4$. There have not as yet been detailed measurements of the spin waves in these systems but some preliminary measurements have been made [29], and better crystals are needed before further progress can be made. The other type of system is $Rb_2Mn_cCr_{1-c}C\ell_4$, which has not only competing anisotropies but also competing exchange constants; the Mn-Mn constant is antiferromagnetic, the Mn-Cr constant weakly antiferromagnetic but the Cr-Cr constant ferromagnetic. At intermediate concentrations there is then a disordered spin glass phase [30]. Detailed measurements have been made of the magnetic excitations in a sample with c = 0.75 [31], which is just antiferromagnetic. The results are shown in fig 5 and there are well defined Ising cluster modes associated with the different possible environments of the Mn ions, and at lower frequencies a propagating spin wave branch. These results are qualitatively similar to those on the diluted Mn systems, figs 2 and 3, possibly because the Cr ions are only relatively weakly coupled to the Mn ions. Further work with different concentrations and in particular in the spin glass phase will be very useful and informative.

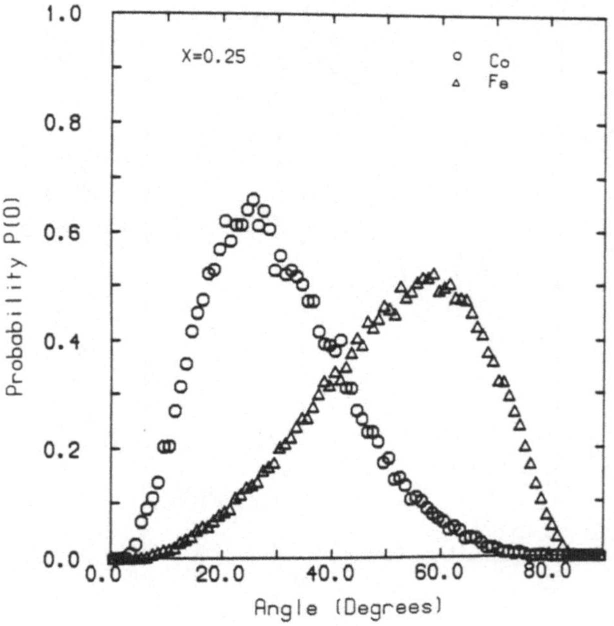

Fig. 4. Distribution of the angle of Co and Fe spins from the c-axis in $K_2Co_{0.25}Fe_{0.75}F_4$ [28].

The theory of the excitations in these highly disordered systems is complex. As yet the analytic theories have used some form of averaging to replace the highly disordered ground state, fig 4, with a more ordered and tractable state. Numerically the most detailed calculations were performed for the $K_2Fe_cCo_{1-c}F_4$ system [28,32] by first finding the equilibrium direction of each of the spins, fig 4, by an iterative procedure, and then using this disordered ground state as the starting point for the equation of motion method of calculating the spin waves. The results are disappointing in that qualitatively they are very similar to those of the mixed antiferromagnets shown in fig 1.

III.A. Theory

In this section we consider the phase transitions in systems for which there is no frustration and the ground state is well defined. The exchange or anisotropy constants of the constituents have the same sign but may differ in magnitude. Harris [33] considered whether the disorder would destroy the sharp phase transition by comparing the fluctuation in transition

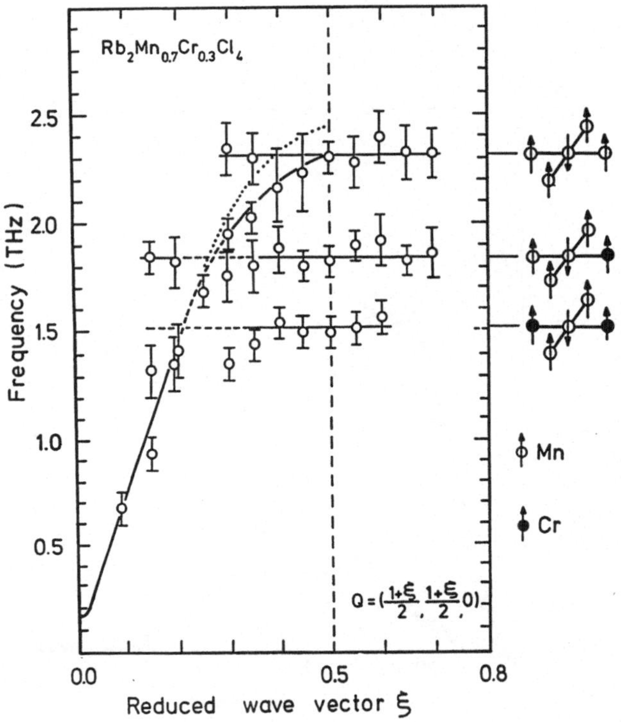

Fig. 5. The frequencies of peaks in the scattering from $Rb_2Mn_{0.75}C\ell_{0.25}F_4$ [31]. Note the Ising cluster modes and the propagating spin waves.

temperature within a correlation volume $\frac{\Delta T_c}{T_c} \sim \zeta^{-\frac{d}{2}}$ with $\frac{T - T_c}{T_c} \sim \zeta^{-\frac{1}{\nu}}$.

The phase transition is then well defined if $\Delta T_c < T - T_c$, giving $\frac{d}{2} > \frac{1}{\nu}$ or using the hyperscaling relation $d\nu = 2 - \alpha$ if $\alpha < 0$. The transition is then uneffected by the statistical fluctuations in the concentration if $\alpha < 0$, and so the exponents are expected to be the same for the disordered and pure systems.

If $\alpha > 0$, the behaviour is unknown, the transition might be smeared, of first order, or the exponents might change to new values for which $\alpha < 0$. Amongst the readily accessible system only the d = 3 Ising model has $\alpha > 0$, and for this system renormalisation group theory [34] has shown that there is a new random Ising fixed point, the universal properties of which differ from those of the pure system. These predictions have now been tested in detail.

III.B. The d = 3 Ising Model

Initially it was very difficult to test the above predictions on real systems because macroscopic fluctuations in the concentration gave a smearing of the critical properties in bulk crystals. Since 1982 the improved quality of the single crystals has enabled detailed experiments to be performed. Experiments have now been performed on two crystals in the $Fe_cZn_{1-c}F_2$ system [35,36] and on two crystals in the $Mn_cZn_{1-c}F_2$ system [37].

The critical neutron scattering was measured and the results fitted at each temperature to a Lorentzian form in wavevector so as to obtain the inverse correlation range, $\kappa(T)$ and the relative susceptibility, $\chi(T)$. The results were then fitted to;

$$\kappa(T) = \kappa_o^+ \;(T - T_N)^{\nu} \qquad T > T_N$$

$$= \kappa_o^- \;(T_N - T)^{\nu} \qquad T > T_N$$

$$\chi(T) = \chi_o^+ \;(T - T_N)^{-\gamma} \qquad T > T_N$$

$$\chi(T) = \chi_o^- \;(T_N - T)^{-\gamma} \qquad T < T_N$$

Some typical fits are shown in fig 6 for the susceptibility. In table 2 we collect together the final results of the different experiments and compare the results with the best theoretical results [38]. Clearly there are small but measurable differences in the exponents between the pure and random fixed points, but very much larger differences in the amplitude ratios.

Despite the apparent agreement between these results and theory, there are still two puzzling features. Firstly, the cross-over exponent from the pure to random behaviour is predicted to be $\frac{1}{\alpha}$ so that the reduced temperature range over which random Ising behaviour can be observed is expected to be $\left(\frac{\Delta J}{J}\right)^{\frac{1}{\alpha}} \sim 10^{-5}$. It is therefore surprising that the random behaviour is observed over a wide temperature range, fig 6. The second puzzling feature is that the random exchange constants give rise to fluctuations in the ordered moment below T_c, [39]. A simple theory then suggests that these will give elastic scattering of a Lorentzian squared form in wavevector. A Lorentzian squared term was not observed in the experiments [37].

III.C. The d = 2 Ising Model

The pure d = 2 Ising model has the specific heat exponents $\alpha = 0$. Consequently the critical phenomena of a random d = 2 system is expected to be the same as that of a pure d = 2 Ising system although one might expect different logarithmic corrections to the exponents. Calculations [40] do indeed give logarithmic corrections and possibly unusual behaviour of the susceptibility. Experiments have been performed on samples of

Table 2. Exponents and Amplitude Ratios for Random Ising Systems

		ν	κ_0^+/κ_0^-	γ	χ_0^+/χ_0^-
$Mn_{0.75}Zn_{0.15}F_2$	[37]	0.715±0.035	0.71±0.02	1.364±0.076	2.56±0.15
$Mn_{0.5}Zn_{0.5}F_2$	[37]	0.755±0.05	0.65±0.05	1.56±-0.16	2.2±0.2
$Fe_{0.46}Zn_{0.54}F_2$	[36]	0.69±0.1	0.69±0.02	1.31± 0.03	2.8±0.4
$Fe_{0.5}Zn_{0.5}F_2$	[35]	0.73±0.03	0.73±0.02	1.45± 0.06	2.2±0.1
Random Ising	[38]	0.70	0.83	1.39	2.2
Pure Ising		0.63	0.51	1.24	5.1

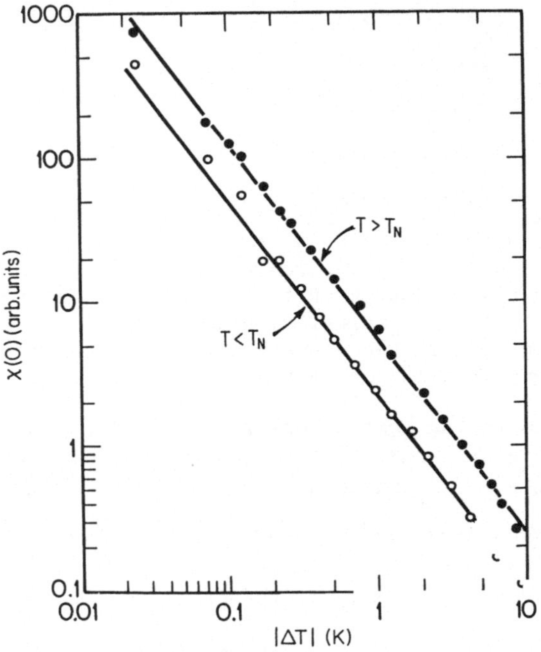

Fig. 6. The susceptibility $\chi(T)$ for $Mn_{0.75}Zn_{0.25}F_2$ [37]. The fits are to single power laws and give the exponents γ = 1.364 above and below T_N.

$Rb_2Mn_{0.5}Ni_{0.5}F_4$ [40] and on $K_2Co_{0.7}Mg_{0.3}F_4$ [41,42]. The initial experiments [40, 41] showed that the critical exponent, γ and ν, measured above T_c and the order parameter exponent β were consistent with the pure d = 2 Ising behaviour. They did not however study the critical scattering below T_c. In view of the large difference in the amplitude ratios between the pure and Ising d = 3 systems, table 2, the critical scattering in the d = 2 system was also measured below T_c [42]. When the results were analysed using the normal Lorentzian form of the scattering below T_c, amplitude ratios were obtained which were very different from those measured or calculated for the pure d = 2 system. This is because for the d = 2 Ising system the critical scattering below T_c has a different form from the usual Lorentzian form as discussed theoretically [43], and as found to be essential for a satisfactory analysis of experiments [44]. Unfortunately the shape of the critical scattering below T_N,

$$\chi(q,T) = \frac{\chi(0,T)(1+\phi y^2)^{\eta/2}}{(1-\lambda+\lambda(1+y^2)^{\frac{1}{2}})^2}$$

with $y = q/\kappa(T)$, contains the non-universal parameters, ϕ and λ. When these were held fixed at the values 0.45 and 0.16 obtained for the pure system, the amplitude ratios were significantly different from those of the pure material. When, however ϕ and λ were altered to 0.4 and 1.0 respectively very good agreement was obtained with the expected amplitude ratios. Clearly without a detailed and numerical value for the non-universal constants we can only conclude that the results are consistent with the theory of the d = 2 Ising model. In particular the exponents found were $\nu = 1.05\pm0.03$, $\gamma = 1.81\pm0.03$ and $\beta = 0.13\pm0.02$ in good agreement with theory: 1.0, 1.75 and 0.125 respectively.

More worrying is the dependence of the results on the form of the critical scattering. The failure of the Lorentzian form has been investigated in detail only for the d = 2 Ising model. Similar work, both theoretical and experimental, is needed for other systems to check if the Lorentzian form is indeed an adequate approximation for analysing critical scattering data.

IV. PERCOLATION

IV.A. Introduction

The subject of the percolation problem is illustrated in fig 7, which shows a square lattice with 50% of the sites occupied and only nearest neighbours connected by bonds. The system forms only finite clusters: the percolation concentration is reached when one of the clusters has infinite extent and this occurs when $c = c_p = 0.59$. In addition fig 7 shows that many of the clusters are highly ramified, and have weak one-dimensional links on the paths connecting different parts of the same cluster.

The percolation point is discussed as a multicritical point at which the long range magnetic order can be destroyed either thermally, by heating the system up, or geometrically, by further diluting the system. The theory of multicritical phenomena [45] then suggests that at T = 0:

$$\kappa \sim (c_p - c)^{\gamma_G} , \quad \chi \sim (c_p - c)^{-\gamma_G}$$

while for $c = c_p$:

$$\kappa \sim (\mu(T))^{\nu_T} , \quad \chi \sim (\mu(T))^{-\gamma_T}$$

where $\mu(T)$ is the appropriate one-dimensional temperature scaling field for the spin system with $\mu(T) = 2\exp(-2JS/k\,T)$ for Ising systems and

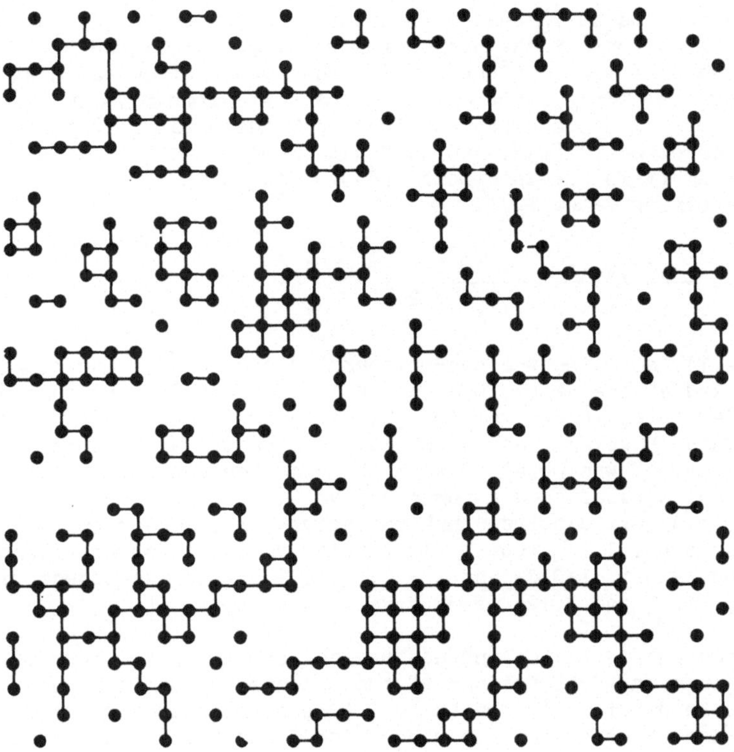

Fig. 7. Simulation of a d = 2 random square lattice with 50% of occupied sites. The nearest neighbour bonds are drawn in to show the nearest neighbour clusters.

$\mu(T) = \frac{k_B T}{JS}$ for isotopic systems. The theory of multicritical phenomena then suggests that the exponents ν_G and γ_G are related to the thermal exponents, γ_T and ν_T by a crossover exponent, ϕ, such that $\gamma_T = \gamma_G/\phi$ and $\nu_T = \nu_G/\phi$. The exponent ϕ was calculated [45] to be 1 for Ising systems.

IV.B. Experiments for $c > c_p$

The theory described briefly above was developed at least in part as a result of a series of experiments performed to measure the percolation phenomena in one-dimensional [47], two-dimensional [17, 48], and three-dimensional [49] systems. The results for the systems without long range order were analysed by fitting the critical scattering to the ubiquitous Lorentzian profile, and extracting the inverse correlation length and amplitude. It was unfortunately too difficult to grow crystals with the concentrations known sufficiently accurately to test the geometric exponents, ν_G and γ_G. Since, however, these were well known from series expansions and analytic theories, most of the experiments concentrated on measuring the temperature dependence. In fig 8 the results are shown for the inverse correlation length for several different concentrations c of $Rb_2Co_cMg_{1-c}F_4$. They clearly suggest that the inverse correlation length can be written as the sum of geometrical and thermal part:

$$\kappa(T, c_p - c) = \kappa_T(T) + \kappa_G(c_p - c).$$

This result was first deduced by Thorpe [50] for one-dimensional systems and since our experiments have been justified for higher dimensions by Stinchcombe [51]. The results also gave $\nu_T = 1.32\pm0.04$ in excellent accord with $\nu_G = 1.35\pm0.01$ and $\phi = 1.0$.

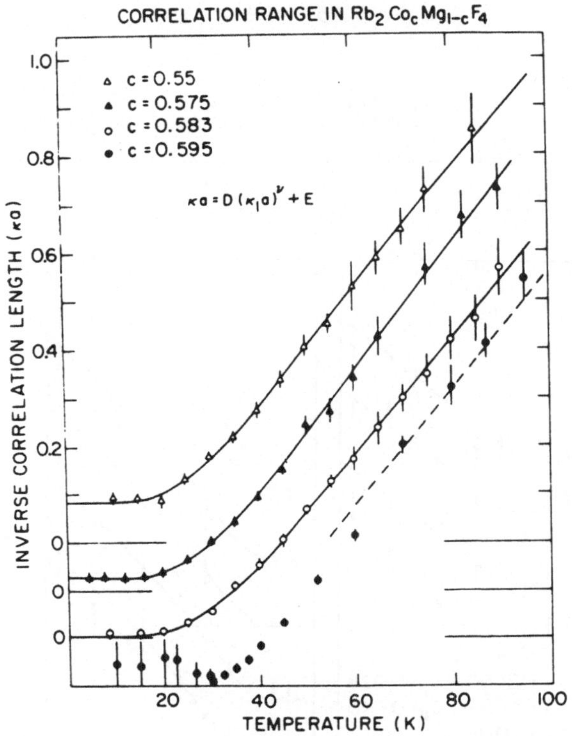

Fig. 8. The inverse correlation length κ for four different concentrations of $Rb_2Co_cMg_{1-c}F_4$. The fits are to the sum of a geometric and thermal term given the temperature dependence of the one-dimensional weak links $\mu(T)$ or $\kappa_1(T)$ [17].

The results for the more nearly isotropic system, $Rb_2Mn_cMg_{1-c}F_4$, were somewhat different [48]. The inverse correlation length was at least approximately the sum of a geometric and thermal part, but the exponent, ν_T, was 0.9 ± 0.03, leading to a crossover exponent $\phi = 1.5\pm0.1$. For several years this result was unexplained, but then Coniglio [52] pointed out that whereas in Ising models at low temperature the one-dimensional weak links do indeed dominate the temperature dependence, in isotropic systems the effects of two weak links in parallel yields a similar contribution, and the cross-over exponent is given by the electrical resistance behaviour from which he deduces that $\phi = 1.43$ in two dimensions in agreement with experiment.

Similar results were obtained for three dimensional systems for $c < c_p$ as shown in fig 9. The results for the different exponents are collected in table 3.

Table 3. Exponents for the Percolation Problem [49].

	d = 2	d = 3
ν_G	1.356±0.005	0.85±0.02
γ_G	2.43±0.03	1.66±0.07
Ising		
ν_T	1.32±0.05	0.85±0.10
γ_T	2.41±0.15	1.7±0.20
Heisenberg		
ν_T	0.90±0.05	0.95±0.10
γ_T	1.50±0.15	1.73±0.15

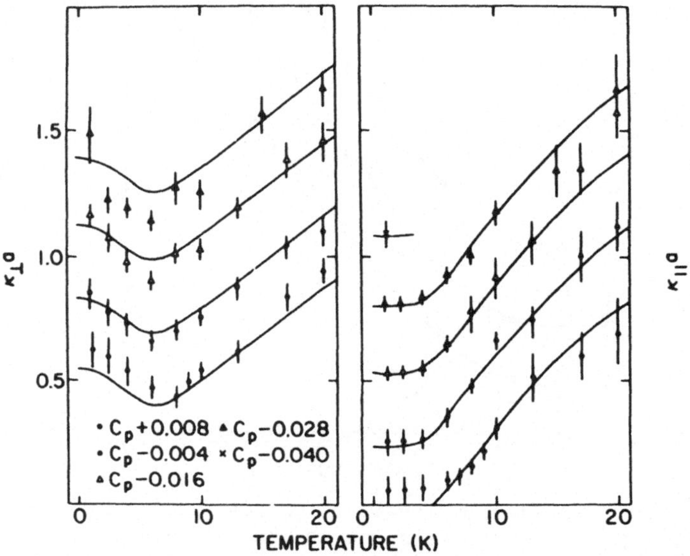

Fig. 9. The inverse correlation length of the transverse and longitudinal fluctuations for $Mn_cZn_{1-c}F_2$. Note that 0.2 has been added to $c = c_p$ -0.004, 0.4 to $c = c_p$ = -0.016, 0.6 to $c = c_p$ - 0°28 and 0.8 to $c = c_p$ -0.040. The solid lines are given by the sum of a geometric and thermal part where the latter is given by $\mu(T)^{0.85}$ [49].

IV.C. Experiments for c > cp

The multicritical theory of the percolation point was shown in the preceding section to give a good description of the measurements for $c < c_p$. The situation is much less satisfactory for $c > c_p$ both theoretically and experimentally. For $c > c_p$ there is a transition to long range order, which of necessity means that there must be many paths connecting different parts of the infinite cluster, and the asymptotic exponents are expected to be those of the appropriate random system, section 3. This implies that the thermal scaling is not dominated by only the one-dimensional weak links, and even above $T_c(c)$ the structure factor will be determined by more than one length scale and have a non-Lorentzian form [53]. At temperatures below $T_c(c)$, the long range order will generate a random field due to the randomness of the interactions and as described in section V this in mean field theory produces a Lorentzian squared term in the scattering cross-section.

Experimentally the situation is also unclear. In $Rb_2Co_xMg_{1-x}F_4$ [17] the behaviour is qualitatively as expected. Long range order sets in at a fairly well defined temperature, $T_c(c)$, at which $\kappa = 0$ within experimental error fig 8, and the scattering profiles are consistent with the Lorentzian form.

The situation is more surprising in $Mn_cZn_{1-c}F_2$ and $RbMn_cMg_{1-c}F_3$ [49]. In both of these cases long range order sets in at a well defined temperature, but the inverse correlation length is not then zero. In the latter, $T_c(c) \sim 14$ K, fig 10, but below 6 K the Bragg intensity or the long range decreases while the inverse correlation length continues to decrease. This behaviour is not yet understood.

Unexpected behaviour may also have been observed in $Co_cZn_{1-c}F_2$ [54]. The inverse correlation length decreased on cooling but was not zero at the onset of long range order. This may be due to concentration fluctuations in the sample, or to the problem of achieving thermodynamic equilibrium at low temperatures in an Ising system. Clearly at low temperatures relatively small long range interactions such as dipolar interactions may play an important role. More work is needed before the behaviour of systems for concentrations close but above the percolation concentration are understood.

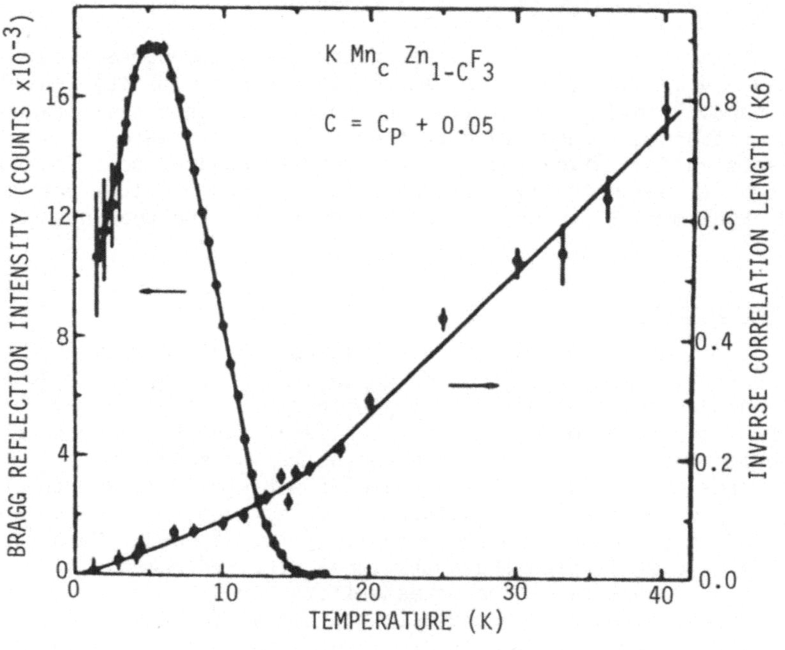

Fig. 10. Bragg scattering and inverse correlation length in $KMn_{0.33}Zn_{0.67}F_3$ [49].

V. THE RANDOM FIELD PROBLEM

V.A. Introduction

The application to a magnetic system of magnetic fields which are randomly directed at each site causes a drastic change in the properties. Imry and Ma [55] suggested that an Ising system might break up into domains if the gain in the energy by taking advantage of the statistically random fields $\sim HL^{\frac{d}{2}}$ for a domain of size L, outweighs the cost in the domain wall energy $\sim JL^{d-1}$. These results then suggest that for Ising systems, the lower critical dimension is 2 and below d_L the random field will necessarily break up the long range order. In systems with continuous symmetry the domain wall energy $\sim L^{d-2}$ gives $d_L = 4$.

After Imry and Ma's pioneering work there were a variety of theories developed [56] based on the ε-expansion about 6 dimensions which agreed with Imry and Ma's result for d_L for continuous systems, but for Ising systems gave $d_L = 3$. The experiments described below were initiated to investigate this discrepancy experimentally, but have over the past 7 years shown that the random field problem is considerably more subtle than initially envisaged by either the theorists or the experimentalists. Meanwhile, however, it is now generally accepted that d_L in equilibrium is 2 [57].

Random fields are often produced when defects or impurities are present in crystals, consequently they occur in most if not all real systems. The effects of impurities are notoriously difficult to control and in particular to eliminate, and so real progress was only possible after Fishman and Aharony [58] showed that the application of a uniform field to a random antiferromagnet generates a random staggered field, partly through the disorder in the direct Zeeman term and partly through the randomness in the exchange interactions. This result then enabled the strength of the random field to be controlled experimentally and externally, and so enabled much more informative experiments to be performed.

A field applied to a site i, H_i, generates a moment $\langle S_j \rangle$ on the site j, where $\langle S_j \rangle = \chi_{ij} H_i$ and χ_{ij} is the appropriate susceptibility. Averaging over the random field on each site, the scattering from these random moments for a wavevector q is then given by $S(q) = |\chi(q)|^2 H^2$, which if $\chi(q)$ is given by a Lorentzian form gives rise to a Lorentzian squared contribution to the scattering. As shown in fig 11, Lorentzian squared profiles are often observed experimentally when systems are cooled in the presence of a random field.

V.B. Metastability

In all of the d = 3 systems investigated experimentally $Co_cZn_{1-c}F_2$ [54], $Fe_cZn_{1-c}F_2$ [60] and $Mn_cZn_{1-c}F_2$ [59,61], it is found that when the samples are cooled in the magnetic field (to produce a random field), FC, long range order is not established, and the scattering cross-section is well described by a Lorentzian squared form. In contrast if the systems are cooled in the absence of a field, ZFC, and the field applied, long range order is retained until the sample is heated above a well defined temperature at which ergodicity is restored. This behaviour is illustrated in fig 11 which shows the large difference between FC and ZFC experiments at 43.4 K below the metastability line, but no difference at 44.0 K above the line. Differences in the behaviour of FC and ZFC measurements are also observed in macroscopic measurements such as thermal expansion and magnetisation [62].

The occurrence of metastability in $Mn_cZn_{1-c}F_2$ at 40 K is initially surprising because the spin wave gap is ~ 1 K, so that the metastability cannot arise from the thermal freezing, which always occurs at sufficiently low temperatures in Ising systems. There are many excited spin waves below the metastability boundary, and so a more sophisticated explanation of the metastability is needed.

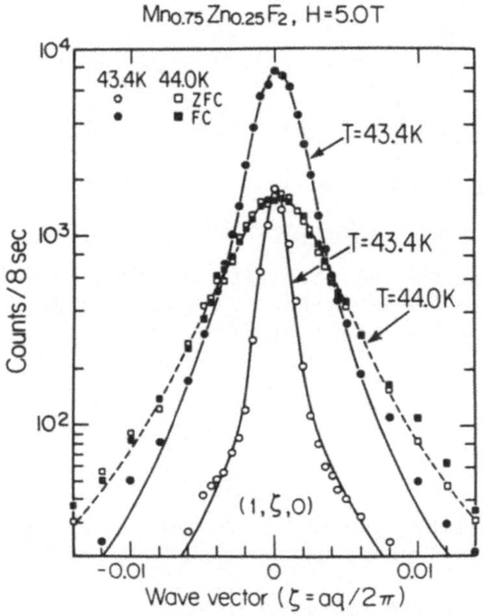

Fig. 11. Scattering as a function of wavevector transfer from $Mn_{0.75}Zn_{0.25}F_2$. The ZFC and FC results differ at 43.4 K but are the same at 44.0 K. The solid lines are fits to a Lorentzian squared profile with in addition a resolution limited Gaussian component for ZFC at 43.4 K [59].

There have now been a variety of experiments to determine the nature of the metastability. The results have shown:

i) The metastability boundary $T_M(H)$ found by neutron scattering measurements occurs at nearly the same temperature as the peak in the temperature derivative of the birefringence of ZFC samples heated in the same field [63]. Detailed macroscopic measurements suggest that the peak is at a temperature slightly, $\Delta T/T_c \sim 5 \times 10^{-3}$, below the metastability boundary.

ii) The difference between $T_M(H)$ and T_c scales as $T_M(H) - T_c + bH^2 = -CH^{2/\phi}$, fig 12, where b is a small mean field correction and ϕ is the random field crossover exponent which was predicted for random fields to be γ [58] and more recently by including random exchange effects as well to be ~1.05γ [64]. These predictions are in agreement with the experimental results, fig 12 and [36].

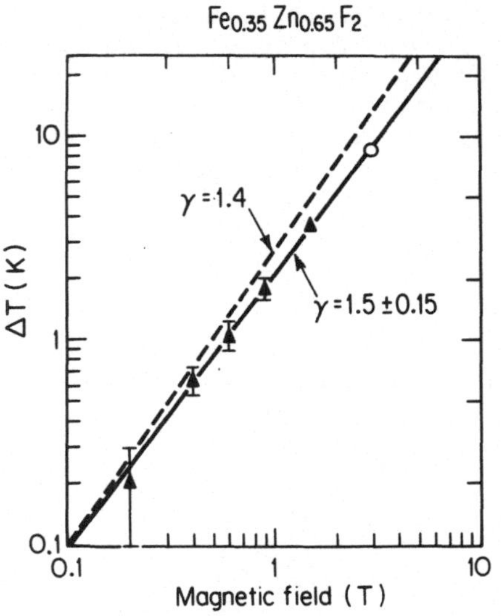

Fig. 12. The change in the metastability temperature $\Delta T = T_M(0) - T_M(H)$ with magnetic field for $Fe_{0.35}Zn_{0.65}F_2$ [60]. The solid and dashed lines have slopes of $2/\gamma$.

iii) Below the metastability boundary all domain states are metastable if they have domains of size larger than the FC domain state [59-61].

iv) On reducing the field the systems relax to a more ordered state, but on raising the field little change occurs. Domain walls cannot be generated below the metastability boundary.

v) At low temperatures the inverse correlation length of the Lorentzian squared profile scattered by the FC domain state scales as H^{ν} with $\nu = 2.1 \pm 0.1$ for $Fe_cZn_{1-c}F_2$ [60], fig 13, but $\nu = 3.5 \pm 0.5$ for $Co_cZn_{1-c}F_2$ [54] and $Mn_cZn_{1-c}F_2$. This difference may arise from the proximity of the spin-flop boundary in the latter materials.

This behaviour was initially discussed in terms of the barrier heights for the motion of domain walls. Villian and others [65] showed that as a domain becomes larger, there are increasingly larger barriers to its motion. In particular it was predicted [65] that a time t after a quench the domain size would be of order

$$L \sim \frac{T}{H^2} \ln\left(\frac{t}{\tau}\right) ,$$

where τ is a microscopic time. Several experiments have now been performed to probe this predicted time dependence. Typical results [59] are shown in table 4 for $Mn_cZn_{1-c}F_2$ at 40.6 K close to the metastability boundary. The shortest reasonable time for τ is the reciprocal of a zone boundary magnon $\sim 10^{-12}$ sec, so the ratio of the domain sizes at 600 secs and 54000 secs is expected to be 1.14 whereas experimentally it is 1.01±0.03. Any relaxation is therefore considerably slower than logarithmic.

Table 4. FC Domain Sizes in $Mn_{0.75}Zn_{0.25}F_2$ [59]

Inverse Correlation Length (10^{-3} r.l.u.)	Time (sec)
1.46 ± 0.07	140
1.39 ± 0.02	600
1.37 ± 0.02	1300
1.35 ± 0.02	4300
1.38 ± 0.02	54000

reciprocal of a zone boundary magnon ~ 10^{-12} sec, so the ratio of the domain sizes at 600 secs and 54000 secs is expected to be 1.14 whereas experimentally it is 1.01±0.03. Any relaxation is therefore considerably slower than logarithmic.

A further difficulty with this theory is to explain the rapid onset of the metastability . Villain and Fisher [66] have now shown that just above a possible transition to long range order at $T_c(H)$, the barrier heights increase with a power of $T - T_c(H)$. This has the effect of explaining the apparently sharp metastability boundary on the time scales applicable for experiments. It is nevertheless still difficult to understand the absence of any time dependence below the boundary, table 4, although measurements of the capacitance of $Fe_cZn_{1-c}F_2$ when FC just below the metastability boundary [67] did show a logarithmic time dependence. The origin of this time dependence and the failure to observe any comparable time dependence in other quantities, in particular the domain size is unknown. King et al [68] have measured the ac magnetic susceptibility and found it is frequency dependent at anomalously low frequencies close to the metastability boundary. We conclude that anomalously slow dynamics can account for the metastability boundary.

Despite this success, the theory of the metastable phase is still not completely understood. Fig 13 shows that the FC correlation length continues to decrease below the metastability boundary: the system is not completely frozen. The asymmetry in the effects of raising and lowering the field (iii) and (iv) above have not been explained in detail, and finally the way in which the same FC state can be reached from several different disordered configurations even when the metastability line is crossed at different fields.

Similar results have been obtained for the two-dimensional systems $Rb_2Co_cMg_{1-c}F_4$ [69,70] and $Rb_2Mn_{0.7}Mg_{0.3}F_4$ (unpublished). There is a metastability line below which a range of different domain sizes are stable, and this line scales with temperature and field as predicted by scaling theories [70]. The only qualitative differences with the d = 3 results are that at low temperatures the FC scattering profiles are the sums of Lorentzian and Lorentzian squared terms or possible Lorentzian to the power 1.5, while the inverse correlation lengths scale as H^{ν} with $\nu = 1.6\pm0.2$ [69]. In $Rb_2Mn_{0.7}Mg_{0.3}F_4$ experiments performed using the ZFC procedure showed that the long range order was stable only up to a critical field, $T_L(H)$, considerably below the metastability boundary, $T_M(H)$. Presumably this difference from the behaviour of the d = 3 systems arises because d = 2 is the lower critical dimension. The metastability theories [65,66] are equally applicable in d = 2 as in d = 3.

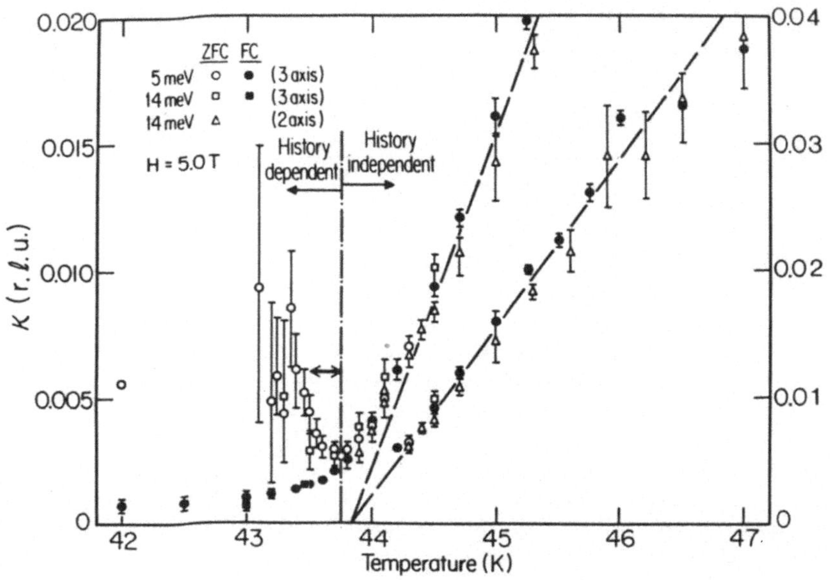

Fig. 13. The inverse correlation length of $Mn_{0.75}Zn_{0.25}F_2$ in a field of 5 T for ZFC and FC experiments in the neighbourhood of the metastability boundary at 43.75 K [59].

V.C. The Phase Transition

In three dimensions theory predicts that the ground state has long range order, and so on cooling a phase transition between a paramagnetic and a long range ordered state is expected. Unfortunately the metastability boundary, $T_M(H)$, occurs just above the phase transition, $T_c(H)$, but nevertheless the scattering in the equilibrium phase above $T_M(H)$ may still provide information about the nature of the phase transition at $T_c(H)$.

Figs 14 and 15 show measurements of the inverse correlation length observed above $T_M(H)$ for various fields in $Mn_cZn_{1-c}F_2$ and in $Fe_cZn_{1-c}F_2$. The results are clearly similar. For H = 0 the exponent $\kappa \sim H^{\nu}$ gives $\nu \sim 0.7$ as described in section II. However, fits to the data at higher fields give exponents between 1.0 and 2.0 depending on the field and the range over which the data are fitted. It is clearly very difficult if not impossible to extract reliable exponents without a detailed knowledge of the appropriate crossover functions.

The amplitude of the Lorentzian squared for one field is shown in fig 16, and similar results were obtained for other fields. If there is to be a continuous phase transition, this Lorentzian squared term must evolve continuously into the Bragg reflection for long range order. If $\beta > 0$ this requires the amplitude to be zero at $T_c(H)$ in evident disagreement with the steady increase on cooling shown in fig 6. Clearly this conclusion depends on the assumption of a Lorentzian squared profile as critical effects may give a different profile, but in the absence of an alternative form these results suggest that the transition is probably of first order [71] even if very weakly so.

The properties of the ZFC state have also been studied close to the metastability boundary. At low temperatures the scattering consists solely of a resolution limited Bragg reflection, but on heating a Lorentzian

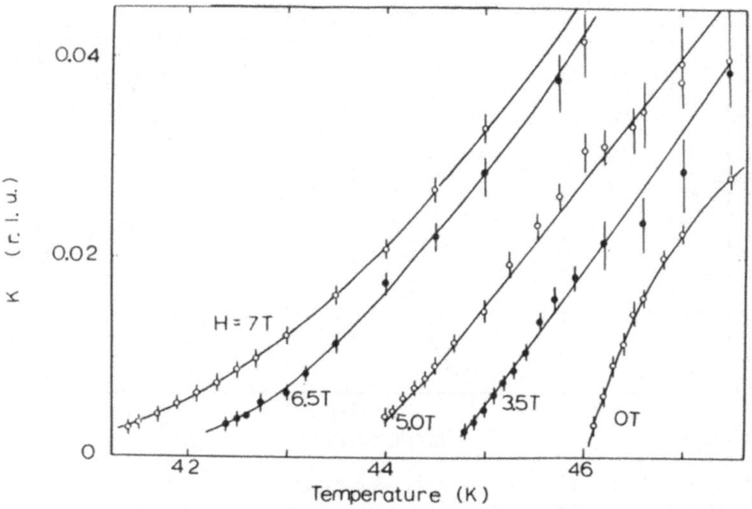

Fig. 14. The inverse correlation length of $Mn_{0.75}Zn_{0.25}F_2$ in various fields above the metastability temperature [59].

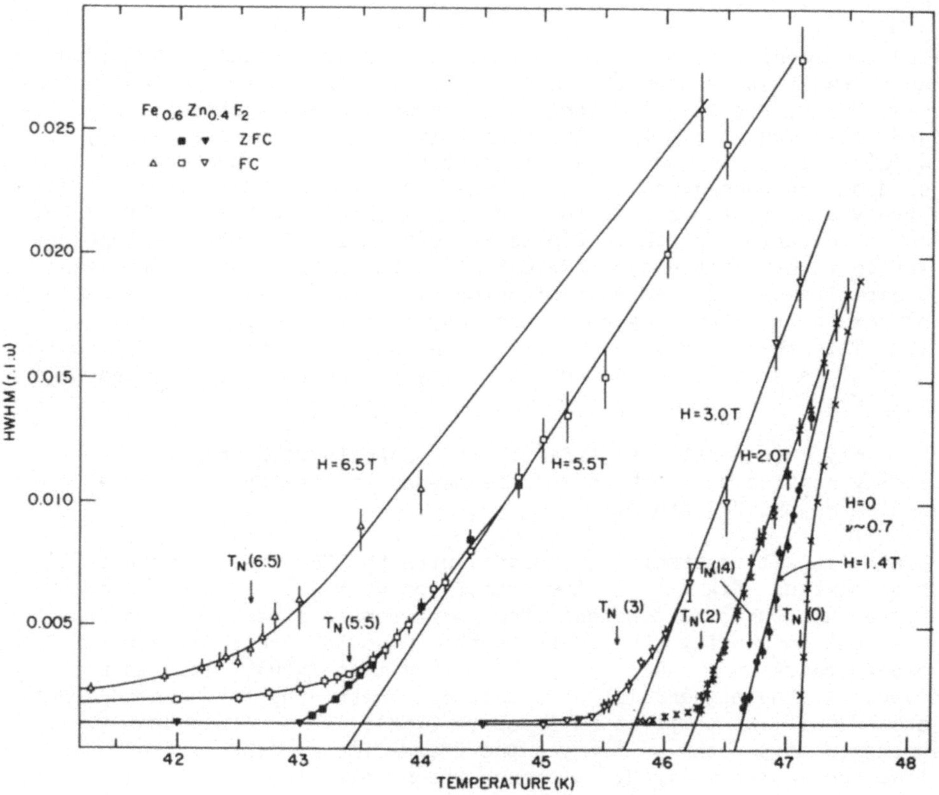

Fig. 15. The inverse correlation length of $Fe_{0.6}Zn_{0.4}F_2$ in various fields - note that T_N is the metastability temperature $T_M(H)$ [60].

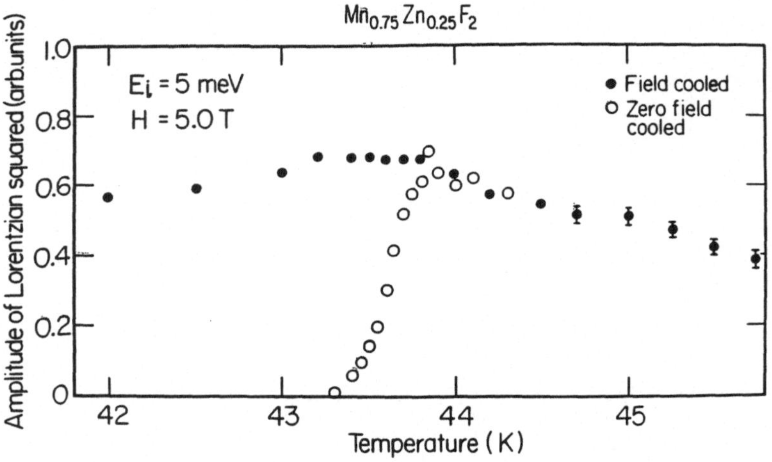

Fig. 16. The amplitude of the Lorentzian squared component of the scattering in $Mn_{0.75}Zn_{0.25}F_2$ as a function of temperature [59].

squared component is also observed whose intensity increases and width decreases until it reaches the FC correlation length when the temperature reaches, $T_M(H)$, fig 13. The long range order decreases rapidly over ~0.4 K at $T_M(H)$ as shown in fig 17. The slight increase in the intensity just below $T_M(H)$ is due to the relief of extinction in these perfect crystals, and so does not correspond to an increase in the long range order. The relatively abrupt decrease in the long range order and the fact that the inverse correlation length is finite at $T_M(H)$, fig 13, suggests that the transition is of first order. Nevertheless the results of birefringence [63], capacitance [68] and magnetisation [72] measurements show a sharp peak in the same temperature region as the long range order is decreasing, fig 18. This peak is well explained as depending on $\ln|T - T_c'(H)|$, $\alpha = 0$ where $T_c'(H)$ is just below the metastability temperature, $T_M(H)$, and higher than, $T_c(H)$.

Clearly this result is difficult to reconcile with the neutron scattering results and further work is needed to clarify the nature from the transition of the ZFC state.

The critical scattering associated with the ZFC state decreases in width on heating, fig 13. If the transition at $T_c(H)$ was not observed solely because the time constants for large domain sizes were too slow, we might expect on heating the ZFC state that κ would tend to diverge at $T_c(H)$, but become constant at some temperature less than $T_c(H)$. This is not observed. At the higher fields no critical scattering is observed for the ZFC state until the temperature is above $T_c(H)$. We conclude that the measurements at present do not yet provide definitive information about the critical properties at $T_c(H)$, although the results shown in fig 16, favour a first order transition in contrast to many of the recent theories [73].

Less detailed measurements have been made on two-dimensional systems using neutron scattering techniques [60,70]. Measurements of the temperature dependence of the birefringence on $Rb_2Co_{0.85}Mg_{0.15}F_4$ [74] showed that the sharp peak observed when H = 0, was broadened in larger fields, and this was interpreted as evidence for the destruction of the phase transition in two dimensions. Experimental evidence for metastability has been observed in a number of other systems [75], but has not been investigated in the detail described above.

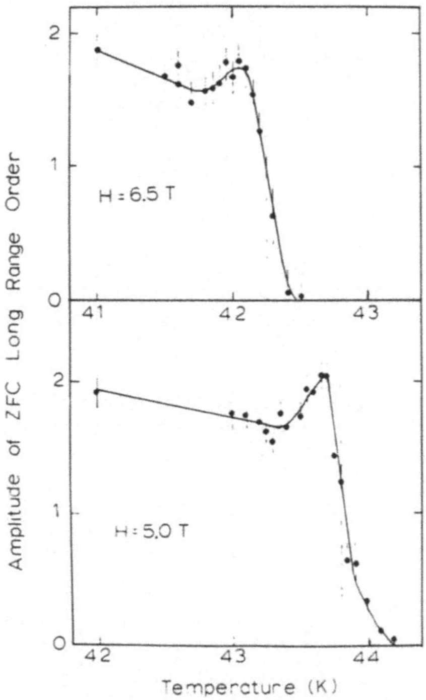

Fig. 17. Amplitude of the long range order when heating the ZFC state in $Mn_{0.75}Zn_{0.25}F_2$ [59].

VI. COMPETING INTERACTIONS AND CONCLUSIONS

VI.A. Competing Anisotropies

The phase transitions in materials such as $K_2Fe_cCo_{1-c}F_4$ in which the Fe moments tend to align perpendicular to the c-axis and the Co moments along the c-axis, were discussed first using mean field theory [76]. There are four phases; a paramagnetic phase, one with the spins aligned parallel to the c-axis, one with them aligned perpendicular and one where both components of the spins are aligned. The critical phenomena and properties of the tetracritical point where these four phases meet were discussed by Aharony and Fishman [77]. The most detailed experiments have been performed using the $Fe_cCo_{1-c}C\ell_2$ system [78] and the results gave agreement with the theory for the lines between the paramagnetic and the ordinary phases in

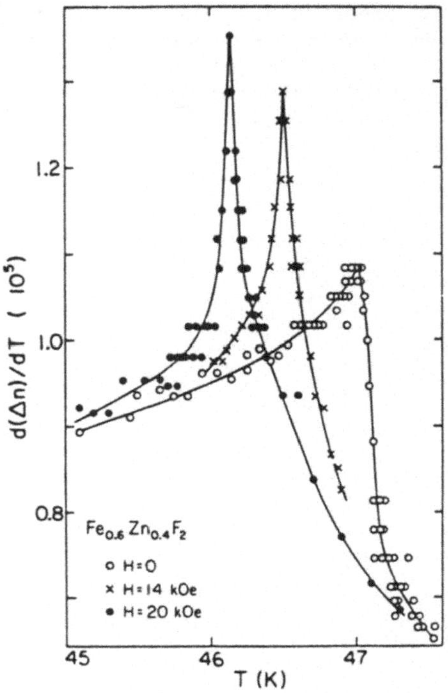

Fig. 18. The temperature derivative of the birefringence showing the critical scattering observed on heating the ZFC state of $Fe_{0.6}Zn_{0.4}F_2$ [63].

which one component of the spin was ordered, but disagreement with the properties of the phase boundaries between the different ordered phases. This is now believed to arise [78], because the ordering of one spin component generates a random field on the other spin component which since the experiments are necessarily performed in a field cooled manner inhibits the ordering of the second component.

These effects are expected to be absent in the higher symmetry $K_2Fe_cCo_{1-c}F_4$ system, and experiments do suggest [79] that the theory works better in this case. A detailed test is not however possible until better samples are available with more precisely known concentrations.

VI.B. Competing Exchange Constants

When exchange constants are of different sign, there are regions of the phase diagram in which spin glass phases are obtained. These will be reviewed at this school by Sherrington [80] and so they are only briefly mentioned here. Many of the experiments have been performed on dilute metallic alloys but there are also a number of simpler insulating systems with competing exchange constants. One system is $Rb_2Cr_cMn_{1-c}C\ell_4$ in which the Cr-Cr interactions are ferromagnetic while the Mn-Mn ones are antiferromagnetic [30]. There have not yet been detailed studies of the properties in the spin glass phase of this system.

EuO and $FeC\ell_2$ have second nearest neighbour interactions which oppose the ordering preferred by the nearest neighbour interactions. Consequently on diluting they form spin glass phases. Extensive work on the properties of the $Eu_cSr_{1-c}O$ system has been performed by Maletta and collaborators [81], and on $Fe_cMg_{1-c}C\ell_2$ [82]. The results show qualitatively similar effects to the random field problem; mestastability, long time constants and Lorentzian squared profiles. We shall not however discuss these results in detail.

VI.C. Conclusions

In these lectures I have described the experimental results available on the excitations and phase transitions of disordered magnets. In systems where the ground state is co-linear and ordered both the properties of the excitations and the phase transitions are basically understood. The excitations can be calculated using molecular dynamics techniques with good agreement with experiment, but the analytic theories such as the coherent potential approximation are both complicated and unsatisfactory. The phase transitions are well defined and have exponents in good agreement with theory.

The properties close to the percolation point are also well understood in the disordered phase. The multicritical picture of geometric and thermal lengths with the latter dominated by the one dimensional weak links describes the results very well. The properties of the ordered phases close to percolation are not understood. It is not yet known if the fracton picture is useful in describing the excitations, and there are a number of experimental results which suggest that the structure and phase transitions of these ordered phases are not as expected. Further work is needed on both the excitations and phase transitons close but above the critical concentration for percolation.

The behaviour of systems with competing interactions is altogether more subtle. There are undoubted problems of metastability and of long time constants. In the case of the random field problem considerable progress has been made but there is still not a detailed understanding of the stability of the different structures in the metastable region, and also not even a complete description of the observed behaviour in the equilibrium paramagnetic phase. These failings of the theory are important because nearly all real materials contain defects and these produce random fields. Consequently the study of phase transitions in most real materials corresponds to performing a random field, FC experiment and so will not give strictly long range order. Further work is needed and is being performed to study the random field problem and the other systems with competing interactions.

ACKNOWLEDGEMENTS

I am grateful to all my collaborators for helping discussions and enjoyable progress on disordered magnetic systems, and in particular to R.J. Birgeneau and G. Shirane. Financial support for the work in Edinburgh has been provided by the Science and Engineering Research Council.

REFERENCES

1. O. Nikotin, P. A. Lindgard and O. W. Dietrich, J. Phys. C 2, 1168 (1969).
2. P. Martel, R. A. Cowley and R. W. H. Stevenson, Can. J. Phys. 46, 1355 (1968) and J. Phys. C 6 , 2997 (1973).
3. R. J. Birgeneau, H. J. Guggenheim and G. Shirane, Phys. Rev. B 8, 304 (1973).

4. D. J. Breed, K. Gilijamse and A. R. Miedema, Physica 45, 205 (1969).
5. M. P. H. Thurlings, E. Frikee and H. W. de Wijn, Phys. Rev. B 25, 4750 (1982).
6. M. T. Hutchings, J. Als-Nielsen, P. A. Lindgard and P. J. Walker, J. Phys. C 14, 5327 (1981).
7. W. Marshall and S. Lovesey, Theory of Thermal Neutron Scattering, Oxford Univ. Press (1971).
8. T. M. Holden, R. A. Cowley, W. J. L. Buyers and R. W. H. Stevenson, Solid State Commun. 6, 154 (1968). W. J. L. Buyers, T. M. Holden, E. C. Svensson, R. A. Cowley and R. W. H. Stevenson, Phys. Rev. Lett. 27, 1442 (1972) and E. C. Svensson, S. M. Kim, W. J. L. Buyers, S. Rolandson, R. A. Cowley and D. A. Jones, AIP Conf. Proc. 24, 161 (1975).
9. E. C. Svensson, W. J. L. Buyers, T. M. Holden, R. A. Cowley and R. W. H. Stevenson, Can. J. Phys. 47, 1983 (1969).
10. J. Als-Nielsen, R. J. Birgeneau, H. J. Guggenehim and G. Shirane, Phys. Rev. B12, 4963 (1975).
11. S. A. Higgins, R. A. Cowley, M. Hagen, J. K. Kjems, U. Durr and K. Fendler, J. Phys. C17, 3235 (1984).
12. G. J. Coombs, R. A. Cowley, D. A. Jones, G. Parisot and D. Tochetti, AIP Conf. Proc. 29, 254 (1976).
13. G. J. Coombs, R. A. Cowley, W. J. L. Buyers, E. C. Svensson, T. M. Holden and D. A. Jones, J. Phys. C 9, 2167 (1976).
14. O. W. Dietrich, G. Mayer, R. A. Cowley and G. Shirane, Phys. Rev. Lett. 35, 1735 (1975).
15. Y. Uemura and R. J. Birgeneau, Phys. Rev. Lett. 57, 1947 (1986).
16. R. A. Cowley, O. W. Dietrich and D. A. Jones, J. Phys. C 8, 3023 (1975).
17. H. Ikeda and G. Shirane, J. Phys. Soc. Japan 46, 30 (1979); R. A. Cowley, R. J. Birgeneau, G. Shirane, H. J. Guggenheim and H. Ideda, Phys. Rev. B21, 4038 (1980).
18. R. A. Cowley, G. Shirane, R. J. Birgeneau and H. J. Guggenheim, Phys. Rev. B15 4292 (1977).
19. M. F. Thorpe and R. Alben, J. Phys. C 9, 2555 (1976).
20. R. Alben, S. Kirkpatrick and D. Beeman, Phys. Rev. B15, 346 (1977).
21. W. J. L. Buyers in Excitations in Disordered Systems, ed. M. F. Thorpe, NATA ASI Series B 78, 411 Plenum (1982).
22. L. J. Clarke (to be published).
23. J. W. Halley and W. K. Holcomb, J. Phys. C 11, 753 (1978).
24. R. J. Elliott, J. A. Krumhansl and P. L. Leath, Rev. Mod. Phys. 46, 45 (1974), W. J. L. Buyers, D. E. Pepper and R. J. Elliott, J. Phys. C 5, 2611 (1972).
25. W. J. L. Buyers, D. E. Pepper and R. J. Elliott, J. Phys. C 6, 1953; (1973); G. J. Coombs and R. A. Cowley, J. Phys. C 8, 1889 (1975).
26. S. Alexander and R. Orbach, J. de Physique Lett. 43, L625 (1982).
27. A. Aharony, S. Alexander, O. Entin-Wohlman and R. Orbach, Phys. Rev. Lett. 58, 132 (1987).
28. S. A. Higgins, Ph.D. Thesis, Edinburgh University (1985).
29. S. A. Higgins, V. H. M. Vlak, M. Hagen, R. A. Cowley, A. F. M. Arts and H. W. de Wijn, J. Phys. C (in press).
30. N. Kohles, H. Theurerkaufer, K. Strobel, R. Geick and W. Treutmann, J. Phys. C 15, L137 (1982); K. Katsumata, T. Nire, M. Tanimoto and H. Yoshizawa, Phys. Rev. B 25, 428 (1982).
31. D. Sieger, H. Tietze, R. Geick, S. Bates, R. A. Cowley, W. Treutmann and U. Steingenburger, Solid State Commun. (in press).
32. S. A. Higgins and M. Hagen (to be published).
33. A. B. Harris, J. Phys. C 7, 1671 (1974).
34. D. E. Khmelnitzkii, Sov. Phys. JETP 41, 981 (1976).
35. R. J. Birgeneau, R. A. Cowley, G. Shirane, H. Yoshizawa, D. P. Belanger, A. R. King and V. Jaccarino, Phys. Rev. B27, 6747 (1983).
36. D. P. Belanger, A. R. King and V. Jaccarrino, Phys. Rev. B34, 452 (1986).
37. P. W. Mitchell, R. A. Cowley, H. Yoshizawa, P. Boni, Y. J. Uemura and R. J. Birgeneau, Phys. Rev. B 34, 4719 (1986).

38. G. Jug, Phys. Rev. B 27, 609 (1983); K. E. Newman and E. K. Reidel, Phys. Rev. B 25, 264 (1982); I. O. Mayer and A. I. Sokolov, Sov. Phys. Solid State 26, 2076 (1984); A. Newlove, J. Phys. C 16, L423 (1983).
39. R. A. Pelcovits and A. Aharony, Phys. Rev. B31, 350 (1985).
40. V. S. Dotsenko and V. S. Dotsenko, J. Phys. C 15, 495 (1983); Ibid J. Phys. C15, L557 (1983).
41. H. Ikeda, M. Suzuki and M. T. Hutchings, J. Phys. Soc. Japan 46, 1153 (1979).
42. M. Hagen, R. A. Cowley and R. M. Nicklow, Phys. Rev. (in press).
43. H. B. Tarko and M. E. Fisher, Phys. Rev. B 11, 1217 (1975); C. A. Tracy and B. M. McCoy, Phys. Rev. B12, 368 (1975).
44. R. A. Cowley, M. Hagen and D. P. Belanger, J. Phys. C17, 3763 (1984).
45. D. Stauffer, Zeit Phys. B22, 161 (1976); T. C. Lubensky, Phys. Rev. B15, 311 (1977).
46. D. J. Wallace and A. P. Young, Phys. Rev. B17, 2384 (1978).
47. Y. Endoh, G. Shirane, R. J. Birgeneau and Y. Ajiro, Phys. Rev. B19, 1476 (1979).
48. R. J. Birgeneau, R. A. Cowley, G. Shirane, J. A. Tarvin and H. J. Guggenheim, Phys. Rev. B21, 317 (1980).
49. R. A. Cowley, G. Shirane, R. J. Birgeneau, E. C. Svensson and H. J. Geggenheim, Phys. Rev. Lett. 39, 894 (1977; Phys. Rev. B22, 4412 (1980).
50. M. F. Thorpe, J. de Physique 36, 117 (1975).
51. R. B. Stinchcombe, J. Phys. C13, 3723 (1980).
52. A. Congilio, Phys. Rev. Lett. 46, 250 (1981).
53. A. Aharony in Multicritical Phenomena, ed. by R. Pynn and A. Skeltorp, NATO ASI Series B, Vol. 106, Plenum p. 307.
54. M. Hagen, R. A. Cowley, S. Satija, G. Shirane, H. Yoshizawa, R. J. Birgeneau and H. J. Guggenheim, Phys. Rev. B28, 2602 (1983).
55. Y. Imry and S. K. Ma, Phys. Rev. Lett. 35, 1399 (1975).
56. A. Aharony, Y. Imry and S. K. Ma, Phys. Rev. Lett. 37, 1367 (1976); A. P. Young, J. Phys. C10, L257 (1977); G. Parisi and N. Soulas, Phys. Rev. Lett. 43, 744 (1979).
57. J. Z. Imbrie, Phys. Rev. Lett. 53, 1747 (1984).
58. S. Fishman and A. Aharony, J. Phys. C12, L279 (1979).
59. R. J. Birgeneau, R. A. Cowley, G. Shirane and H. Yoshizawa, Phys. Rev. Lett. 54, 2147 (1985); R. A. Cowley, R. J. Birgeneau and G. Shirane, Physica 140A, 285 (1986).
60. R. A. Cowley, H. Yoshizawa, G. Shirane and R. J. Birgeneau, Zeit, Phys. B. 58, 15 (1984); D. P. Belanger, A. R. King and V. Jaccarino, Phys. Rev. B31, 4538 (1985); H. Yoshizawa, R. A. Cowley, G. Shirane and R. J. Birgeneau, Phys. Rev. B31, 4548 (1985).
61. R. A. Cowley, H. Yoshizawa, G. Shirane, M. Hagen and R. J. Birgeneau, Phys. Rev. B30, 6650 (1984).
62. Y. Shapiro, N. F. Oliveira, Jr. and S. Foner, Phys. Rev. B30, 6639 (1984); A. R. King, V. Jaccarino, D. P. Belanger and S. M. Rezende, Phys. Rev. B32, 503 (1985).
63. D. P. Belanger, A. R. King, V. Jaccarino and J. Cardy, Phys. Rev. B31, 4538 (1983).
64. A. Aharony, Europhysics Letters 1, 617 (1986).
65. J. Villain, Phys. Rev. Lett. 54, 1543 (1984); G. Grinstein and J. Fernandez, Phys. Rev. B29, 6389 (1984).
66. J. Villain, J. de Physique 46, 1843 (1985). D. S. Fisher, Phys. Rev. Lett. 56, 416 (1986).
67. D. P. Belanger, S. M. Rezende, A. R. King and V. Jaccarino, J. Appl. Phys. 57, 3294 (1985).
68. A. R. King, J. A. Mydosh and V. Jaccarino, Phys. Rev. Lett. 56, 2525 (1986).
69. R. J. Birgeneau, H. Yoshizawa, R. A. Cowley, G. Shirane and H. Ideda, Phys. Rev. B28, 1438 (1983).
70. D. P. Belanger, A. R. King and V. Jaccarino, Phys. Rev. Lett. 54, 577 (1985).

71. A. P. Young and M. Nauenberg, Phys. rev. Lett. 54, 429 (1985).
72. W. Kleeman, A. R. King and V. Jaccarino, Phys. Rev. B34, 479 (1986).
73. A. J. Bray and M. A. Moore, J. Phys. C18, L927 (1985).
74. I. B. Ferriera, A. R. King, V. Jaccarino, J. Cardy and H. J. Guggenheim, Phys. Rev. B28, 5192 (1983).
75. H. Ideda, J. Phys. C16, L1033 (1983).
P. Wong and J. W. Cable, Phys. Rev. B28, 5361 (1983).
P. Wong and J. W. Cable, Phys. Rev. B30, 485 (1984).
76. P. A. Lindgard, Phys. Rev. B14, 4074 (1976),Ibid B16, 2168 (1978).
77. S. Fishman and A. Aharony, Phys. Rev. B18, 3507 (1978).
78. P. Wong, P. M. Horn, R. J. Birgeneau, C. R. Safinya and G. Shirane, Phys. Rev. Lett. 45, 1974 (1980); Ibid Phys. Rev. B27, 428 (1983).
79. W. A. H. M. Vlak, E. Frikkee, A. F. M. Arts and H. W. de Wijn, J. Phys. C16, L1015 (1983).
S. A. Higgins, R. A. Cowley, M. Hagen, J. K. Kjems, U. Durr and K. Fendler, J. Phys. C17, 3235 (1984).
80. D. Sherrington (this school).
81. H. Maletta and W. Xinn, in Handbuch on the Physics and Chemistry of Rare Earths, ed. K. A. Gschneidner and L. Eyring, Vol. 12, North Holland (1986).
82. P. Wong, S. von Molnar, T. T. M. Palstra, J. A. Mydosh, H. Yoshizawa, S. M. Shapiro and A. Ito, Phys. Rev. Lett. 55, 2043 (1985).

SPIN GLASSES

D. Sherrington

Department of Physics
Imperial College
London SW7 2BZ
England

ABSTRACT

A review is presented of the key concepts and techniques of the theory of spin glasses, together with their applications to problems of complex optimization and to model neural networks.

I. INTRODUCTION

The expression "spin glass" was originally employed as a description of a state in some magnetic alloys in which the orientations of magnetic moments (or "spins") "freeze" over experimental observation times, but without spatial periodicity (hence "glass"). Subsequent study has exposed many new and subtle concepts and shown them to be quite ubiquitous. As a consequence, more recently "spin glass" has come to describe a wide class of situations in which a combination of conflicting ordering instructions and quenched disorder leads to cooperative behaviour of a novel kind, to be discussed in detail below. As well as magnetic materials, this class encompasses many other solid state systems, and also, via mathematical mappings, includes problems in operational research (complex optimization) and in model neural networks. In all of these areas, conceptualization born in disordered magnets has cast new light. In these lectures I shall introduce and partially overview the key concepts and theoretical techniques, with only limited discussion of experiments and history, and no attempt at completeness of cover or attribution - for a review of early history see Coles [1], for recent extensive reviews see Binder and Young [2], Malletta and Zinn [3], and Fischer [4]. For simplicity, initially I shall employ the language of magnetism and use magnetic examples.

In common with all cooperative magnetism, the spin glass problem lies within the field of the statistical mechanics of strongly interacting units. However, unlike conventional magnets, the interactions between the spin variables are disordered. One is concerned with ordering (of spin orientations) in the presence of disorder (of interactions). This disorder is quenched, or fixed, a constraint on the orientational ordering.

$$H = -\frac{1}{2}\sum_{ij} J_{ij}\, \underline{S}_i . \underline{S}_j \tag{1}$$

A simple model of magnetism is the Heisenberg model, of Hamiltonian where the variables are the orientations of the spin vectors $\{\underline{S}_i\}$ and the $\{J_{ij}\}$ are quenched (constrained) interactions. In a conventional magnet the spin locations $\{\underline{R}_i\}$ are periodically arranged, as a crystal, and J_{ij} depends on the labels i,j only as a function of $(\underline{R}_i-\underline{R}_j)$; the interactions between the spins are ordered. Disorder arises either if the spin locations are not periodic or if the J_{ij} are random, rather than depending only on $(\underline{R}_i-\underline{R}_j)$.

Furthermore, in a system characterized by eqn (1), irrespective of whether the interactions are ordered or not, if all the J_{ij} are non-negative all pair interaction energies

$$E_{ij} = - J_{ij}\, \underline{S}_i . \underline{S}_j \quad ; \text{ all } i,j \tag{2}$$

can be simultaneously minimized (by orienting all the $\underline{S}_i$ in the same direction - ferromagnetic order). If, however, some of the J_{ij} are negative, conflicts may arise and simultaneous minimization of all pair energies may be impossible. Such a situation is known as frustration [5].

Spin glasses are characterized by interactions which are both disordered and frustrated. Disorder alone can lead to such interesting features as localization [6] and percolation [7]. Frustration alone is responsible for helical spin ordering in pure systems whose spins have continuous orientational degrees of freedom [8], and for rich sequences of phases in systems with discretely restricted spin or pseudospin variables [9]. The combination can be even more fascinating.

Before proceeding, let us note a fundamental difference between spin glasses and conventional (atomic) glasses. In window glass, for example, the variables which freeze are the atomic positions. The interactions

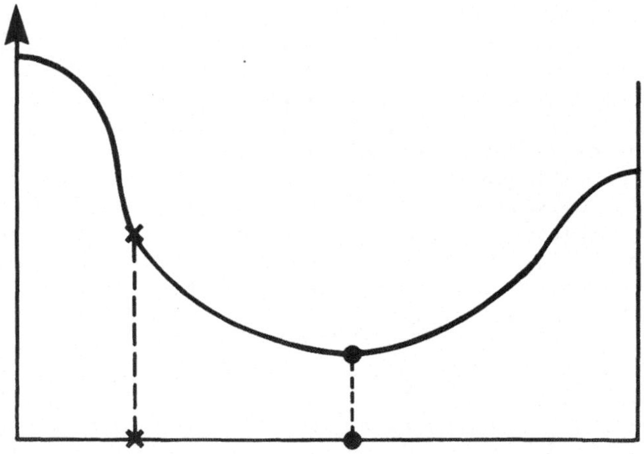

Fig. 1. Schematic plot of free energy as a function of position in macroscopic phase space for an unfrustrated system; also energy or cost as a function of position in microscopic phase space.

between the atoms are translationally invariant, so that the true minimum energy state is a crystal. The glassy state is a long-lived metastable excited state. In spin glasses the minimum energy state is not a periodically oriented spin state. Furthermore, although there are spin glass systems in which the atomic structure on which the spins reside is amorphous, this is not necessary for the appellation.

Disorder and frustration clearly have the potential to destroy periodic magnetic order. They also have the potential to replace it by some non-periodic or amorphous order. Indeed, it was the observation of an apparent sharp phase transition [10] without any corresponding onset of periodic magnetic order which was the driving force for the theoretical work [11-14] which opened the subject as a new area of statistical mechanics. The greater modern interest lies, however, in another consequence of disorder and frustration, a complex chaotically evolving quasi-fractal structure to the free energy surface in an appropriate parameter space. Whereas conventional unfrustrated systems can be considered to have smooth energy and free energy structures, as illustrated schematically in Fig. 1, a spin glass is believed to have a structure with a multiplicity of hills and valleys, effectively on all scales, like a fractal mountain landscape. This is illustrated in Fig. 2. The multipeaked structure hinders motion in phase space and casts doubts on the relevance of the ergodic theorem on practical timescales, suggesting a re-assessment of the normal tenets of statistical mechanics. Furthermore, perturbing the system, for example by the application of a magnetic field, modifies the energy landscape, normally in a non-trivial fashion, and leads to a difference in behaviour between systems prepared in different ways.

The application of techniques and concepts from the theory of spin glasses to complex optimization and to models of neural memory will be explored later in these lectures, but let us now just note a hint of the potential. Many hard optimization problems are characterized by having as their objectives the minimization of cost functions which, in their parameter spaces, have complex multivalley structures as in Fig. 2. In

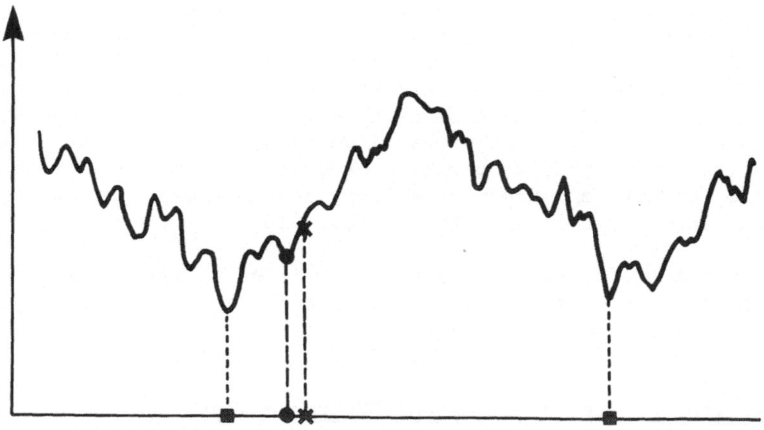

Fig. 2. Schematic plot of free energy as a function of position in macroscopic phase space for a frustrated or NP-complete system; also energy or cost as a function of position in microscopic phase space.

neural systems, memory states are believed to correspond to self-consistent patterns of activity of the neurons, stable against finite disturbances of these activities. This is reminiscent of the energetic stability against fluctuations in phase space of the minima of Figs. 1 and 2. Clearly, a useful memory will have many metastable memory states, suggesting that the analogues of disorder and frustration, found respectively in the dendritic nature of the synaptic connections and in the combination of excitatory and inhibitory character of the synapses, are essential to the function of the brain.

The theory of spin glasses is much more difficult than that of periodic magnets, and even qualitative questions of great consequence remain incompletely answered. The struggle to develop the theory has, however, produced new techniques, new concepts and much interest, with potentially wide implications. The exposure of several of these will be my aim in the rest of these lectures.

I.A. Frustration

Frustration refers to the inability of a system to satisfy simultaneously all its ordering instructions. Its potential relevance is seen readily [5] by considering a set of four Ising spins ($\sigma_i = \pm 1$) located on the corners of a square and interacting only with their nearest neighbours via pairwise energies

$$E_{ij} = - J_{ij}\, \sigma_i\, \sigma_j, \tag{3}$$

with the J_{ij} either + J (ferromagnetic) or -J (antiferromagnetic). If the number of antiferromagnetic bonds around the square is even, the $\{\sigma_i\}$ can be chosen so as to minimize simultaneously all four exchange bonds, giving a total energy -4J. However, if there is an odd number of negative bonds around the square, not all bonds can be simultaneously satisfied; at least one must be dissatisfied, giving a total energy of -2J as the minimum achievable. Furthermore, whereas the only ground state degeneracy in the former (unfrustrated) case is associated with global spin inversion ($\sigma_i \to -\sigma_i$; all i), in the latter (frustrated) case it is four times greater, since there is degeneracy associated with which of the four bond-ordering instructions to disobey. Generally, frustration increases the energy and the degeneracy of the minimum energy states.

Not surprisingly, frustration can have major consequences when it occurs throughout an interacting network. For example, consider a square lattice of Ising spins with nearest neighbour $\pm$J bonds. If all the fundamental cells, or plaquettes, are unfrustrated, there is a paramagnet-ordered phase transition at a finite temperature, and the ground state is only doubly degenerate (due to global inversion). If, however, all the plaquettes are frustrated, cooperative ordering is suppressed, and the ground state has both an extensive entropy and an energy extensively greater than that for the unfrustrated case [15].

The concept of frustration is not restricted to systems all of whose non-zero bonds are of the same strength, nor to Ising systems [16]. It does, however require antiferromagnetic interactions; any closed loop having an odd number of antiferromagnetic bonds is frustrated.

I.B. Prototypes, Characteristics and "Universality"

As I have noted already, disorder and frustration are necessary ingredients for spin-glass behaviour. The details of their origin appear to be of secondary importance, the main qualitative features being common to all. To illustrate this "universality" and to expose some of the experimental characteristics, in this section I shall introduce three prototypical classes of system, note the nature of disorder and frustration in each, and indicate some of the common consequences.

The first class of note is that of the original experimental manifestations, the canonical metallic spin glasses. These are substitutional metallic alloys of a non-magnetic host, such as Cu or Au, and a local-moment-bearing impurity, such as Mn or Fe, at finite but not too great concentrations (less than order 15%). Here the disorder lies in that the magnetic atoms are distributed randomly on the sites of the underlying lattice. As well as possible short-range direct exchange, the localized spins interact indirectly through the conduction electrons, by the Ruderman-Kittel-Kasuya-Yosida (RKKY) mechanism. This RKKY interaction is of long range and is oscillatory in sign as a function of separation. In the nearly-free electron approximation, appropriate to an approximately spherical Fermi surface, the wavelength of oscillation is π/k_F where k_F is the Fermi wave-vector. Within this approximation the effective exchange has the asymptotic form

$$J(R) = \frac{J_o}{(k_F R)^3} \cos(2k_F R + \emptyset) \; . \tag{4}$$

The frustration is caused by the oscillation in sign of J(R). Furthermore, because of the combination of random site occupation and the variation of the sign of J(R) with separation R, pairs of occupied sites receive different mutual ordering instructions by different paths and, because of their different environments, even pairs of spins with the same separation are non-equivalent. It is this combination of frustration and random non-equivalency which is at the root of the spin glass problem.

A second experimental class is that of insulating alloys, such as $Eu_x Sr_{1-x} S$ and $Cd_{1-x} Mn_x Te$, in which only the Eu and Mn are magnetic. In these alloys the exchange interactions are of shorter range than in the metallic alloys, have a different origin (superexchange), but are still frustrated by having a next nearest neighbour antiferromagnetic exchange of sufficient magnitude to interfere with the nearest neighbour contribution over concentrations up to order x = 0.6 (but greater than an appropriate percolation value). The nearest neighbour interaction can be of either sign - it is ferromagnetic for $Eu_x Sr_{1-x} S$ but antiferromagnetic for $Cd_{1-x} Mn_x Te$.

A third class has spins on all the sites of a lattice but with exchange bonds J_{ij} randomly positive or negative. This is the canonical theoretical model, due to Edwards and Anderson [12] who chose the exchange interactions to be quenched independent random variables symmetrically distributed about $J_{ij} = 0$, thereby giving a random arrangement of frustrated and unfrustrated plaquettes and excluding the possibility of any conventional periodic magnetic order. A suitable bias to the J_{ij} distribution allows also the possibility of periodic order [17], in much the same way as does increasing the concentration of magnetic sites in the previous two classes. If the range of the interactions is taken to be infinite, still with each J_{ij} equally probable and independently chosen [13], then the random bond model becomes exactly soluble (in principle) and has been the guide to much of the progress in the subject.

All these classes, and others, show a number of characteristic experimental* consequences within the parameter range for spin-glass

* Experiments on the third class, random-bond models, are computer simulations.

behaviour and over finite (and perhaps infinite) life-times. One is the occurence of a cusp in the a.c. susceptibility at a characteristic temperature T_g, with no apparent periodic freezing of spin orientations beneath this temperature (as shown, for example, by neutron diffraction). This cusp is rounded out in an applied static magnetic field. A second is a difference in the d.c. susceptibility for $T < T_g$ depending on whether the measuring field is applied after cooling in zero field or the cooling is done with the measuring field already applied. The field-cooled (FC) susceptibility is greater than the zero-field-cooled (ZFC) susceptibility, the latter being similar to the a.c. susceptibility in the limit of low field. Another characteristic preparation-dependent feature concerns remanent magnetization beneath T_g; the isothermal remanent magnetization σ_{IRM} obtained by cooling the system in zero field, then applying the field for a macroscopic period of time, and finally switching off the field, is smaller than the thermoremanent magnetization σ_{TRM}, in which the field is applied at a temperature greater than T_g, the system is cooled to the required temperature and then the field is removed. Dynamic relaxation to equilibrium is slow. Examples of these features are shown in Figs. 3-6.

The cusp in the susceptibility suggests a phase transition to a frozen but non-ferromagnetic magnetic phase. The lack of periodic order shows it to be non-conventional. The preparation dependent effects

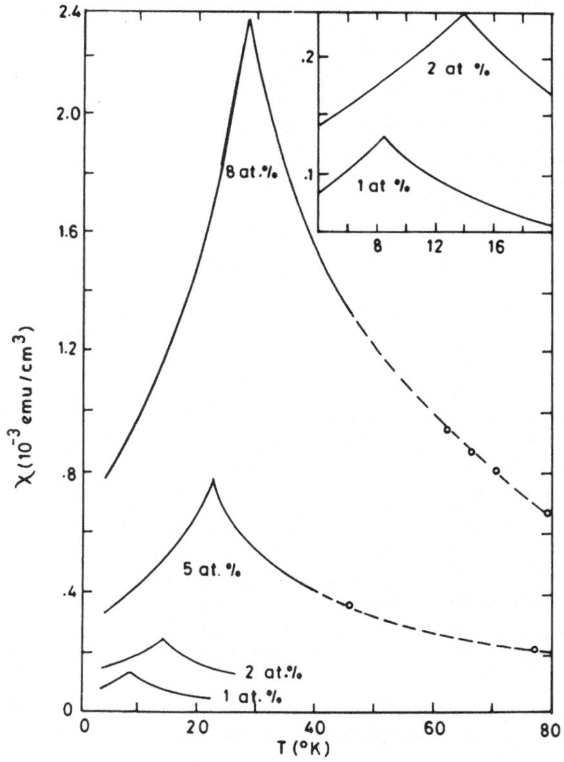

Fig. 3. a.c. susceptibility of the canonical spin glass <u>Au</u> Fe; from Ref. 10.

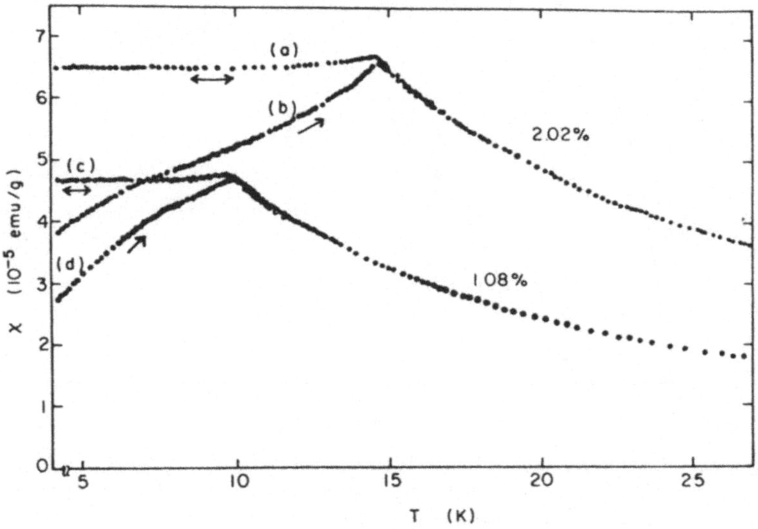

Fig. 4. d.c. susceptiblity results for two Cu Mn alloys with 1.08 and 2.02 at % Mn. Curves (a) and (c) are obtained by cooling in the measuring field (FC), (b) and (d) are the results of zero field cooled (ZFC) experiments; from S. Nagata, P.H. Keesom and H.R. Harrison, Phys. Rev. B 19, 1633 (1979).

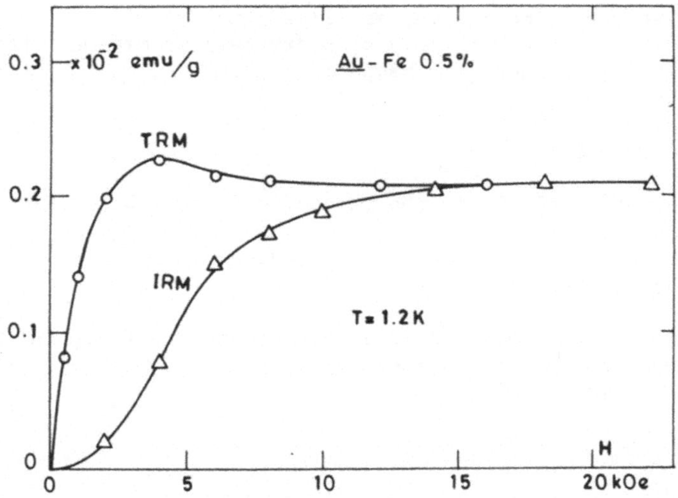

Fig. 5. Remanent magnetizations as measured in Au -0.5% Fe at 1.2K. IRM denotes isothermal remanent magnetization, TRM denotes thermoremanent magnetization; from J.L. Tholence and R. Tournier, J. de Physique 35, C4-229 (1974).

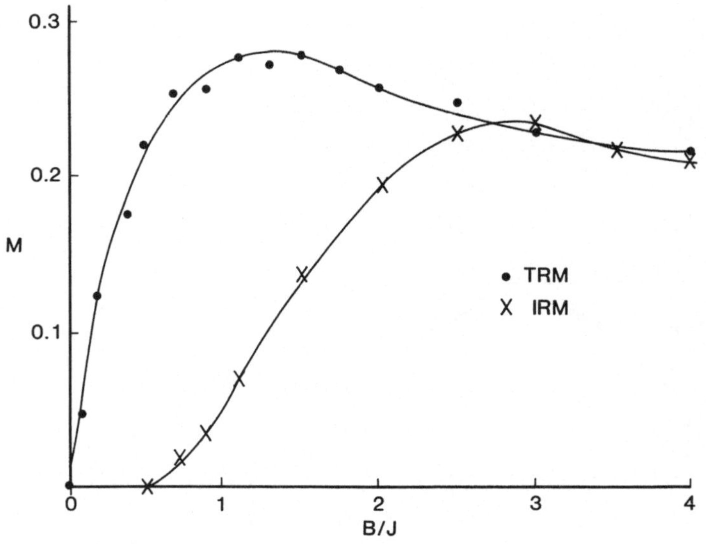

Fig. 6. Remanent magnetizations as obtained in a computer simulation of a model two-dimensional Ising spin glass with Gaussian distributed nearest neighbour exchange and at a temperature one quarter of the Gaussian exchange width; from W. Kinzel, Phys. Rev. B 19, 4595 (1979).

suggest that the free energy structures for temperatures less than T_g have many non-symmetry-related minima, as illustrated in Fig. 2, in contra-distinction to the forms with only symmetry-related minima, characteristic of simple conventional periodic magnetic phases.

II ANALYTIC THEORY

Let us focus initially on the random bond model with Ising spins; ie.

$$H = - \sum_{(ij)} J_{ij} \sigma_i \sigma_j \qquad ; \sigma_i = \pm 1 \tag{5}$$

with the spins on all the sites of a lattice and the $\{J_{ij}\}$ independent random exchanges having the same distribution $P(J_{ij})$ for all pairs of sites (ij) the same distance apart.

II.A. Possibility of a New Phase

Let us first note that if $P(J_{ij})$ is symmetric,

$$P(J_{ij}) = P(-J_{ij}), \tag{6}$$

then even a very crude mean field theory [18] indicates that although ferromagnetism is impossible, a new kind of order with non-periodically frozen spins may occur.

Allowing for non-equivalency of sites, naive mean field theory gives

$$\langle\sigma_i\rangle = \tanh(\sum_j \beta J_{ij}\langle\sigma_j\rangle) \quad (7)$$

where < > denotes a thermodynamic average,

$$\langle 0\rangle = \frac{\mathrm{Tr}\ 0\ \exp(-\beta H)}{\mathrm{Tr}\ \exp(-\beta H)}, \quad (8)$$

$$\mathrm{Tr} \equiv \sum_{\{\sigma_i\} = \pm 1} \quad (9)$$

and

$$\beta \equiv (k_B T)^{-1}. \quad (10)$$

Conventionally, the critical temperature T_c for the onset of order is given by the existence of a non-trivial solution ($\langle\sigma\rangle \neq 0$) to the linearized mean field equation. In the present case this corresponds to

$$\langle\sigma_i\rangle = \sum_j J_{ij} \langle\sigma_j\rangle/kT_c. \quad (11)$$

Averaging over sites and bonds and ignoring correlations between these averages yields

$$[\langle\sigma_i\rangle] = \sum_j [J_{ij}][\langle\sigma_j\rangle]/k_B T_c \quad (12)$$

where the [ ] brackets denote such an average over the spatial disorder. For a symmetric distribution $[J_{ij}]$ is zero so that there is no non-trivial solution to (12) at finite temperature. If, however, (11) is first squared and then averaged, a similar ignoring of correlations gives

$$[\langle\sigma_i\rangle^2] = \sum_j [J_{ij}^2][\langle\sigma_j\rangle^2]/(k\ T_g)^2, \quad (13)$$

which has a non-trivial solution at

$$(kT_g)^2 = \sum_j [J_{ij}^2], \quad (14)$$

with "order parameter" $[\langle\sigma_i\rangle^2]$; that is, below T_g one can have a "frozen-spin" phase with zero overall magnetization. Of course, this crude theory has several questionable features, ignoring correlations and reaction fields and implicitly assuming all the eigenstates of J_{ij} to be extended, but we shall see that a more sophisticated mean field theory also leads to the prediction of a new cooperative magnetic phase without periodic long range order.

There are two main mean field theories which deal carefully with the subtleties of the spin glass. I shall concentrate on that known as replica mean field theory [13,14]. The other is based on taking account of reaction or cavity field modifications to eqn (7); in its initial form this approach is associated with Thouless, Anderson and Palmer [19] but it has reached great sophistication in the hands of Mézard, Parisi and Virasoro [20]. The replica and cavity methods give the same final results.

In all the theories the concern is with large systems, in the thermodynamic limit as the number of spins N approaches infinity, and in statistically relevant quantities, averages or distributions over ensembles of systems with statistically equivalent constraints.

II.B. Replica Theory

1. Replicas in Lieu of Disorder. When one is interested in statistically representative quantities, rather than specific instances, it is often convenient to average over the quenched disorder at an early stage. This is the philosophy we shall follow, averaging over all representations with the same exchange distributions $P(J_{ij})$. It is however, crucial that one averages only extensive observables, which on macroscopic scales depend only on the preparation rules and not on the detailed resulting disorder [21]. Thus, in time-dependent statistical mechanics, it is legitimate to average the free energy, $F = -k_B T \ln Z$, but not the partition function Z itself. Unfortunately, whereas Z, being a sum of exponentials,

$$Z = \mathrm{Tr} \exp(-H(\{J_{ij}\})/k_B T), \tag{15}$$

is relatively easy to average, ln Z is much harder. A mathematical artifice employed to overcome this difficulty is to use the identity

$$\ln Z = \lim_{n\to 0} \frac{1}{n}(Z^n - 1) \tag{16}$$

to write Z^n as

$$Z^n = \mathrm{Tr}_n \exp(-\sum_{\alpha=1}^{n} H^\alpha(\{J_{ij}\})/k_B T), \tag{17}$$

where

$$H^\alpha(\{J_{ij}\}) = -\sum_{(ij)} J_{ij}\, \sigma_i^\alpha \sigma_j^\alpha \quad ; \; \sigma_i^\alpha = \pm 1, \tag{18}$$

corresponding to n replicas of the system with identical $\{J_{ij}\}$. Tr_n is a trace over all nN spins,

$$\mathrm{Tr}_n \equiv \prod_{\alpha=1}^{n} \sum_{\{\sigma_i^\alpha\}=\pm 1} . \tag{19}$$

ln Z is thus replaced by the limiting value of the partition function of a system with n independent Ising spins on each site, but with the interactions between sites identical for each group of spins. The disorder average is now readily performed;

$$[Z^n] = \mathrm{Tr} \exp\{\sum_{(ij)} \sum_r [J_{ij}{}^r]_c/(k_B T)^r$$

$$\sum_{\alpha,\beta,..} (\sigma_i^\alpha \sigma_i^\beta \ldots \sigma_i^{\gamma r})(\sigma_j^\alpha \sigma_j^\beta \ldots \sigma_j^{\gamma r})\} \tag{20}$$

where $[J_{ij}{}^r]_c$ indicates the rth cumulant average of J_{ij}, and α,β... indicate r replica labels, each ranging from 1 to n.

If we follow Edwards and Anderson [12] and restrict ourselves to

Gaussian $P(J_{ij})$ then $[J_{ij}^{\ r}]_c$ is zero for all $r > 2$. Further restricting to nearest neighbour interactions, we have for the average free energy

$$[F] = -k_B T \lim_{n \to 0} \frac{1}{n} \left\{ \mathrm{Tr}_n \exp\left(\sum_{(ij)} \left[\beta \tilde{J}_o \sum_\alpha \sigma_i^\alpha \sigma_j^\alpha + (\beta \tilde{J})^2 \sum_{\alpha\beta} \sigma_i^\alpha \sigma_i^\beta \sigma_j^\alpha \sigma_j^\beta \right]\right) - 1 \right\} \tag{21}$$

where $\tilde{J}_o$ is the mean and $\tilde{J}$ the standard deviation of $P(J_{ij})$. Thus we have replaced the disordered system of (3), averaged over the J_{ij}, by one with a periodic (ordered) effective Hamiltonian,

$$H_{eff} = -\sum_{(ij)} \left(\tilde{J}_o \sum_{\alpha=1}^{n} \sigma_i^\alpha \sigma_j^\alpha + \beta \tilde{J}^2 \sum_{\alpha,\beta=1}^{n} \sigma_i^\alpha \sigma_i^\beta \sigma_j^\alpha \sigma_j^\beta \right), \tag{22}$$

with a more complicated interaction involving higher-dimensional spins and requiring analysis in the limit as $n \to 0$.

2. Replica Mean Field Theory. In general (21) cannot be evaluated exactly. A natural approximation is a mean field theory in which, by analogy with coventional mean-field theory, one converts the problem to a self-consistantly determined single-site problem by the replacements

$$\sum_{(ij)} \sigma_i^\alpha \sigma_j^\alpha \to \sum_{(ij)} (\sigma_i^\alpha m^\alpha - (m^\alpha)^2/2); \quad m^\alpha = \langle \sigma_i^\alpha \rangle_n \tag{23a}$$

$$\sum_{(ij)} \sigma_i^\alpha \sigma_i^\beta \sigma_j^\alpha \sigma_j^\beta \to \sum_{(ij)} (\sigma_i^\alpha \sigma_i^\beta q^{\alpha\beta} - (q^{\alpha\beta})^2/2); \quad q^{\alpha\beta} = \langle \sigma_i^\alpha \sigma_i^\beta \rangle_n; \ \alpha \neq \beta \tag{23b*}$$

where $\langle\ \rangle_n$ refers to a thermodynamic average against the effective Hamiltonian and the order parameters m^α, $q^{\alpha\beta}$ are to be determined self-consistently. For convenience we are assuming J_o to be non-negative so that $\langle \sigma_i^\alpha \rangle$ is the same on all sites⁺. $\langle \sigma_i^\alpha \sigma_i^\beta \rangle$ is also site-independent.

With these substitutions the evaluation of $[F]$ becomes a single-site problem

$$[F] = -Nk_B T \lim_{n \to 0} \frac{1}{n} \left\{ \mathrm{Tr}_n \exp\left[\beta \tilde{J}_o z \sum_\alpha (\sigma^\alpha m^\alpha - (m^\alpha)^2/2) + (\beta \tilde{J})^2 z \left(n + 2 \sum_{(\alpha\beta)} (\sigma^\alpha \sigma^\beta q^{\alpha\beta} - (q^{\alpha\beta})^2/2)\right)\right] - 1 \right\} \tag{24}$$

where z is the coordination number and the trace is now single-site. m^α, $q^{\alpha\beta}$ are given by

$$m^\alpha = \frac{\mathrm{Tr}_n\, \sigma^\alpha \exp[-\beta \tilde{H}_n]}{\mathrm{Tr}_n \exp[-\beta \tilde{H}_n]} \tag{25}$$

---

* For $\alpha = \beta$, $\sigma^\alpha \sigma^\beta = 1$.

\+ Were J_o negative, we should need to employ a staggered magnetization as the $\langle \sigma_i \rangle$ order parameter, as in a conventional antiferromagnet.

and

$$q^{\alpha\beta} = \frac{Tr_n\ \sigma^\alpha\ \sigma^\beta\ \exp[-\beta\tilde{H}_n]}{Tr_n\ \exp[-\beta\tilde{H}_n]}, \tag{26}$$

where $-\beta\tilde{H}_n$ is the argument of the exponential in eqn (24), or, equivalently, by the extremal equations

$$\frac{\delta\tilde{F}_n}{\delta m^\alpha} = \frac{\delta\tilde{F}_n}{\delta q^{\alpha\beta}} = 0\ , \tag{27}$$

where

$$\tilde{F}_n = -k_B T\ \ln\ Tr_n\ \exp(-\beta\tilde{H}_n). \tag{28}$$

These mean field equations are believed to be exact in the thermodynamic limit ($N \to \infty$) for the Sherrington-Kirkpatrick infinite-range spin glass [13] in which the interactions extend over all pairs of spins, with the J_{ij} independent random parameters, equally distributed with mean and variance each scaling as N^{-1}, where N is the total number of spins; ie.

$$\tilde{J}_o = J_o N^{-1}, \tag{29}$$

$$\tilde{J} = J\ N^{-1/2}. \tag{30}$$

One is still, however, faced with the limiting procedure $n \to 0$. This is greatly simplified in the replica symmetric ansatz in which one assumes m^α to be independent of α and all the $q^{\alpha\beta}$ ($\alpha \neq \beta$) to be identical;

$$\begin{aligned} m^\alpha &= m \quad ; \quad \text{all } \alpha, \\ q^{\alpha\beta} &= q \quad ; \quad \text{all } \alpha \neq \beta. \end{aligned} \tag{31}$$

This ansatz is natural, since the replicas are mathematical artifices and indistinguishable. In fact, however, the true situation is more subtle than (31), but we shall defer discussion until later.

Within the replica-symmetric ansatz the exponent in eqn (24) may be linearized in σ^α by means of the identities

$$\exp(2\lambda \sum_{(\alpha\beta)} \sigma^\alpha\sigma^\beta) = \exp(\lambda((\sum_\alpha \sigma^\alpha)^2-1)) \tag{32}$$

and

$$\exp(\lambda x^2) = (2\pi)^{-1/2} \int dy\ \exp(-y^2/2 + (2\lambda)^{1/2} yx). \tag{33}$$

We then readily obtain

$$[F] = N \left\{ \frac{J_o m^2}{2} - \frac{\beta J^2}{4} (1-q)^2 - k_B T \int dh\ P(h)\ \ln\ 2\cosh(\beta h) \right\}, \tag{34}$$

$$m = \int dh\ P(h)\ \tanh(\beta h), \tag{35}$$

$$q = \int dh\ P(h)\ \tanh^2(\beta h), \tag{36}$$

where

$$P(h) = (2\pi J q^2)^{-1/2} \exp(-(h-J_o m)^2/2Jq^2), \tag{37}$$

$$J_o = \tilde{J}_o z \quad \text{and} \quad J = \tilde{J} z^{1/2}. \tag{38}$$

Thus, in this approximation, m and q behave as the averages of a paramagnetic magnetization and its square against an effective field h, of self-consistently determined distribution given by eqn (37).

Before noting the further consequences of these equations, let us note that m and q can be directly related to the original problem as follows. We may write

$$\langle\sigma_i\rangle^r = \frac{\mathrm{Tr}_r\, \sigma_i^{\alpha_1}\sigma_i^{\alpha_2}\ldots\sigma_i^{\alpha_r}\exp(-\beta(H^{\alpha_1}+H^{\alpha_2}+\ldots+H^{\alpha_r}))}{\mathrm{Tr}_r\, \exp(-\beta(H^{\alpha_1}+H^{\alpha_2}+\ldots+H^{\alpha_r}))} \tag{39}$$

where the α's are (real) replica labels and the H^{α_r} all have the same disorder. Multiplying numerator and denominator by Z^{n-r}, taking the limit $n \to 0$, and utilizing the fact that $\lim_{n\to 0} Z^n = 1$, there results

$$\langle\sigma_i\rangle^r = \lim_{n\to 0} \mathrm{Tr}_n\, \sigma_i^{\alpha_1}\ldots\sigma_i^{\alpha_r} \exp(-\beta\sum_{p=1}^{n} H^{\alpha_p}). \tag{40}$$

Disorder averaging then yields

$$[\langle\sigma_i\rangle^r] = \lim_{n\to 0} \langle\sigma_i^{\alpha_1}\ldots\sigma_i^{\alpha_r}\rangle_n \quad ; \alpha_1 \neq \alpha_2 \ldots \neq \alpha_r. \tag{41}$$

In particular, therefore,

$$[\langle\sigma_i\rangle] = \lim_{n\to 0} m^{\alpha}, \tag{42}$$

$$[\langle\sigma_i\rangle^2] = \lim_{n\to 0} q^{\alpha\beta}. \tag{43}$$

Hence, in the limit $n \to 0$, m is identified as the average magnetization and q as the average squared local magnetization.

From this identification it is clear that non-zero q implies a cooperatively ordered magnetic state, non-zero m implies a ferromagnetic component to that order. We thus identify solutions to eqns (35)-(38) as (i) paramagnetic if $m = q = 0$, (ii) ferromagnetic if $m \neq 0$, $q \neq 0$, and (iii) spin glass if $m = 0$, $q \neq 0$. The resultant phase diagram is shown in Fig. 7a, whilst for comparison a typical experimental phase diagram is shown in Fig. 7b. When account is taken of the different scalings with concentration of effective interactions J_o and J, there is qualitative accord between these figures.

Let us now concentrate on the spin glass phase. Beneath the spin glass transition temperature $T_g = J/k_B$, replica mean field theory predicts that q grows continuously as (T_g-T). Using the correlation function expression for the differential susceptibility*,

$$\chi = (k_B T)^{-1} \sum (\langle\sigma_i\sigma_j\rangle - \langle\sigma_i\rangle\langle\sigma_j\rangle), \tag{44}$$

---

* This expression follows directly by formal differentiation of $M = \sum_i \langle\sigma_i\rangle = \sum_i \mathrm{Tr}\,\sigma_i \exp(-\beta H)/\mathrm{Tr}\exp(-\beta H)$.

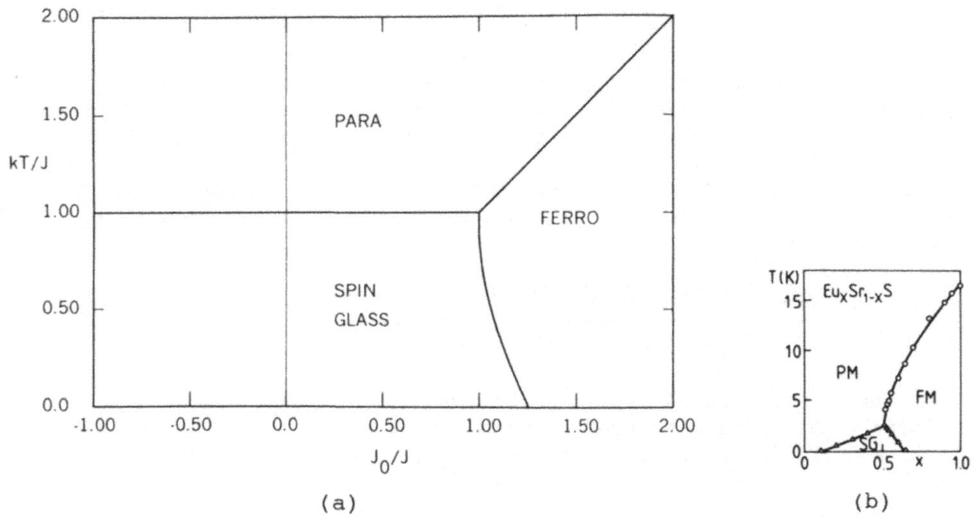

Fig. 7. (a) Phase diagram obtained in replica-symmetric mean field theory for a random bond Ising spin glass; from Ref. 13.
(b) Phase diagram of $Eu_xSr_{1-x}S$; from H. Maletta and P. Convert, Phys. Rev. Lett. 42, 108, (1979).

we obtain, in the case of $J_o = 0$ and no external field,

$$\chi = \frac{N}{k_B T}(1-[\langle\sigma_i\rangle^2]) \tag{45a}$$

$$= \frac{N}{k_B T}(1-q) \ . \tag{45b}$$

Thus for $T > T_g$ one has Curie behaviour, but beneath T_g the onset of q gives rise to a reduction and a cusp in $\chi(T)$ at T_g. In the presence of an external field b, eqn (37) is modified by $h \rightarrow h - b$, and hence χ can also be obtained from $\chi = N\partial m/\partial b$, yielding again eqn (45b) for $J_o = b = 0$. Finite J_o leads to enhancement (by $(1-\beta J_o)^{-1}$ for b = 0). In Fig. 8 are shown the predictions for a variety of values of J_o,b, showing the cusp for b = 0, rounded by a finite field. Again, there is qualitative accord with the experimental a.c. or ZFC susceptibilities (see Figs. 3 and 4).

In fact, however, the replica-symmetric ansatz is not everywhere stable [13, 22-24]. The problem is clearly exposed [22] by expanding $q^{\alpha\beta}$ and m^α about their replica symmetric values

$$q^{\alpha\beta} = q + \eta^{\alpha\beta}, \ m^\alpha = m + \varepsilon^\alpha, \tag{46}$$

expanding [F] to second order in the fluctuations, and analysing the resultant normal modes in replica space. In the limit $n \rightarrow 0$ the ansatz is found to be unstable against replica-symmetry breaking η- modes beneath a surface in (T/J, b/J, J_o/J) space, as shown in Fig. 9. This instability surface is known as the de Almeida-Thouless (AT) surface, its projections as de Almeida-Thouless lines. Beneath the surface a more subtle ansatz is required. That due to Parisi [14] has proven to be at least marginally stable against all fluctuations tested to date. Parisi's ansatz is based on a continuous hierarchy of symmetry breakings.

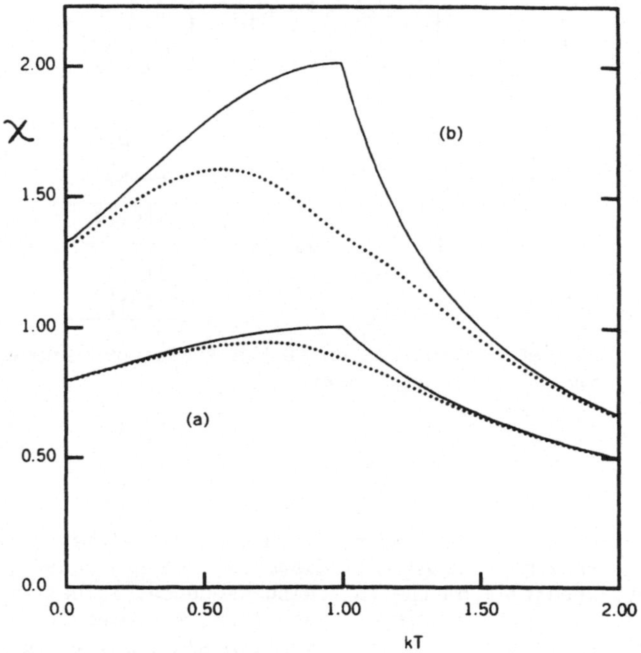

Fig. 8. Differential susceptibility of a random-bond spin glass as given by replica-symmetric mean field theory. Solid curves are for zero field, dotted curves for b = 0.1J. Curves (a) are for $J_o/J = 0$, curves (b) are for $J_o/J = 0.5$; from Ref. 13.

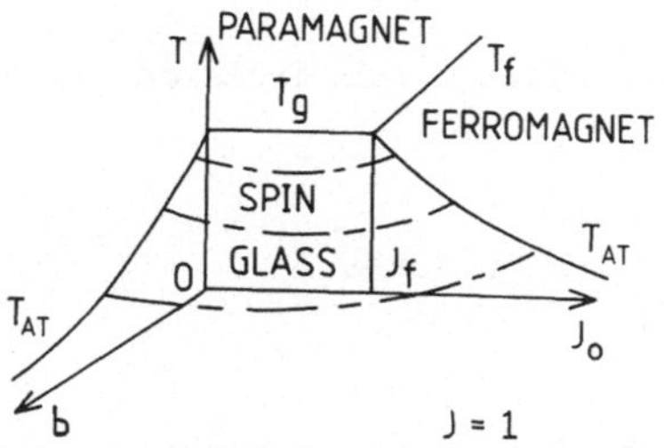

Fig. 9. de Almeida-Thouless surface for the limit of stability of the replica-symmetric ansatz for mean field theory for a random-bond Ising model. The ansatz is unstable on the side of the surface closer to the origin.

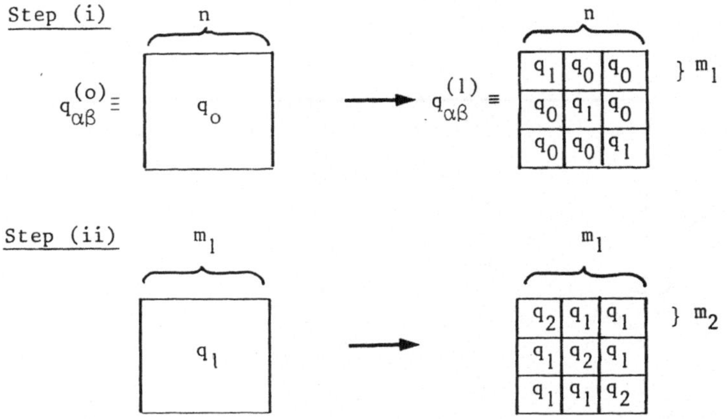

Fig. 10. Schematic illustration of the Parisi procedure for replica-symmetry breaking.

Considering $q^{\alpha\beta}$ as an (nxn) matrix with diagonal elements zero, the replica-symmetric ansatz takes all off-diagonal elements equal (to q). The Parisi ansatz results from the following sequence, illustrated in Fig. 10. First divide the (nxn) matrix $q^{\alpha\beta}$ into $(n/m_1)^2$ blocks of $(m_1 x m_1)$ matrices. Take the off-diagonal blocks to have all their elements equal to q_o. Sub-divide the diagonal $(m_1 x m_1)$ blocks into $(m_1/m_2)^2$ sub-blocks of $(m_2 x m_2)$ matrices. Take the off-diagonal $(m_2 x m_2)$ sub-blocks to have all their elements equal to q_1. Subdivide further the $(m_2 x m_2)$ sub-blocks into $(m_2/m_3)^2$ sub-sub-blocks of $(m_3 x m_3)$ matrices. Take the off-diagonal $(m_3 x m_3)$ sub-sub-blocks to have their elements equal to q_2. And so on, sub-dividing diagonal blocks, with

$$n \geq m_1 \geq m_2 \ldots \geq 1. \tag{47}$$

Now take the limit $n \to 0$, reversing the inequalities to

$$0 \leq m_1 \leq m_2 \ldots \leq 1 \tag{48}$$

and pass to the limit of infinite sub-division

$$\left.\begin{array}{l} \dfrac{m_k}{m_{k+1}} \to 1 - \dfrac{dx}{x} \\ q_k \to q(x) \end{array}\right\} \quad 0 \leq x \leq 1 \ . \tag{49}$$

$q(x)$; $0 \leq x \leq 1$ is now the Parisi order function. The average free energy can be expressed as a functional of $q(x)$ and the order function itself obtained by taking its extremum. The order parameter m remains replica-symmetric.

Before discussing the character of the solutions emerging from the Parisi treatment, it is perhaps helpful to indicate the significance of $q(x)$. To this end, let us introduce the concept of overlap between states. The overlap between two microstates s, s' is defined by

$$\tilde{q}^{ss'} = N^{-1} \sum_i \sigma_i^s \sigma_i^{s'} \tag{50}$$

where $\{\sigma_i^s\}$ is the spin orientation in microstate s. A distribution of overlaps can then be defined by

$$\tilde{P}(q) = \sum_{ss'} p_s \, p_{s'} \, \delta(q - \tilde{q}^{ss'}) \tag{51}$$

where p_s, $p_{s'}$ are the probabilities of microstates s, s'. In thermodynamic equilibrium

$$p_s = Z^{-1} \exp(-\beta E_s) \tag{52}$$

where E_s is the energy of state s. The average overlap distribution in equilibrium is directly related to q(x) by

$$[\tilde{P}(q)] = \int dx \, \delta(q-q(x)) = dx/dq. \tag{53}$$

Similarly, we may define an overlap between thermodynamic macrostates S, S' by

$$q^{S,S'} = N^{-1} \sum_i m_i^S \, m_i^{S'} \tag{54}$$

where

$$m_i^S = \langle \sigma_i \rangle^S \tag{55}$$

is the local magnetization averaged over thermodynamic state S. Again, a distribution of overlaps may be defined by

$$P(q) = \sum_{SS'} P_S \, P_{S'} \, \delta(q-q^{SS'}) \tag{56}$$

where P_S, $P_{S'}$ are the probabilities of macrostates S, S'. In Gibbs thermodynamic equilibrium

$$P_S = Z^{-1} \exp(-\beta F_S) \tag{57}$$

where F_S is the free energy of state S. These distributions are related in the thermodynamic limit by

$$P(q) = \tilde{P}(q). \tag{58}$$

Thus dx/dq gives the average distribution of overlaps between thermodynamic states in equilibrium. q(x) independent of x yields [P(q)] with a single delta function, corresponding to a single thermodynamic state*. q(x) varying with x gives more structure to [P(q)], indicating the existence of multiple states.

Clearly, the average of overlap-related quantities follows directly. For example,

$$[\langle \sigma_i \rangle^2] = \int dq \, q[P(q)] = \int_0^1 dx \, q(x), \tag{59}$$

* q(x) is normally restricted to positive values. Global inverse states are given by adding its reflection q(-x). Interest is, however, in states not related by global symmetry operations.

$$[\langle\sigma_1\,\sigma_2\,\ldots\,\sigma_k\rangle^2] = \int dq\; q^k[P(q)] = \int_0^1 dx\; q(x)^k. \tag{60}$$

Rather than deriving eqn (53) directly here, let us reconsider eqn (39) allowing for replica-symmetry breaking. In this case one must average the right hand side over explicit identifications of the α_p,

$$[\langle\sigma_i\rangle^r] = \lim_{n\to 0} \frac{1}{n(n-1)\ldots(n-r+1)} \sum_{\alpha_1\neq\alpha_2\ldots\neq\alpha_r} \langle\sigma^{\alpha_1}\sigma^{\alpha_2}\ldots\sigma^{\alpha_r}\rangle \tag{61}$$

Noting that the Parisi procedure yields,

$$\lim_{n\to 0} \frac{1}{n(n-1)} \sum_{\alpha\neq\beta} (q^{\alpha\beta})^k = \int_0^1 dx\; q(x)^k \tag{62}$$

the result (63) follows immediately. Result (64) follows from a related analysis. Averages with $r > 2$ can also be related to $q(x)$ but will not be discussed explicitly.

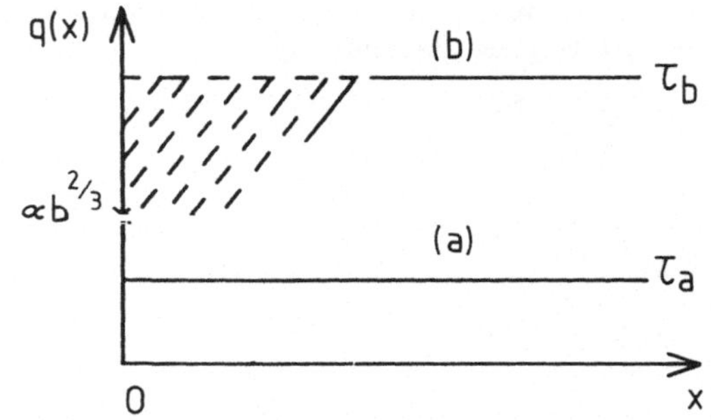

Fig. 11. Parisi order function $q(x)$ (a) above the de Almeida-Thouless surface (b) below the AT surface, with parameters shown for $J_o = 0$. τ_a, τ_b are the corresponding reduced temperatures; $\tau = (T_g - T)/T_g$; for τ small.

Turning now to the solutions of Parisi mean field theory, let us first comment that, above the AT surface $q(x)$ is independent of x, implying not only replica symmetry but also a single thermodynamic state. Beneath the AT surface $q(x)$ is found to have a form as illustrated in Fig. 11, with an upper plateau $q(1)$ for $1 \geq x \geq x_2$, a possible lower plateau $q(0)$ for $0 \leq x \leq x_1$, and a monotonic variation in between. In the absence of an external field or spontaneous magnetization, x_1 and $q(0)$ are both zero, but in general they increase monotonically with effective field. $q(1)$ and x_2 increase with decreasing temperature beneath T_g. The Almeida-Thouless surface corresponds to the coalescence of the plateaux. Turning to $[P(q)]$ this form of $q(x)$ implies that there is a delta function peak at the maximum overlap of q(1) and, for q(0) ≠ 0, another at the minimum overlap q(0), with a continuous spread of overlaps between these limits. The peak at q(1) is interpreted as reflecting the overlap of a state with itself, S = S' in eqn (56), whilst the continuous region between q(0) and q(1) is due to overlaps between a range of non-equivalent thermodynamic states. This can be taken as evidence for a set of non-equivalent global minima in the free energy (cf. Fig. 2).

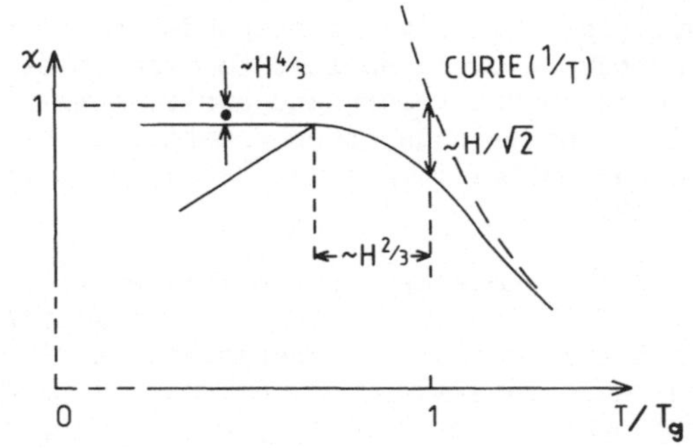

Fig. 12. Susceptibility of an Ising spin glass in an applied field H, as predicted by Parisi's mean field theory. The upper solid curve shows the full Gibbs average, obtained from the full q(x) and interpreted as the field-cooled (FC) susceptibility. The lower curve shows the result of restricting to one thermodynamic state, as obtained from q(1) and interpreted as the zero-field-cooled (ZFC) susceptibility.

Turning to the susceptibility, and specialising to $J_o = b = 0$, the Gibbs-averaged susceptibility

$$\chi = N(k_B T)^{-1}(1 - [\langle\sigma_i\rangle^2]) \tag{45a}$$

becomes now, using eqn (59),

$$\chi = N(k_B T)^{-1}\left(1 - \int_0^1 q(x)dx\right). \tag{63}$$

For $T > T_g$, where q(x) is zero, this gives the usual Curie 1/T behaviour, but beneath T_g

$$1 - \int_0^1 q(x)\, dx = T/T_g \tag{64}$$

and

$$\chi = N/k_B T_g \ ; \text{all } T < T_g. \tag{65}$$

Thus for $T < T_g$ the prediction is a plateau, comparable with the field-cooled susceptibility exhibited in Fig. 4. On the other hand, q(1) behaves essentially as q in the replica-symmetric theory (and is often referred to as q_{EA}, after Edwards and Anderson, the inventors of the replica-symmetric spin-glass mean-field theory), so that

$$\chi_{EA} = N(k_B T)^{-1}(1 - q(1)) \tag{66}$$

behaves analogously to the a.c. or zero-field-cooled susceptibilities. Recalling the interpretation of q(1) as self-overlap, the interpretation is made of the replacement of q(x) by q(1) as corresponding to a system restrained to a single thermodynamic state, as might be expected, on the basis of the interstate barrier picture of Fig. 2, to occur in an a.c. or ZFC experiment, while the full q(x) corresponds to having access to all states. This philosophy is often applied to other observables, giving a "folklore mapping" between the theory and experimental results corresponding to different methods of preparation or of time-scale. The predictions for the susceptibility in the presence of an applied field are illustrated in Fig. 12.

A note of caution is, however, appropriate at this stage. The Parisi theory is only a mean field theory, albeit a very sophisticated one. Mean field theory does not take account of thermodynamic fluctuations in systems with finite-range interactions*, nor can it predict critical dimensions+. On the other hand, it does give a number of predictions remarkably close, at least qualitatively, to experimental observations, it has been shown to be in quantitative accord with computer simulations of the infinite-range spin glass§, and it is believed to be of direct relevance to certain models of neural memory and to the partitioning of random graphs.

Another point worth mentioning, since it runs counter to naive expectation, is that the free energy of the finite q(x) mean field theory beneath T_g is higher than the analytic continuation of the paramagnetic free energy and the Parisi result is higher than that of replica symmetric theory. In fact, naive choice of the minimum free energy is erroneous. The correct result is the minimum free energy <u>subject to stability</u>. The requirement of even replica symmetric stability beneath T_g is sufficient to eliminate the continuation of the paramagnetic phase. The instability of replica-symmetry breaking modes in the replica-symmetric phase beneath the AT surface leads to its elimination in favour of the Parisi phase.

Another interesting feature exhibited by the Parisi solution is ultrametricity [27,28]. A space is described as "ultrametric" if distances in that space obey the following property: given three points a,b,c separated by "distances" in some space d_{ab}, d_{bc}, d_{ca} and labelled such that

$$d_{ab} \geq d_{bc} \geq d_{ca}, \tag{67}$$

* A clear manifestation of deficiency is that replica mean field theory predicts a cusp in the specific heat at the temperature of the susceptibility cusp, whereas experimentally it is continuous at this temperature and rounded at a higher temperature.

+ The lower critical dimension is that beneath which there is no phase transition at finite temperature. The upper critical dimension is that at which mean field critical exponents become exact.

§ For example the transition temperature [23,24], the susceptibilities [23,25] and the average overlap distribution [26].

then

$$d_{ab} = d_{bc} \; ; \tag{68}$$

i.e., the two largest distances between three points are equal. Distances are a measure of separation. Overlap is a measure of similarity, the converse of separation. Thus, a space of overlaps is ultrametric if for three states S, S', S'', labelled so that

$$q^{SS'} \leq q^{S'S''} \leq q^{S''S}, \tag{69}$$

one has

$$q^{SS'} = q^{S'S''}; \tag{70}$$

i.e., the two smallest overlaps are equal. With the overlap definition of eqn (54)*, Parisi theory predicts such ultrametricity of a non-trivial kind; trivial ultrametricity is when all q (or d) are equal. The interest in this result is its implication of hierarchical order. This implication is apparent if one considers an evolutionary tree, as in Fig. 13, takes its end-points as the states and their pairwise overlap to be determined by how far back one needs to go to find a common ancestor, the overlap being smaller the further back one needs to go. One readily sees that for any three states, the two smallest overlaps are equal. There is a similar hierarchy in the Parisi spin glass, with the overlap definition as given by eqn (54), and for some other measures [29,30]. If the overlap is expressed by the height in a generalised pairwise phase space one obtains a fractal mountainous "landscape".

The Parisi theory predicts other unusual features, borne out by simulations on the infinite-range model [26], including a lack of self-averaging+ of the overlap distribution, the corresponding self-averaging quantity being the distribution functional of the distribution of overlaps, P(P(q))[27]. We shall not, however, pursue these aspects further.

Another interesting association which will not be pursued is between the parameter x, of Parisi's q(x), and the timescale τ [31]. This correspondence has

$$\tau(x + \delta x) << \tau(x). \tag{71}$$

Beyond mean field theory, in finite dimensions greater than unity, there are no comparable exact theoretical results and controversy still reigns. Renormalization group analysis [35] suggests that the upper critical dimension is six but detailed theoretical analysis is complicated by uncertainties as to whether the Parisi mean field theory is an appropriate basis around which to perform expansions. Computational investigations concerned with lower dimensions will be discussed in the next section.

---

* Several measures of distance are possible. A simple one is $d_{SS'} = 1-q^{SS'}$. One which obeys triangular inequalities is

$$d_{SS'} = \{N^{-1} \Sigma ((m_i^S) - (m_i^{S'})^2)\}^{1/2}$$
$$= (q^{SS} + q^{S'S'} - 2q^{SS'})^{1/2}$$

\+ A quantity is self-averaging if in the thermodynamic limit it takes the same value(s) for all specific realisations of the quenched disorder.

Finally, in this section, we note that the replica theory discussed above for Ising spins can be extended to vector and other generalized spins, but refer the reader to refs. 33-35 for reviews.

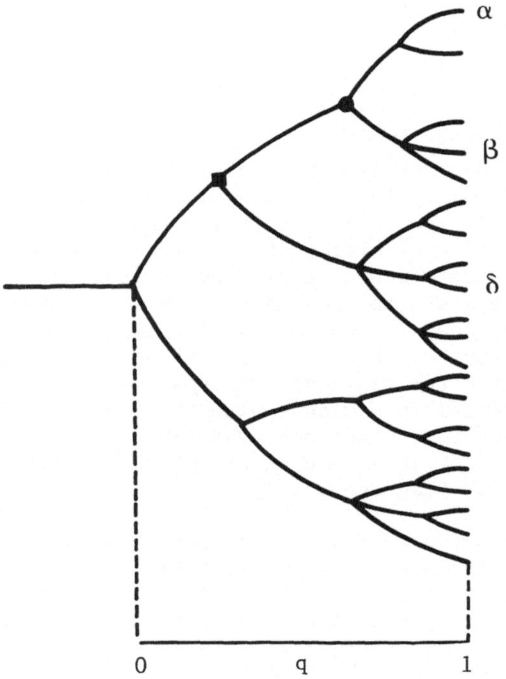

Fig. 13. An evolutionary tree, illustrating the occurence of ultrametricity. If the overlap between two final states is measured by the degree of evolution of their nearest common ancestor, as shown, then for any group of three final states the two smallest overlaps are equal.

III. COMPUTER SIMULATIONS

A complete and reliable analytic study of finite-range spin glasses is still lacking, although a number of qualitative and semiquantitative discussions and arguments have been given, with conclusions which have not always remained time-invariant. Computer simulations have provided valuable supplementation and, for example, the most convincing evidence concerning lower critical dimensionality. In fact, spin glasses are a particularly notable example of the use of computational physics as a third major mode of study in physics (as well as analytic theory and conventional experiment). In this section are described some of those simulational studies, involving procedures for which no conventional experimental analogues exist.

III.A. Zero Temperature Scaling

The signature of a macroscopically frozen phase is the presence of long-range order. The first study of long-range correlation relevant to spin glasses was by Morgenstern and Binder [36] who used transfer matrix techniques to study $[\langle\sigma(\underline{r}')\sigma(\underline{r}+\underline{r}')\rangle^2]$, the average of the square of the correlation between Ising spins on sites a distance $\underline{r}$ apart. They found that in two dimensions this correlation function decayed with r even for the lowest temperatures studied, although the decay was not explicitly followed to zero for low temperatures. They deduced that spin glass order was absent in two dimensions.

A more recent study relies on the scaling with distance of an effective exchange interaction between blocks of spins [37]. It systematizes the idea that only a macroscopically ordered system is affected by details of its boundary conditions, a concept also used to discuss localization of electronic states. An effective interaction between blocks of size L is given by (one-half of) the energy difference between a block of size L with periodic boundary conditions and the same block with antiperiodic boundary conditions. For a spin glass with individual bonds randomly positive or negative this energy difference is randomly of either sign but its modulus is the relevant measure. The average of this modulus over many realizations of the microscopic disorder will be denoted J(L). Its behaviour as $L \to \infty$ determines the existence or absence of spin glass order;

$$\lim_{L\to\infty} J(L) = \begin{cases} \infty & : \text{order} \\ 0 & : \text{no order} \end{cases} \tag{72}$$

In fact, in studies of an Ising spin glass, Bray and Moore [37] found J(L) to scale as

$$J(L) \sim JL^{y} \tag{73}$$

with

$$\begin{aligned} d = 2 &: \quad y = -\,0.291 \pm 0.002 \\ d = 3 &: \quad y = +\,0.19 \pm 0.01 \end{aligned} \tag{74}$$

demonstrating that there is order in three dimensions but not in two.

An extension of this study demonstrated the chaotic evolution discussed in section 1. Bray and Moore [38] now took the microscopic exchange interactions first according to

$$J_{ij} = J_{ij}^{(1)} \tag{75a}$$

and then

$$J_{ij} = J_{ij}^{(1)} + J_{ij}^{(2)} \tag{75b}$$

where the $J_{ij}^{(1)}$, $J_{ij}^{(2)}$ are drawn randomly from Gaussian distributions of width J, $\varepsilon \ll J$ respectively. The resultant effective block spin exchanges for each case (based on the energetic difference between periodic and anti-periodic boundary conditions) were used to generate effective block spin exchange widths J(L), ε(L) and the scaling of ε(L)/J(L) was studied. It was found that

$$\frac{\varepsilon(L)}{J(L)} \sim L^{\xi} \tag{76}$$

with $\xi > 0$. Small deviations in the initial J_{ij} distribution lead to diverging ground state differences. That is, the ground states evolve chaotically. A similar chaotic evolution is believed to result from perturbations of other parameters, such as the temperature and the applied field.

These scaling conclusions obtained numerically are in accord with heuristic droplet arguments [39].

III.B Monte Carlo Simulation

The value of Monte Carlo simulation was recognized early in the current wave of interest in spin glasses [18, 40], and the technique has been widely employed since. However, here we shall concentrate on more recent applications directed towards determining phase transition temperatures and critical behaviour.

First, we describe briefly the Monte Carlo principles and procedures [41]. Monte Carlo simulation is in a sense an inverse of the usual ergodic transformation from an average of a real system over real time to a Gibbs average over an ensemble of systems,

$$\langle A\rangle = \lim_{\tau\to\infty} \frac{1}{\tau}\int_{t_o}^{t_o+\tau} A(t)dt \rightarrow \langle A\rangle = \frac{\sum_s A_s \exp(-\beta E_s)}{\sum_s \exp(-\beta E_s)} \tag{77}$$

where on the left the time evolution A(t) is that of nature and on the right s labels the energy eigenstates, the microscopic configurations. In principle, this Gibbs average could be evaluated directly numerically by calculating A_s and E_s for all microstates and summing as in eqn (77). However, this rapidly becomes prohibitive since even for an Ising system the number of microstates for N spins is 2^N. For continuous spins it is infinite and even if the allowable orientations are discretized it is totally impractical. A naive Monte Carlo procedure would pick states s randomly with equal probability and sum $A_s\exp(-\beta E_s)$ and $\exp(-\beta E_s)$ over the choices to approximate the expression on the right in eqn (77). Except at very high temperatures (which is not the region of interest), this would be wasteful too because $\exp(-\beta E_s)$ would be small for many of the states chosen. All practical Monte Carlo procedures employ a technique known as "importance sampling", which chooses states not equally randomly but with a probability reflecting their importance according to the Boltzmann factor $\exp(-\beta E_s)$. In such procedures one therefore evaluates ⟨A⟩ from

$$\langle A\rangle \simeq \frac{1}{\tau}\int_{t_o}^{t_o+\tau} A(t)dt \tag{78}$$

where A(t) now follows a stochastically determined dynamics designed so that it samples states of the system corresponding to the probability $\exp(-\beta E_s)/\sum_s \exp(-\beta E_s)$; the probability of transitions from state s to s', $W_{ss'}$, in such a dynamics must satisfy the detailed balance condition

$$W_{ss'}/W_{s's} = \exp\{-\beta(E_{s'} - E_s)\} \; . \tag{79}$$

Time in eqn (78) is measured in simulation steps, Monte Carlo steps.

Provided that a dynamics satisfies eqn (79) it will lead in the long time limit to the Gibbs distribution, irrespective of whether it corresponds to the physical dynamics (which also satisfies eqn (79)). For the evaluation of static thermodynamic averages all choices are equivalent fundamentally, although some may be more practical in the time needed to approach Gibbs distribution. Clearly, however, if one is interested in dynamic correlations, the choice of simulational dynamics is relevant.

The most common algorithm used to implement eqn (79) is that due to Metropolis et al. [42]. In the Metropolis algorithm a change in microstate is considered, the energetic consequence of its implementations calculated, and the change effected if either (i) its implementation would lead to an energetic reduction or (ii) if it would lead to an energetic increase ΔE <u>and</u> $\exp(-\beta\Delta E)$ is greater than a number chosen randomly between zero and one; otherwise the move is rejected. These random number choices are made independently at each step. The choice of microstate change tested may be made randomly or systematically, but more normally the former.

An alternative algorithm used reasonably frequently is that known as the heat bath or Glauber algorithm. In this algorithm a move is accepted with probability $\exp(-\beta\Delta E)/(\exp(-\beta\Delta E)+\exp(\beta\Delta E))$, irrespective of the sign of ΔE. This algorithm has the virtue of being closer to physical dynamics.

A systematic procedure for Monte Carlo simulation of an equilibrium property of a spin glass would proceed as follows:

(i) Choose any external parameters, such as temperature, field etc.

(ii) Choose the size of system (and boundary conditions).

(iii) Choose randomly, from the distribution to be considered, the positions and/or magnitudes of the bonds.

(iv) Choose a random starting configuration for the spin orientations.

(v) Choose randomly or systematically a change in local spin orientation (normally of a single spin).

(vi) Investigate the energetic consequence of the change in (v).

(vii) Accept or reject the change according to the algorithm being used.

(viii) Return to (v) and repeat, while monitoring some measure of equilibration.

(ix) Continue this cycle while collecting data, making averages over times greater than the equilibration time.

(x) Repeat from (iii) for different quenched disorders, average or look at distributions as appropriate.

(xi) Repeat from (ii) for different sizes to eventually use scaling to extrapolate to the thermodynamic limit or make other size-dependence investigations.

(xii) Repeat from (i) for different temperatures, external fields, interaction distribution probabilities, concentrations etc.

In some cases one also runs simultaneously but independently multiple copies with the same interactions and external parameters and cross-correlates between the copies. For example, a cross-correlation between two replicas of their instantaneous overlap

$$q^{(1,2)}(t) = N^{-1} \sum_i \sigma_i^{(1)} \sigma_i^{(2)} \tag{80}$$

enables one to determine the overlap probability distribution $\tilde{P}(q)$, of eqn (51) [26];

$$\tilde{P}(q) = \frac{1}{\tau} \int_{t_o}^{t_o+\tau} \delta(q - q^{(1,2)}(t))dt \tag{81}$$

This is an example of a measurement which can be readily performed in a computer simulation but for which no experimental analogue is known. An extension to study correlations between three or more independently evolving replicas is relevant to an examination of ultrametricity.

The list (i)-(xii) above includes a crucial but highly non-trivial step, the test for equilibration, necessary since both t_o and τ in eqns (78), (81) must be greater than the equilibration time. In spin glasses equilibration times become very long as the temperature is lowered and simply guessing when t_o and τ are great enough can be very misleading. A systematic measure is needed. One simple procedure is to monitor the decay of a suitable quantity to its equilibrium value. For example, one might monitor

$$q(t) = N^{-1} \sum_i \sigma_i(t+t_o)\sigma_i(t_o) \ . \tag{82}$$

At t=0, q(t)=q, but for t greater than the sample equilibration time, $\sigma_i(t+t_o)$ has lost memory of $\sigma_i(t_o)$. Thus, if t is less than the time for global inversions, [q(t)] settles to

$$q = N^{-1} \sum_i [\langle\sigma_i\rangle^2] \tag{83}$$

and the time to do so yields a measure of the characteristic equilibration time $\tau_{eq}(N,T)$. t_o must be checked a posteriori to be greater than τ_{eq}. τ_{eq} grows rapidly as N is increased and T decreased.

In fact, if t is taken much greater than τ_{eq}, global inversions of the spin system occur and q(t) changes sign. The global inversion time is much greater than the equilibration time and diverges with N. Potential problems associated with global inversion are avoided by monitoring a quantity invariant to such changes, such as

$$q_U^{(2)}(t) = N^{-2} \sum_{ij} \sigma_i(t+t_o)\ \sigma_j(t+t_o)\ \sigma_i(t_o)\ \sigma_j(t_o) \tag{84a}$$

$$= (N^{-1} \sum_i \sigma_i(t+t_o)\ \sigma_i(t_o))^2. \tag{84b}$$

For t=0, $q_U^{(2)}(t) = 1$, but $[q_U^{(2)}(t)]$ tends for $t > \tau_{eq}$ to

$$q^{(2)} = N^{-1} \sum_{ij} [\langle\sigma_i\sigma_j\rangle]^2 = \chi_{SG}\ , \tag{85}$$

the spin glass susceptibility measuring $\langle\sigma\rangle^2$ response to a random field or the cubic response to a uniform field.

$[q_L^{(2)}(t)]$ approaches $q^{(2)}$ from above. A useful supplement is to take a second measure approaching from below, with coalescence as the best measure of the equilibration time [43]. Such a measure is provided by monitoring an appropriate cross-correlation between two replicas having identical quenched disorder, but evolving independently, such as

$$q_L^{(2)}(t) = \frac{1}{t}\int_{t_o}^{t_o+t} dt'\, N^{-2} \sum_{ij} \sigma_i^{(1)}(t')\, \sigma_j^{(1)}(t')\, \sigma_i^{(2)}(t')\, \sigma_j^{(2)}(t') \quad (86a)$$

$$= \frac{1}{t}\int_{t_o}^{t_o+t} dt'\, (N^{-1} \sum_{i} \sigma_i^{(1)}(t')\, \sigma_i^{(2)}(t'))^2 \quad (86b)$$

where superscripts 1,2 on σ label the replicas. For t=0, $q_L^{(2)}(t) = 0$, but as $t \to \tau_{eq}$ $[q_L^{(2)}(t)]$ approaches $q^{(2)}$ from below.

Simulations are neccessarily done on finite-sized systems, whereas in thermodynamics one is interested in the limit $N \to \infty$. Clearly, one needs a systematic extrapolation procedure. Also, of particular concern is the fact that a phase transition only occurs in the limit $N \to \infty$, non-analyticities being smoothed for finite systems. If one's interest is the critical region near a continuous phase transition, a valuable tool for data analysis is finite-size scaling [44-46]. If one assumes a single relevant length-scale in the critical region, given by the correlation length ξ, scaling with the reduced temperature $\tau = (T_g - T)/T_g$ as

$$\xi \alpha\, \tau^{-\nu}, \quad (87)$$

then the order parameter distribution over systems of size L is given by

$$P(q) = L^{\beta/\nu}\tilde{P}(qL^{\beta/\nu}, L^{1/\nu}\tau), \quad (88)$$

where $\tilde{P}(x,y)$ is a universal function and β, are critical exponents. To estimate the transition temperature a particularly useful measure is

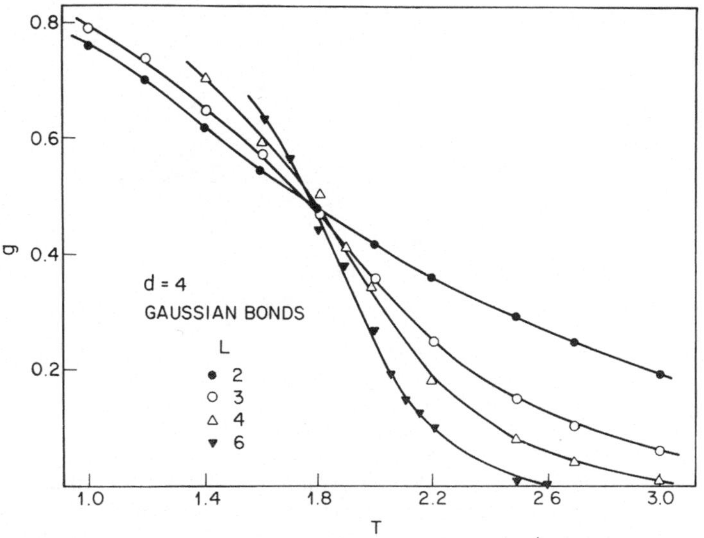

Fig. 14. $g_L(T)$ curves for a nearest neighbour Ising spin glass in four dimensions; from Ref. 47.

$$q_L(T) = \frac{1}{2}\left\{3 - \frac{[\langle q^4\rangle]}{[\langle q^2\rangle]^2}\right\} \tag{89}$$

where q is obtained by cross-correlation between two replicas. Using eqn (88) g takes the form

$$g_L(T) = \tilde{g}(L^{1/\nu}\tau) \tag{90}$$

so that at $\tau = 0$ g is independent of L and T_g is given by the coalescence of g_L for all L. In fact, in the thermodynamic limit $L \to \infty$ g_L is expected to have a sharp discontinuity at $T = T_g$ since the function is constructed to be zero if q is Gaussian-distributed, as expected above T_g, and unity if q has a single dominant peak at $q \neq 0$, as would be the case below T_g if there is a single thermodynamic phase. Even if P(q) has more structure below T_g, as in the Parisi solution, g_∞ will have a discontinuous jump as the temperature is reduced through T_g. The greater rounding-off of the discontinuity as L is reduced, again leads to a picture of crossings of $g_L(T)$ for a different L at $T = T_g$. Fitting to the scaling form (90) provides a further check.

This technique has been used by Bhatt and Young [43, 47] to examine the phase transition of the Ising spin glass for nearest neighbour interactions for d = 2, 3 and 4 and for infinite-ranged interactions. The infinite-range results confirm the transition temperature obtained by mean field theory. Fig. 14 reproduces their results for d = 4, showing a clear crossing and a low temperature spin-glass phase. For d = 2, the results

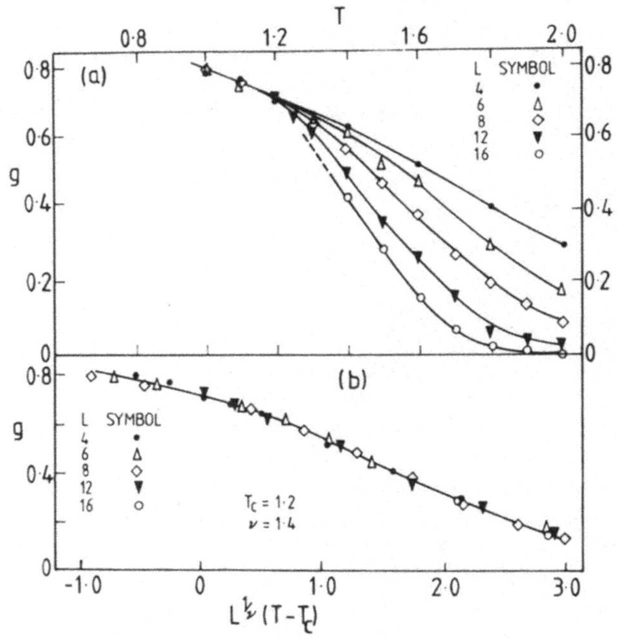

Fig. 15. $g_L(T)$ for the three-dimensional $\pm J$ Ising spin glass (a) versus T to show the coming together at $T = T_c$ and (b) scaling plot versus $L^{1/\nu}(T - T_c)$ demonstrating that all L fall on a universal curve for $T > T_c$; from Ref. 47.

are consistent with a zero-temperature phase transition, although the systems cannot be equilibrated to this temperature. For d = 3 the values of $g_L(T)$ come together as T is reduced (at $T_g/J \simeq 1.2$ for $\pm$ J interactions, at $T_g/J \simeq 0.9$ for Gaussian interactions) but thereafter remain together; see Fig. 15. Interpreting the temperature of this coming-together as T_g gives good scaling curves but the implication of the subsequent sticking-together of $g_L(T)$ for all L for $T < T_g$ remains unclear. However, it has been noted [43, 47] that this behaviour is that to be expected of a low temperature phase with power-law decay of spin correlation functions as a function of separation, suggestive of a critical phase at all temperatures less than T_g.

Vector models can be studied by an extension of the above Monte Carlo method [48, 49]. The indication is that in three dimensions the short-range Heisenberg spin glass has a zero-temperature transition in the spin variables, although it may have a higher-temperature transition to chiral ordering. A finite-temperature spin-glass ordering, as indicated by the analogue of $g_L(T)$, is, however, induced by random anisotropy [50]. This may well be the explanation as to why experimental systems with Heisenberg spins regularly show well-defined spin glass transitions with critical exponents closer to those of Ising spin glass models than those of isotropic Heisenberg models.

IV. ANALOGIES AND APPLICATIONS

Frustration and disorder are not restricted to spin glasses and conceptualization and techniques developed for spin glasses are now finding application in problems apparently very removed physically or materially. The links are mathematical and conceptual. In this section an attempt is made to illustrate and partially to expose some of those links.

Two areas of application will be discussed, (i) hard optimization problems, as experienced in computer science and operational research, and (ii) neural networks for memory, as believed relevant to an understanding of the human brain and the possible development of a new class of content-addressable computers with noise-tolerant distributed storage.

IV.A. Optimization

In operational research, or various aspects of planning and design, one is often faced with the problem of drawing an optimal balance between conflicting ideals. We shall be concerned here only with situations where this problem can be quantified and formulated as that of minimizing some cost (or measure) function with respect to controllable parameters, subject to fixed constraints. A classic simple example of such a problem is that of routing a salesman through a given set of (effectively randomly located) cities so as to visit each city once whilst travelling the minimum distance. In this travelling salesman problem the cost function is the tour length, the variables the order of visiting the cities and the constraints their locations (and separations) [51]. Another example, which we shall discuss in particular detail later, concerns the partitioning of the vertices of a randomly connected graph into equal-sized sub-sets with the minimum number of connections between them [52,53], relevant to practical problems such as microchip design and processor load balancing.

In general, these problems are very hard, usually what is known as NP-complete [53,54], soluble in a time polynomial in their size only on a non-deterministic Türing machine (or, in practice, requiring a time scaling faster than polynomial in their size for complete solution on a conventional deterministic computer). In general, finding the ground state of a spin glass is an NP-complete problem*.

It is clear that as the number of elements in these problems becomes large it rapidly becomes impossible to test all possible solutions to look for the optimal one. For example, in the graph bi-partitioning problem the number of possibilities is $2^{(N-1)} \simeq \exp(N \ln N)$, which is highly non-polynomial, whilst for the travelling salesman who wishes to visit N cities and then return home there are even more possibilities, $N! \simeq \exp(N \ln N)$!

One technique is to guess a solution, calculate its cost and also its gradient with respect to changes of a few variables, and thence iterate through such few-variable changes to the lowest cost this procedure allows. The result is illustrated in Figs 1 and 2 where now the ordinate is to be interpreted as cost, the abscissa as measuring the position in a space of such few-variable changes, the crosses denoting initial guesses and the dots the results of the corresponding iterative improvements. For a simple system with a smooth cost function landscape, as Fig. 1, this procedure rapidly yields the optimal solution, but when the cost function landscape has a more fractal character, as in Fig. 2, iterative improvement yields only local and not global optima. Fig. 2 is believed to be more representative of NP-complete problems.

Clearly, it would be useful to have a procedure not only to descend into valleys, but also to surmount hills and pass to lower valleys. Drawing on experience in trying to find spin glass ground states by computer simulation, and guided by conceptual ideas from statistical mechanics and the principles of metallurgical annealling, Kirkpatrick et. al. [56] proposed the technique of optimization by simulated annealling (OSA). This technique extends the analogy of cost to energy by adding also an analogue of temperature, replacing iterative improvement (rapid quenching in metallurgy) by computer simulated annealing. One starts a minimum cost (ground state) search by performing a Monte Carlo simulation (as in the previous section, but without necessarily waiting for equilibration) commencing at a fairly high "temperature" and gradually reducing the "temperature". The analogue of the test for a change in local spin orientation is a few-variable change (as in the iterative improvement), such as an exchange of the order of visiting two cities, the probability of its acceptance being determined by its resultant cost change, the temperature and the algorithm (eg. Metropolis). Clearly, an infinitely slow cool will lead to the ground state, but even a finite rate of cooling leads to a significant improvement over the descent-only algorithm for systems with quasi-fractal cost landscapes. The experimental analogue is well-known to the metallurgist who anneals to remove the defects which are a feature of rapid quenching, temperature providing the energy to enable them to overcome the energy barriers which impede their motion.

OSA is now a valuable tool in the designer's toolbox. There remain, however, several options for its practical implementation, different choices for variable changes to test, different cooling rate schedules, types of problem most suited to its use, etc. Some of these require experience from conventional approaches, such as the choices of variable changes, some can be monitored during a simulated cool, such as the measurement of the "specific heat" from the energy fluctuations to ascertain temperatures of cooperative ordering (such as associated with

---

* An isolated exception is the Ising nearest neighbour spin glass in two dimensions with no external field [55].

phase transitions or clustering) at which cooling should be slowed; some involve trial and error, such as whether to cool at a constant rate or with a geometrical reduction of temperature; but others can be investigated analytically using tools developed for spin glasses.

Let us now turn, therefore, to analytic applications to optimization. In keeping with the philosophy of the earlier sections we restrict the discussion to statistical properties, such as the average cost over an ensemble of systems with the same construction rules, but different explicit realizations. Techniques developed for spin glasses have been used to study travelling salesman problems, graph matching and colouring problems and several others [57, 58]. We shall discuss here only the equi-bipartitioning of random graphs [59].

Consider a graph having N vertices {i} connected by edges characterized by an NxN symmetric connectivity matrix $\underline{\underline{a}}$ whose elements a_{ij} are either 1 or 0 depending upon whether edge (ij) is present or absent. The issue we address is the partitioning of the vertices into two groups, each containing half the vertices, so as to minimize the number of edges between the groups. Our particular interest will be in the case of random connectivity.

Denoting the location of a vertex by $\sigma_i = 1$ if the vertex is in one group, $\sigma_i = -1$ if it is in the other, we have for the number of cross edges

$$N_c = \sum_{(ij)} a_{ij}(1-\sigma_i\sigma_j)/2, \tag{91}$$

whilst the equipartitioning requirement imposes the constraint

$$\sum_i \sigma_i = 0. \tag{92}$$

Interpreting the σ_i as Ising spins, $\sigma = \pm 1$ corresponding to spin up or spin down, the optimization requirement is equivalent to finding the ground state energy of a random magnet of Hamiltonian

$$H = -J \sum_{(ij)} a_{ij}\sigma_i\sigma_j, \tag{93}$$

subject to a constraint of zero overall magnetization. Although the interactions in eqn (93) are purely ferromagnetic the zero-magnetization constraint makes the problem frustrated.

The statistical mechanical techniques developed earlier for spin glasses may be applied to the evaluation of the cost/ground state energy averaged over an ensemble of systems of equivalent connectivities. Again it is convenient to introduce an artificial temperature $T(=(k_B\beta)^{-1})$ to yield a partition function

$$Z = \mathrm{Tr}' \exp(-\beta H), \tag{94}$$

where the trace is restricted by the constraint (92), and thence an average free energy

$$[F] = -k_B T[\ln Z] \tag{95}$$

and, finally, the average ground state energy

$$[E] = \lim_{T\to 0} [F]. \tag{96}$$

We shall be interested in the thermodynamic limit, as the number of vertices N becomes very large.

As noted earlier, of particular interest are random graphs, the average being performed over the class of equivalent graphs. Again the replica procedure is useful in averaging ln Z. If wished, the zero-magnetization constraint can be eliminated in favour of an extra cost penalty

$$\delta H = \lambda(\sum_i \sigma_i)^2. \tag{97}$$

These graph partitioning problems are closely related to spin glass problems as characterized by

$$H = - \sum_{(ij)} J_{ij}\, a_{ij}\, \sigma_i\, \sigma_j \tag{98}$$

where the J_{ij} are randomly quenched $\pm J$.

It is interesting to consider two limits of the extent of connectivity; one, referred to as extensive, in which the average valence (vertex coordination number) is proportional to N; the other, referred to as intensive, in which the average valence is independent of N. In the case of extensive valence the average minimal cost can be expressed in terms of the average ground state energy of the SK model and results from the latter applied directly [60, 61]. When the valence is intensive the thermodynamics of (98) is again exactly soluble in principle, but is more complicated than for the SK model; for independently random a_{ij} this model was originally introduced by Viana and Bray (VB) [62], whilst for fixed local valence it is closer to the problem of a spin-glass on a Bethe lattice [63-65]. The bipartitioning of random graphs of intensive valence, both fixed and average, as well as corresponding spin glasses, have been studied by the analytic techniques of section 2 [66-68], and fixed intensive valence has also been studied simulationally [69]. The main analytic complication of the case of intensive valence is that in place of a single order function $q^{\alpha\beta}$ are needed a hierarchy, $q^{\alpha\beta}$, $q^{\alpha\beta\gamma}$, $q^{\alpha\beta\gamma\delta}$ .... The consequent subtleties remain incompletely resolved at this time.

IV.B. Neural Networks

The brain offers major challenges to scientists of many disciplines. As the controlling organism for human memory and behaviour, its understanding is clearly of importance to biological and behavioural scientists. Its ability to perform easily and quickly recognition and deductive tasks far beyond the capabilities of conventional computers, despite the relative slowness of its operational units, makes it of growing interest to computer scientists. Its high connectivity and clearly cooperative operation pose challenges for the application of the art of the statistical mechanist. All in all, it is hardly surprising that neural networks are a rapidly growing field of activity. In this sub-section we discuss some idealizations of the problem of neural memory from the point of view of a spin glass theorist.

The brain contains many neurons (of order 10^{10}) connected by many synapses (of order 10^{14}-10^{15}), with the synaptic connections highly dendritic and mostly non-local. In an idealized model a neuron can be considered at any time to be in one of two states, firing or not firing [70]. Labelling a neuron by i we refer to its state at time t as

$$\sigma_i(t) = 1 \quad ; \text{ firing} \tag{99}$$

or

$$\sigma_i(t) = -1 \quad ; \text{ non-firing.}$$

If a neuron fires it sends signals down all its outgoing synapses to other neurons. These signals may be excitatory or inhibitory, considered as giving a positive or a negative potential at the next neuron. Assuming only pairwise interactions, the potential at neuron j due to neuron i can be expressed as

$$V_{ji}(t) = J_{ji}\,(\sigma_i(t)+1)/2, \tag{100}$$

where J_{ji} measures the strength and sign of the contribution if i has fired, whilst $(\sigma_i(t)+1)/2$ takes the value unity if i is firing, zero if it is not firing. The total input potential at j is

$$V_j(t) = \sum_i J_{ji}\,(\sigma_i(t)+1)/2\ . \tag{101}$$

This input potential affects the subsequent action of neuron j, which now has a probability to fire determined by the difference between $V_j(t)$ and a "threshold" W_j. As a function of $(V_j - W_j)$ this probability is believed to have a sigmoid shape, with firing probability p_j zero if $V_j \ll W_j$, unity if $V_j \gg W_j$ and passing continously, monotonically and quite rapidly between these values as $(V_j - W_j)$ passes through zero. Let us initially consider this to be further idealized to

$$p_j = \Theta(V_j - W_j) \tag{102}$$

where $\Theta(x)$ is the Heaviside step function

$$\Theta(x) = 0 \quad ; x < 0$$

$$\ 1 \quad ; x > 0 \tag{103}$$

If neuron j fires it now passes on signals to the neurons reached by its outgoing synapses, and, in turn, each such neuron's subsequent activity is determined by the totality of signals received. After a time, any initial pattern of activity may evolve to a stable pattern or sequence of patterns of activity. Such stable patterns or sequences correspond to retrieval states. To operate as a memory these retrieval states should be related to stored information, whilst their achieval from an initially noisy pattern is the recall process. Memorized information corresponds to activity patterns or sequences, whilst the storage of this information is in the synapses [71], since it is they which determine the stable states. From eqns (101) and (102) it follows that the condition for a pattern of activity to be stable is

$$\sigma_i = \text{sign}\,\{\sum_j J_{ij}\,(\sigma_j + 1) - 2W_i\} \tag{104}$$

An important advance in the realization of the applicability of techniques of statistical mechanics to the analysis of such neural networks came from J.J. Hopfield [72] who noted that eqn (104) is closely analogous to the condition for a state of an Ising model to be stable against single spin flips. Explicitly, consider a Hamiltonian

$$H = \sum_{(ij)} J_{ij}\, \sigma_i\, \sigma_j - \sum_i b_i\, \sigma_j \quad ; J_{ij} = J_{ji}. \tag{105}$$

States stable against single spin-flips satisfy

$$\sigma_i = \text{sign}\, \{\sum_j J_{ij}\, \sigma_j - b_i\}. \tag{106}$$

If b_i is taken as

$$b_i = 2W_i - \sum_j J_{ij} \tag{107}$$

and J_{ij} is assumed symmetric in the model brain, then eqns (104) and (106) are equivalent. Although there is no biological evidence for

$$J_{ij} = J_{ji}, \tag{108}$$

its assumption makes the analysis of stable activity patterns amenable to the techniques of static statistical mechanics. Eqn (108) has now been superseded, but this observation was a psychological milestone.

In the analysis below, we concentrate initially on this Hamilitonian formulation. With no further analysis, we already see, by analogy with the discussions in earlier sections, that in order to get many solutions to eqn (106), corresponding to many possible memorized patterns, it is necessary that the J_{ij} are frustrated, some J_{ij} positive, and excitatory, some negative and inhibitory, as observed in the brain. Similarly, the quasi-random connectivity of the dendritic synapses is surely also relevant.

As noted, memory information is stored in the J_{ij}. In the Hopfield model J_{ij} is taken to have the form proposed by Hebb [71],

$$J_{ij} = N^{-1} \sum_{\mu=1}^{p} \xi_i^\mu\, \xi_j^\mu \tag{109}$$

where the $\{\xi_i^\mu\}$; $\xi_i = \pm 1$ are memorized patterns, p in number.

If p is intensive*, independent of N, the complete square nature of

$$H = N^{-1} \sum_{\mu=1}^{p} \sum_{(ij)} \xi_i^\mu\, \xi_j^\mu\, \sigma_i\, \sigma_j \tag{110}$$

$$= \frac{1}{2} N^{-1} \sum_{\mu=1}^{p} (\sum_i \xi_i^\mu\, \sigma_i)^2 \tag{111}$$

permits a straightforward steepest descents analysis [73,74], yielding the self-consistency equations for the stable states [75]

$$m^\mu = N^{-1} \sum_i \xi_i^\mu \langle\sigma_i\rangle \tag{112}$$

---

* The reason for this qualification is that only if the number of squared terms is intensive is the simplest functional integral formulation extremally dominated [74].

$$\langle\sigma_i\rangle = \tanh(\beta \sum_{=1}^{p} m^{\nu} \xi_i^{\nu}). \tag{113}$$

If p is extensive, proportional to N, the more sophisticated techniques developed for the SK model can be applied successfully to averages over ensembles of systems with different uncorrelated random patterns [76].

Note that in these statistical mechanical analyses the stable states can be investigated at any temperature T. In the limit $T \to 0$ they correspond to the idealization (105). However, for finite T, the equivalence is to a more rounded firing probability/input potential curve, closer to the believed-true sigmoid curve. Temperature is thus a measure of threshold noise.

Already for finite p, the solutions to (112) and (113) demonstrate the value of noise. For all $T < 1$ the embedded patterns are solutions; that is

$$\langle\sigma_i\rangle = \langle\sigma\rangle \xi_i^{\mu} \quad ; \text{any one } \mu, \text{ all } i \tag{114}$$

are solutions, or equivalently, there are solutions

$$m^{\mu} = m \quad ; \text{any one } \mu \tag{115}$$

$$0 \quad ; \text{all other } \mu,$$

corresponding to finite overlap with one memorized pattern, zero with all others. For $1 > T > 0.46$ these are the only solutions, but for $T < 0.46$ there are additionally solutions which are spurious mixtures of memorized patterns. Thus noise smooths the free energy surface, eliminating spurious recall. For $T > 1$ the "noise" is too great and no non-trivial stable solutions remain ($\langle\sigma\rangle = 0$ is the only solution).

Let us now turn to extensive p, $p = \alpha N$, with α independent of N. For this case perfect retrieval is not possible with Hebb synapses. The maximum p for perfect retrieval is $p_{max} = N/2\ln N$. If, however, one accepts a finite fraction of errors, then good retrieval (with an error of less than 2%) is achieved at $T = 0$ up to $\alpha_c = 0.138$; the error fraction is given by

$$\frac{N_e}{N} = \frac{(1 - m)}{2}, \tag{116}$$

where m is given by eqn (114) with μ corresponding to the memory whose retrieval is sought, averaged over memories. Unfortunately, however, at $\alpha = \alpha_c$ there is an unphysical catastrophe and no recall at all is possible for $\alpha > \alpha_c$. Fig. 16 shows this retrieval quality as a function of α for $T = 0$. Fig. 17 shows the phase diagram predicted for general α, T. Beneath the curve labelled T_M, states having good overlaps with the stored patterns are at least metastable, although only beneath curve T_c are they global minima. In the region between T_M and T_g the minima are spurious spin-glass states, whilst above T_g only the paramagnetic solution survives. T_R bounds a region of replica-symmetry breaking.

The importance of this analysis lies partly in its specific predictions for the Hopfield-Hebb model, but more particularly in the demonstration that the application of statistical mechanical techniques can provide quantitative and qualitative information concerning neural memory. In fact, many of the specific restrictions imposed above can now

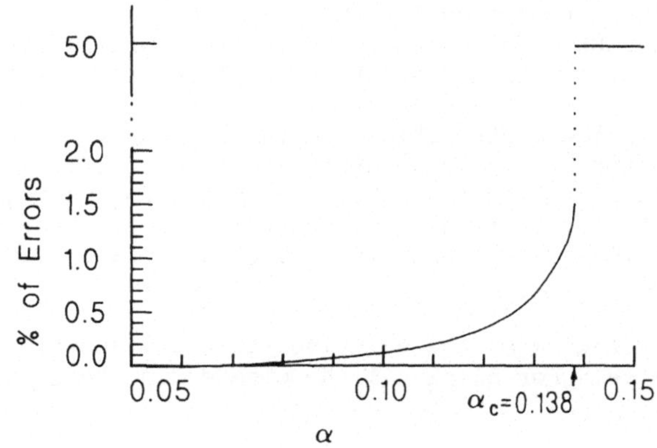

Fig. 16. Percentage of errors as a function of α at T = 0 for a Hopfield model with Hebb synapses storing p = αN uncorrelated patterns; from Ref. 76.

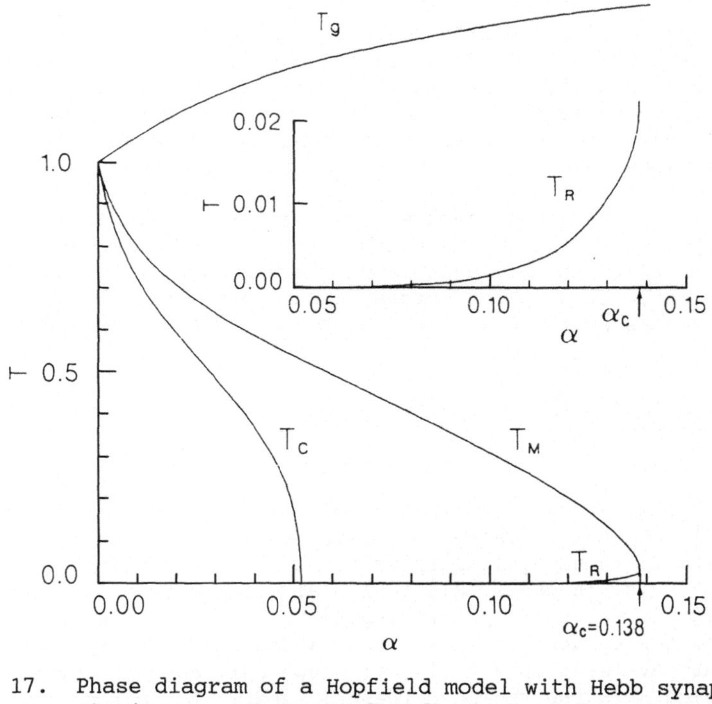

Fig. 17. Phase diagram of a Hopfield model with Hebb synapses storing p = αN uncorrelated patterns; from Ref. 76.

be removed and analysis has been extended to include dilution, synaptic clipping, asymmetric synapses, correlations between memories, storage and retrieval of sequences, and other more realistic or alternative features. Some of these extensions can be analyzed within static statistical mechanics, others require dynamic theory. Furthermore, statistical mechanics can be applied in a matter inverse to that discussed above. This was beautifully demonstrated by Gardner [77] who examined the storage limits of eqn (104) for the perfect recall of random memories of arbitrary average correlation, allowing for all possible $\{J_{ij}\}$. To this end she took the patterns to be retrieved as the quenched parameters and the $\{J_{ij}\}$ as the variables, using replica analysis to treat the quenched average of the optimization problem. For

$$W_i = \frac{1}{2} \sum_j J_{ij} \tag{117}$$

the maximum storage is given by

$$\begin{aligned} \alpha_c &= 2(1 + 2m^2/\pi) && ; \text{ small m} \\ &\simeq - 1/(1 - m)\ln(1 - m) && ; \text{ large m} \end{aligned} \tag{118}$$

where m is the average overlap between memories,

$$m = [N^{-1} \sum_i \xi_i^\mu \xi_i^\nu] \quad ; \mu \neq \nu. \tag{119}$$

V. CONCLUSION

In these lectures I have attempted to expose the concepts underlying the spin glass problem, to indicate some of the techniques which have been developed to understand it, and to illustrate some of the applications of both the concepts and the techniques to problems in areas widely different in their physical character but related conceptually and mathematically.

Among the messages I have tried to promulgate are the following:

(i) Disorder and frustration are not nuisances, simply complicating analysis; rather they are key ingredients to conceptually new phenomena with both rich structure and purpose.

(ii) True experiment, analytic theory and computer simulation are, all three, important methods of study, symbiotically related and all invaluable to understanding complex physical problems. In particular, computer simulation is not simply a procedure for effectively mimicking the procedures of conventional experiment on possibly model systems; it also provides probes for which no experiment exists.

(iii) Transfers from one area of science to another, or from pure science to application, need not be through similarity of materials or even of physical properties, but can also be carried by their mathematical formulation and conceptualization in abstractly related spaces. Fundamental science can indeed have unexpected applications which could not be anticipated at the outset.

ACKNOWLEDGEMENTS

The author's understanding of, and excitement by, the subject of spin glasses has been greatly aided by discussion and advice from many friends, colleagues, associates and students. He thanks them all. He also acknowledges the support of the Science and Engineering Research Council of the United Kingdom, who have provided him with several research grants since the inception of his interest in the subject, as also several other agencies who have given partial support at various stages.

REFERENCES

1. B. R. Coles, "The Origins and Influences of the Spin-Glass Problem" in "Multicritical Phenomena", R. Pynn and A. Skjeltorp, eds., Plenum, New York (1983).
2. K. Binder and A. P. Young, Revs. Mod. Phys. 58, 801 (1986).
3. H. Maletta and W. Zinn, "Spin Glasses" in "Handbook on the Physics and Chemistry of Rare Earths", K. A. Gschneider Jr. and L. Eyring, eds., North Holland, Amsterdam (1986).
4. K. H. Fischer, Phys. Stat. Sol. b, 116, 357 (1983); 130, 13 (1985).
5. G. Toulouse, Commun. in Phys. 2, 115 (1977).
6. Y. Nagaoka and H. Fukuyama (eds.), "Anderson Localization", Springer, Heidelberg (1982).
   P. Lee and T. V. Ramakrishnan, Revs. Mod. Phys. 57, 287 (1985).
   D. M. Finlayson (Ed.), "Localization and Interaction", Edinburgh Univ. Press, Edinburgh (1986).
7. D. Stauffer, "Introduction to Percolation Theory", Taylor and Francis, London (1985).
8. R. J. Elliott (Ed.), "Magnetic Properties of Rare Earth Metals", Plenum, New York (1972).
9. J. Yeomans, to be published in "Solid State Physics", Vol. 41, ed. H. Ehrenreich and D. Turnbull, Academic Press, New York (1987)
10. V. Cannella and J. A. Mydosh, Phys. Rev. B 6, 4220 (1972).
11. K. Adkins and N. Y. Rivier, J. Physique Colloq. C 4-25, 237 (1974).
12. S. F. Edwards and P. W. Anderson, J. Phys. F 5, 965 (1975).
13. D. Sherrington and S. Kirkpatrick, Phys. Rev. Lett. 35, 1972 (1975).
14. G. Parisi, Phys. Rev. Lett. 43, 1754 (1979).
15. J. Villain, J. Phys. C 10, 1717 (1977).
16. J. Villain, J. Phys. C 10, 4397 (1977).
17. D. Sherrington and B. W. Southern, J. Phys. F 5, L49 (1975).
18. D. Sherrington, AIP Conf. Proc. 29, 224 (1975).
19. D. J. Thouless, P. W. Anderson and R. Palmer, Phil. Mag. 35, 593 (1977).
20. M. Mézard, G. Parisi and M. A. Virasoro, Europhys. Lett. 1, 77 (1986); "Spin Glass Theory and Beyond", World Scientific, Singapore 1987.
21. R. Brout, Phys. Rev. 115, 824 (1959).
22. J. R. de Almeida and D. J. Thouless, J. Phys. A 11, 983 (1978).
23. S. Kirkpatrick and D. Sherrington, Phys. Rev. B 17, 4384 (1978).
24. R. N. Bhatt and A. P. Young, Phys. Rev. Lett. 54, 924 (1985).
25. G. Parisi, J. Phys. A 13, 1887 (1980).
26. A. P. Young, Phys. Rev. Lett. 51, 1206 (1983).
27. M. Mézard, G. Parisi, N. Sourlas, G. Toulouse and M. Virasoro, Phys. Rev. Lett. 52, 1156 (1984).
28. R. Rammal, G. Toulouse and M. A. Virasoro, Revs. Mod. Phys. 58, 765 (1986).
29. M. Thomsen, M. F. Thorpe, T. C. Choy, D. Sherrington and H-J. Sommers, Phys. Rev. B 33, 1931 (1986).
30. G. G. Athanasiu, C. P. Bachas and W. F. Wolff, Phys. Rev. B 35, 1965 (1987).

31. H. Sompolinsky, Phys. Rev. Lett. 47, 935 (1981).
32. A. B. Harris, T. C. Lubensky and J. H. Chen, Phys. Rev. Lett. 36, 415 (1976).
33. D. Sherrington, "The infinite-ranged m-vector spin glass", in "Proceedings of the Heidelberg Colloquium on Spin Glasses", J. L. van Hemmen and I. Morgenstern, Eds., Springer, Heidelberg (1983).
34. D. Elderfield and D. Sherrington, J. Phys. C 16, 1233 (1983).
35. D. Sherrington, Prog. Theor. Phys. (Japan) Supp. 87, 180 (1986).
36. I. Morgenstern and K. Binder, Phys. Rev. Lett. 43, 1615 (1979).
37. A. J. Bray and M. A. Moore, J. Phys. C 17, L463 (1984).
38. A. J. Bray and M. A. Moore, Phys. Rev. Lett. 58, 57 (1987).
39. D. S. Fisher and D. A. Huse, Phys. Rev. Lett. 56, 1601 (1986).
40. K. Binder and K. Schröder, Phys. Rev. B 14, 2142 (1976); Solid State Commun. 18, 1361 (1976).
41. K. Binder, Ed., "Monte Carlo Methods in Statistical Physics", Springer, Berlin (1979).
42. N. Metropolis, A. W. Rosenbluth, M. N. Rosenbluth, A. H. Teller and E. Teller, J. Chem. Phys. 21, 1087 (1953).
43. R. N. Bhatt and A. P. Young, Phys. Rev. Lett. 54, 924 (1985).
44. M. E. Fisher, "The Theory of Critical Point Singularities", in "Proc. Int. Summer School "Enrico Fermi" Course LI", ed. M. S. Green, Academic Press, New York (1971).
45. K. Binder, Z. Phys. B 43, 119 (1981).
46. M. N. Barber, "Finite Size Scaling", in "Phase Transitions and Critical Phenomena, Vol. 8", C. Domb and J. L. Lebowitz, eds., Academic Press, New York 1983.
47. R. N. Bhatt and A. P. Young, "Numerical Studies of Spin Glasses", in "Heidelberg colloquium on Glassy Dynamics", J. L. van Hemmen and I. Morgenstern, eds., Springer, (1986).
48. J. A. Olive, A. P. Young and D. Sherrington, Phys. Rev. B 34, 6341 (1986).
49. A. Chakrabarti and C. Dasgupta, Phys. Rev. Lett. 56, 1404 (1986).
50. A. Chakrabarti and C. Dasgupta, Phys. Rev. B 36, 793 (1987).
51. E. L. Lawler, J. K. Lenstra, A. H. G. Rinnoy Kan and D. B. Shmoys, "The Travelling Salesman Problem", Wiley, New York (1985).
52. N. Christofides, "Combinatorial Optimization", Wiley, New York (1979).
53. C. H. Papadimitriou and K. Steglitz, "Combinatorial Optimization", Prentice-Hall, Englewood Cliffs (1982).
54. M. R. Garey and D. S. Johnson, "Computers and Intractibility", Freeman, San Francisco (1979).
55. F. Barahona, R. Maynard, R. Rammal and J. P. Uhry, J. Phys. A 15, 673, (1982).
56. S. Kirkpatrick, C. D. Gelatt and M. P. Vecchi, Science 220, 671 (1983).
57. S. Kirkpatrick and G. Toulouse, J. Physique 46, 1277 (1985).
58. M. Mézard, "Spin Glasses and Optimization", in "Heidelberg Colloquium on Glassy Dynamics", eds, J. L. van Hemmen and I. Morgenstern, Springer, Heidelberg (1986).
59. D. Sherrington, "Graph partitioning as a spin glass problem", in "Time-dependent effects in disordered systems", eds, R. Pynn and A. Skjeltorp, Plenum, New York (1987).
60. Y. Fu and P. W. Anderson, J. Phys. A 19, 1605 (1986).
61. W. Wiethege and D. Sherrington, J. Phys. A 20, L9 (1987).
62. L. Viana and A. J. Bray, J. Phys. C 18, 3037 (1985).
63. D. R. Bowman and K. Levin, Phys. Rev. B 25, 3438 (1985).
64. D. J. Thouless, Phys. Rev. Lett. 56, 1082 (1986).
65. D. Sherrington and K. Y. M. Wong, J. Phys. A 20, L785 (1987).
66. I. Kanter and H. Sompolinsky, Phys. Rev. Lett. 58, 164 (1987).
67. M. Mézard and G. Parisi, Europhys. Lett. 3, 1067 (1987).

68. K. Y. M. Wong and D. Sherrington, J. Phys. A 20, L793 (1987).
69. J. R. Banavar, D. Sherrington and N. Sourlas, J. Phys. A 20, L1 (1987).
70. W. McCullough and W. Pitts, Bull. Math. Biophys. 5, 115 (1943).
71. D. Hebb, "The Organization of Behaviour", Wiley, New York (1949).
72. J. J. Hopfield, Proc. Natl. Acad. Sci. USA 79, 2554 (1982); 81, 3088 (1984).
73. D. Sherrington and J. F. Fernandez, Phys. Lett. 62A, 457 (1977).
74. J. F. Fernandez and D. Sherrington, Phys. Rev. B 18, 6270 (1978).
75. D. J. Amit, H. Gutfreund and H. Sompolinsky, Phys. Rev. A 32, 1007 (1987).
76. D. J. Amit, H. Gutfreund and H. Sompolinsky, Phys. Rev. Lett. 55, 1530 (1985); Ann. Phys. 173, 30 (1987).
77. E. A. Gardner, Europhys. Lett. 4, 481 (1987).

OPTICAL TECHNIQUES AND EXPERIMENTAL INVESTIGATION OF DIFFUSION PROCESSES IN DISORDERED MEDIA

P. Evesque

Laboratoire d'Optique de la Matière Condensée
Université P. et M. Curie, 4 place Jussieu
75252 Paris Cedex 05

C. Boccara

Laboratoire d'Optique Physique, E S P C I
10, rue Vauquelin, 75231 Paris Cedex 05

ABSTRACT

Different diffusion processes in disordered media will be investigated using optical methods and interpreted in the light of new theoretical approaches based on fractals:

We will investigate heat diffusion processes in disordered materials by heating the surface of the sample with a pulsed laser and by determining the time dependence of the surface temperature. Much care will be taken to describe the properties of the surface temperature when this surface is either tortuous or fractal. New theoretical results will be given.

In the case of naphthalene D_8 crystal doped with naphthalene H_8, two triplet excitations which meet together fuse in a singlet state which instantaneously luminesces. Time resolved spectroscopy will allow us to determine the time dependence of the rate of fusion. The results will be interpreted in the light of the percolation theory inside and outside the critical region.

Time resolved spectroscopy and transient grating experiments will be used to investigate the geometry of the pore space of a porous material which is called vycor. We will discuss the efficiency fo each of these two methods to determine the tortuosity of the pore space.

Determination of the state density of an amorphous semiconductor will be performed by a photothermal detection of absorption. This technique is more sensitive than classical ones when absorption is small. It is then peculiarly efficient in the case of the state density in the forbidden gap.

I. INTRODUCTION

Optical methods are efficient tools in order to investigate physical properties of disordered materials. In one hand, it is well known that absorption and emission spectra are of prime interest in determining the density of states and energy levels of doped crystals and glasses. Moreover these techniques can also bring information on the relaxation processes existing in the materials especially when these techniques are associated with time resolved detections and pulsed excitations. In the same way, the invention of picosecond and femtosecond lasers which are characterized by very short resolution times and high power emission intensities had enabled to develop a new set of experimental techniques based on non-linear optics ; for instance they had allowed to get information on new parameters using either Raman emission or photon echo experiments as well as four wave mixing and hole burning techniques and so on ...

On the other hand, the physical nature of the different processes which can be investigated with optical techniques in disordered materials are many fold : it goes from coherent propagation, collision and diffusion of electronic excitations to energy transfer processes between donor and acceptor molecules. It also concerns heat diffusion and sound propagation. Spin glass dynamics can also be studied through optical methods based on Faraday effect.

We will not intend to give a complete overview of the subject but rather we will focus our attention on a few examples which will illustrate the wide range of this topic and the efficiency of the methods.

In the first part of this article, we will pay attention to a few cases of heat diffusion in disordered media. The basic idea of these experiments is to heat with a pulsed laser the surface of the disordered medium under investigation and to determine the time dependence of heat dissipation inside the material. This can be achieved either by measuring directly the temperature of the surface with an infrared detector or by measuring the temperature gradient which exists near the surface either inside or outside the material. This last method uses the so called "mirage" effect.

The second part of this article will be devoted to a problem of diffusion in a percolating case. This is the case of a binary naphthalene crystal. We will see that time resolved spectroscopy technique applied on the triplet and singlet states of naphthalene can bring information on the rate of fusion and trápping of excitations. As these processes strongly depend on the mean number of sites visited by the excitation, we will show that such an experiment relates the reaction kinetics to the anomalous geometry that a random walker can explorate at a given time. The experiment will then consist in exciting the triplet state at time $t = 0$ and to look at the fusion rate at time t through the singlet state luminescence.

In a third part we will use different techniques based either on energy transfer processes and time resolved spectroscopy, or diffusion of molecules associated with transient grating experiment in order to elucidate the intimate structure of a porous medium : the vycor.

In the last part we will see that photothermal detection of absorption can be much more sensitive than classical methods ; this will be applied to the determination of the density of states in the forbidden gap of an amorphous semiconductor.

II. A FRACTAL APPROACH OF HEAT DIFFUSION IN ROUGH SURFACES AND HETEROGENEOUS MEDIA

Heat sources and samples with usual geometrical shapes such as planes, points, cylinders..., have been mostly considered [1]. When highly complex systems have to be studied, it is difficult to account for the individual properties of each component to get the global properties of the system. For instance when one deals with heat diffusion one can account for a perturbated sample surface by introducing an "equivalent layer" whose thermal properties are different from the bulk ones [2]. Or one can try to modelize the complex sample by an assembly of elements of simple geometrical shapes (e.g., spheres, ...[3]).

Recently, macroscopic self similar objects have been considered in the framework of fractal theory developed by Mandelbrot [4]. We would like, at this point, to suggest how such geometrical approach can be a great help to understanding the thermal behavior of rough surfaces and heterogeneous samples. These problems are of great interest both from a fundamental point of view (physics of disordered system) and for practical applications to characterize divided materials of high technological interest (ceramics, sintered materials...).

Let us recall that a structure is self similar if we cannot tell the difference in the structure as we change the scale. Among the fractal objects one can distinguish two kinds of self similar structures, the geometrical ones which we will use as an example to define the fractal dimension $\bar{d}$ and the random self similar structures which are of actual importance to describe physical systems. It is obvious that in real physical systems the scale range at which this self similarity occurs is limited both in the small dimension limit (e.g., grains or atoms...) and in the large dimension one where the objects behave like Euclidean entities.

Fig. 1 shows an example of self similar geometrical construction called the Sierpinki gasket. We can see that if we change the unit length by 1/2 we have 3 equal pieces. Let us recall that for a straight line interval, a square or a cube if we change the scale by 1/2 we have $(1/2)^{-d}$ equal pieces of the initial object, d being equal to 1, 2 or 3 respectively. Here d is the usual Euclidean dimension. We can use this difinition to get the dimension of the Sierpinski gasket.

$$(1/2)^{-\bar{d}} = 3 \qquad \text{thus} \qquad \bar{d} = 1.58$$

$\bar{d}$ is called the fractal dimension.

Now for application to heat diffusion it is worth focussing our attention on random self similar structures. For such structures the above properties must be considered as an average. As an example Fig.2 shows a so-called Brownian motion curve whose fractal dimension is 1.5.

We will label d the dimensionality of the Euclidean space in which the stucture is dived (d = 3 in the general case) ; $\bar{d}$ will be the notation of the fractal dimension and $\bar{\bar{d}}$ the spectral one. The diffusion coefficient will be labelled D.

We will first consider the case of a fractal heat source embedded in an Euclidean space of finite diffusion coefficient. However in some cases of random media, one can only heat a small part of the surface of these media. In this case the part of the structure which is really heated can not reflect the self similar properties of the whole structure ; in order to take into account this problem, we will consider in the second part the case of a fractal source of dimentionality $\bar{d}_s$, embedded in a fractal conducting structure of dimensionality $\bar{d}$.

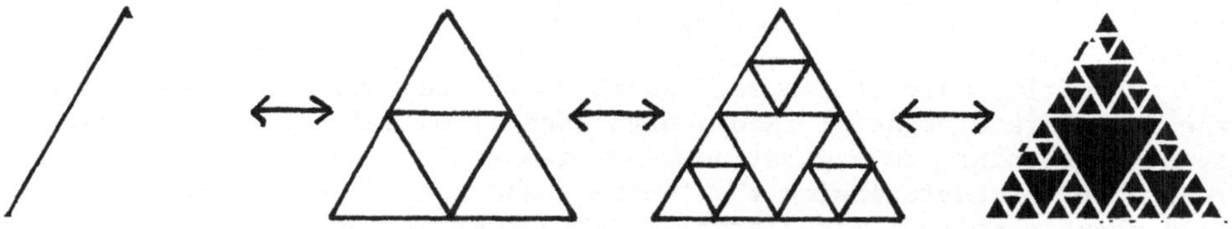

Fig. 1. The Sierpinski gasket

II.A. Euclidean and Fractal Heat Source

In order to get a first insight into the behavior of fractals let us consider the simple following situation : when a point is heated at time $t = 0$ by a fast heat pulse, the temperature in the surrounding three dimensional medium whose diffusivity is D is given by :

$$T\,(r,\,t) = \frac{e^{-r^2/4Dt}}{(4Dt)^{3/2}}$$ (geometrical dimension of the source : $d = 0$, unit source excitation)

Integrating this contribution over a line source will lead to :

$$T\,(r = 0,t) \sim (Dt)^{-1} \quad ; \quad (d = 1)$$

for a point located on the line source.

And over a plane source to :

$$T\,(r = 0,t) \sim (Dt)^{-1/2} \quad ; \quad (d = 2)$$

for a point located on the plane source.

More generally assuming a source at time $t = 0$ with a fractal dimension $\bar{d}_s$ and a diffusion in a 3-dimensional space one finds that the temperature of source decreases as :

$$T\,(t) \sim (Dt)^{(\bar{d}_s-3)/2}$$

The generalization of this equation to a $\bar{d}_s$ fractal source embedded in d-dimensional Euclidean space is :

$$T(t) \sim (Dt)^{(\bar{d}_s - d)/2}$$

As an illustration of this result we have used the curve of Fig.2 as a unit source at time $t = 0$ and compute the temperature of the central point of this source. (In fact we have averaged the results over 20 curves). The source being limited in space we expect a crossover at long times, and because of quantification (1000 points to built the curve) crossover at short times... In between the slope is close to the expected value $-(3/2 - \bar{d}/2) = -0.75$.

II.B. Diffusion within Random Fractal Structures

In some cases, the structure in which heat diffusion occurs exhibits a strong disorder and can be mapped on a fractal structure of fractal and fracton dimensions $\bar{d}$ and $\bar{\bar{d}}$ [5]. The heat source is obviously a part of this structure which will be assumed fractal of fractal dimensionality $\bar{d}_s$ ($\bar{d}_s < \bar{d}$). From Rammal and Toulouse [6], one knows that $\bar{\bar{d}} < \bar{d}$.

$\bar{\bar{d}}$, sometimes called the spectral dimension is introduced in order to account for the peculiar diffusion processes [5].

Indeed looking at a diffusion process implies to consider the mean square displacement $\langle r^2 \rangle$ of the diffusers (phonons,...).

In an Euclidean space $\langle r^2 \rangle \sim t$ and thus the diffusion coefficient $D = d/dt\,(\langle r^2 \rangle)$ is a constant.

It has been demonstrated that in a fractal space $\langle r^2 \rangle \sim t^{\bar{\bar{d}}/\bar{d}}$ and the diffusion coefficient D is time dependent, thus the usual heat diffusion solutions are no more valid.

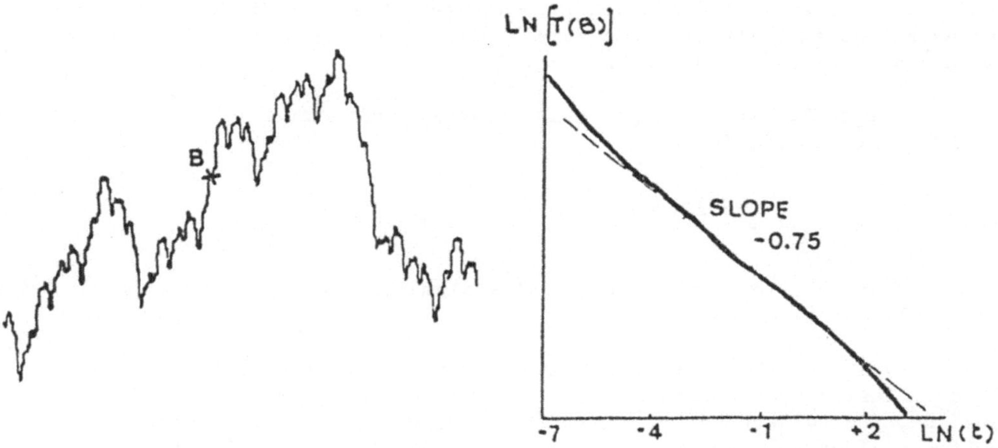

Fig. 2. Brownian scalar motion curve ($\bar{d} = 1.5$) and the temperature at B as a function of time after a pulsed excitation.

Among the random fractal structures a particular attention has been devoted to the so-called "percolation network" [7]. To create a percolation network, each intersection of a d-dimensional grid is e.g., occupied at random with probability p. A critical probability p_c is found such as for $p > p_c$ a connected cluster will cross the grid (infinite cluster in the case of an infinite grid). Such structure, which can be easily generated by computer calculation [8] is found to exhibit a fractal structure [9] ($\bar{d} = 1.9$ for d=2 and $\bar{d} = 2.6$ for d=3). Moreover, Alexander and Orbach have conjectured that $\bar{\bar{d}} = 4/3$ [10] for such percolating network.

Let us go back to heat diffusion and suppose that at time t=0 the sample surface is (uniformly) heated by a short pulse of heat, the heat source dimensionality being $\bar{d}_s$ (e.g. $\bar{d}_s = 0,1,2$ for a point, line and a plane source respectively). After a time t, heat has diffused over a volume V defined by $\langle r^2 \rangle \sim t^{\bar{\bar{d}}/\bar{d}}$; with $V \sim \langle r^2 \rangle^{\bar{d}/2}$. As the energy deposited at time t=0 in this volume is proportional to the surface volume, this energy is $E \sim \langle r^2 \rangle^{\bar{d}_s/2}$ and the temperature T(r) is proportional to E/V :

$$T(t) = E/V \sim \langle r^2 \rangle^{(\bar{d}_s - \bar{d})/2} = (t^{\bar{\bar{d}}/\bar{d}})^{(\bar{d}_s - \bar{d})/2}$$

$$T(t) \sim t^{-\bar{\bar{d}}/2 + \bar{d}_s \bar{\bar{d}}/(2\bar{d})}$$

One can verify that for an Euclidean sample of dimension 3 ($\bar{d} = \bar{\bar{d}} = d = 3$) excited by a point, a line or a plane ($\bar{d}_s = 0,1,$ or 2) one finds as in section II.A.

$$T \sim t^{(\bar{d}_s - d)/2}$$

III. EXPERIMENTAL RESULTS

We have checked the time behavior of the surface temperature of various optically opaque samples (both in the visible and in the IR).. The sample surface is heated by a short (~ 10 ns) light pulse (0.53 nm) and its average temperature is monitored by a fast IR dectetor after being collected by an elliptic mirror. The signal is recorded and averaged with a digital oscilloscope Lecroy 7600.

For an Euclidean sample and a fractal source such as a rough surface $3 > \bar{d}_s > 2$, formula leads to :

$$T \sim t^{-3/2 + \bar{d}_s/2} \sim t^{-\alpha} \qquad \text{with } 0 < \alpha < 1/2$$

Thus the slope in log-log scales is smaller than the usual 1/2 for a plane excitation.

In a real physical situation we expect a cross-over corresponding to a diffusion over a distance equal to the deepest structures of the surface. Indeed such behavior has been observed by group on opaques rough surfaces of carbon samples. The crossover between the two lines varying towards the short time scales as the polishing is improved (smaller grain size of the polishing paper).

For a fractal sample, compact enough to assimilate its surface to a plane ($\bar{d}_s = 2$) one gets :

$$T \sim t^{-(\bar{\bar{d}}/2) - (\bar{\bar{d}}/\bar{d})}$$

To give an order of magnitude of $\bar{\bar{d}}/2 - \bar{\bar{d}}/\bar{d}$ let us consider the case of the 3-d percolating network ($\bar{d} = 2.6$; $\bar{\bar{d}} = 4/3$) which has been used many times [9] as a model for disordered systems. For such system :

$$\bar{\bar{d}}/2 - \bar{\bar{d}}/\bar{d} = 0.154$$

The exponent is thus much lower than 0.5 which is expected in Euclidean case. We have looked at a larger variety of random disordered structures and have found in many occasions a power law over a large time scale with an exponent typical in the range 0.15 - 0.25. As an example Fig.3 shows the result obtained for a weakly bounded assembly of copper spheres.

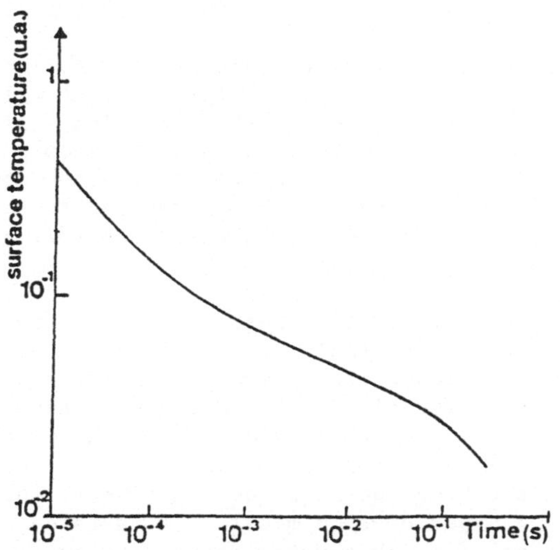

Fig. 3. Surface temperature of an assembly of copper spheres.

We do not claim that the structures that we have studied are fractal systems ; but now that it is well established that nature very often exhibits a fractal geometry [4], it appears interesting to examine heat diffusion in the framework of such theory.

The important point which has to be underlined is that, when diffusion processes takes place in random media (random structures, random bonds...), the average square length covered by the diffuser during its random walk $\langle r^2 \rangle$, is a power function of the time whose exponent account for the dimensionality of the walk.

IV. REACTION KINETICS IN A REAL SYSTEM CLOSE TO PERCOLATION. THE CASE OF A BINARY CRYSTAL OF NAPHTHALENE

One subject of great interest during these past few years has been the problem of the energy migration in doped crystals. A common approach towards this problem has been the study of naphthalene D_8 (ND_8) crystals doped with naphthalene H_8 (Nh_8) containing traces of bethamethyl-naphthalene (BMN) as supertrap at low temperature ($T < 4$ K). The great advantage of this crystal is that a random distribution can be achieved and that the molecular, crystallographic structures, and the coupling strengths are very well known [11-20]. This material seems to be the simplest one which can be used for studying the kinetics of the energy migration in random structure. However, a controversy still exists on the interpretation of the experimental results obtained by Kopelman in the case of the triplet states [21,22]. Does the energy migration come from a dynamic percolation [23,24] an Anderson transition [25-27], or are the experiments correctly described by the model developed by Fayer et al. [28-30].

Our tentative approach to this problem has been to use laser selective excitation and time resolved techniques. We have performed these experiments in order to determine the time dependence of the Nh_8 (BMN) triplet state luminescence. Our experiments have also been concerned with the singlet state luminescence created by the fusion of two triplet excitations.

One of the goal, here, will be to demonstrate that an anomalous behaviour can be generated by a long range migration. It will be due to some special geometrical properties of the diffusion : if energy migration exists between donors and if these donors are randomly distributed, one should expect that the number of paths connecting two donors far away from each other will be strongly reduced compared to that expected in a classical 3-D case. This problem looks very much like a percolation one.

In a first step, we will briefly describe the system and give its different physical characteristics. This will enable us to compute the different times for a triplet excitation to jump from one naphthalene H_8 to a neighbouring one as function of the distance between these two naphthalenes. As these jumptimes strongly depend on the distance, we will assume a cutoff in the distribution of the transfer rates. This cut-off will depend on the time scale of the experiment. This enables us to map the problem on a percolation model. For instance, we report in figure 4, a simulation of a 2-D crystal doped with one impurity at a concentration of 8 %. The donors have been connected together by taking into account only the shortest possible jumps which will allow the migration. One can see in this picture that the way to go from one part of this crystal to another one is much more 1-dimensional than two-dimensional. This randomness reduces the effective dimensionality for energy migration.

In order to determine the dynamics of trapping and fusion in this system we have first to describe the migration in one cluster this will be done by computing the average number N(t) of sites visited by a random walker between time t=0 and time t. Since clusters at the percolation thresholds are fractals, this will require to introduce the spectral dimension $\bar{\bar{d}}$ of Alexander and Orbach [6,10]

$$N(t) \sim t^{\bar{\bar{d}}/2}$$

We will then average N(t) over the different cluster size and relate it to the dynamics of the trapping and the fusion. For instance we will demonstrate that the number I of fusions per unit of time follows

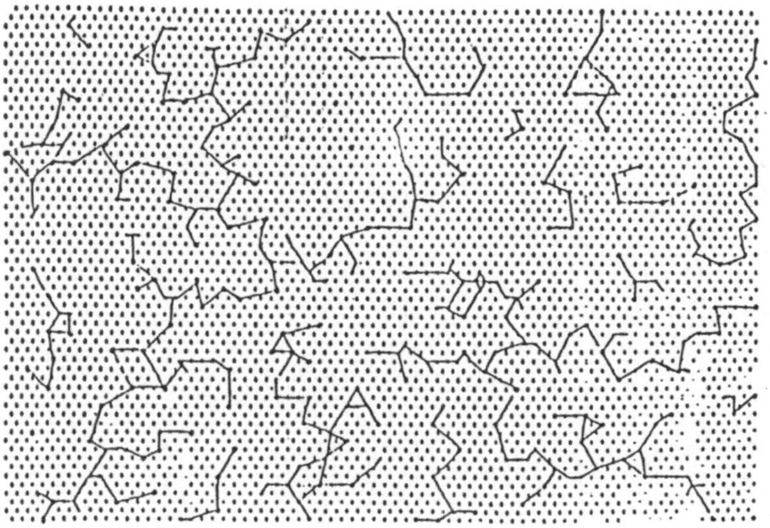

Fig. 4. Simulation of a crystal (1) doped with 8% impurities (8) randomly distributed. The bars indicate all possible jumps that an excitation can make. The jump lengths are smaller or equal to fourth nearest neighbours.

at short time the law :

$$I \sim t^{-\alpha} \qquad \text{with} \qquad \alpha = 1 - \bar{\bar{d}}\,(1-\varepsilon)/2$$

where ε is related to the other classical percolation exponents [32] through $\varepsilon = \beta/(\beta+\gamma)$. α should lay around 0.37 in 2-D percolation case.

We will then seek for a kinetics equation. Labelling n_T the density of excited triplet states and assuming a random distribution of these states, we will prove that at short time the kinetics equation can be written as

$$\frac{d}{dt}(n_T) = \frac{d}{dt}(\langle N(t)\rangle) \quad n_T \quad \text{with} \quad \langle N(t)\rangle = t^{\bar{\bar{d}}(1-\varepsilon)/2}$$

where $\langle N(t)\rangle$ is the average number of distinct sites visited by a random walker between times $t = 0$ and t.

Time resolved experiments [31,33] have been used in order to confirm or infirm this theoretical approach and we have studied the migration of triplet excitations at low temperature ($T < 4.2$ K) in naphthalene D_8 crystals doped with 5% to 12% of naphthalene H_8 containing a small amount of betamethylnaphthalene. The analysis of the time dependence of the fusion of two triplet excitations leads to clearly demonstrate the fractal behavior of the exploration at short time and the Euclidean behavior at longer times. The $\bar{\bar{d}}$ spectral dimension has been measured for the largest studied concentrations ($c > 10\%$) : it lays very near 1.33 which is the predicted value for a percolation model (Fig.5). Deviations from $\bar{\bar{d}} = 1.33$ have been detected for the smallest concentrations $c = 5\%$ and 8% ; we have also determined the time T_0 at which a crossover between a fractal and an Euclidean exploration occurs for the different studied concentrations. Taking into account the repetition rate of the exciting laser and the T_0 value, we will justify the experimental deviations of $\bar{\bar{d}}$ as induced by a non random distribution of excitations in the clusters.

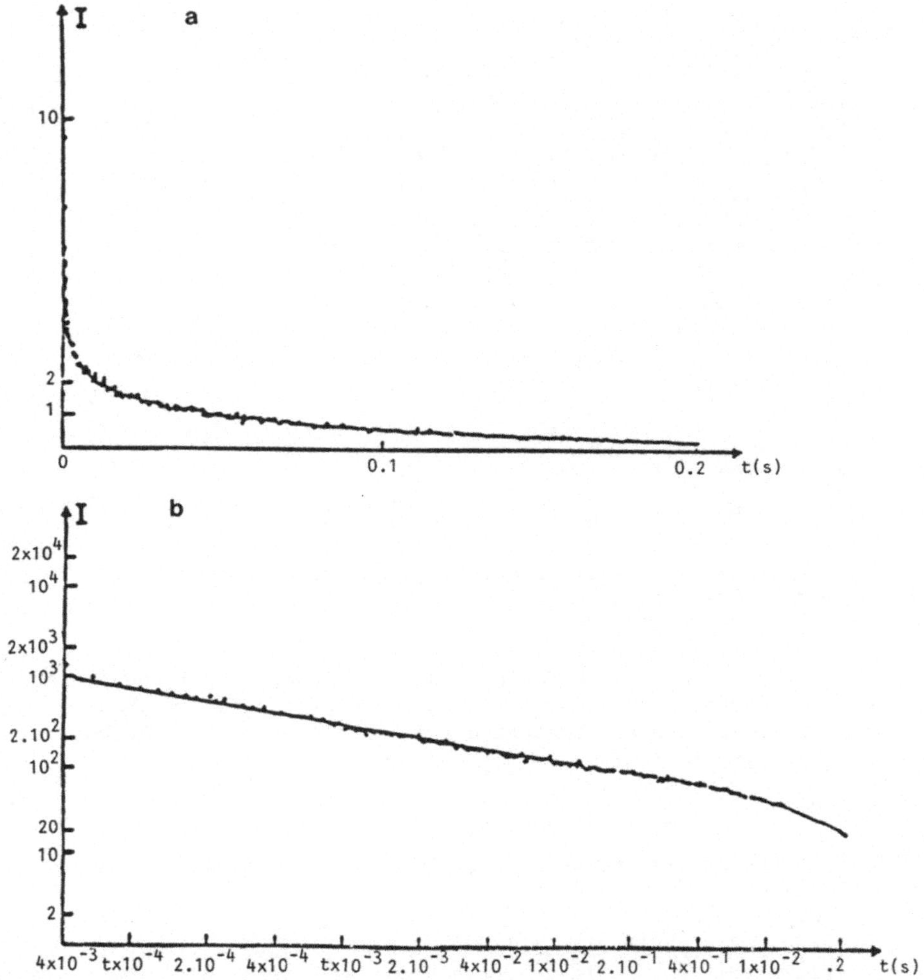

Fig. 5. Time dependence of the singlet luminescence I of naphthalene H_8 in a crystal of naphthalene D_8 doped at 10%. This luminescence is created by the fusion of two triplet excitations. a) linear scale, b) log-log scale. I decreases as t^{-} with 0.37 which confirms the percolation model (see text).

It turns out that a 20% sample does not exhibit this fractal behavior on the studied time scale ; this is also in agreement with the percolation model and with the computed transfer times. Finally, these experimental results demonstrate that the time dependence of the fusion of two excitations gives a better insight of the energy migration processes than the time dependence of the supertrapping process.

V. OPTICAL STUDY OF THE GEOMETRY OF THE PORE SPACE OF VYCOR

Nowadays a great interest has been devoted to the study of porous materials and recent works [34] on rocks structures seem to indicate that the geometry of some pore spaces is fractal [35]. On the other hand, one of the most intensively studied porous materials is the Corning vycor glass (PVG), (7930), the structure of which seems to be well represented through a compact packing of glass spheres of about 200 Å diameter so that the pore radius is about 20 Å and the porosity about 28% [35] ; this is confirmed by electron microscope photographs [35]. In the same way, this description quite agrees with recent results on superfluid helium embedded in this porous material [36].

However different authors have tentatively suggested a fractal structure for PVG in view of their recent results concerning one-step energy transfer [37] and diffusion of excited molecules [38] in this porous material. For instance, Even and coworkers [37] have used a picosecond laser and time resolved spectroscopy experiment to study the time and concentration dependences of the energy transfer occuring between an excited donor molecules (the rhodamine B) and a distribution of acceptor (malachite green) embedded in vycor. We will recall that they have interpreted their results in the light of the Blumen and Klafter theory [39]. This one is a generalization of the theoretical prediction of the time dependence of a one step trapping process due to a dipole-dipole interaction to the case of a $\bar{d}$ fractal space.

If one labels t the time, τ_0 the radiative lifetime of the donor, the donor luminescence intensity I decays as :

$$I(t) \sim \exp -\left\{\frac{t}{\tau_0} + \alpha\left(\frac{t}{\tau_t}\right)^{\bar{d}/6}\right\}$$

where α is a constant which depends on the trap concentration and τ_t the characteristic time of the dipole-dipole interaction.

In view of their experimental results, Even et al. [37] have concluded to a 1.75 fractal dimension of the pore structure.

Following Yang and coworkers [40] we will see, however, that such a transfer strongly depends on the variations of the local environment at the typical length scale at which the transfer occurs. So this experimental set-up turns out to be unable to test a real dilation symmetry and the fractal character of the space. Moreover the main features of the experimental results can be obtained assuming only the previously mentioned structure made of a compact packing of shperes, which makes really doubtfull the fractal interpretation. This will be proved using computer calculations.

From an other point of view, it is now well known that optical transient grating experiments allows to measure diffusion coefficients of excited states by measuring the variation of the relaxation time T of the gratings with its interfringe Λ. In a classical Euclidean space T should vary as Λ^2. We had suggested [41] that this technique be applied to detect an anomalous geometry and predict the behavior of the gratings extinction when considering a fractal or a percolating system.

In particular, we have found that in the large interfringe Λ limit which assumes that the length scale ξ at which the disorder can be averaged is much smaller than the interfringe Λ of the gratings the diffracted signal shall decrease exponentially with a time constant T given by :

$$T = T_o \, (\xi / a)^{(2\bar{d}/\bar{\bar{d}})-2} \, (\Lambda / a)^2$$

where T_o is the jump time for a random walker to jump to a distance a and a is the smallest distance at which the space looks fractal. The fractal space is characterized by its fractal ($\bar{d}$) and spectral ($\bar{\bar{d}}$) dimensions.

The greatest difficulty which had to be solved in order to perform this experiment in vycor was to find an excited state which last a time longer than the time needed to perform a random walk on a distance larger than the interfringe Λ (Λ>1 μm so that T>1s).

This has been achieved by Dozier et al. [38] who have performed transient grating experiments on a solution of a dye molecule (azobenzene) which undergoes a photoisomerization. After much care has been taken in order to try to avoid any sticking of azobenzene on the surface of the pores and after carefully matching the refractive index of the liquid to the vycor one, they could measure the diffusion coefficient in both the bulk solution and the solution in PVG and have found these two diffusion coefficients to differ from each other. They have interpreted this result by assuming a large tortuosity of the paths connecting the pores and a small connectivity of the pores at a small length scale compared to the interfringe of the gratings. Assuming that these geometrical features can be mapped on a fractal or a percolation model, the diffusion coefficient can be calculated [41] ; it does not only depend on the fractal dimension but also on the spectral one [10] so that a whole class of different fractal spaces with different fractal dimension is possible [38] without taking into account any sticking effect.

In order to definitely prove the fractal nature of the PVG pore space, we wanted to demonstrate that diffusion was not modify by stickings of azobenzene molecules on the PVG surface. We have then performed transient grating experiments on a picosecond time scale and measure typical reorientation times of the molecule. The basic idea underlaying this work can be summarized as follows : since the works of lord Rayleigh, Einstein, Langevin and Perrin on Brownian motion, it is well established that rotational and translational diffusions are related to each other. So, assuming that the translational diffusion coefficient measured by the transient grating method is not disturbed by sticking effects implies also that the rotational relaxation time of the molecule is independent of the fact that the liquid is or is not embedded in PVG. On the contrary, if we find two different rotational relaxation times for the molecule in the bulk solution and in the liquid embedded in PVG, this will be induced by a very local change of the translational and rotational diffusion properties (i.e. at a much smaller length scale than the 40 Å diameter of the pore size since the hydrodynamic radius of azobenzene is about 6.5 Å). In turn, this will probably mean that the molecule sticks on the wall of the pores.

It is this last phenomenon which has been detected with this last study. It is corroborated with optical density measurements which definitely prove that the equilibrium concentration of azobenzene in the bulk liquid is smaller than the one in the liquid embedded in the pores.This finally means either that azobenzene sticks spontaneously on the vycor walls or that chemical properties of the liquid inside and outside the vycor are drastically different, which is an other possibility since this liquid spontaneously wet the porous materials and due to the large surface area of this solid material.

VI. ABSORPTION SPECTRA OF AMORPHOUS SEMICONDUCTORS BY PHOTOTHERMAL DEFLECTION SPECTROSCOPY (MIRAGE EFFECT)

The nature of the optical absorption in semiconductors at and below the absorption edge is of interest, particularly in the case of amorphous materials. The sensitivity of conventional transmission techniques are limited by the requirement of measuring the difference between two nearly equal signals. Furthermore, such techniques are highly sensitive to scattering. While adequately sensitive, the disadvantages of photoacoustic detection [42] are that it is highly sensitive to scattered light reaching the detectors and requires either a reliable coupling between the transducer and the sample or to cut the sample in order to put it in a closed cell. We have been able to obtain absorption spectra by "Mirage" [43,44] (photothermal deflection) spectroscopy of hydrogenated amorphous silicon (a-Si:H) films in the range of 2.1 to 0.6 eV. These spectra would be difficult to obtain by other methods.

When an intensity-modulated light beam (pump beam) is absorbed by a medium, heating will ensue. This heating causes a periodic index of refraction gradient in a thin layer adjacent to the sample surface. A second beam (probe beam), propagating through this thin layer, will then experience a periodic deflection which can be quantitatively related to the optical absorption. The magnitude of the deflection ϕ is related to optical absorption in the following manner

$$\phi \sim (1-e^{-\alpha \ell}) \sim \frac{L}{n}\,\frac{dn}{dT}\,\frac{dT}{dx}$$

where ℓ is the film thickness, L the probe beam optical path and $\frac{dT}{dx}$ the thermal gradient.

Fig.6 shows a mirage set-up which is very compact ($\sim$20 cm long) and whose sensitivity in difraction is 10^{-10} rd/$\sqrt{Hz}$ [45].

Our pump beam was the monochromatized output of HgXe arc lamp (0.01 eV bandwidth), and the deflection of the He-Ne laser probe beam was monitored with a conventional position sensor whose output was detected with a lock-in amplifier and normalized for the intensity variations of the pump beam as the wavelength was changed. Since the dn/dT of liquids is typically an order of magnitude larger than that for gases, we immersed the sample in filtered CCl_4. The experimental setup with the exception of the pump source, was enclosed to eliminate temperature gradient caused by air currents.

Absolute absorption coefficient (α) can be determined in one of two ways : 1) For large absorptions, the signal saturates

$$\phi = A\,(1-e^{-\alpha \ell}) \Rightarrow \quad \phi_{sat} = A$$

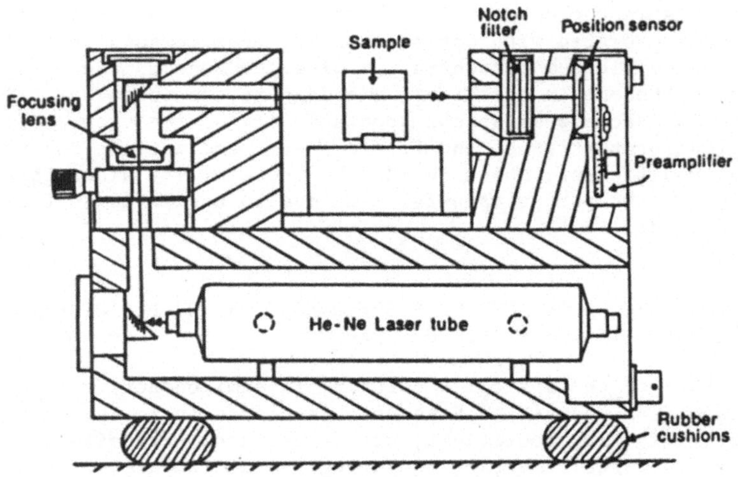

Fig. 6. Experimental arrangement of the compact mirage cell.

The value of the saturated signal can be used to determine $\alpha\ell$ from the equation

$$\alpha\ell = -\ln\left[1 - (\phi/\phi_{sat})\right]$$

or, 2) the reflection and transmission of the samples are measured at a photon energy where absorption is significant. The equations for reflection and transmission are then solved for the index of refraction and the absorption coefficient using a numerical routine. The detailed treatment is given by M.L. Theye group papers [46] on amorphous semiconductors. In Fig. 7 we give the absorption edge and tail of a-Si:H films ($\sim 1\mu$m thick) deposited under various conditions. Our measurements extend the measured values of α by two orders of magnitude over those obtained

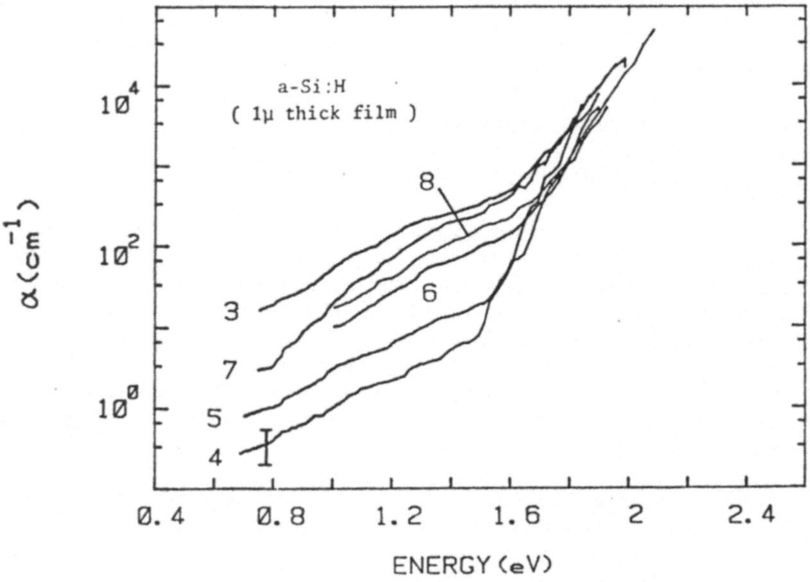

Fig. 7. Absorption coefficient of amorphous Si semiconductor measured with a photothermal technique.

by other techniques. The noise level at 0.7 eV corresponds to an $\alpha \ell = 10^{-5}$ for a 1 mW beam and a few seconds averaging time. Hence, the limiting sensitivity is 10^{-8} W of absorbed power.

Due to its high sensitivity, this technique is now widely used for a-Si and other amorphous semiconductor. The 1 eV absorption has been correlated to the number of dangling silicon bonds by comparing photoacoustic signals to EPR signals [47]. More details will be found in the paper by Adler for the analysis of gap states in amorphous semiconductor. We would like to underline that a photothermal experiment integrates the absorption of the surface and of the bulk of the thin film, whereas photoconductivity which is a sensitive compeeting technique, only deals with photoinduced carriers and ignores localized surface effects.

REFERENCES

1. H.S. Carlsraw, and J.C. Jaeger, "Conductor of heat in solids", Oxford Clarendon (1959)
2. B.K. Bein, S. Krieger and J. Pelge, Can. J. Phys. 64, 1208 (1986)
3. M. Hlavacek, Arch. Mech. (Varszawa), 32, 491 (1980)
4. B. Mandelbrot, "The fractal geometry of nature", Freeman, New York (1983) and "Les objets fractals", Flammarion, Paris (1975)
5. R. Orbach, Science, 231, 814 (1986)
6. R. Rammal, G. Toulouse, J. de Physique (Paris) Lettres, 44, L-13 (1983)
7. R. Pynn, and A. Skjeltorp, "Scaling phenomena in disordered systems", edit. NATO, ASI series Plenum (1985)
8. H.E. Stanley, and N. Ostrowsky, "On growth and form" edit. NATO ASI series, Nijhoff Amsterdam (1985)
9. R. Zallen, "The Physics of amorphous materials", John Wiley (1983)
10. S. Alexander, and R. Orbach, J. de Physique (Paris) Lettres, 43, L-625 (1982)
11. C.A. Hutchinson Jr, and B.W. Magnum, J. Chem. Phys. 34, 908 (1961)
12. M. Schwoerer and H.C. Wolf, Mol. Cryst. 3, 177 (1967)
13. B.J. Botter, C.J. Monhof, J. Schmidt, and J.H. van der Waals, Chem. Phys. Lett. 43, 210 (1976)
14. D.W.J. Cruickshank, Acta Crystallogr. 10, 504 (1957)
15. P. Reineker, D. Richard, and U. Schmid, J. Chem. Phys. 76, 5245 (1982)
16. U. Doberer and H. Port, Chem. Phys. Lett. 85, 253 (1982)
17. D.M. Hanson, J. Chem. Phys. 52, 3409 (1970)
18. C.L. Braun and H.C. Wolf, Chem. Phys. Lett. 9, 260 (1971)
19. Ph. Pee, J.P. Lemaistre, F. Dupuy, R. Brown, and Ph. Kottis, Chem. Phys. 64, 389 (1982)
20. B.J. Botter, A.J. van Strien, and J. Schmidt, Chem. Phys. Lett. 49, 39 (1977)
21. R. Kopelman, E.M. Monberg, F.W. Ochs, and P.N. Prasad, J. Chem. Phys. 62, 292 (1975)
22. D.C. Ahlgren, E.M. Monberg, and R. Kopelman, Chem. Phys. Lett. 64, 122 (1979)
23. R. Kopelman, E.M. Monberg, and F.W. Ochs, Chem. Phys. 19, 413 (1979)
24. E.M. Monberg and R. Kopelman, Chem. Phys. Lett. 58, 492 (1978)
25. J. Klafter, and J. Jortner, J. Chem. Phys. 71, 2210 (1979)
26. J. Klafter, and J. Jortner, Chem. Phys. Lett. 60, 5 (1978)
27. J. Klafter, and J. Jortner, Chem. Phys. Lett. 49, 410 (1977)
28. G.R. Gochanour, H.C. Andersen, and M.D. Fayer, J. Chem. Phys. 70, 4254 (1979)
29. R.F. Loring, M.C. Andersen, and M.D. Fayer, J. Chem. Phys. 76, 2015 (1982)
30. R.F. Loring and M.D. Fayer, Chem. Phys. 70, 139 (1982)
31. P. Evesque, J. Phys. (Paris), 44, 1217 (1983)
32. P.G. de Gennes, C.R. Acad. Sc. Ser. B 296, 881 (1983)

33. P. Evesque, and J. Duran, J. Chem. Phys. 80, 3016 (1984)
34. J. Kats, and A.H. Thompson, Phys. Rev. Lett. 54, 1325 (1985)
35. K. Kadukora, Ph.D. Dissertation, University of California, Los Angeles (1983)
36. J.R. Beamish, A. Hikata, and C. Elbaum, Phys. Rev. B 27, 5848 (1983)
37. - U. Even, K. Rademann, J. Jortner, N. Manor, and R. Reisfeld, Phys. Rev. Lett. 52, 2164 (1984)
- P. Levitz and J.M. Drake, Phys. Rev. Lett. 58, 686 (1987)
38. W.D. Dozier, J.M. Drake, and J. Klafter, Phys. Rev. Lett. 56, 197 (1986)
39. J. Klafter, and A. Blumen, J. Chem. Phys. 80, 875 (1984)
40. C.L. Yang, P. Evesque, and M.A. El-Sayed, J. Phys. Chem. 89, 3442 (1985)
41. - P. Evesque, J. Duran, and A. Bourdon, J. Phys. C 18, 2643 (1985)
- P. Evesque, J. Duran, and A. Bourdon, J. Phys. (Paris) C7, 45 (1985)
42. A. Rosencwaig, "Photoacoustic and photothermal spectroscopy", John Wiley and Son, New York (1980)
43. A.C. Boccara, D. Fournier, and J. Badoz, Appl. Phys. Lett. 36, 130 (1979)
44. W.B. Jackson, N.M. Amer, A.C. Boccara, and D. Fournier, 20, 1333 (1981)
45. F. Charbonnier, and D. Fournier, Rev. Sci. Instrum. 57, 1126 (1986)
46. M.L. Theye, A. Georghia, K. Driss-Khodja, and A.C. Boccara, "11th Conf. on Amorphous and Liquid Semiconductors", Rome (1985)
47. W. Jackson, N. Amer, D. Fournier and A.C. Boccara, "Technical Digest 2nd Int. topical mentions on photoacoustic spectroscopy" (1981)

DETECTION OF PROTONIC PERCOLATION ON HYDRATED LYSOZYME POWDERS

G. Careri and A. Giansanti

Dipartimento di Fisica
Università di Roma I
Roma 00185, Italy

and

J. Rupley

University Department of Biochemistry
Biological Sciences West
University of Arizona, Tucson, AZ 85721, U.S.A.

ABSTRACT

Powders of the protein Lysozyme at low hydration display proton conductivity. The conduction process follows the percolation 2D model for both the critical water fractional coverage of the macromolecule surface and the critical exponent. This critical hydration threshold is close to the onset of the catalytic activity. We view the proton conduction as a proton transfer process along threads of Hydrogen bonded water molecules.

I. INTRODUCTION

The aim of this work is to establish the presence of non localized and random pathways for proton transfer on the surface of an enzyme, to help concerted action among H-bonded side chains responsible for catalysis. As it will appear in the following, the protein Lysozyme is a suitable candidate for such experimental study.

Over the past years we have studied several thermodynamic and dynamic properties of the hydration of Lysozyme powders, because here hydration can be varied by adding water to dry macromolecules until a dilute solution is obtained. The advantage of this system is that several measurements can be made at all system compositions, using different techniques. This makes it possible to study the subsequent events induced by hydration in a series of increasing complexity.

Lysozyme is a comparatively simple enzyme, and almost everything is known about its hydrated powders, thanks to I.R. spectroscopy, E.P.R. relaxation, heat capacity and other thermodynamic and dynamic properties, and especially the enzyme activity towares appropriate substrates [1]. We may summarize these data by saying that the hydration-stepwise process consists of three well-defined stages: i) from 0 to about 60 H_2O molecules/macromolecule, dominated by the interaction of water with the charged groups of the protein; ii) from 60 to about 200 H_2O molecules/macromolecule, where some major changes in surface water arrangements take place; iii) from 220 to about 300 or more H_2O molecules/macromolecule, where the enzymatic activity starts and grows with increasing hydration, together with the condensation of water molecules onto weakly interacting unfilled patches of surface where the molecules are in rapid motion. It is important to note that no structural transitions in the absorbed water or in the protein itself have been detected in the range from 60 to 220 H_2O molecules/macromolecule (or hydration h included between 0.07 and 0.25 g H 0/g.dry weight).

II. EXPERIMENTAL RESULTS

Since the water molecules carry a strong dipole moment, we have recently investigated the dehydration process of Lysozyme powders, using a gravimetric dielectric technique at Megahertz frequencies [2]. When the dielectric losses were measured in H_2O - or D_2O - isopiestic hydrated samples, the relaxation showed a strong isotope effect, indicating that the inferred conductivity was mainly protonic. Moreover, the combination of pH dependence and an effect produced by complexation with a substrate indicate the existence of a proton conduction that involves ionizable side-chain groups of the protein surface, particularly of the active site of the enzyme.

More recently, the analysis [3] of the hydration dependence of the capacitance in the low hydration limit suggested a new interpretation of the onset of dielectric properties within the framework of the percolation model. This general physical model has been proven applicable to a broad range of processes, particularly to study the electrical conductivity of a network of conducting and non-conducting elements [4]. One of the most appealing aspects of the percolation transition is the presence of a threshold, where a long-range connectivity among the elements of the system suddenly appears at a critical concentration of carriers.

In native Lysozyme powders the dielectric capacitance displayed a sharp increase at a water content threshold h = 0.150 ± 0.016 g/g, followed by saturation at increased hydration as shown in Fig. 1a. From the capacitance data at different frequencies one can derive the d.c. conductivity σ, which, as shown in Fig. 1b, displays a similar sharp increase. Since the hydration of one monolayer is h = 0.38 $\pm$ 10% g/g, the experimental volume ratio for surface percolation is 0.40 10%, a value very close to the 0.45 $\pm$ 0.03 predicted by theory [3]. Notice that for three-dimensional networks, regardless of their structure, the conduction threshold predicted by theory is 0.16 $\pm$ 0.02, and this rules out connectivity through the protein interior, where water molecules are known to be very sparse. Moreover, the threshold h was found to be constant from pH 3 to pH 8, indicating that the local geography of water clusters about ionizable sites of the protein surface is not of primary importance. Thus only the number of water molecules acting as interconnected conductivity sites is relevant; and as a matter of fact, the same threshold is found for both H_2O - D_2O-hydrated samples.

From percolation theory [4], one can easily derive that above the threshold the conductivity σ must follow the power law

$$\sigma(h) - \sigma(h_c) = (h_c - h_c)^{\tau} \quad (1)$$

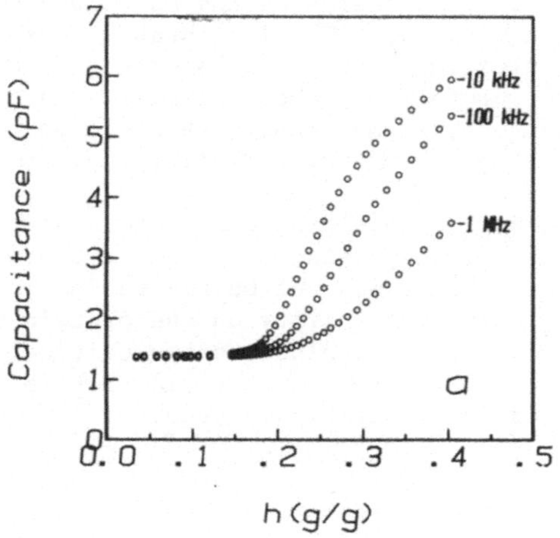

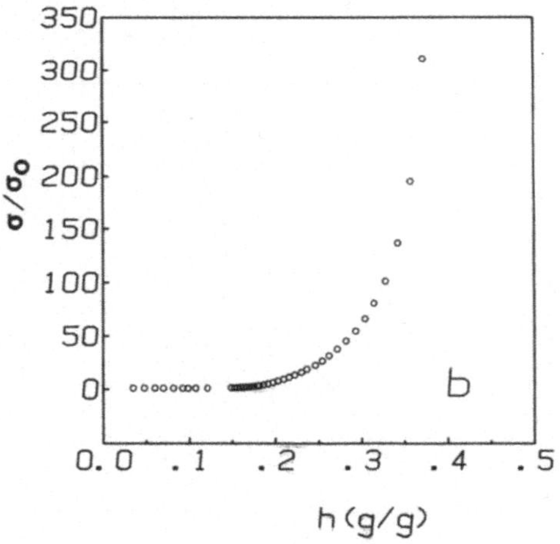

Fig. 1. a. Capacitance of the composite capacitor containing a sample of Lysozyme powder of ph. 5.28 as a function of hydration level of the protein. The capacitance data are given for three frequencies. The hydration level was decreased from the high hydration limit of 0.35 h to the low hydration limit of near 0.07 h by a passage of a stream of dry air through the apparatus (see ref. [2]).
b. Normalized conductivity as a function of the hydration level h. σ_o is the limiting low hydration conductivity of the sample.

where τ depends on the dimensionality of the system. Results of the analysis are shown in Fig. 2 and are in very good agreement with the theoretical prediction for a 2D conduction process. A more complete study of the critical exponent is in progress at the present time.

We picture the percolative path as a proton transfer along a random but continuous thread of hydrogen-bonded water molecules absorbed on the protein surface, the water molecules acting as valves in the proton flow along these fluctuating threads. The long statistical path of interconnected water molecules thus acts as a "short-cut" to bypass the local complex details of the protein surface. Finally, since the percolation theory is size-independent, we may expect that the percolation threshold will be similar for all proteins which display similar sorption isotherms.

For lysozyme-saccharide complexes a higher value of the percolation threshold, has been found, suggesting that the presence of a "foreign body", where the water bridges may not be favorable for proton transfer, must affect the long-range connectivity on the protein surface. This hydration level, $h_c = 0.25$, is so close to the critical level for the onset of enzyme activity in Lysozyme powders [1] that it suggests protonic percolation is involved in Lysozyme catalysis.

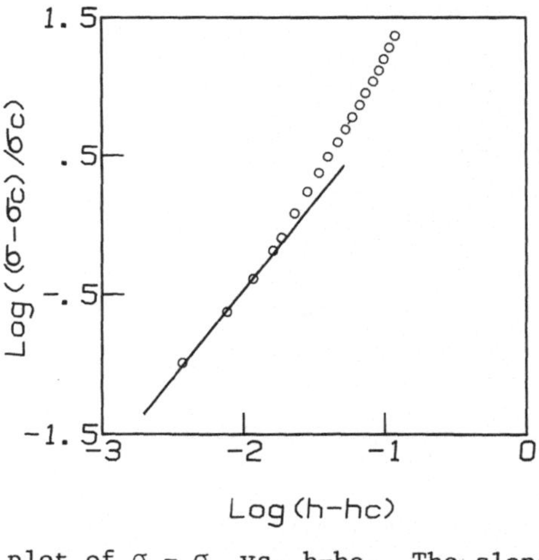

Fig. 2. Log-Log plot of $\sigma - \sigma_c$ vs. h-hc. The slope is 1.25 ± 0.02.

III. DISCUSSION

The percolative transition we considered above reflects a change in the onset of a long-range connectivity developed stochastically in a randomly-structured system of many elements. In the particular case of Lysozyme powders a 2% change in the surface water content shifts the system from a non-conductive to a proton-conductive mode. The threshold for this process, the proton conduction, is discontinuous, whereas the physical or chemical structure of the system is not. The event which suddenly occurs is a large-scale one, because it requires a long-path proton movement over a non-uniform surface.

In our opinion, we may extend the above considerations to other biological processes where a large-scale process can occur if we perform the appropriate correlation of several smaller-scale events [5]. For instance, we may describe the conductive regime of hydrated powders as a series of correlated single-proton transfers along a random thread of water molecules. Since each single-proton transfer can be considered a small-scale event caused by a local fluctuation, the large-scale event results from the correlation of a great many of these small-scale events. Actually, a similar process is observed in laser physics, where several microscopic events (the transitions among atomic states) become cooperatively time-correlated to produce one macroscopic event (the building of the coherent wave).

REFERENCES

1. J.A. Rupley, E. Gratton, & G. Careri Trends Biochem. Sci 8, 18 (1983).
2. G. Careri, M. Geraci, A. Giansanti & J.A. Rupley Proc. Natl. Acad. Sci. USA 82, 5342 (1985).
3. G. Careri, A. Giansanti and J.A. Rupley, Proc. Natl. Acad. Sci. USA 83, 6810 (1986).
4. R. Zallen, "The Physics of Amorphous Solids", Wiley, New York (1983).
5. G. Careri, "Order and Disorder in Matter", Benjamin Cummings, Palo Alto, (1983), Ch. 5.

FLUCTUATIONS IN SYSTEMS SUBJECT TO EXTERNAL FORCES

I. Oppenheim

Department of Chemistry
Massachusetts Institute of Technology
Cambridge, MA 02139

ABSTRACT

We consider a system initially at equilibrium. An external force, $\tilde{F}$, is switched on at t=0 such that F(t) = F (independent of time) for t>0. At long times, the system either appears to come to equilibrium or to a steady-state; i.e. the macroscopic properties of the system are time-independent and, in equilibrium, there are no fluxes and, in the steady-state, there are time-independent macroscopic fluxes. For systems which appear to come to equilibrium, the fluctuations are qualitatively the same as in true equilibrium. However, in systems which appear to come to a steady-state, the fluctuations may be qualitatively different from those in equilibrium. A mode-coupling theory for these results will be presented and the qualitative difference between steady-state and equilibrium fluctuations described.

I. INTRODUCTION

We utilize mode-mode coupling techniques to describe time independent and time dependent fluctuations in equilibrium and steady state systems. The fluctuations in steady state systems are qualitatively different from those in equilibrium systems.

We consider a classical system of N identical point particles in a volume V whose equilibrium properties are obtained from the grand canonical distribution function. We first introduce the multimode quantities, Q, made up of products of slow dynamical variables. Next, we formulate the general Langevin equations describing the time dependence of these quantities and use these equations to obtain generalizations of Onsager's hypothesis. Whereas the equilibrium distribution function for this system is well known, the appropriate distribution function for steady-state systems is not known. However, if we assume that the distribution function for computing averages of $\tilde{Q}$ can be written as $\rho(Q)$, we can use the requirement that the average of $\tilde{Q}(t)$ is independent of time to obtain the appropriate steady state form, ρ_{ss}. Once this is done, the computation of steady state averages of $\tilde{Q}$ and $\tilde{Q}(t)\tilde{Q}$ can be performed.

II. MULTIMODE QUANTITIES

The slow linear dynamical variables of the system will be denoted by the vector $\underset{\sim}{A}$. The components of $\underset{\sim}{A}$, A_α, each have the property that $\dot{A}_\alpha \tau_m << A_\alpha$, on the average, where τ_m is a molecular time of order 10^{-12} sec. and the dot indicates a time derivative. Some of the components of $\underset{\sim}{A}$ correspond to the conserved variable densities, i.e. number density, momentum density and energy density, but there may be nonconserved slow variables as well, e.g. the stress tensor in solids. The vector $\underset{\sim}{A}$ depends on the phase point of the system, $X = r^N, p^N$, and on a position in space $\underset{\sim}{r}$. Here r^N and p^N denote the positions and momenta of all particles in the system. The form of $\underset{\sim}{A}(X,\underset{\sim}{r})$ is

$$\underset{\sim}{A}(X,\underset{\sim}{r}) = \sum_{j=1}^{N} \underset{\sim}{A}_j \, \delta(\underset{\sim}{r}-\underset{\sim}{r}_j) \tag{1}$$

and

$$\underset{\sim}{A}(X(t),\underset{\sim}{r}) \equiv \underset{\sim}{A}(\underset{\sim}{r},t) = e^{iLt} \underset{\sim}{A}(X,\underset{\sim}{r}) \tag{2}$$

where iL is the Liouville operator for the system. For example, the dynamical variable, $N(\underset{\sim}{r})$, whose average is the number density, is given by

$$N(\underset{\sim}{r}) = \sum_{j=1}^{N} \delta\,(\underset{\sim}{r}-\underset{\sim}{r}_j) \tag{3}$$

and

$$N(\underset{\sim}{r},t) = \sum_{j=1}^{N} \delta\,(\underset{\sim}{r}-\underset{\sim}{r}_j(t)) \ . \tag{4}$$

The macroscopic number density is given by

$$n(\underset{\sim}{r},t) = \overline{N(\underset{\sim}{r},t)} = \mathrm{Tr}\,(\rho(0)N(\underset{\sim}{r},t)) \tag{5}$$

where $\rho(0)$ is the distribution function of the system at time zero and the trace involves a sum over the number of particles in the system and an integral over r^N and p^N. The variables $\underset{\sim}{A}$ have the property that $\dot{\underset{\sim}{A}} \sim \lambda$ where λ is small. It is convenient to deal with the spatial Fourier transforms of $\underset{\sim}{A}$, i.e.

$$\underset{\sim}{A}_k = \int e^{i\underset{\sim}{k}\cdot\underset{\sim}{r}} \underset{\sim}{A}(\underset{\sim}{r})d\underset{\sim}{r} \tag{6}$$

and

$$\underset{\sim}{A}_k(t) = \int_V e^{i\underset{\sim}{k}\cdot\underset{\sim}{r}} \underset{\sim}{A}(\underset{\sim}{r},t)d\underset{\sim}{r} \ . \tag{7}$$

We note that if A is a slow variable, then products of A's are also slow variables. Thus, all of the slow variables of the system are included in the multimode quantities $\underset{\sim}{Q}(t)$ given by

$$Q_k^{(0)} = V\delta_{ko}$$

$$\underset{\sim}{Q}_k^{(1)}(t) \equiv \hat{\underset{\sim}{A}}_k(t) \equiv \underset{\sim}{A}_k(t) - \langle \underset{\sim}{A}_k \rangle$$

$$Q^{(2)}_{k-k',k'}(t) = \hat{A}_{k-k'}(t)\hat{A}_{k'}(t) - \langle\hat{A}_{k-k'}\hat{A}_{k'}\rangle$$

$$-\langle\hat{A}_{k-k'}\hat{A}_{k'}\hat{A}^*_{k}\rangle \cdot \langle\hat{A}_{k}\hat{A}^*_{k}\rangle^{-1} \cdot \hat{A}_{k}(t) \qquad (8)$$

•
•
•

Here, the symbol <> denotes an equilibrium average and the superscript * the complex conjugate. For Q to be a slow variable all of its wave vectors must be less than some cutoff, k_c, such that $k_c\xi \ll 1$ where ξ is the equilibrium correlation length in the system. Whereas there are a small number of linear slow variables there are essentially an infinite number of nonlinear slow variables.

An important property of the Q's is that they are orthogonal in the sense that

$$\langle Q^{(\ell)} Q^{(m)*}\rangle = \delta_{\ell m} \langle Q^{(\ell)} Q^{(\ell)*}\rangle \qquad (9)$$

and only Q's with the same total wave vector couple. At a later stage we will need estimates of the order of magnitude of various correlation functions. We find that

$$\langle Q^{(\ell)} Q^{(\ell)*}\rangle \leqslant N^{\ell}$$

and (10)

$$\langle \dot{Q}^{(\ell)} Q^{(\ell')*}\rangle \leqslant N^{\min(\ell,\ell')}$$

Since the Q's span all of the slow variables of the system, any dynamical variable which is orthogonal to that set must be fast. Thus, if

$$\langle I(t)Q^*\rangle = 0 \qquad (11)$$

for all Q, I(t) must be a fast variable.

III. LANGEVIN EQUATIONS

We can write an equation for the time derivative of Q(t) in the form

$$\dot{Q}(t) \equiv \underline{\underline{M}}(t)*Q(t) + I(t) \qquad (12)$$

where $I(t)$ is defined by Eq. (6) and $\underline{\underline{M}}(t)$ is defined by

$$\underline{\underline{M}}(t) \equiv \langle\dot{Q}(t)Q^*\rangle * \langle Q(t)Q^*\rangle^{-1} . \qquad (13)$$

Here, the * between the brackets denotes a sum over the components of the vectors as well as a sum over the repeated wave vectors on each side of the star. Note first that $I(t)$ has the property

$$\langle I(t)Q^*\rangle = 0 \qquad (14)$$

and is a fast variable in the sense that $\langle I(t)I^*\rangle$ decays to zero on a molecular time scale.

The matrix $\underline{\underline{M}}(t)$ can be rewritten as

$$\underline{\underline{M}}(t) = \langle\dot{Q}Q^*\rangle * \langle QQ^*\rangle^{-1} + \int_0^t d/d\tau\, \underline{\underline{M}}(\tau)d\tau$$

$$= \langle\dot{Q}Q^*\rangle * \langle QQ^*\rangle^{-1} - \int_0^t \langle I(\tau)I^*\rangle * \langle Q(\tau)Q^*\rangle^{-1} d\tau \quad . \tag{15}$$

Eq. (15) is exact but it can be approximated by

$$\underline{\underline{M}}(\infty) \equiv \underline{\underline{M}} = [\langle\dot{Q}Q^*\rangle - \int_0^\infty \langle I(\tau)I^*\rangle d\tau] * \langle QQ^*\rangle^{-1} \tag{16}$$

for $t > \tau_m$. Eq. (16) is correct through $O(\lambda^2)$, the smallness parameter characterizing the time derivative of Q. The first term on the right-hand side of Eq. (16) is a streaming or Euler term $\sim\lambda$ and can be evaluated purely from thermodynamic considerations. The second term on the right-hand side of Eq. (16) is a Navier-Stokes or dissipative term $\sim\lambda^2$ and involves transport coefficients like thermal conductivity as well as thermodynamic quantities.

The orders of magnitude of the various elements of the matrix $\underline{\underline{M}}$ are readily found from Eqs. (10). The element $\underline{\underline{M}}_{\ell\ell'}$ which couples $\dot{Q}^{(\ell)}$ to $Q^{(\ell')}$ is of order

$$\underline{\underline{M}}_{\ell\ell'} \lessdot N^0 \qquad \text{for } \ell \gtrdot \ell'$$
$$\lessdot N^{\ell-\ell'} \qquad \text{for } \ell' \gtrdot \ell \quad . \tag{17}$$

Eq. (12) can now be rewritten as:

$$\dot{Q}(t) = \underline{\underline{M}} * Q(t) + I(t) \tag{18}$$

which is of the form of a generalized Langevin equation since it is linear in Q, has a time independent coefficient $\underline{\underline{M}}$ and $I(t)$ has the properties of a rapidly fluctuating force.

The formal solution of eq. (18) is:

$$Q(t) = \underline{\underline{e}}^{Mt} * Q + \int_0^t \underline{\underline{e}}^{M(t-\tau)} * I(\tau)d\tau \tag{19}$$

where $Q = Q(t=0)$. Eq. (18) immediately leads to the results:

$$\langle Q(t)Q^*\rangle = \underline{\underline{e}}^{Mt} * \langle QQ^*\rangle \tag{20}$$

$$\overline{Q(t)} = \underline{\underline{e}}^{Mt} * \overline{Q} \tag{21}$$

$$\overline{Q(t)Q^*} = \underline{\underline{e}}^{Mt} * \overline{QQ^*} \tag{22}$$

where the bar denotes an average over an arbitrary distribution function of the form $\rho(Q)$. Eqs. (21) and (22) follow from the fact that products of $\underset{\sim}{Q}$'s can always be written as linear sums of $\underset{\sim}{Q}$'s since the $\underset{\sim}{Q}$'s form a complete set. Eqs. (21) and (22) are generalizations of the Onsager hypothesis.

Note that for equilibrium averages

$$\langle B(t)\rangle = \langle B\rangle \text{ and } \langle \dot{B}(t)\rangle = 0 \tag{23}$$

for all dynamical variables since $iL\rho_{eq} = 0$. The criterion for a steady-state system is that

$$\overline{\underset{\sim}{Q}(t)}^{ss} = \overline{\underset{\sim}{Q}}^{ss} \tag{24}$$

for all $\underset{\sim}{Q}$. This can be true only if

$$\underline{\underline{M}} * \overline{\underset{\sim}{Q}} = 0 \tag{25}$$

which follows from Eq. (21). In the next section we shall use Eq. (25) to determine the form of $\rho(\underset{\sim}{Q})$ for steady-state systems.

It is clear from Eqs. (20) and (22) that once the equal time averages of $\underset{\sim}{Q}\underset{\sim}{Q}^*$ are found, the time-dependent averages follow fairly easily.

IV. EQUAL TIME FLUCTUATIONS

The equilibrium distribution function for the system is given by

$$\rho_{eq} = \frac{\frac{1}{N!h^{3N}} e^{N\beta_o\mu_o} e^{-\beta_o H(X)}}{\Xi} \tag{26}$$

where the grand canonical partition function is

$$\Xi = \sum_{N=0}^{\infty} \int dX \frac{1}{N!h^{3N}} e^{N\beta_o\mu_o} e^{-\beta_o H} \quad . \tag{27}$$

Here $\beta_o = 1/k_B T_o$, T_o is the equilibrium temperature and μ_o is the equilibrium chemical potential.

The equilibrium equal time correlation functions can be written

$$\langle \hat{\underset{\sim}{A}}_k \hat{\underset{\sim}{A}}_k^* \rangle \simeq \langle \hat{\underset{\sim}{A}}\hat{\underset{\sim}{A}} \rangle \, (1 + O(k^2\xi^2)) \tag{28}$$

where $\hat{A} = \hat{A}_{k=o}$. These are easily obtained using thermodynamic derivatives. For example

$$\begin{aligned} \langle \hat{N}\hat{N} \rangle &= (\frac{\partial n}{\partial \beta\mu})_\beta V \\ \langle \hat{N}\hat{E} \rangle &= -(\frac{\partial n}{\partial \beta})_{\beta\mu} V = (\frac{\partial e}{\partial \beta\mu})_\beta V \; . \end{aligned} \tag{29}$$

Here n is the equilibrium number density and e the equilibrium energy density. We have dropped the subscript o momentarily. In equilibrium, all pertinent bilinear (and higher order) time independent fluctuations can be expressed in terms of thermodynamic quantities.

A frequently used generalization of the equilibrium distribution function to nonequilibrium systems is the local equilibrium distribution function

$$\rho_{LE}(t) = \frac{\rho_o\, e^{\underset{\sim}{A}^* * \underset{\sim}{\Phi}(t)}}{\langle e^{\underset{\sim}{A}^* * \underset{\sim}{\Phi}(t)} \rangle} \tag{30}$$

Here the $\underset{\sim}{\Phi}$ are conjugate forces which depend on $\underset{\sim}{k}$ and t but <u>not</u> on X. More explicitly

$$\underset{\sim}{A}^* * \underset{\sim}{\Phi} = \sum_{\underset{\sim}{k}} \underset{\sim}{A}_{-k} \cdot \underset{\sim}{\Phi}_k(t) \tag{31}$$

and, for example

$$\Phi_N(\underset{\sim}{r},t) = \beta(\underset{\sim}{r},t)[\mu(\underset{\sim}{r},t) - \frac{1}{2} m\, v^2(\underset{\sim}{r},t)] - \beta_o\mu_o$$

$$\Phi_E(\underset{\sim}{r},t) = -\beta(\underset{\sim}{r},t) + \beta_o \tag{32}$$

$$\underset{\sim}{\Phi}_{\underset{\sim}{P}}(\underset{\sim}{r},t) = \beta(\underset{\sim}{r},t)\underset{\sim}{v}(\underset{\sim}{r},t)$$

where $\underset{\sim}{v}(r,t)$ is the local space and time dependent velocity. Clearly by adjusting $\Phi(\underset{\sim}{r},t)$ appropriately, all averages of the linear dynamical variables can be written

$$\underset{\sim}{a}_k(t) \equiv \overline{\hat{\underset{\sim}{A}}_k(t)} = \mathrm{Tr}\,(\rho_{LE}(t)\hat{\underset{\sim}{A}}_k) \quad . \tag{33}$$

However, this is not a suitable distribution function for computing

$\overline{Q^{(\ell)}(t)}$ for $\ell > 1$.

For systems linearly displaced from equlibrium, Eq. (30) becomes

$$\rho_{LE} = \rho_{eq}(1 + \hat{\underset{\sim}{A}}^* * \underset{\sim}{\Phi}) \quad . \tag{34}$$

A suitable generalization of this formula is

$$\rho = \rho_{eq}\Big(1 + \hat{\underset{\sim}{A}} * \underset{\sim}{\Phi} + \sum_{\ell=2}^{\infty} \underset{\sim}{Q}^{(\ell)*} * \underset{\sim}{\Phi}^{(\ell)}\Big) \tag{35}$$

where $\underset{\sim}{\Phi}^{(\ell)}$ is the conjugate force for $\underset{\sim}{Q}^{(\ell)}$ and is a macroscopic quantity which does not depend on X. We shall now use Eq. (35) together with Eq. (25) to determine the appropriate forms for Φ and $\Phi(\ell)$ for the steady-state distribution function. In the development below, we drop the ss notation.

Using Eq. (35) we find

$$\overline{\underset{\sim}{Q}^{(\ell)}} = \langle \underset{\sim}{Q}^{(\ell)} \underset{\sim}{Q}^{(\ell)*} \rangle * \underset{\sim}{\Phi}^{(\ell)} \tag{36}$$

Since $\overline{\underset{\sim}{Q}^{(\ell)}} \sim$ N for systems linear displaced from equilibrium we find that $\underset{\sim}{\Phi}^{(\ell)} \sim \tilde{N}^{1-\ell}$. Substitution of Eq. (36) into Eq. (25) we find

$$\underline{\underline{M}}*\overline{Q} = \underline{\underline{M}}*\langle QQ^*\rangle*\Phi = 0 \tag{37}$$

which is a set of equations to determine all of the Φ's. If we write $\underline{\underline{M}} = \underline{\underline{N}}*\langle QQ^*\rangle^{-1}$ where

$$\underline{\underline{N}} = \langle \dot{Q}Q^*\rangle - \int_o^\infty \langle I(\tau)I^*\rangle d\tau \quad , \tag{38}$$

Eqs. (37) become

$$\begin{aligned}
\underline{\underline{N}}_{11}*\Phi + \underline{\underline{N}}_{12}*\Phi^{(2)} + \underline{\underline{N}}_{13}*\Phi^{(3)} + \cdots &= 0 \\
\underline{\underline{N}}_{21}*\Phi + \underline{\underline{N}}_{22}*\Phi^{(2)} + \underline{\underline{N}}_{23}*\Phi^{(3)} + \cdots &= 0 \\
\underline{\underline{N}}_{31}*\Phi + \underline{\underline{N}}_{32}*\Phi^{(2)} + \underline{\underline{N}}_{33}*\Phi^{(3)} + \cdots &= 0
\end{aligned} \tag{39}$$

Using the fact that $\underline{\underline{N}}_{\ell\ell'} \sim N^{\min(\ell,\ell')}$ and $\Phi^{(\ell)} \sim N^{1-\ell}$, these eqs. simplify to

$$\underline{\underline{N}}_{11}*\Phi = 0 \tag{40a}$$

$$\underline{\underline{N}}_{21}*\Phi + \underline{\underline{N}}_{22}*\Phi^{(2)} = 0 \tag{40b}$$

$$\underline{\underline{N}}_{31}*\Phi + \underline{\underline{N}}_{32}*\Phi^{(2)} + \underline{\underline{N}}_{33}*\Phi^{(3)} = 0 \tag{40c}$$

neglecting terms of order $(k_c\xi)^3$. Eq. (40a), together with the boundary conditions on the system, determine Φ; i.e.

$$[\langle \dot{\hat{A}}\hat{A}^*\rangle - \int_o^\infty \langle I_A(\tau)\ I_A^*\rangle d\tau] * \Phi = 0 \quad ; \tag{41}$$

Eq. (40b) determines $\Phi^{(2)}$ in terms of Φ and Eq. (40c) determines $\Phi^{(3)}$ in terms of Φ and $\Phi^{(2)}$. It is clear that while Φ involves only thermodynamic quantities, $\Phi^{(\ell)}$, $\ell > 1$, will involve transport coefficients as well.

We now must specify the boundary conditions in order to determine Φ. We consider a system with a temperature gradient in the z direction with boundary planes at z = -L and L. The boundary conditions are:

$$T(z = -L) = T_{-L},\ T(z = L) = T_L,\quad v(\text{boundary}) = 0. \tag{42}$$

The solution of eq. (40a) yields:

$$v = 0$$

$$T(x,y,z) = T_{-L} + \frac{T_L - T_{-L}}{2L}(z+L) \tag{43}$$

$$\nabla p_h = 0$$

where p_h is the hydrostatic pressure. The equilibrium values in Eq. (32) are taken at $z = 0$.

Eq. (40b) can be written more explicitly in $\underset{\sim}{r}$ space as:

$$\int \underset{\sim}{N}_{\substack{ac\\b}} (\underset{\sim}{r}_1,\underset{\sim}{r}_2;\underset{\sim}{r}')\cdot\underset{\sim}{\Phi}_c(\underset{\sim}{r}')d\underset{\sim}{r}'$$

$$+ \int \underline{\underline{N}}_{\substack{a\ a'\\b\ b'}} (\underset{\sim}{r}_1,\underset{\sim}{r}_2;\underset{\sim}{r}_1',\underset{\sim}{r}_2') * \underset{\sim}{\Phi}^{(2)}_{\substack{a'\\b'}}(\underset{\sim}{r}_1,\underset{\sim}{r}_2)d\underset{\sim}{r}_1'd\underset{\sim}{r}_2' = 0 \ . \tag{44}$$

Here a,b etc label the slow linear variables. The leading contribution to the second term in Eq. (44) can be obtained by writing

$$\underline{\underline{N}}_{\substack{aa'\\bb'}}(\underset{\sim}{r}_1,\underset{\sim}{r}_2;\underset{\sim}{r}',\underset{\sim}{r}_2') = \underline{\underline{N}}_{aa'}(\underset{\sim}{r};\underset{\sim}{r}_1')\langle\hat{A}_b(\underset{\sim}{r}_2)\hat{A}_{b'}(\underset{\sim}{r}_2')\rangle$$

$$+ \underline{\underline{N}}_{bb'}(\underset{\sim}{r}_2;\underset{\sim}{r}_2')\langle\hat{A}_a(\underset{\sim}{r}_1)\hat{A}_{a'}(\underset{\sim}{r}_1')\rangle \ . \tag{45}$$

We now take the Fourier transform of Eq. (44), fixing the center of mass variable $\underset{\sim}{R} = \underset{\sim}{r}_1+\underset{\sim}{r}_2/2$ at 0.

The quantity of interest is

$$\chi_{ab}(\underset{\sim}{q}) = \frac{1}{(2\pi)^3}\int d\underset{\sim}{k}\ [\ \overline{A_{a(\underset{\sim}{q}+\underset{\sim}{k}/2)}A_{b(\underset{\sim}{q}-\underset{\sim}{k}/2)}}^{ss} - \langle A_{a(\underset{\sim}{q}+\underset{\sim}{k}/2)}A_{b(\underset{\sim}{q}-\underset{\sim}{k}/2)}\rangle_L\ ] \tag{46}$$

where the subscript L denotes an average over the local equilibrium distribution function Eq. (34). The quantity $\chi_{ab}(\underset{\sim}{q})$ is the difference between the true steady state fluctuation and that which would be obained using the local equilibrium function. This difference exists because of the necessity of introducing $\Phi^{(2)}$ to insure that $\overline{Q^{(2)}(t)}^{ss} = \overline{Q^{(2)}}^{ss}$.

So far, we have not explicitly described our system of interest. It could be a fluid, a crystal or a disordered solid. In order to illustrate the importance of χ, we shall write it explicitly for fluid systems, though essentially the same results hold for the other systems as well.

The results are that the χ for correlations which exist in equilibrium are unimportant, i.e. χ_{NN}, χ_{NE}, χ_{EE}, $\chi_{\underset{\sim}{P}\underset{\sim}{P}}$, but the χ for correlations which are zero in equilibrium or in the nonconvecting local equilibrium steady state are appreciable. Thus, for example

$$\underset{\sim}{\chi}_{PN}(\underset{\sim}{q}) = \frac{-k_BTn}{2q^2\Gamma_s}\hat{q}\,\hat{q}\cdot\frac{\nabla T}{T} + \frac{k_BTn(T\gamma_T)}{q^2(\nu+\Gamma_T)}\hat{q}x(\hat{q}x\frac{\nabla T}{T}) \tag{47}$$

$$\underset{\sim}{\chi}_{PE}(\underset{\sim}{q}) = \frac{-k_B T h}{2q^2 \Gamma_s} \hat{q}\hat{q} \cdot \frac{\nabla T}{T} + \frac{k_B T(h - nC_p/\gamma_T)T\gamma_T}{q^2(\nu + \Gamma_T)} \hat{q} x (\hat{q} x \frac{\nabla T}{T}) \tag{48}$$

where the thermodynamic quantities and transport coefficients are measured at $z = 0$, h is the enthalpy density, γ_T is the thermal expansivity, Γ_s is the sound attenuation coefficient, ν is the dynamic shear viscosity and Γ_T is the thermal diffusivity. Finally $\hat{q}$ is the unit vector in the direction of $\underset{\sim}{q}$.

The correlation between $\underset{\sim}{P}$ and N or E is surprising since in these nonconvecting systems $\overline{P}^{ss} = \underset{\sim}{0}$. The correlation is produced by the presence of the vectorial heat flux. The apparent divergence as $q \to 0$ is spurious becuase we have assumed that $qL \gg 1$ in order to obtain Eqs. (47) and (48). The expressions in these equations have simple physical interpretations. For example the first terms can be written schematically as

$$\begin{aligned} \underset{\sim}{\chi}_{PN} &\sim - \frac{\langle P^2\rangle^{1/2} \langle \hat{N}^2\rangle^{1/2}}{V} \frac{c_o}{q^2\Gamma_s} \hat{q}\hat{q} \cdot \frac{\nabla T}{T} \\ \underset{\sim}{\chi}_{PE} &\sim - \frac{\langle P^2\rangle^{1/2} \langle E^2\rangle^{1/2}}{V} \frac{c_o}{q^2\Gamma_s} \hat{q}\hat{q} \cdot \frac{\nabla T}{T} \end{aligned} \tag{49}$$

where c_o is the adiabatic sound velocity. The correlation exists for the component of momentum in the direction of the heat flux and has a magnitude proportional to the ratio of how far a sound mode propagates before it decays to the distance over which the temperature varies. The other terms in Eqs. (47) and (48) arise from coupling of transverse momentum and heat modes.

Note that the q^{-2} dependence implies that

$$\overline{\underset{\sim}{P}(\underset{\sim}{r})N(\underset{\sim}{r}')}^{ss} \sim \frac{1}{|\underset{\sim}{r}-\underset{\sim}{r}'|} \sim \overline{\underset{\sim}{P}(\underset{\sim}{r})E(\underset{\sim}{r}')}^{ss} ; \tag{50}$$

the correlations decay as $1/r$ for intermediate distance, $r \ll L$. Thus in this simple steady state, long range correlations are set up due to the presence of the heat flux.

V. TIME DEPENDENT FLUCTUATIONS

Once the time independent fluctuations have been obtained we can use Eqs. (20) and (22) to obtain the time dependent fluctuations. We shall focus on the density-density fluctuations since they can be measured directly by light scattering.

In the equilibrium system we find from Eq. (20)

$$\langle \hat{N}_k(t)\hat{N}_{-k}\rangle = \underline{\underline{e}}^{M_k t}_{NN} \langle \hat{N}_k \hat{N}_{-k}\rangle + \underline{\underline{e}}^{M_k t}_{NE} \langle \hat{E}_k \hat{N}_{-k}\rangle \quad . \tag{51}$$

In the steady state system we find

$$\overline{\hat{N}(t)\hat{N}}^{ss} = \underline{\underline{e}}^{Mt}_{N\ell} * \overline{\underset{\sim}{Q}^{(\ell)}\hat{N}}^{ss}$$

$$= \underset{\approx N\ell}{e^{Mt}} * \langle \underset{\sim}{Q}^{(\ell)}\hat{N}\underset{\sim}{Q}^{(j)*}\rangle * \underset{\sim}{\Phi}^{(j)} \tag{52}$$

Because of the orthogonality properties of the Q's and the facts that $Q^{(\ell)}\hat{N}$ can be expressed as a linear sum of Q's up to order $\ell+1$ and that $\hat{Q}^{(j)}\hat{N}$ can be expressed as a linear sum of Q's up to order $j+1$, we find that $j = \ell+1, \ell, \ell-1$ and Eq. (52) can be written

$$\overline{\hat{N}(t)\hat{N}}^{ss} = \sum_{\ell=1}^{\infty} e^{Mt}_{N\ell} * [\langle \underset{\sim}{Q}^{(\ell)}\hat{N}\underset{\sim}{Q}^{(\ell-1)*}\rangle * \Phi^{(\ell-1)} + \langle \underset{\sim}{Q}^{(\ell)}\hat{N}\underset{\sim}{Q}^{(\ell-1)*}\rangle * \Phi^{(\ell)}$$

$$+ \langle \underset{\sim}{Q}^{(\ell)}\hat{N}\underset{\sim}{Q}^{(\ell+1)*}\rangle \cdot \Phi^{(\ell+1)}] \quad . \tag{53}$$

Note that $\Phi^{(0)} = 1$ and $\underset{\sim}{\Phi}^{(1)} = \underset{\sim}{\Phi}$. We now use the facts that

$e^{Mt}_{N\ell} \sim \langle N^{1-\ell})$, $\Phi^{(\ell)} \sim N^{1-\ell}$, $\langle \underset{\sim}{Q}^{(\ell)}\hat{N}\underset{\sim}{Q}^{(\ell-1)*}\rangle \langle N^{\ell}$, $\langle \underset{\sim}{Q}^{(\ell)}\hat{N}\underset{\sim}{Q}^{(\ell)*}\rangle \langle N^{\ell}$ and

$\langle \underset{\sim}{Q}^{(\ell)}\hat{N}\underset{\sim}{Q}^{(\ell+1)*}\rangle \langle N^{\ell+1}$ to write the leading terms in Eq. (53) as

$$\overline{\hat{N}(t)\hat{N}}^{ss} = \underset{\approx N1}{e^{Mt}} \cdot \langle Q^{(1)}\hat{N}\rangle + \underset{\approx N1}{e^{Mt}} \cdot \langle \underset{\sim}{Q}^{(1)}\hat{N}\underset{\sim}{Q}^{(1)*}\rangle * \underset{\sim}{\Phi}^{(1)}$$

$$+ \underset{\approx N1}{e^{Mt}} \cdot \langle \underset{\sim}{Q}^{(1)}\hat{N}\underset{\sim}{Q}^{(2)*}\rangle \cdot \underset{\sim}{\Phi}^{(2)}$$

$$+ \underset{\approx N2}{e^{Mt}} * \langle \underset{\sim}{Q}^{(2)}\hat{N}\underset{\sim}{Q}^{(1)*}\rangle \cdot \underset{\sim}{\Phi}^{(1)} \tag{54}$$

The first term on the right-hand side is the equilibrium result; the second term is the local equilibrium correction to the equal time correlation function and is small; the third term is the steady state correction to the equilibrium correlation function and is important; the fourth term is the local equilibrium correction to the propagator and is again small.

Using the results of the last section, we obtain the following approximate result for the dynamic structure factor

$$S_{NN}(\underset{\sim}{q},\omega) = \frac{2k_BTn}{mc_o^2}\{(\frac{C_p}{C_v}-1)\frac{\Gamma_T q^2}{\omega^2+(\Gamma_T q^2)^2}$$

$$\frac{\Gamma_s q^2}{(\omega-qc_o)^2 + [\Gamma_s q^2(1-\varepsilon^+(\underset{\sim}{q}))]^2} + \frac{\Gamma_s q^2}{(\omega+qc_o)^2 + [\Gamma_s q^2(1-\varepsilon^-(\underset{\sim}{q}))]^2}\}$$

For equilibrium, $\varepsilon^+ = \varepsilon^- = 0$, and for the steady state

$$\varepsilon^{\pm} = \pm \frac{c_o}{2q^2\Gamma_S}\hat{q} \cdot \frac{\underset{\sim}{\nabla} T}{T} \quad . \tag{56}$$

The scattering spectrum consists of a Rayleigh peak centered at $\omega = 0$ which is unmodified by the linear steady state and two Brillouin peaks centered at $\omega = \pm qc_0$. The steady state corrections modulate the sound attenuation coefficients Γ_s increasing it for sound modes going opposite to the heat flux and therefore decreasing the intensity and decreasing Γ_s for sound modes going in the direction of the heat flux and increasing their intensity.

The techniques presented here are described in more detail in reference [1]. The results for fluids have been obtained by a number of investigators using a number of different techniques and references to their work is given in [1]. Similar results for solids are given in [2]. Experimental confirmations of these results have been reported by Beysens [3] and by others [4]. Extensions of these techniques to the nonlinear regime are reported in [5] and some of the results have appeared in references [6]. Finally, the effects of boundaries have been discussed in [7].

REFERECNES

1. J. Machta and I. Oppenheim, Physica A 112:361 (1982).
2. A. Griffin, Can. J. Phys. 46:2843 (1968).
3. D. Beysens, Y. Garrobos and G. Zalcer, Phys. Rev. Lett. 45:403 (1980).
4. R. Penney, H. Kiefte and M.J. Clouter, Bull. Can. Assoc. Phys. 39:888 (1983); G.H. Wegdam, N.M. Keulen, and J.C.F. Michielsen, Phys. Rev. Lett. 55:630 (1985).
5. I. L'Heureux, Thesis M.I.T. (1987).
6. D. Ronis and I. Procaccia, Phys. Rev. A 26:1812 (1982); T.R. Kirkpatrick, E.G.D. Cohen and J.R. Dorfman, Ibid. 995 (1982).
7. G. Satten and D. Ronis, Phys. Rev. A 27:2577 (1983).

SPECTROSCOPY OF INORGANIC GLASSES

W. M. Yen

Department of Physics and Astronomy
University of Georgia
Athens, Georgia 30602, USA

ABSTRACT

We present in this paper a brief review of the status of our understanding of the optical properties of inorganic glasses activated with various impurity ions and/or centers. We sketch the historical evolution of this discipline and outline recent developments based largely on relatively new laser spectroscopic techniques. These experimental methods have allowed us to suppress inhomogeneous features of the optical spectra of crystalline and disordered materials alike and have permitted us a more detailed look at the structure of all classes of solids and at the dynamical processes which affect the optically excited state of individual centers.

I. INTRODUCTION

This review concerns itself with a subset, albeit the most common one, of the large variety of disordered systems which is the subject of this course. We will focus our attention on common insulating glasses; more specifically, we are interested here in those glasses which have been optically activated through the introduction of an impurity of defect so that they show coloration when viewed by the naked eye. The insulating glasses of interest are encountered in a multitude of common, every day uses as well as an increasing number of technical applications.

Glass has a long and very interesting history which can be traced to 2,000BC and perhaps even earlier. The discovery of glass has been attributed by Pliny to the Phoenicians and the craft flourished principally in the Middle East as raw materials for glass making, i.e., sand, soda, or potash and fuel, are in good supply in that region. There are a number of excellent histories of glass and glass technologies available which make fascinating reading [1,2], consequently, we will not belabor these historical facts. Because the batch materials were not purified, various contaminants were invariably present in most of the early surviving glasses; this result in the coloration of glass, with the color characteristics being peculiar to the locale of manufacture because of the impurities which were prevalent there. The precise ingredients making up a particular glass were then considered to be secrets of the craft and were guarded jealously by the ancient craftsmen. The glass industry as we know it today grew from these origins driven by various consumer and technological needs. Much effort was spent, for example, in the development of clear glasses for windowpane use and undoubtedly the initial correlation between impurity content of the batch material and the coloration of the glass was established then. Attempts to actually develop a rigorous physical understanding of the details of the spectra of impurity doped glasses. however, did not commence until the late 1950's [3], shortly after the seminal articles by Sugano and Tanabe [4] on the nature of the spectra of transition metal ions in crystals.

The insulating glasses of interest here are typical of materials in the vitreous state in that there is no long range positional order of the glass constituents over macroscopic distances. The insulating nature of these materials implies that their ground states consist of filled valence bands and that forbidden gaps in energy intervene before additional electronic states (conduction bands) are encountered. These gaps are in most cases sizable and in their inactivated state the glasses will look clear to the eye. In other words, the glass matrix acts as an optically inert host to the impurity centers; by using the dopant ion as a probe, we then can derive information on the structure and the nature of interactions which affect the glassy host material.

Because of the wide variety of glassy and vitreous materials, and because in general their very nature is evidence of a precarious state of metastability, the terminology used to describe the glassine state is not very exact. For example, it is sometimes difficult to distinguish between a real glass and a supercooled liquid. Definitions, nevertheless, have evolved by convention: a glass is required to possess those mechanical properties that define a solid, and though the glass state is understood to be a non-equilibrated state, glasses are required to be stable against crystallization over some lengthy period of time. Finally, the state is to be reproducible upon recycling of external parameters such as temperature.

A somewhat more quantitative parameter may be assigned to the glassy state, i.e., the glass transition temperature, T_g. In these terms, a material is said to undergo a glass transition if the amorphous condensed phase shows a sudden or discontinuous change in the derivative of some thermodynamic property from crystal to liquid-like values as a function of increasing temperature. The temperature at which the discontinuity occurs is then defined as T_g. There are some intrinsic difficulties with this definition, one of which is illustrated in Figure 1 [5]. The currently most commonly accepted value of T_g is associated with the liquid viscosity, η of the compound. Specifically, T_g is the point at which η attains a value of 10^{+13} poise. Because of the uncertainties connected with this definition, it is generally accepted that T_g cannot be calculated and that this quanitity is mostly a convenient descriptive parameter.

The thermodynamic conditions which lead to the formation of glasses viz crystallization have been considered by a number of authors beginning with Zachariesen [6]. There is a complex interplay between the nature of the bonding, the kinetics of crystal formation and the rates of energy extraction from the liquid state. This subject is well beyond the scope of this presentation and reference is made to a number of past and recent reviews [5,7,8].

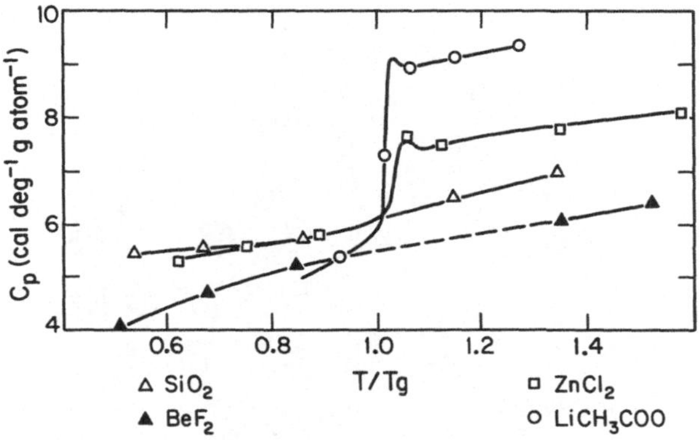

Fig. 1. Changes of the heat capacity observed in the vicinity of the glass transition temperature for a number of NWF of the Mx_2 type. The transition temperatures are: SiO_2, T_g = 1176°C; BeF_2, T_g = 319°C, and $ZnCl_2$, T_g = 975°C. The behavior of an ionic lithium acetate is shown for comparison T_g = 125°C [5].

We have organized the discussion as follows: we review briefly the terminology utilized when dealing with glass of the type of interest to us and the microscopic structure of this type of materials. We will then consider the nature of the activating centers and the types of spectra and transitions these impurities produce when introduced into insulating glasses. Finally, we will present advances attained in understanding both the static and dynamic spectroscopic properties of doped glass derived through the use of laser spectroscopy. We conclude with some general remarks on future prospects and developments in this area of research.

II. TERMINOLOGY AND STRUCTURE

As the name implies, an amorphous or disordered material is not expected to contain any local or lattice symmetry extending over macroscopic distances. The glass equivalent of a crystalline lattice is known as a network or matrix; though single atomic entities can combine to form a network, the most common glasses generally contain molecular units such as $(SiO_4)^{-2}$ tetrahedra which arrange themselves in ways paralleling their crystalline counterparts. The principal difference in these terms between a glass and a crystal of the same composition is that in the glass the molecular bond lengths and bond angle can vary in a random way. Cations in a simple molecular crystal are known as network formers (NWF). To maintain neutrality, anions will bond and connect NWF's in sufficient numbers to saturate their valence; bonds of this type are known as bridging bonds. Bridging bonds may be altered to non-bridging bonds by introducing another type of cation, the network modifier (NWM). These latter cations form bonds with the anions and thus modify or interrupt the interconnections of the NWF cations. Alkaline or alkaline-earth ions are commonly used as NWM, and the amount which may be substituted into a given network depends on the nature of the molecular building blocks forming the original network. Such considerations delimit the compositional ranges over which glasses may

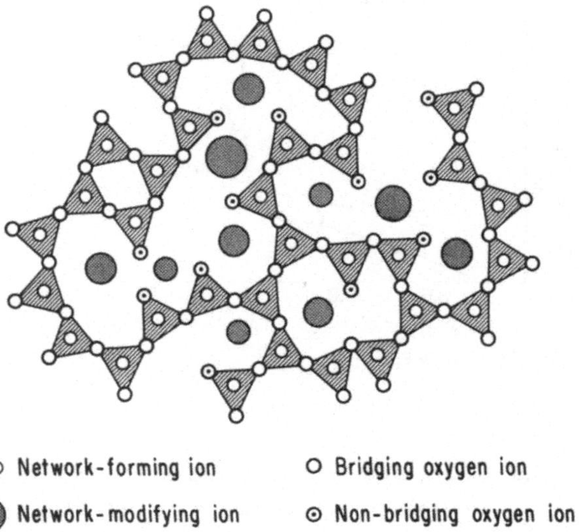

Fig. 2. Two dimensional representation of an oxide glass showing NWF, NWM cations and bridging and non-bridging bonds as depicted by Zachariasen [6].

be formed. Ternary or more complex glasses have been designed to accommodate higher valence NWM such as the trivalent rare earths which are optically active in the visible and which are of interest to us here. In the more complex glasses, the distinction between NWF and NWM becomes rather fuzzy and both types of cations can sometimes be considered as genuine constituents of the glass [9]. Figure 2 illustrates various definitions introduced above for a two-dimensional representation of an oxide glass [6].

The local coordination number of charges surrounding the NWF cations vary depending upon the size of the cation and are related in part to the coordination normally encountered by these cations when they form crystals. The coordination of NWM on the other hand can vary and there is evidence that these cations can have anywhere from four to nine fold coordination [9,10,11]. The coordination surrounding the NWF implies that a certain degree of local order exists in the microscopic domain, consistent with the concept of the formation of networks. It is not surprising then that diffuse structure is observed when X-rays are diffracted from glasses. The patterns observed and the radial distributions derived from these patterns are reminiscent of those observed in liquids and they generally serve to confirm the model of glass structure discussed above [6,12].

Coloration of glasses is accomplished by the intentional introduction of various types of positively charged ions [3]. Metal ions of the transition metal series, $(3d)^n$, have been most commonly employed for these purposes. Certain heavier ions, such as Pb, U, Tl, have also been used. The use of lanthanides $(4f)^n$ as a coloration agent, though now widespread, has been a relatively recent development. The transition metal ions enter into glasses both as NWF and NWM, and roughly speaking, concentrations in the neighborhood of 1 wt% may be incorporated into most glasses without affecting the properties of the host glass. Lanthanides cannot enter simple glasses as NWF and their incorporation requires additional NWM for charge compensation and other purposes. Certain heavier and more complex glasses, such as the lanthanum crown glass, have been developed to fulfill specific optical instrumentation requirements; in these cases, of course, rare earths can and do serve as NWF.

The intrinsic instability of the glassy state invariably leads to devitrification, however, this process may take extended periods of time. Eqyptian glass beads dating to approximately 500BC, for example, show no evidence of recrystallization. For binary and more complex glasses, phase separation is found to occur; under the proper circumstances, one or the other phase may devitrify leading to the formation of ceramics. The latter form, needless to say, is another class of materials with important technical applications [13].

III. OPTICAL PROPERTIES OF IMPURITY CENTERS IN GLASSES

Spectroscopic studies of glasses initially concentrated on those which contained transition metal ions. These studies sought to determine the absorptive optical properties which led to their coloration. Emissive properties were treated only in a cursory way as interest in the luminescence of disordered materials did not arise until the advent of lasers.

A sequence of early workers attempted to interpret the spectra of glasses using atomistic models which have long since been discredited [3]. It was not until 1951 that Hartmann and Ilse [14] were successful in applying the theories of Bethe [15] and Van Vleck [16] to the spectra of Ti^{3+} in a crystal; this led to the subsequent publication of the works of Sugano, Tanabe and Kanimura [4,17] which established the behavior of 3d ions in solids. Bates and Douglas [18] first extended the work in crystal to glasses. As it is to be expected, there is a one-to-one correlation between the spectra of ions in crystalline and in amorphous materials. This is most apparent when the energy levels involved are well isolated as is the case in the Mn^{3+} spectra shown in Figure 3. Bates [19] was able to deduce average coordination ıbers

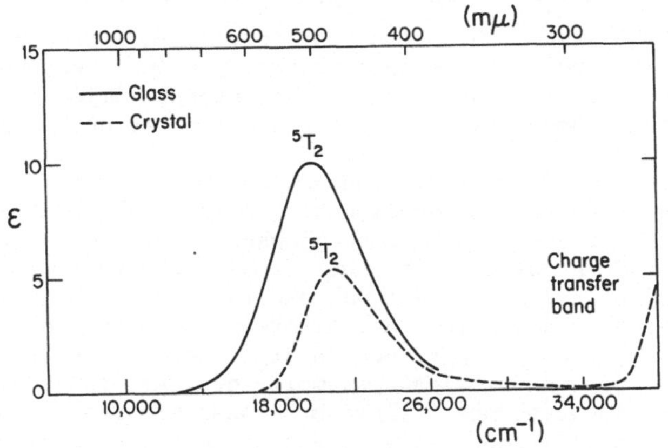

Fig. 3. Comparison of the spectra of Mn^{3+} in a soda lime silicate glass and in a hydrated crystal. The shift of the $^5T_2$ absorption towards lower energies implies a smaller field in the glass [19].

for the active centers in various glasses through this type of analysis. He was also able to extract a measure of the crystalline field parameters, E/B, for the transition metal ions by fitting the peaks of the absorption bands. Generally, these field values for glasses are 5-10% lower than those found in similar coordinated ions in crystals; this decrease in the average field can be understood in terms of the lower packing densities encountered in the disordered networks.

Historically, the spectra of the other important class of glass dopants, i.e., the lanthanides or rare earths, was not fully unraveled until the mid 1950's. It follows that no comprehensive effort was made to understand the properties of these glasses until then [20]. In the past two decades, however, there has been considerable interest in the optical properties of this class of material related in part to effort to develop high power laser devices. An extremely comprehensive data base has been developed on a large variety of glasses containing the majority of the rare earths, special emphasis has been placed on those ions which have useful luminescence properties [21,22]. Again, there is an exact correlation between the rare earth energy levels in glass and their crystalline counterparts, and the weak field approximation applicable in the latter is also found to be effective in the former.

The principal difference to be found between the spectra of ions in crystals and in glasses arises not in the intrinsic nature of the electronic states, but in the considerable broadening which is introduced in glasses by the disorder which is prevalent in these systems [23]. This broadening is extraneous to the ions and is said to be inhomogeneous; in the case of amorphous materials, its physical origin can simply be traced to the random variations in the local coordination and bonding at each individual ion site. The concept of homogeneous and inhomogeneous broadening and the manifestations of this broadening on the optical spectra of a trivalent rare earth ion are illustrated in Figure 4.

The inhomogeneous broadening in glasses is considerable. In the case of 4f doped materials, this broadening ranges up to $200 cm^{-1}$ and it is sufficient to obscure the Stark splittings produced by the local fields in the L•S states. In 3d materials, the inhomogeneities not only lead to broadening of the electronic states, but also are capable of affecting the energy of the multiplets so as to produce overlapping spectra; this is the case, for example, with the familiar ^2E and $^4T_1$ states of Cr^{3+} when this ion is introduced into a glass [24]. Conventional spectroscopic methods generally cannot deconvolute out the inhomogeneous contributions, hence, the interpretation of glass spectra obtained through these means is difficult to carry out in any detail.

The disorder induced variation of crystal field parameters from site to site leads to some additional universal observations concerning the optical properties of activated glasses. Because of the presence of molecular building blocks in the glass network, glasses contain relatively high lying vibrational excitations which, along with other excitations such as acoustical and disorder modes of the network, can couple to optically excited states and produce non-radiative de-excitation processes. An example of the consequences is the quenching of luminescence. These processes are faster in glasses than in crystals, and it is generally true that emissive states in glasses are fewer than those found in similarly activated ordered systems. Additionally, since oscillator strengths and, hence, lifetimes of states depend on the crystal field parameters, a random variation of the latter results in a spread of the values of the lifetime across the distribution. Thus, whenever the luminescence decay time of an optical

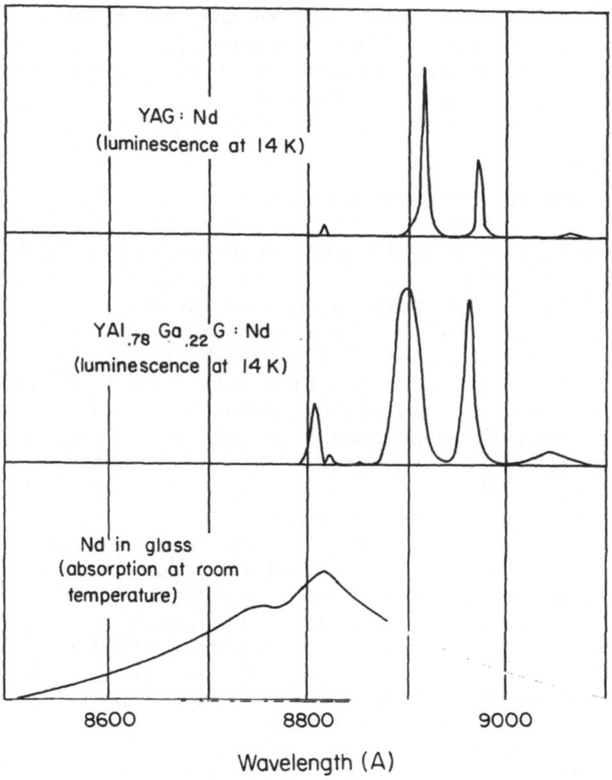

Fig. 4. Spectra of the $^4F_{3/2}$ level of Nd^{3+} in a crystal, a mixed crystal and in a glass showing varying degrees of inhomogeneous broadening [23].

transition in glass is measured by conventional methods, the observed lifetime is highly non-exponential as a consequence of the randomness of the system [21].

All dynamical processes observed in the excited states of ions in crystals have been also documented in glasses and a considerable volume of work exists in this area [25]. It is relatively easy to establish, through conventional spectroscopy, that energy transfer occurs in glasses doped with various centers. Phenomena such as concentration quenching, sensitized lumunescence, and optical energy trapping are all well documented experimentally in amorphous systems. Analysis of these processes, however, is a formidable problem, as it involves averaging procedures over several sets of random variables which may be correlated in some instances. Conclusions may, nevertheless, be drawn based on our understanding of the dynamics of ordered materials, and as a whole, interactions appearing in crystals are paralleled by those which are effective in glasses, provision being made for the additional complications produced by the large inhomogeneous broadening existing in the latter [26].

A number of reviews addressed to the conventional spectroscopic properties of activated glasses have appeared [5,26,28], and the reader is referred to the literature for a more detailed description of the optical processes affecting these systems.

IV. LASER SPECTROSCOPY OF IONS IN GLASSES

The availability of tunable laser sources have provided us with a versatile and powerful source with which to conduct optical spectroscopic studies. The methodology and techniques involved in laser spectroscopic have been reviewed by several authors [29,30], and thus are not repeated here. Of special interest to us is the fact that these experimental methods provide us with ways with which we can suppress some of the inhomogeneous spectral contributions, and under the proper circumstance extract the homogeneous or intrinsic spectra of individual ions situated at specific disordered sites. The spectra thus derived not only provides information on the nature of and interactions affecting the electronic states of the center, but also yields data on the surroundings of the excited ion. Thus, these techniques can be used to probe into the microscopic structure of glasses; similarly, laser based time resolved spectroscopies can provide us with information on the elementary excitations prevalent in glasses. Fluorescence line narrowing (FLN) and its time resolved version (TRFLN) have been the most common techniques used in the study of optical properties of glasses, though other methods such as hole burning have also been used. Reviews of various aspects of laser spectroscopy of ordered and disordered solids are also available [26].

IV.A. Structural Studies

FLN using lasers was first demonstrated by Riseberg [31] and by Motegi and Shionoya [32] in Eu doped glasses. Since then, these techniques have been applied extensively in the study of doped glasses, principally 4f activated systems. Comprehensive studies have been carried out on trivalent Nd, Eu and Yb glasses [26].

In experiments on Eu containing glasses, for example, the $^7F_0$ to $^5D_0$ inhomogeneously broadened transition is selectively excited with a laser with a frequency width which is narrower than the breadth of the transition. The normal broadened fluorescence to the $^7F_{0,1,2}$ ground states is then observed to show various degrees of narrowing. As illustrated in Figure 5, dramatic changes occur in the FLN spectra of the Eu ion as the laser is tuned across the inhomogeneously broadened absorption. These variations are obviously connected to the large site to site variations of the local electrostatic fields at the ion location. Since the splittings of the Stark manifolds are a function of the symmetries of the "crystal" field, the energy level diagrams obtained through the use of FLN have been used to derive the appropriate parameters for each site [10] as is also shown in Figure 5. These parameters can be used to derive geometrical models of the nearest neighbor environment of the impurity ion [33].

Parallel efforts using computer simulation methods of the formation of glasses have been carried out by Brawer [34] and others [35]. Using molecular dynamic simulations, these workers have been able to follow the dynamics of a finite number of molecules through a heating and cooling cycle. Upon cooling, the ensemble of molecules, which includes a single dopant ion, condense into a disordered structure. Many such simulations are carried out, the collection of results is then taken to

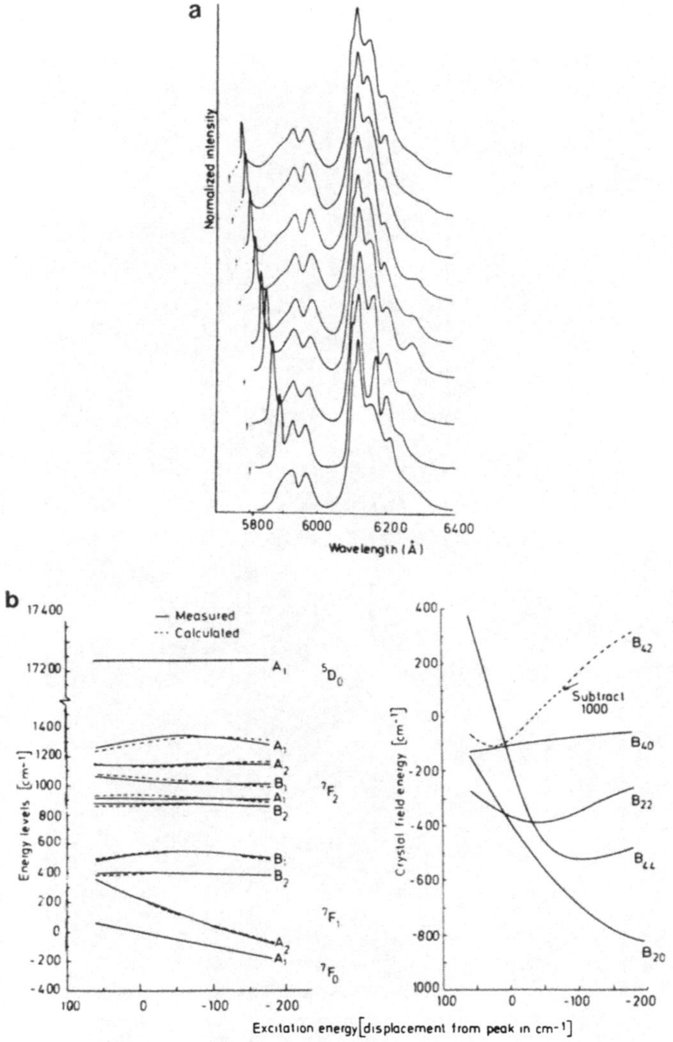

Fig. 5. (a) FLN spectra of Eu^{3+} in a silicate glass showing the large distributions of fields affecting the $^5D_0 \rightarrow {}^7F_1$ transitions. The arrow points at the laster excitation wavelengths within the inhomogeneously broadened absorption $^7F_0 \rightarrow {}^5D_0$ in this glass. (b) Variations of the $^7F_0$, $^7F_1$ and $^7F_2$ observed in the silicateglass with $^5D_0$ energy held as a constant. B^k_q calculated to yield the ^7F variations are shown to the right. The variation of the A_2 level of $^7F_1$ (C_{2v} symmetry) is due to the large B_{44} variation as a function of $^5D_0$ pumping energy. From Brecher and Riseberg [10].

represent the actual distribution of local environments surrounding the impurity. Examples of the local symmetries thus derived are shown in Figure 6 for a Eu doped BeF_2 glass. The results obtained through the simulations are generally consistent with experimental observations in X-ray scattering and in FLN spectra [14]. Though simulations have well-understood limitations and the structures generated cannot be

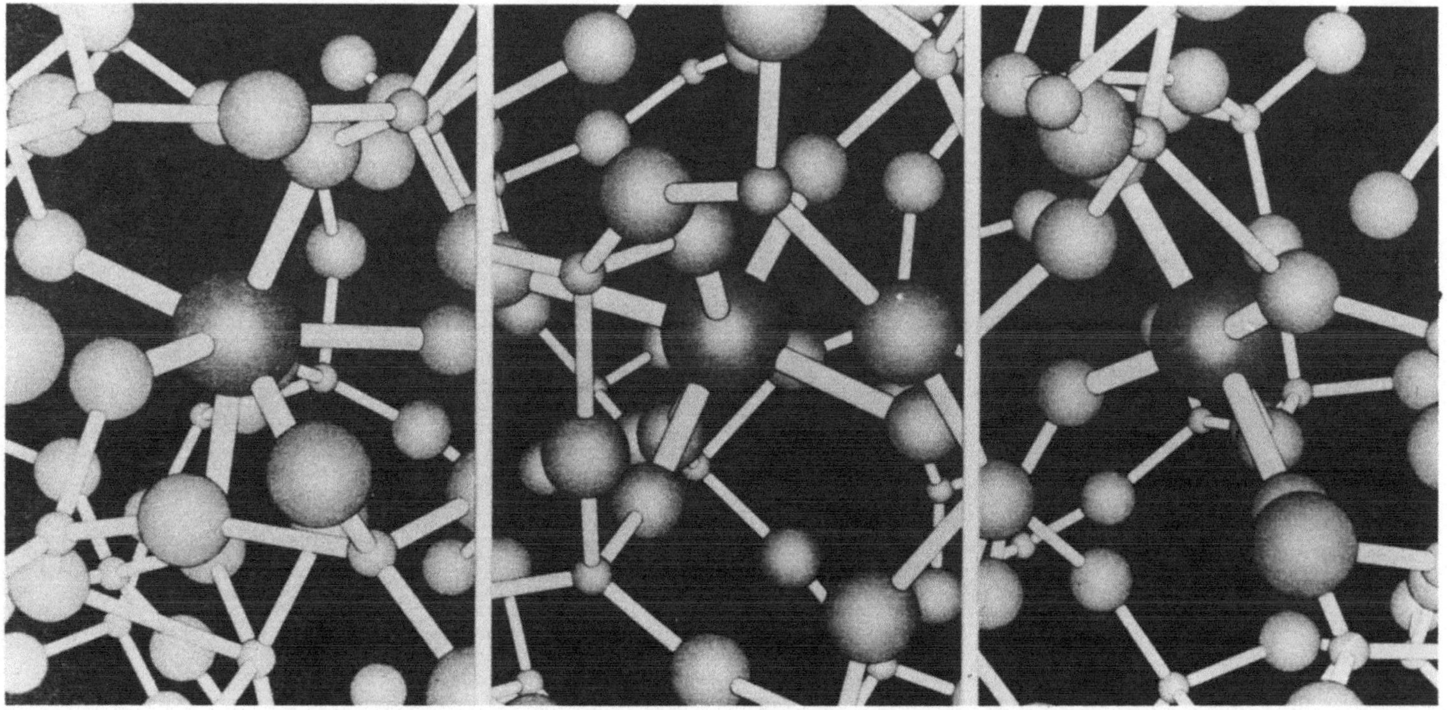

Fig. 6. Graphic display of three sites surrounding 4f ions in BeF_2 glass obtained in simulation. The large spheres in the center represent the rare earth ion; 5, 6 and 7 coordinated ions are shown in figure [35].

proven to be unique, the information derived has proven to be extremely useful. Indeed, these developments signal that we have the capabilities to replicate disordered structures in a computational way.

Hole burning techniques can also be employed to probe into the inhomogeneously broadened profile of optical transitions in glasses. Because of the limitations imposed thermal broadening, these studies have predominantly been done at low temperatures and have been complementary to FLN studies. In hole burning, only two absorptive levels are involved; consequently, these cases are somewhat easier to interpret and more levels are amenable to study. Of interest is the observation that in certain rare earth doped inorganic glasses, permanent holes may be burned into the absorption profile. These holes are permanent if the low network temperatures are maintained. In certain instances, the hole may return after temperature recycling. It is thought that for inorganic activated glasses the hole burning process in photo-physical, i.e., the actual environment surrounding the excited ion is actually altered by the absorption process. Additional work is clearly required in this area as it is technically of some importance [36].

If the electronic transition can be made to narrow, then its associated structure will show equivalent narrowing. These effects have been demonstrated for phonon and magnon sidebands in crystals. Narrowing of sidebands associated with a rare earth no-phonon transition in inorganic glasses has also been investigated by Hall and co-workers [37]; the shielding of 4f electrons, however, makes assisted transitions in these systems relatively weak, thus, no detailed analysis could be carried out in the case of Gd^{3+} work. More recently, Begin and co-workers [38] have carried FLN experiments on the ^2E state of a Cr^{3+} doped silicate glass and observed the sideband shown in Figure 7. As

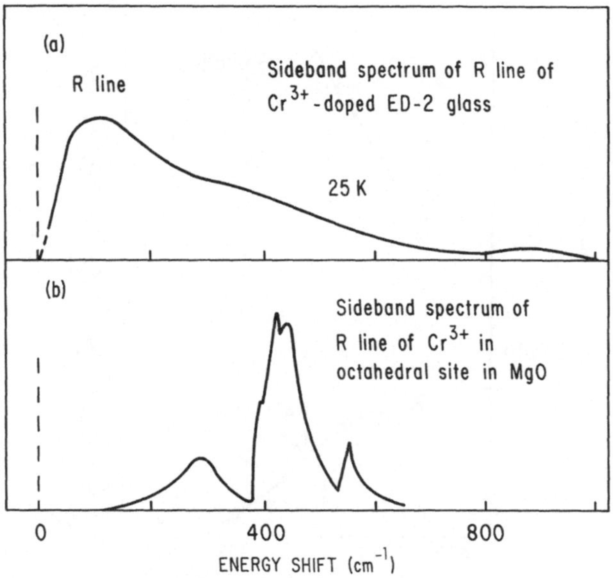

Fig. 7. (a) Phonon/vibrational sidebands observed in a FLN experiment done on the ^2E state of Cr^{3+} in ED-2 glass. Note the change in distribution relative to the crystalline sidebands (b) [38].

can be seen there, the distribution of vibrational frequencies which couple to the no-phonon electronic transition differ considerably from those encountered in crystals being generally shifted to lower energies. These results have already been proven to be crucial to our understanding of relaxation processes in the excited state [39]. Studies of this type will likely play an important role in allowing us to probe deeper into the nature of the elementary excitations of disordered systems, specially the low energy disorder modes [40].

IV.B. Radiative and Non-radiative Relaxation

As has already been noted, the distribution of sites in a disordered structure leads to a correlated distribution of lifetimes and results in a highly non-exponential decay of metastable states of active centers. Pulsed FLN techniques allow us to determine the lifetimes/oscillator strengths of any subset of ion sites within the inhomogeneous absorption. When this is done, in the absence of inter-ion interactions, the subset lifetimes then become pure exponentials as would be expected. Similarly, the large variation of fields also lead to changes in the ion-phonon coupling, and the non-radiative relaxation rates are expected to change when measured across the inhomogeneously broadened profile. Site selective laser spectroscopic techniques once again have allowed us to deconvolute these variations as is illustrated in Figure 8 [41].

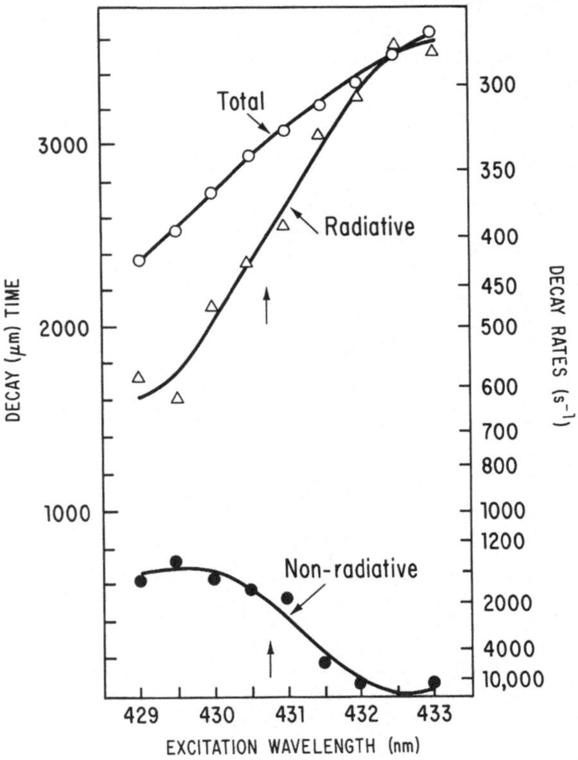

Fig. 8. Observed variation of the radiative and non-radiative lifelines of the state of Nd^{3+} as measured across the inhomogeneous absorption indicating variations in coupling strengths [41].

The thermal dependence of the optical linewidth of transitions of ions in glasses has been a subject of considerable current interest [42]. The study of the resonant FLN signal in a variety of 4f doped glasses by Selzer and others [43,44], led to the observation that the linewidths showed an almost universal T^n dependence on the temperature down to moderately low temperatures (approximately 5K) and that there also appears to be an anomalously large contribution to the dephasing at low temperatures. A typical example is shown in Figure 9 [44]. The so called disorder modes or two level systems (TLS) were suspected to play a role in the additional dephasing observed at low temperatures and a number of theoretical explanations ensued. Difficulties persisted in the explanation of these properties principally because of the absence of a change in the temperature dependence. This has been remedied in more recent work by Macfarlane and Shelby [45], Hegarty, Broer and co-workers [46,47], and by Brundage and Yen [48]. It is very likely that TLS play an important role in dephasing at very low temperatures, and that a Raman-like scattering process of acoustic excitations, such as those observed by Bergin et al. [38], dominate the linewidth process throughout the range of investigations.

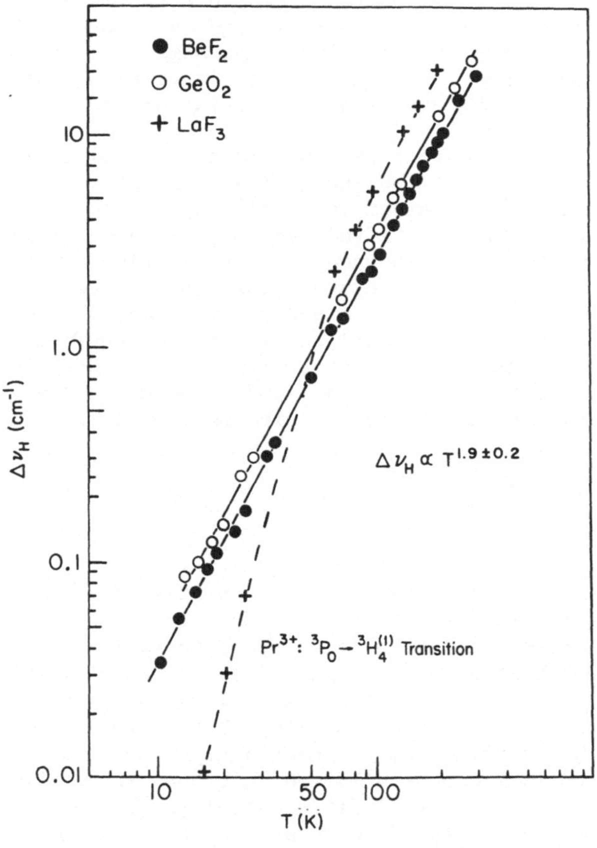

Fig. 9. Temperature dependence of the homogeneous width of the $^3P_0 \rightarrow {}^3H_4$ transition of Pr^{3+} doped glasses and in a similarly coordinated $YAlO_3$ crystal. Note T^2 dependence and the anomalously large rate of relaxation observed in the glasses at low temperatures [53].

IV.C. Optical Energy Transfer

If the concentration of dopant ions is increased, additional dynamical processes occur which can be traced to the inter-ionic interaction. These interactions lead to the transfer and diffusion of optical energy and are manifested by a rich assortment of phenomena documented in ordered and in disordered materials [49,50]. The preponderant bulk of extant studies in glasses have been carried out using conventional as opposed to laser spectroscopic techniques. Because of this, though the existence of dynamical effects in glasses is well established, no detailed analysis of these processes has been carried out. This is because conventional spectroscopy is not able to provide a measure of the dynamics occuring within a system when the ions are identical. The situation is partially remedied by the time resolved version of FLN or TRFLN which allows us to probe spectral changes within an inhomogeneous distribution as a function of time [51]. Hegarty and co-workers [52] have shown that TRFLN combined with conventional measurements of Donor (D) to Acceptor (A) dynamics are necessary to completely resolve the problem. Transfer processes observed in doped glasses mimic the behavior seen in crystals; successful analysis of data of TRFLN results would yield information on the state of the donor system prior to the transfer of energy to some other spectral feature. Several laser based studies of transfer processes have been conducted beginning with the work by Motegi and Shionoya [32]. However, complications resulting from the large inhomogeneous broadening, which require averaging over several random variables, have prevented a detailed analysis of the macroscopic behavior of the D-A dynamics.

Some general conclusions can be made regarding the microscopic interactions which exist between ions in glasses. Recent results by Brundage and Yen [53], obtained on a Yb^{3+} doped silicate glass, indicate that the processes which are effective in glasses are similar to those observed in crystals where phonon mediated multipolar interactions [54] are thought to be dominant. This subject has also been reviewed in a number of places [50], but clearly it remains as an area where additional theoretical and experimental work is required. Understanding of optical energy transfer processes in glasses, since they are readily accessible to experimental measurements, could serve as a precursor to the understanding of the general problem of transport in any disordered system.

V. CONCLUSIONS

Our intention has been simply to describe briefly an area of research which is extremely rich in content both in the historical and in the scientific senses. The specific systems dealt with in this review are the prototypes of all other disordered systems which are the subject of this course and, of course, are the most common and the most diverse. We have sought to illustrate that spectroscopic studies of glasses can be carried out conveniently and that various types of optically active ions can be utilized to probe into the structure of amorphous systems and to yield information regarding the excitations in these materials. A complete understanding of the structure and the dynamics of the optically excited states of ions in glasses is expected to have immediate scientific and technical consequences. Much of the work on these systems has been fueled by the technical community concerned with high power laser devices. Though these interests are currently on the wane, studies of relaxation and diffusion of energy in disordered structures will certainly have an impact on the general problem of transport in amorphous systems. As noted earlier, the recent development of transparent ceramics which can be activated by impurity ions presents us with an opportunity to investigate a system that is intermediate between crystals and glasses. Results derived in these ceramics are likely to be very interesting and useful in the evolution of various phases of the disorder problem.

ACKNOWLEDGEMENTS

This work is supported under the auspices of the Material Science Program of the U.S. Department of Energy, Grant DE-FG09-87ER45291. The author has also benefited from support from the University of Georgia Research Foundation.

REFERENCES

1. F. J. T. Malone, Glass in the Modern World (Doubleday, New York, 1968).
2. R. W. Douglas and S. Frank, A History of Glassmaking (Fowles, Oxfordshire, 1972).
3. W. A. Weyl, Coloured Glasses (Dawson's of Pall Mall, London, 1959).
4. Y. Tanabe and S. Sugano, J. Phys. Soc. (Jpn) 9, 753 (1954); J. Phys. Soc. (Jpn) 11, 864 (1956) and J. Phys. Soc. (Jpn) 13, 880 (1958).
5. J. Wong and C. A. Angell, Glass: Structure by Spectroscopy (Dekker, New York, 1976).
6. W. H. Zachariasen, J. Amer. Chem. Soc. 54, 3844 (1932); J. Chem. Phys. 3, 162 (1935).
7. J. M. Stevels, "The Structure and Physical Properties of Glass," in Handbuch der Physik, S. Flugge, ed. (Springer-Verlag, Berline, 1962), Vol. XIII.
8. D. R. Uhlmann and H. Yinnon, "The Formation of Glass, " in Glass-Science and Technology, D. R. Uhlmann and N. J. Kreidl, eds., (Academic Press, New York, 1983), Vol. 1, Chap. 1.
9. K. Patek, Glass Lasers (Butterworth, London, 1970).
10. C. Brecher and L. A. Riseberg, Phys. Rev. B21, 2607 (1980).
11. M. J. Weber, J. Non-Cryst. Solids 47, 117 (1982).
12. B. E. Warren, Z. Kristallog. 86, 349 (1933).
13. G. H. Beall and D. A. Duke, "Glass Ceramics Technology, " in Glass-Science and Technology, D. R. Uhlmann and N. J. Kreidl, eds., (Academic Press, New York, 1983) Vol. 1.
14. H. Hartmann and F. E. Ilse, Z. Phys. Chem. 197, 239 (1951).
15. H. Bethe, Ann. Phyk. 3, 133 (1929).
16. J. H. Van Vleck, Phys. Rev. 41, 208 (1932).
17. S. Sugano, Y. Tanabe and H. Kanimura, Multiplets in Transition Metal Ions in Crystals (Academic Press, New York, 1976).
18. T. Bates and R. W. Douglas, Trans. Soc. Glass Tech. 43, 289 (1959).
19. T. Bates, "Liquid Field Theory and Absorption Spectra of Transition Metal Ions in Glasses," in Modern Aspects of the Vitreous State, J. D. Macken, ed., (Butterworth, London, 1962), Vol. 2. Chap. 5.
20. S. Hüfner, Optical Spectra of transparent Rare Earth Compounds (Plenum Press, New York, 1978).
21. S. E. Stakowski, R. A. Saroyan and M. J. Weber, "Neodynium Doped Glass Spectroscopic and Physical Properties," Lawrence Livermore National Laboratory Report No. M-095, 2nd Revision (Livermore, 1981), unpublished.
22. S. E. Stakowski, "Glass Lasers," in Handbook of Laser Science and Technology, M. J. Weber, ed., (CRC Press, Boca Raton, 1982), Vol. 1, p. 215.

23. G. F. Imbusch and R. Kopelman: "Optical Spectroscopy of Electronic Centers in Solids," in Laser Spectroscopy of Solids, W. M. Yen and P. M. Selzer, eds. (Springer-Verlag, Berlin, Second Edition, 1986), Topics in Applied Physics, Vol. 49, Chap. 1.
24. L. J. Andrews, A. Lempicki and B. C. McCollum, J. Chem. Phys. 74, 5526 (1981).
25. See for example: R. Reisfeld, Structure and Bonding 30, 65 (1976).
26. M. J. Weber, "Laser Excited Fluorescence Spectroscopy in Glasses," in Laser Spectroscopy of Solids, W. M. Yen and P. M. Selzer, eds. (Springer-Verlag, Berlin, Second Edition, 1986), Chap. 6.
27. R. Reisfeld, Structure and Bonding 13, 53 (1973).
28. I. Zschökke (ed.), Optical Spectroscopy of Glasses (D. Reidel Publishing, Amsterdam, 1986).
29. P. M. Selzer: "General Techniques and Experimental Methods in Laser Spectroscopy of Solids," in Laser Spectroscopy of Solids, W. M. Yen and P. M. Selzer, eds. (Springer-Verlag, Berlin, Second Edition, 1986), Chap. 4.
30. W. M. Yen: "Laser Spectroscopy," in Handbook on the Physics and Chemistry of Rare Earths, K. A. Gschneider and L. Eyring, eds. (Elsevier Science, Amsterdam, 1987), Vol. 12, Chap. 6.
31. L. A. Riseberg, Phys. Rev. Lett. 28, 789 (1972).
32. N. Motegi and S. Shionoya, J. Lumin 8, 1 (1973).
33. C. Brecher and L. A. Riseberg, Phys. Rev. B13, 81 (1976).
34. S. A. Brawer, Phys. Rev. Lett. 46, 778 (1981).
35. S. A. Brawer and M. J. Weber, J. Chem. Phys. 75, 3572 (1981).
36. R. M. Macfarlane and R. M. Shelby: "Measur4ements of Optical Dephasing by Spectral Holeburning in Rare Earth Doped Glasses," in Coherence and Energy Transfer in Glasses, P. A. Fleury and B. Golding, eds. (Plenum Press, New York, 1984) p. 189.
37. D. W. Hall, S. A. Brawer and M. J. Weber, Phys. Rev. B25, 2828, (1981).
38. F. J. Bergin, J. F. Donagan, T. J. Glynn and G. F. Imbusch, J. Lumin. 34, 307 (1986).
39. D. L. Huber, J. Lumin. 36, 327 (1987).
40. P. W. Anderson, B. I. Halperin and C. M. Varma, Phil. Mag 25, 1 (1972).
41. C. Brecher, L. A. Riseberg and M. J. Weber, Phys. Rev. B18, 5799 (1978).
42. See: M. J. Weber (ed), Optical Linewidths in Glasses, J. Lumin. 36, 179-329 (1987).
43. P. M. Selzer, D. L. Huber, D. S. Hamilton, W. M. Yen and M. J. Weber, Phys. Rev. Lett. 36, 813 (1976).
44. J. Hegarty and W. M. Yen, Phys. Rev. Lett. 43, 1126 (1979).
45. R. M. Macfarlane and R. M. Shelby, J. Lumin. 36, 179 (1987).
46. J. Hegarty, J. Lumin. 36, 273 (1987).
47. M. M. Broer, B. Golding, W. H. Haemmerle, J. R. Simpson and D. L. Huber, Phys. Rev. B33, 4160 (1986).
48. W. M. Yen and R. T. Brundage, J. Lumin. 36, 209 (1987).
49. W. M. Yen: "Experimental Studies of Energy Transfer in Rare Earth Ions in Crystals," in Spectroscopy of Crystals Containing Rare Earth Ions, A. A. Kaplyanskii and R. M. Macfarlane, eds. (Elsevier Science, Amsterdam, 1987), Chap. 4.
50. W. M. Yen: "Experimental Studies of Optical Energy Transfer in Glasses," in Coherence and Energy Transfer in Glasses, P. A. Fleury and B. Golding, eds. (Plenum Press, New York, 1984), p. 145.
51. W. M. Yen, J. Lumin. 18/19, 639 (1979).

52. J. Hegarty, D. L. Huber and W. M. Yen, Phys. Rev. B23. 6271 (1981); Phys. Rev. B25, 5638 (1982).
53. R. T. Brundage and W. M. Yen, Phys. Rev. B34, 8810 (1986).
54. T. Holstein, S. K. Lyo and R. Orbach, "Excitation Transfer in Disordered Systems," in Laser Spectroscopy of SOlids, W. M. Yen and P. M. Selzer, eds. (Springer-Verlag, Berlin, Second Edition, 1986), Chap. 2.

EFFECTS OF DISORDER ON THE SPECTRAL PROPERTIES OF Cr-DOPED GLASSES, GLASSCERAMICS AND CRYSTALS

G. Boulon

Laboratoire de Physico-Chimie des Matériaux Luminescents
Unité Associée au CNRS
Université Claude Bernard de Lyon
69622 - Villeurbanne-cédex, France

ABSTRACT

The illustrations of the role played by the disordered structures on the optical properties of luminescent materials are numerous. Among the different activator ions, it is especially interesting to consider the Cr^{3+} ion. This ion is not only a structural probe but also a nucleating agent which generates nanocrystallites and microcrystallites inside the vitreous phase so that it gives us the possibility to follow the evolution of the structure from the glassy phase to the crystalline phase via the glass-ceramic phase. In all materials, the disordered occupation of Cr^{3+} sites is a general tendency. Systematically regular, perturbed and pair ions are detected by the presence of the $^2E \rightarrow {}^4A_2$ zero-phonon lines associated with vibronic bands. In addition, energy transfers take place between each kind of site and rigorous interpretations are difficult. Anyway the relations between the structural properties and the optical properties of Cr^{3+}-doped compounds have to be studied not only for basic research reasons but also for applications in lasers and luminescent solar collectors. In this course we will see a few examples of Cr^{3+}-doped different host networks like oxide glasses, spinel-type glassceramics and tunable solid-state lasers by using mainly the recent Laser-Spectroscopy techniques.

I. INTRODUCTION

First, let's recall the most important features of the Cr^{3+}-ion spectroscopy.

The usual coordination of the Cr^{3+} ion is the octahedral coordination. In oxide materials the six oxygen ions are located as shown in Fig. 1. The ideal point-group symmetry is O_h. The transitions associated with the unfilled $3d^3$ electronic shell have been elaborated by Tanabe and Sugano [1]. The O_h crystal field splits the free-ion wave functions of the $3d^3$ configuration into crystal-field wave functions having t_{2g}^3 and $t_{2g}^2e_g^1$ electron

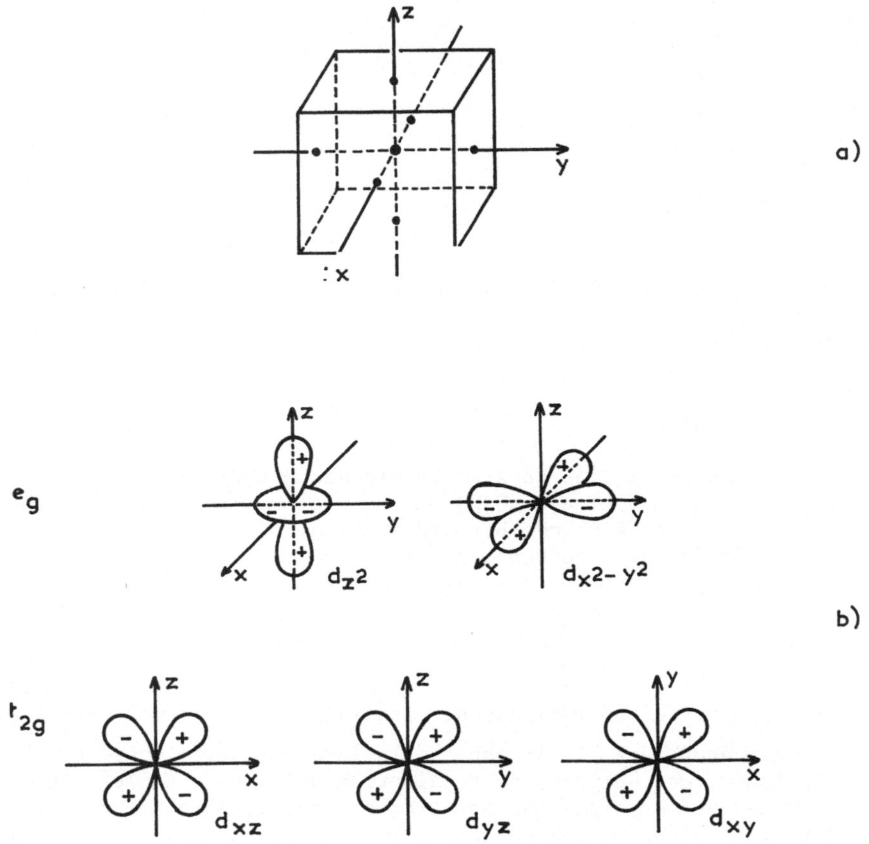

Fig. 1 (a) The octahedral coordination of the Cr^{3+} ion
(b) The t_{2g} and e_g electron configurations of the d orbital.

configurations. In Fig. 1 we have drawn the e_g and t_{2g} orbitals in which the geometry occurs in the luminescence mechanisms. Two types of multiplicities are found: quartet and doublet depending on the spin alignments of the electrons. Fig. 2 presents the evolution of the energy levels as a function of the ligand field strength: on the left the Russel-Saunders terms of the free Cr^{3+} ion and on the right the group theoretical notations. The ligand field strength is evaluated by Δ/B: Δ is measured by 10Dq, the value of the zero-phonon line of the ${}^4A_2 \rightarrow {}^4T_2$ transition in the absorption or excitation spectra, but as it is unusual to observe the structure of this absorption spectrum clearly, the authors take the energy in wavenumber of the maximum of the vibronic broad band. Table 1 gives a few Dq values in different materials. B measures the separation 15B between Russel-Saunders terms such as ${}^4F$ and ${}^4P$.

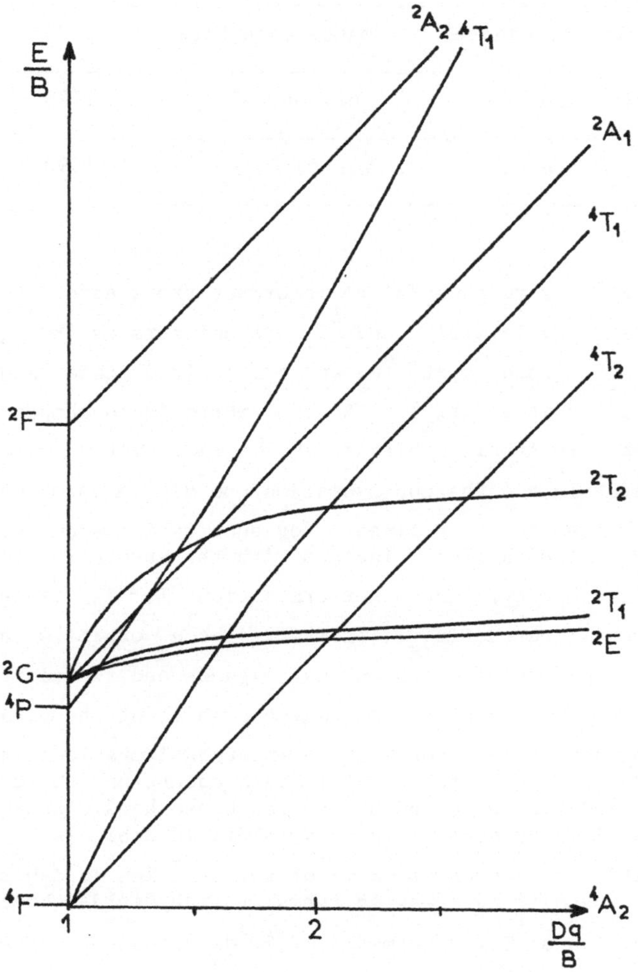

Fig. 2. Tanabe and Sugano's diagram applied to the d^3 orbitals in an octahedral coordination.

Table 1. Average value of the Dq parameter in Cr^{3+}-doped crystals.

host	Structure	Dq (cm^{-1})
$AnAl_2O_4$	spinel	1850
Al_2O_3 (ruby)	corundum	1830
$MgAl_2O_4$	spinel	1800
$LaMgAl_{11}O_{19}$ (hexaaluminate)	magnetoplumbite	1770
$BeAl_2O_4$ (alexandrite)	chrysoberyl	1725
$Gd_3Sc_2Ga_5O_{12}$ (GSGG)	garnet	1590

This diagram is very useful to interpret the absorption transitions but only two excited levels ^2E and $^4T_2$ are emitters by $^3E \to {}^4A_2$ and $^4T_2 \to {}^4A_2$ transitions. The ground state $^4A_2$ and the excited state ^2E arise from the lower set of d^3 orbitals $(t_{2g})^3$. So that there is no change in the geometry of the electronic orbitals of the equilibrium position of both $^4A_2$ and ^2E parabola and the configuration coordinate is practically unchanged giving rise to sharp lines: for such an intraconfigurational transition the emission line coincides with the absorption line as with rare earth spectroscopy. The other transition $^4T_2 \to {}^4A_2$ arises from the $(t_{2g})^2\, e_g{}^1$ orbital to the $(t_{2g})^3$ orbital. The e_g orbitals shown in Fig. 1 point along the direction of the oxygen ligands and the electrostatic repulsion in the $^4T_2$ excited state causes a shift of the parabola in the configuration coordinate: for such an interconfigurational transition, the vibrational modulation of the crystal field occurs and the emission broad band does not coincide with the absorption broad band. Because this transition involved changes in the crystal-field orbitals e_g and t_{2g}, it is very sensitive to the environment of the Cr^{3+} ion as can be seen in Fig. 2 and 3. Usually we speak in terms of high fields above the crossing point between ^2E and $^4T_2$, intermediate fields around the crossing point, and low fields below. The three cases have been summarized in Fig. 3 where we show that the fluorescence spectra change from the sharp line $^2E \to {}^4A_2$ to the broad band $^4T_2 \to {}^4A_2$ depending on the crystal field strength related to the distance between Cr^{3+} and its neighbour ligands. It is easy to understand due to electrostatic interaction that the lowest fields are characterized by the largest distances. Intensive research is performed on such materials to find new tunable solid-state lasers and new luminescent solar collectors in the near infra-red on the $^4T_2 \to {}^4A_2$ broad band emission [2,3].

All transitions should be parity forbidden and $^2E \rightarrow {}^4A_2$ is even spin forbidden whereas $^4T_2 \rightarrow {}^4A_2$ is spin allowed. So in the electric dipole allowed transition approximation, crystal field admixing with odd-parity configurations is required for any optical transitions to be observed. Because the $^4A_2 \rightarrow {}^4T_2$ and $^4A_2 \rightarrow {}^4T_1$ transitions are allowed and as they are observed in solids by vibronic broad bands in the visible range, they are used as the pumping bands of the Cr^{3+} ion. After excitation into $^4T_2$ or $^4T_1$ levels the rapid non-radiative desexcitation occurs followed by the fluorescence from the ^2E and $^4T_2$ relaxed lowest excited levels. Let us

Table 2. The lifetime decreases with increasing site size due to the variation of ε parameter which controls the Boltzman occupation of the $^4T_2$ level [3].

Host	YAG	YGG	GGG	YSGG	GSGG	LLGG
$\varepsilon(cm^{-1})$	1000	650	380	350	50	-1000
$\tau(\mu s)$	—	241	159	139	115	68
$B(cm^{-1})$		656	645	650	658	651
D_q/B		2.30	2.28	2.27	2.20	2.07

add that the oscillator strength in the glassy phase is higher by a factor of about two than in the crystalline phase because in the glassy phase the lowest symmetry allows more transitions. It is known that Cr^{3+} is generally recognized as being O_h symmetry with certain distortions in oxide crystals and obviously in glasses. Axial distorsion of the O_h symmetry give rise to a splitting both of the excited ^2E level and the ground state $^4A_2$ level, the former by some tens of cm^{-1} (29 cm^{-1} in ruby) and the latter by less than 1 cm^{-1}.

We also recall the thermal equilibrium of the two ^2E and $^4T_2$ levels in Fig. 3. The time constant τ is given by the classical formula:

$$\tau^{-1} = \frac{A(^2E) + \frac{g_2}{g_1} A(^4T_2) \exp(-\varepsilon/kT)}{1 + \frac{g_2}{g_1} \exp(-\varepsilon/kT)}$$

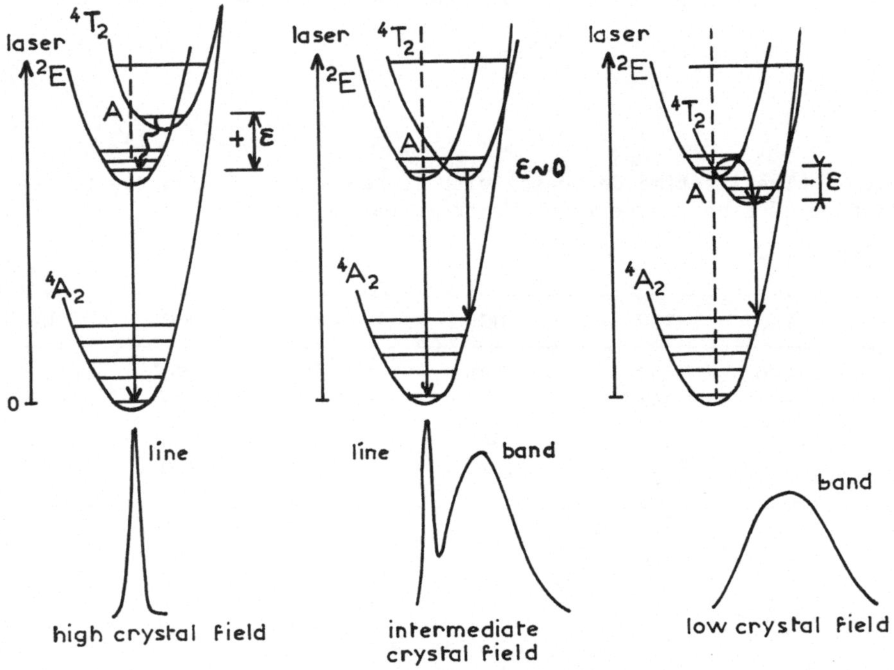

Fig. 3. Configuration coordinate diagrams of the Cr^{3+} levels in the three main cases usually met with experience. The laser pumps one of the vibrational level belonging to the $^4T_2$ electronic excited state. The $^4T_2$ broad band maximum represented by the point A decreases from high to low crystal field and the emission spectrum changes from the $^2E \rightarrow {}^4A_2$ sharp line to the $^4T_2 \rightarrow {}^4A_2$ broad band.

$A(^2E)$ and $A(^4T_2)$ are the emission probabilities of ^2E and $^4T_2$ levels respectively, with $A(^2E) << A(^4T_2)$; g_1 and g_2 are the degree of degeneracy of the levels with $\frac{g_2}{g_1} = 3$ and ε is the energy difference between the lowest vibrational levels of the $^4T_2$ and ^2E excited levels. The experimental results give $\tau(^2E)$ around a few ms whereas for $\tau(^4T_2)$ around a few tens of μs. The total time constant depends strongly on both temperature and the ε parameter. This ε parameter may be evaluated from the temperature dependence either of τ or the fluorescence intensity coming from each excited level (see Table 2). However, in the intermediate crystal field case ($\varepsilon \sim 0$) the decays cannot be used to recognize $^4T_2$ and ^2E levels because the strong mixing of the excited level wavefunctions : τ takes the same value either for $^2E \rightarrow ^4T_2$ or $^4T_2 \rightarrow ^4A_2$ transitions assuming that the non-radiative $^2E \rightarrow ^4T_2$ transitions are rapid. We will note that in most cases the dacay profiles depend on the emission wavelength and they show very often a curvature at short time due to other mechanisms such as, energy transfer between the excited configurations of distinct Cr^{3+} ions.

All these data should be useful in understanding the following spectroscopic results of the Cr^{3+}-doped materials.

II. Cr^{3+} DOPED GLASSES AND GLASSCERAMICS

The inhomogeneous broadening of any optical line is much greater in glasses than in crystals because of the disordered structure of the former which allows a wide distribution of multisites. For example, the two expected broad band of Cr^{3+} ions appear on the absorption spectrum of the cordierite glass 52% SiO_2, 34.7% Al_2O_3, 12.5% MgO with 0.8% of Cr_2O_3, in Fig. 4 at room temperature. One near 442 nm corresponds to the $^4A_2 \rightarrow ^4T_1$ transition, another one near 660 nm is attributed to $^4A_2 \rightarrow ^4T_2$ transition (Dq parameter of 1515 cm^{-1}). We also observe Fano resonances of sharp levels with $^4T_2$ vibronic levels: the former one near 640nm ($^2T_1$ level) and the latter near 692 nm (^2E level). If spinel-like structure microcrystallites are developed within the cordierite glass by heat-treatment, the $^4A_2 \rightarrow ^4T_2$ absorption spectra on Fig. 4 are shifted to shorter wavelengths: 600 nm for the first particles in sample 2 (Dq parameter of 1670 cm^{-1}) and 550 nm for the last stage of the nucleation in sample 3 (Dq parameter of 1800 cm^{-1}). The increasing of the Dz parameter is directly related to the distance between Cr^{3+} ion and the oxygen ligands: this distance ranges from 2.08 Å in $MgCr_2O_4$ (microcrystallites in sample 2) to 2.02 Å in $MgAL_{2-x}Dr_xO_4$ (microcrystallites in sample 3). Furthermore, the position of the $^4A_2 \rightarrow ^4T_2$ absorption band of the glassy phase at lower energy than the maxima of the glassceramics makes it possible to say that the mean distance $Cr^{3+}-O^{2-}$ in glass is relatively large. In addition, we observe the bandwidths are narrower in the glassceramic phase than in the glassy phase as a result of the higher order of the Cr^{3+}-doped microcrystals.

These data have been recently obtained by using the Cr^{3+} nucleating and luminescent probe ion during both the nucleation and the growth of the microcystallites. Different techniques have been studied: Small Angle Neutron Scattering and Small Angle X-4ay Scattering give information on particle sizes between 40 A and 400. By Electron Microscopy and X-Ray Diffraction we were able to know the shape of the particles and to define their crystallographic nature whereas Electronic Paramagnetic Resonance and Laser Spectroscopy Techniques bring information about the local environment of Cr^{3+} ions [4-5].

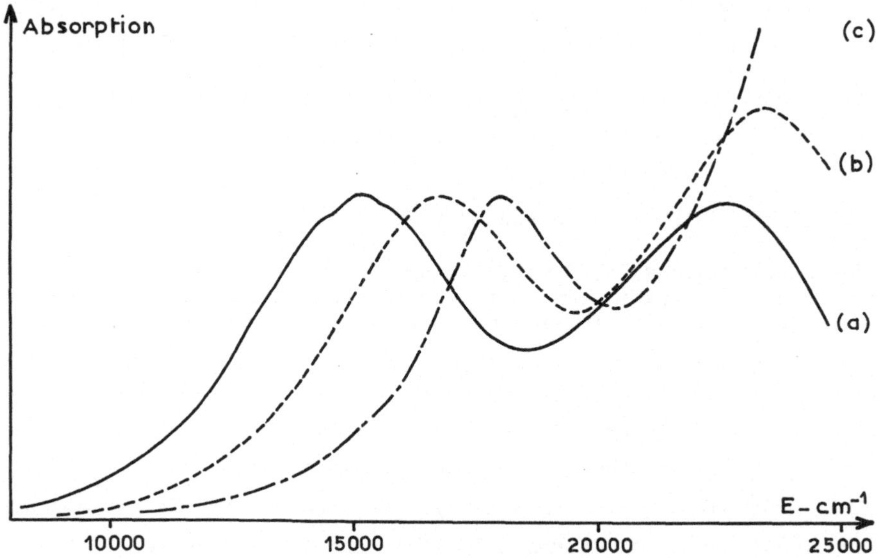

Fig. 4. Absorption spectra at room temperature. The spectra are arbitrary normalized for the same absorption of the $^4A_2$ $^4T_2$ band. (a) Base glass (sample 1), (b) glass heat treated 4H at 875°C and 2h at 900°C (sample 2), (c) glass heat treated 10 mn at 950°C (sample 3).

Under laser excitation into $^4T_2$ level, we observe drastic changes in the emission spectra as it can be seen in Fig. for the three samples. For the glass, the spectrum is composed of two bands: one is very broad peaking near 900 nm ascribed to the $^4T_2 \rightarrow ^4A_2$ transition, the other one at

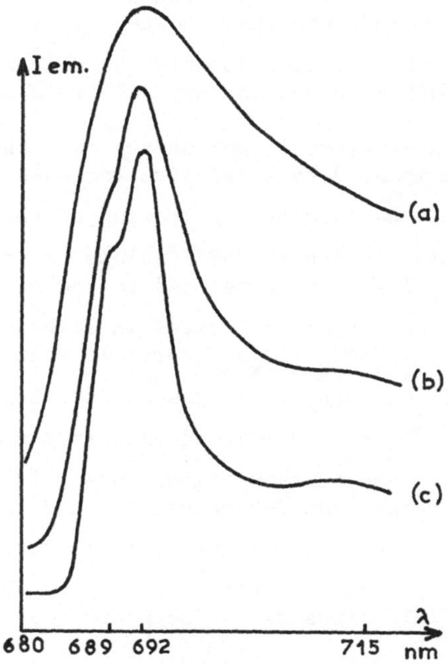

Fig. 5. Emission spectra of cordierite glass. (a) without heat-treatment (sample 1), (b) heat-treatment 4h 875°C + 2h 900°C (sample 2), (c) heat-treatment 10mn 950°C (sample 3), (d) T.R.S. of (b) delay 1ms gate 3 ms.

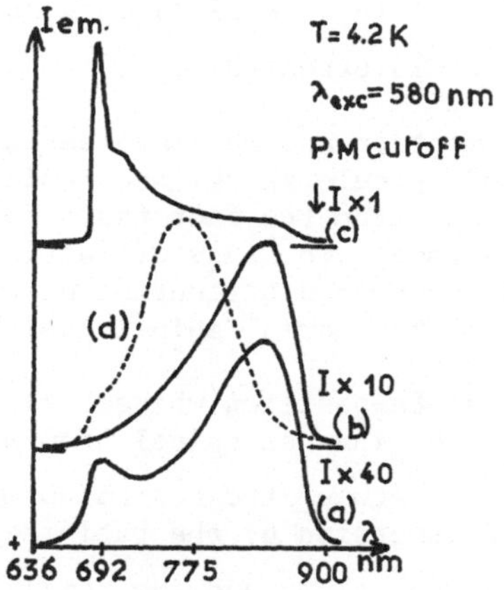

Fig. 6. Detail of the $^2E \rightarrow {}^4A_2$ emission line at 4.2K under 580nm excitation. Once again, each curve is arbitrarily normalized for the same maximum. (a) Base glass (sample 1), (b) glass heat treated 10 mn at 950°C (sample 3), (c) synthetic spinel powder $MgAl_2O_4 : Cr^{3+}$.

692 nm attributed to $^2E \leftarrow ^4A_2$ transition, is narrower but enough large (band-width = 630 cm^{-1}) to traduct the presence of multisites. Time Resolved Spectra clearly indicate that the two fluorescence bands have different decay times and that consequently the Cr^{3+} ions are located in different sites with both high and low crystal field strengths.

With the first heat-treatment, we observe that the crystallisation of the sample 3 is accompanied by a relative increase in emission from the $^2E \rightarrow ^4A_2$ transition as well as by a narrowing of the line from 630 cm^{-1} to 250 cm^{-1} after 10 min heating at 950°C (close to crystallisation temperature at 960°C). This can be related to the higher crystal field and the narrower distribution of Cr^{3+} sites in spinel-like structure microcrystals of the type $MgAl_{2-x}Cr_xO_4$ (particle size = 400 A). However, Cr^{3+} low crystal field ions are still present in the glassy phase as it can be seen by the $^4T_2 \rightarrow ^4A_2$ residual broad band in the near infra-red. The intensity of the latter band is even higher in sample 3 than in sample 1 probably due both to energy transfer mechanisms and the fact that the Cr^{3+}-crystal field strength in spinel crystals is strongly dependent on the initial growth. So, a part of the intensity may be due to Cr^{3+} intermediate crystal field sites as has been explained in the previous chapter.

In sample 2, the particle sizes are 91Å. The time resolved spectra show a broad band at 775 nm with a long life time (~1ms) coming from Cr^{3+} clusters strongly coupled by exchange and superexchange interaction. This conclusion was confirmed by EPR and by the electronic diffraction pattern of the microcrystallites indicating that their structure is close to that of $MgCr_2O_4$, the cell parameter a_o = 8.35 Å being much higher than that of the Cr^{3+}-doped spinel (a_o = 8.09 Å) in good agreement with the evolution of the crystal field strength.

It is worthwhile to mention another interesting point related to the disorder of the elaborated spinel-type glassceramics. One peculiarity of this spinel structure arises from the fact that there are two kinds of cations to be distributed among two kinds of lattice positions. In samples a more or less disordered distribution of the cations of these two kinds of sites exists. The normal spinel $MgAl_2O_4$ is characterized by Mg^{2+} cations in tetrahedral 8a-position whereas Al^{3+} cations occupy octahedral 16d-position. The inverse spinel from synthetic samples is characterized by a disorder between the positions of Mg^{2+} and Al^{2+} cations. One defines the degree of inversion by the ratio between the number of Mg^{2+} cations at octahedral positions and the total number of Mg^{2+} cations both at octahedral and tetrahedral positions. The values of the degree of inversion are 1% in natural samples and between 10% and 30% in synthetic samples like our materials. So, the two broad lines peaking at 688.2 nm and 691.8 nm at 4.4K in Fig. 6 can be associated with the disorder of the two types of Cr^{3+} ions and the inhomogeneous broadening can be associated with the disorder of the second nearest cation neighbours. We did not observe the $^2E \rightarrow ^4A_2$ at 684.8 nm of the regular octahedral sites showing

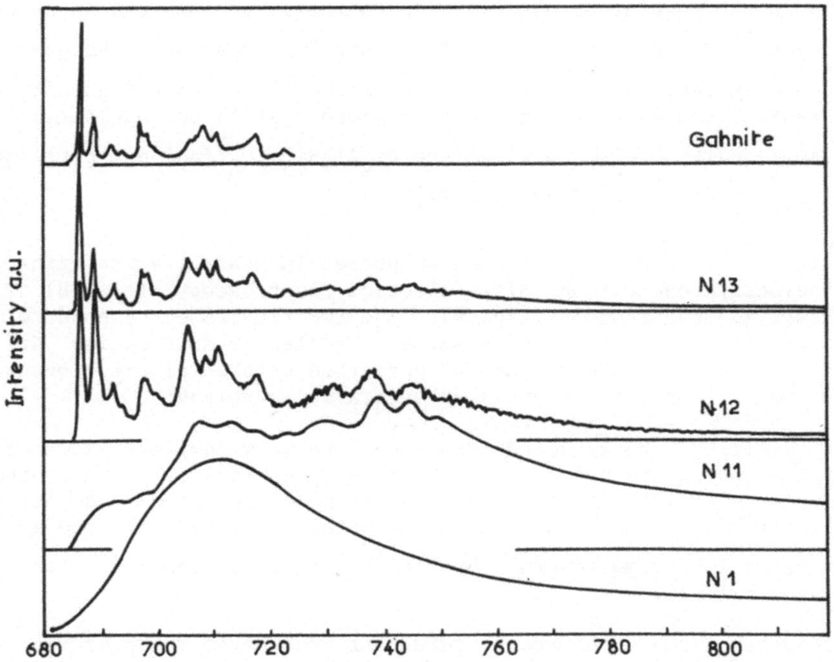

Fig. 7. Emission spectra from 680 and 820 nm at 4.4K of the glass N_1, the glassceramics, N_{12} and N_{13} and chomium-droped gahnite under the 532 nm excitation wavelength of the doubled YAG laser frequency. The spectrum indicates the beginning of a structure which is well resolved for N_{12} and N_{13}. The spectra of N_{13} and gahnite are very similar and the structure between 720 and 800nm is attributable to Cr^{3+} ions in high concentration of the $ZnAl_{2-x}Cr_xO_4$ crystalline phase. This broad band appears in N_{11} du to the weak crystal field of the site.

N_1 : 73.6 SiO_2; 11.8 Al_2O_3; 4.2 Li_2O; 7.0 ZnO

N_{11}: 1.6 TiO_2; 1.5 ZrO_2; 0.3 As_2O_3; 0.024 Cr_2O_3

N_{12}: heat-treatment 10h - 750°C and 2h - 860°C

N_{13}: heat-treatment 10h - 750°C and 1.5h - 900°C and 2h - 910°C.

that the natural disorder of Mg^{2+} and Al^{3+} takes place during the crystallisation of the glassceramics.

Contrary to Mg-Spinel-like structure, the Zn-Spinel-like structure is nearly 100% normal spinel as $ZnAl_2O_4$ gahnite crystals. Recent studies have shown that we are able to follow the crystallization by using the same processes but with much simplification because all Cr^{3+} sites are practically in the regular octahedral site in the final form and we detect the $^2E \rightarrow {}^4A_2$ regular line at 686.2 nm. The Laser-Spectroscopy techniques are even the only techniques to show the presence of Cr^{3+}-doped gahnite type microcrystallites among a lot of other microcrystallites nucleated in glass as β-quartz ZrO_2, and virgilite $Li_xAl_xSi_{3-x}O_6$ (and not petalite $LiAlSi_4O_{10}$ as expected) (see Fig. 7)[6,7].

All Cr^{3+} disorder effects can be probed in glassceramics either by laser selective excitation (Site-Selection Spectroscopy) (Fig. 8) or by Time Resolved Spectroscopy (Fig. 9). The two figures correspond to Mg-Spinel-like structure glassceramics. We clearly distinguish Cr^{4+}-pairs around 705-708 nm at short time and perturbed octahedral sites between 688 and 698 nm at long time under laser excitation wavelength.

In addition, the dynamical processes are very complex. However, two striking features are noted: (i) the non-exponential profiles at short time due to energy transfer between unperturbed, perturbed and pair ions as well for high crystal field as low crystal field and (ii) the exponential part at long time governed by the ^2E longlived sensitizer ions as can be seen in Fig. 10.

An attempt has been made to interpret the energy transfer from the ^2E level to the $^4T_2$ level in borate glasses [8], neglecting the diffusion

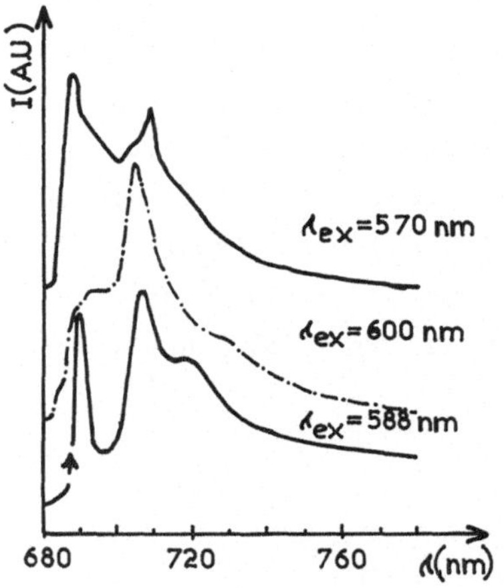

Fig. 8. The emission spectra of Mg-spinel like structure glassceramics at 4.4 K under three laser excitations (λ_{exc}).

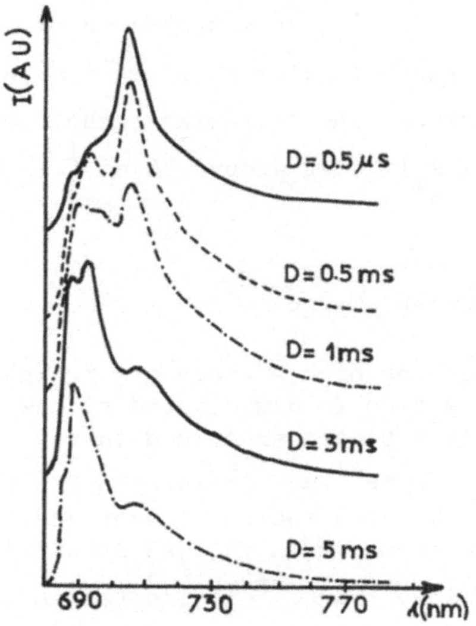

Fig. 9. The emission spectra of Mg-Spinel like structure glassceramics at 4.4K under 570nm laser excitation at different delays (D). The gate width is 100 μs.

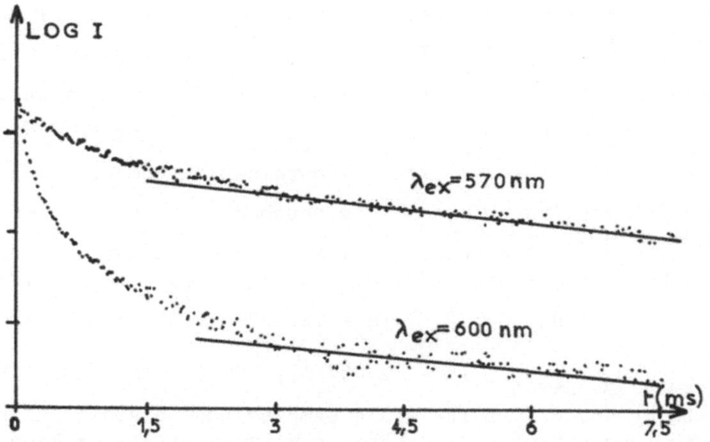

Fig. 10. Decay curves of the 705 nm fluorescence under two laser excitations (λ_{exc}) .σ = 3.4 ms is the time-constant of the linear part at 4.4K.

within the ^2E level and using the well-known formula form Inokuti and Hirayama in the case of electric dipole-dipole interaction by fitting the decay intensity. The critical distance for Cr interaction was found to be 1.4 nm. But this treatment is highly approximative in glasses because the theory assumes a random distribution of Cr^{3+} centers so that effects of perturbed and pairs are not taken into account.

The disorder effect can also be measured on two other spectroscopic parameters: $^4A_2$ ground state splitting called |2D| and ^2E excited level splitting. The fluorescence line narrowing technique yields |2D| from 0.95 to 1.35 cm^{-1} and ^2E splitting around 70 cm^{-1} in Mg-Spinel like structure glassceramics [5].

III. Cr^{3+}-DOPED CRYSTALLINE PHASES

In the research field of high average power high efficiency solid state lasers special attention is now devoted to new chromium/neodymium doped oxide garnets as it has been explained in ref. [9]. Among several requirements, the choice of $Cr^{\#+}$-sensitizer ion for the Nd^{3+}-activator is due to the good matching between the pump light and its broad absorption bands. Moreover failure is met with the YAG standard material because the energy transfer is very low: as Cr^{3+} is located in a strong crystal field site its fluorescence consists in the sharp line $^2E \rightarrow {}^4A_2$ having poor overlap with the Nd^{3+} absorption spectrum and characterized by a long life time (1.5ms). So the objective is to get larger crystalline cell by substituting Y^{3+} and Al^{3+} by larger cations in order to decrease the crystal field at Cr^{3+} site and to observe the broad band $^4T_2 \rightarrow {}^4A_2$ with a short decay time giving a fast and efficient energy transfer. In addition, due to the broad band emission the Cr^{3+}-doped garnets are possible materials for solid-state tunable laser in the near infra-red.

The crystallographic structure is cubic with 8 formula molecules per unit cell. The general composition is $C_3A_2D_3O_{12}$: Y^{3+} or Gd^{3+} are located in the dodecahedral site C (coordination number 8), Al^{3+}, Ga^{3+}, Sc^{3+} and Cr^{3+}-activator ion in the octahedral site A (coordination number 6), Al^{3+}, Ga^{3+} in the tetrahedral site D (coordination number 4). Examples of such materials derived from YAG ($Y_3Al_5O_{12}$) are GGG ($Gd_3Ga_sO_{12}$) (a = 12.376 A), GSGG ($Gd_3Sc_2Ga_3O_{12}$) (a = 12.567 Å) GGG-Ca, Mg, Zr ($Gd_{3-x}Ca_xGa_{s-x-2y}Mg_yZr_{x+y}O_{12}$) (a = 12.497 A), GGG-Ca, Zr without Mg or GGG-Mg, Zr without any Ca. Large scale samples can be grown due both to industrial experience in magnetic bubble memory substrates making and to the development of new solid-state lasers.

III. A. Cr^{3+}-DOPED GGG (Ca, Mg, Zr)

The absorption spectrum shows the broad bands $^4A_2 \rightarrow {}^4T_1$ and $^4A_2 \rightarrow {}^4T_2$ peaking respectively at 470 nm and 640 nm. At room temperature the fluorescence spectrum consists in the $^4T_2 \rightarrow {}^4A_2$ broad band, maximum at 750 nm, with a shoulder at about 695 nm (Fig. 11). The Cr^{3+} time-constant is about 106 μs at low concentration. Both these spectra and the time constant value show that in this compound the Cr^{3+} ions are in weak crystal field sites such as in GSGG. For this reason, we think that this host is a good candidate for near infra-red tunable solid state lasers.

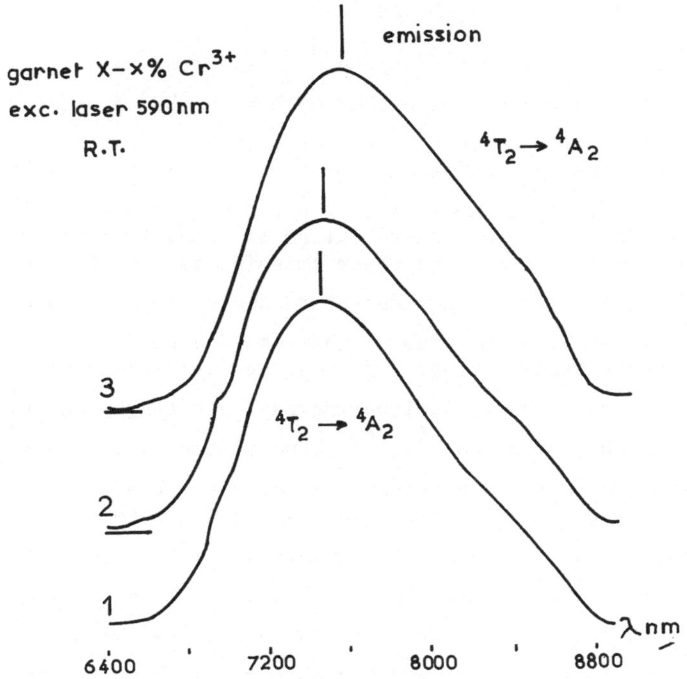

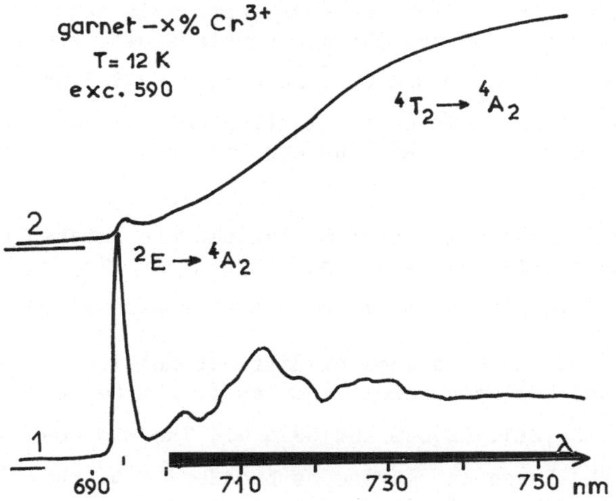

Figs. 11, 12. The concentration dependence of the uncorrected emission spectra of the x Cr^{3+}-doped GGG (Ca,Mg,Zr) samples under 590nm laser excitation. x:(1): 0.1%; (2): 0.38%; (3): 4.6%.

At room temperature we observe the shift of the $^4T_2\rightarrow$ broad band maxiumum to higher wavelengths as a result of the decreasing of the crystal field strength due to the increasing of the unit cell parameters. At 12K we can resolve the $^2E\rightarrow{}^4A_2$ line associated with its phonon-side band at low Cr^{3+} concentration. (1): 0.1%; (2): 4.6%.

The fluorescence spectrum is strongly concentration and temperature dependence. At 4.2 K, a $^2E \rightarrow {}^4A_2$ broad line at 6950 A associated with its vibronic side-bands is superposed to the $^4T_2 \rightarrow {}^4A_2$ broad band (Fig. 12). The excitation spectra registered either with a tunable dye-laser in the ^2E spectral range or with a Xenon lamp in the visible spectral range and the fluorescence decays are different for the line and for the band. This shows that the Cr^{3+} ions are in reality located not in one octahedral site but rather in two different octahedral kinds of sites [9]: some have a rather weak intermediate field and their emission is constituted by the broad band $^4T_2 \rightarrow {}^4A_2$ decreasing as temperature decreases and a broad $^2E \rightarrow {}^4A_2$ line, weak even at 4.2K, which lies on the weak energy side of the main band. Others have a rather strong intermediate field and their emission is constituted by the main $^2E \rightarrow {}^4A_2$ line increasing strongly as temperature decreases and probably of a weak $^4T_2 \rightarrow {}^4A_2$ band hidden by the $^4T_2 \rightarrow {}^4A_2$ band of the other kinds of sites. Moreover each of these kinds of sites has a broad distribution of crystal field values as is indicated by the large inhomogeneous width of the $^2E \rightarrow {}^4A_2$ line (width 40 cm^{-1} whereas in ruby the width is a few cm^{-1}) and the non-exponentiality of all decays even at low concentration. If Cr^{3+} concentration is increased the $^2E \rightarrow {}^4A_2$ line and its vibronic structure disappears at 4.2 K probably due to the formation both of exchange-coupled Cr^{3+} pairs and Cr^{3+} low crystal field sites. However, crystalline growth problems remain in these materials because the Cr^{3+} segregation coefficient and the lattice constant a(A) are different: α=2.96 and a = 12.4931 A for 0.08% Cr^{3+} sample; α = 2.48 and a = 12.4749 A for 4.6% Cr^{3+} sample. In addition the lattice constant a does not increase as is expected by the shift of the fluorescence band towards the long wavelength.

The existence of energy transfer mechanisms can be seen both at short time with a non-exponential decay and at longer time with an exponential part showing that ^2E longlived level governs this process [9].

The hypothesis of the existence of distinct multisites has been confirmed under laser selective excitation as it can be seen in Fig. 14 for Cr^{3+}-doped GGG (Ca,Zr), GS_2G_3G and GS_1G_4G. The two most striking feature lies: (i)the large inhomogeneous broadening of the lines in GGG (Ca,Zr) due to the presence of sites of various crystal field induced by the cations CA^{2+} and Zr^{4+} introduced in the lattice and (ii) the unexpected 4 and even 5 $^2E \rightarrow {}^4A_2$ lines in scandium garnets due to some perturbation effects [10]. We think that the Cr^{3+}-environments are perturbed by the degree of octahedral rare-earth (Gd^{3+}) substitution as it has been mentioned earlier in $RE_3(Ga_{2-x}Cr_x)Ga_3O_{12}$ (RE = rare earth) [11]. With gadolinium gallium garnet the degree of substitution has been evaluated at 3%.

Then we have shown that Cr^{3+}-centers are excellent fluorescent probes to detect special disorder in such garnets.

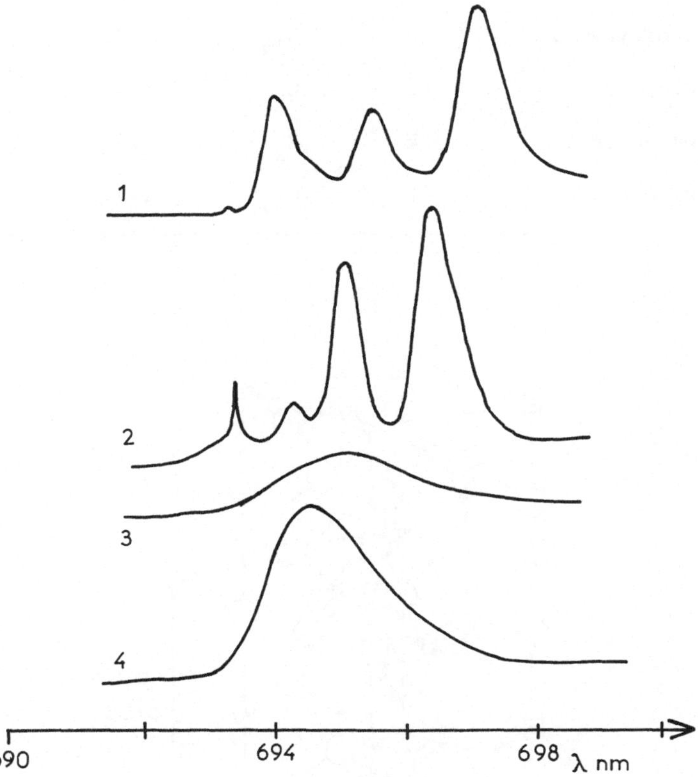

Fig. 13. The emission spectra of the ^{2}E $^{4}A_{2}$ transition at 6K for several Cr^{3+}-doped GGG-type garnets under $^{4}T_{2}$ laser excitation.

(1) GS_1G_4G (a_o = 12.567 Å)

(2) GS_2G_3G (a_o = 12.550 Å)

(3,4): GGG (Ca,Zr) high and low laser excitation (a_o = 12.460 Å).

Table 3

		Wyckoff position	site symmetry	mean Al-O	coordinance
.	regular octahedron	2a	D_{3d}	1.88	6
X	trigonal bipyramid	2b	C_{3v}	1.93	5
.	tetrahedron	4f	C_{3v}	1.85	4
.	distorted octahedron	12k	C_s	1.90	6
o	antiprism	4f	C_{3v}	1.92	6

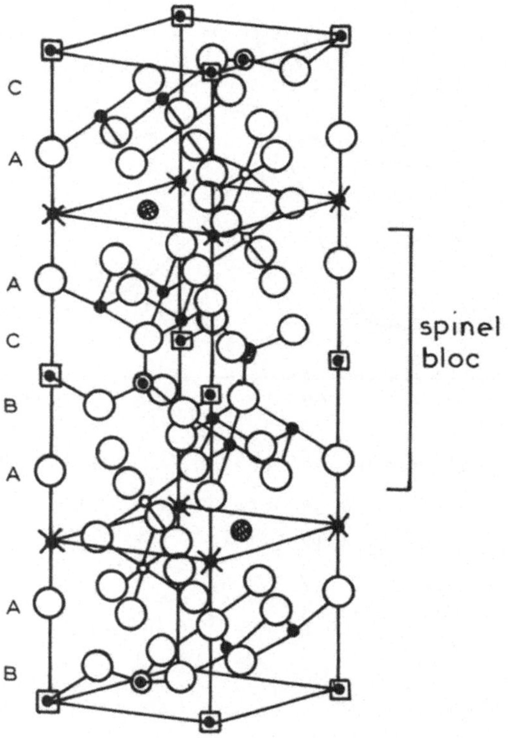

Fig. 14. The unit cell of the magnetoplumbite-like structure (see Table 3).

III.B. Lanthanum Hexaaluminate $LaMgAl_{11}O_{19}$

A whole series of compounds $LnMgAl_{11}O_{19}$ (ln=La,Ce,Pr,Nd,Sm,Eu,Gd) and $LeMAl_{11}O_{19}$ (M=Mg,Mn,Fe,Co,Ni) with magnetoplumbite-like structure has been prepared in the form of large single crystals and their characteristics investigated in France [12-13]. Among these materials $LaMgAl_{11}O_{19}$ appears to be an important crystalline laser host matrix which can be doped with different ions: La^{3+} being substitutable by Nd^{3+}, it forms a new performant laser material and Al^{3+} being substitutable by Cr^{3+}, this compound may be considered as a new potential Cr^{3+} vibronic laser whose present leader is alexandrite $BeAl_2O_4{:}Cr^{3+}$ for which we will give the properties in the next part of this lecture.

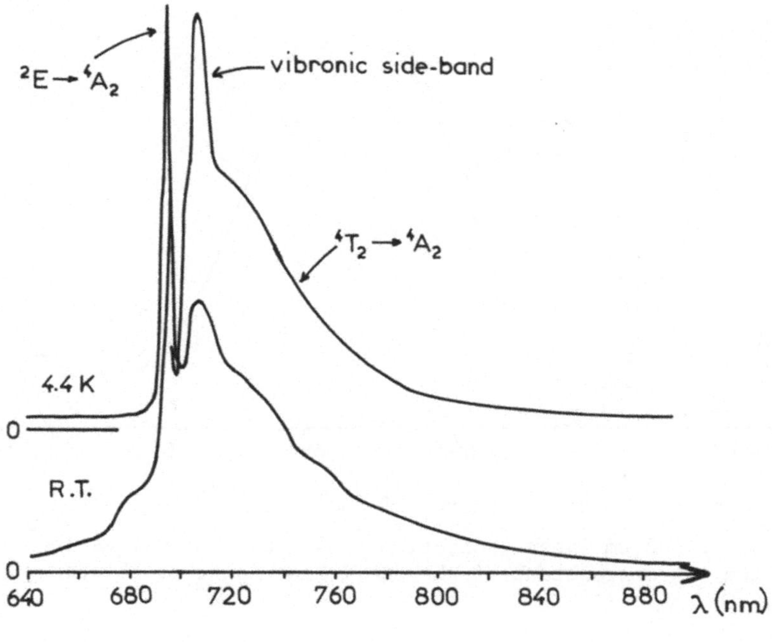

Fig. 15. Temperature dependence of the emission spectra of 0.01 Cr^{3+}-doped $LaMgAl_{11}O_{19}$ under 532 nm laser excitation into the $^4T_2$ excited level.

The magnetoplumbite like structure of the hexaaluminate is hexagonal with space group $P6_{3/mmc}$. The unit cell (Fig. 14) is built of spinel-like blocks containing Al^{3+} and Mg^{2+} cations separated by mirror planes containing Al^{3+} and Mg^{2+} cations. The Wyckoff position of cations, their coordinance and the average oxygen-cation distance are gathered in Table 3.

The position of the maximum of the $^4T_2 \rightarrow ^4T_2$ absorption band 565 nm close to that of alexandriti (580 nm), shows that we may assume as intermediate crystal field for the main Cr^{3+} site yielding the possibility to observe the fluorescent transitions from both ^2E and $^4T_2$ levels. It is confirmed by the emission spectrum in Fig. 15: it consists of a main broad line at 695 nm (bandwidth 64 cm^{-1} even at 4.4 K) due to the $^2E \rightarrow ^4A_2$ transi-

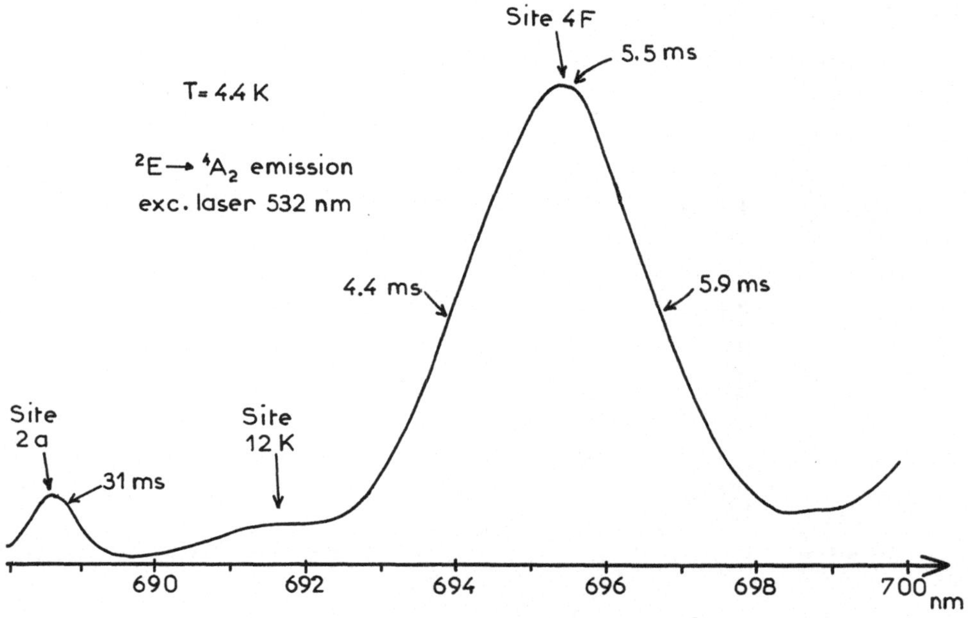

Fig. 16. Resolved emission spectrum of the 0.01 Cr^{3+}-doped $LaMgAl_{11}O_{19}$ under 532 nm laser excitation at 4.4 K. The values indicate the τ time-constant of the quasi-laser part of the decay.

tion with some satellites and of a broad band expanding up to 880 nm due to the $^4T_2 \rightarrow ^4A_2$ transition. When the temperature decreases from 295 K to 4.4 K the resolution of the satellite lines increases (Fig. 16) and we clearly observe three kinds of $^2E \rightarrow ^4A_2$ lines corresponding to three kinds of sites. The main line at 695 nm arises from Cr^{3+} in the 4f site while the weak lines at 691 nm and 688.6 nm may be connected with 12k and 2a sites respectively. The value of the time constant for the 688.6 nm is very high approximately 31 ms; this indicates that the transition corresponds to an inversion center octahedral site, as with the 2a regular octahedron of the magneto plumbite structure. The relatively weak intensity of this fluorescence line may be

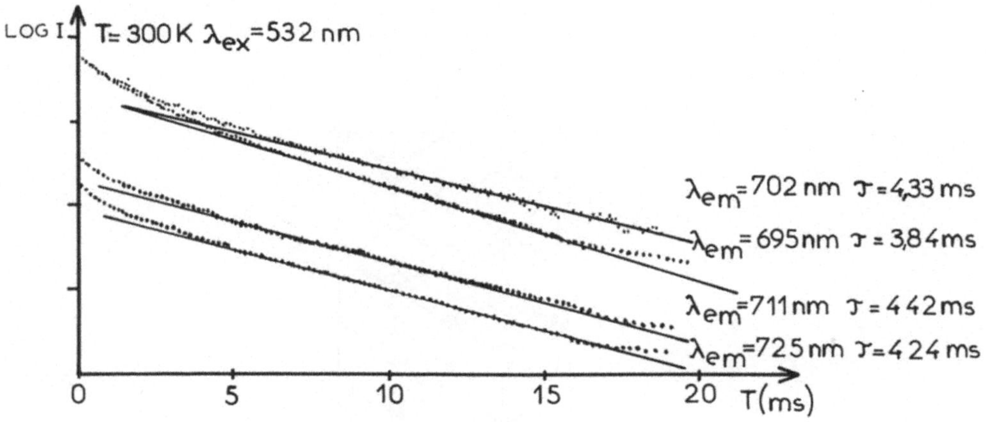

Fig. 17. The decay curves of the different emissions of the 0.01 Cr^{3+}-doped $LaMgAl_{11}O_{19}$.

connected with the very low population of the 2a sites in good agreement with the crystallographic data. The other line at 691 nm probably corresponds to Cr^{3+} ions doped 12k sites. So, the previous assignment is essentially consistent with:

i) the presence of three octahedral sites
ii) the population of each available kind of site in the lattice.
iii) the values of the average time constant
iv) The usual decreasing dependence of the $^2E \rightarrow {^4A_2}$ transition energy with the increasing of the bond length between Al^{3+} (or Cr^{3+}) and O^{2-} ions [14]. However, we need to confirm this hypothesis both from calculations and experiments.

Fig. 17 shows that the fluorescence decays are mostly exponential at 300K for low Cr^{3+} concentrations but not when either the concentration or the temperature increase in relation with the energy transfer mechanisms between each kind of site. Another time, we see that the ^2E longlived level govern the energy transfer not only from the main ^4F sites but also from the 2a sites because a long tail may be recorded in the long time portion with a very weak intensity.

This analysis will be more complete by noting the increasing of the satellite line intensity around 710 nm with the increasing of the concentration. It follows that the satellite peaks are probably due to Cr^{3+} pair ions as usual in the chromium doped crystals. These fluorescence lines also overlap the vibronic spectrum of the zero-phonon line for the $^2E \rightarrow {^4A_2}$ transitions in the three kinds of octahedral sites yielding the still more difficult interpretation. However, ESR experiments show that satellites arise from Cr^{3+}-Cr^{3+} pains when the magnetic field is parallel to C axis as is the case with Cr^{3+} ions in two neighbour 4f antiprism sites which are close enough (2.75 Å) to develop interactions between each other.

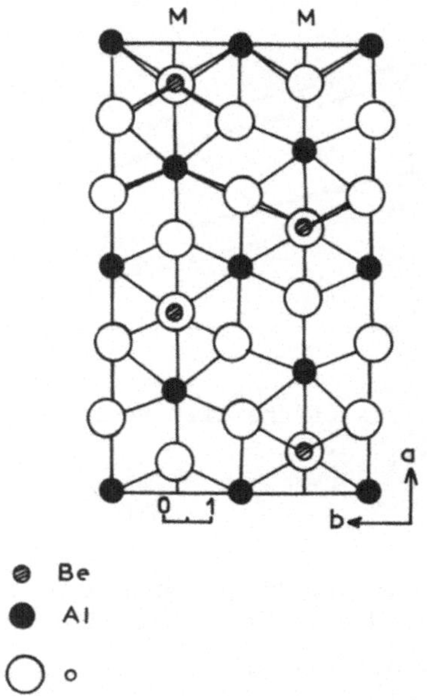

Fig. 18. c-axis view of chrysoberyl structure M denotes mirror planes.

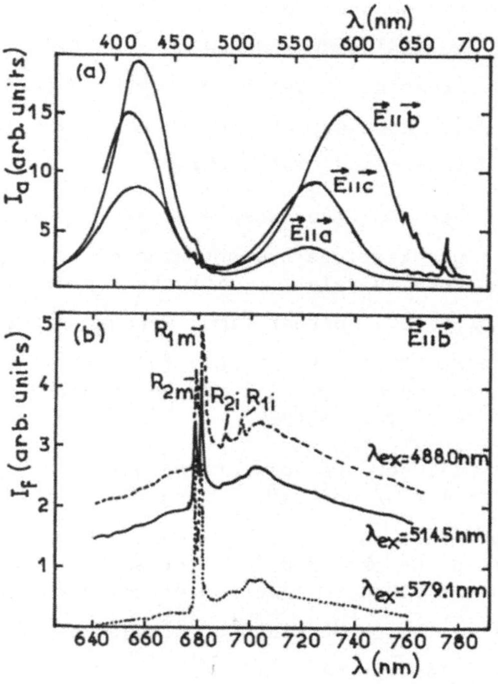

Fig. 19. Absorption and fluorescence spectra of $BeAl_2O_4 : Cr^{3+}$.

III.C. Alexandrite $BeAl_2O_4$

It is well known that Alexandrite has become a commercial tunable solid-state laser material. The results of a series of different types of laser-spectroscopy measurements have been reported providing new information concerning the crystal-field energy levels, ion-ion interaction and electron-phonon interaction properties of this material [15-16].

The most crystal has the chrysoberyl hexagonal-close-packed structure shown in Fig. 18. The space group is orthorhombic P_{nma} with four molecules per unit cell and lattice parameters a=9.40Å, b=5.476Å and c=4.427Å. The Al^{3+} ions are octahedrally coordinated by the oxygen ions and occur in two unequivalent crystal field sites. The Al^{3+} sites lying in the mirror-symmetry planes of the lattice have the site symmetry of the C_s point group, while the other Al^{3+} sites possess inversion symmetry and belong to the C_i point group. The Cr^{3+} ions enter the crystal substitutionally for the Al^{3+} ions 78% replacing Al^{3+} in the mirror sites and the rest going into the inversion sites [17].

Fig. 19 shows the absorption spectra at room temperature. It consists of two broad bands centered at about 420 and 480 nm associated with transitions from the $^4A_2$ ground state to the $^4T_2$ and $^4T_1$ excited states as expected. Three sets of sharp lines centered near 680, 650 and 470 nm are associated with transitions from the $^4A_2$ ground state to the split components of the ^2E (680nm), $^2T_1$ (650nm) and $^2T_2$ (470nm) levels. All these lines are associated with mirror-site ions and it has been shown that they are consistent with the polarization selection rules for the $^4A_2 \rightarrow ^2E$ lines [16]. The lines associated with the inversion-site ions are very weak and difficult to observe in this absorption band due to the strong overlapping with the broad bands of the mirror-site ions. However, the resolution of the two octahedral sites can be seen in the fluorescence spectra of the Fig. 19. At room temperature we observe both $^2E \rightarrow ^4A_2$ lines R_{1m} and R_{2m} of the mirror-site Cr^{3+} ions with a lifetime of 290 μs and R_{1i} and R_{2i} of the inversion-site Cr^{3+} ions with a lifetime of 48ms. Contrary of the cases of garnet and hexaaluminate materials, the homogeneity is greater in alexandrite as it is reflected by the small bandwidth (~ 2 cm^{-1}) of this R lines. The broad band peaking at lower energies is due to the $^4T_2 \rightarrow ^4A_2$ transition superposed with the vibronic spectra from each kind of site.

Obviously, ion-ion interaction occurs in alexandrite as in the other materials and it was found energy transfer between Cr^{3+} ions in mirror sites to Cr^{3+} ions in inversion sites. The mechanism has been treated by an expression derived by Inokuti and Hirayama [18] assuming electric dipole-dipole interaction. The classical critical interaction distances R_o are 35 A in mirror site at 300K and 40 A in inversion site at 10K [16]. In addition strong interaction produces coupled pairs with energy levels and transitions much different than those of isolated ions. Six different types of pairs were observed: two are ferromagnetically coupled and four antiferromagnetically coupled, but the spectrum is too coupled to identify the exact energy diagrams. However the lattice structure of chrysoberyl provides a variety of different combinations of close-neighbour pair sites involving both mirror and inversion locations. The authors of this excellent work mention that normal assumptions concerning energy transfer between randomly distributed sensitizer and activator ions are not strictly valid for Cr^{3+} ions in alexandrite crystals because there is a region around each sensitizer within which strong coupling produces pair spectra instead of energy transfer. The knowledge of the dynamics of the Cr^{3+} ions is very important to laser characteristics: an uniform distribution of Cr^{3+} ions in the crystal will minimize the presence of exchange-coupled pairs which do not contribute to laser emission. In addition, the energy transfer between the mirror sites and the inversion sites must be eliminated by minimizing the percentage of Cr^{3+} ions in inversion sites which might be done through co-doping with ions which preferentially occupy these sites.

IV. CONCLUSION

The aim of this lecture was to show how the spectral and dynamical properties of Cr^{3+}-doped solid state hosts reveal different aspects of the disorder. This activator ion plays both the role of a nucleating agent and a fluorescent structural probe so that the crystallization process of the spinel-type structure glassceramics may be analyzed by using laser-spectroscopy techniques. The disorder of the wide distribution of multisites in glasses is probed from the highest to the lowest crystal field strengths connected with the continuous variation of lowest crystal field strengths connected with the continuous variation of the Cr^{3+}-O^{2-} distance. In crystalline phases distinct and slightly different sites are easily detected: they are regular octahedral sites, perturbed octahedral sites and pairs. Such studies allow us to confirm structural disorder of the host matrix due to the inversion degree between tetrahedral and octahedral site cations of the spinel-like structure or between the dodecahedral and octahedral site cations of the garnet-like structure. The population of the Cr^{3+}-doped octahedral sites may be estimated but we have shown that the fluorescence intensity can sometimes be strongly perturbed by energy transfer mechanisms governed by the ^2E longlived excited levels. Such dynamical properties are very difficult to interpret due to the absence of random distribution of only one kind of site and the presence of several distinct emitting centers. Theoretical developments are needed to reach a better understanding of the disorder effects on the optical properties of luminescent materials. A very promising application concerns a new class of laser hosts for tunable operation at room temperature in the near infrared spectral range.

REFERENCES

1. S. Sugano, Y. Tanabe, H. Kamimura, Multiplets of Transition Metal Ions in Crystals, Academic Press--New York (1970).
2. R. Reisfeld, Glass Lasers and Solar Applications in Spectroscopy of Solid State Laser Type Materials, ed. B. DiBartolo--Plenum Press, New York-London, ASI Series (1987).
3. B. Sturn, G. Huber, J. Appl. Phys. 57 (1), 45 (1985).
4. F. Durville, B. Champagnon, E. Duval, G. Boulon, F. Gaume, A. F. Wright and A. N. Fitch, Physics and Chemistry of Glasses, 25 126 (1984).

5. F. Durville, B. Champagnon, E. Duval, G. Boulon, J. Phys. Chem. Solids, 46, 701 (1985).
6. V. Poncon, M. Bouderbala, G. Boulon, A. M. Lejus, R. Reisfeld, A. Buch and M. Ish-Shalom, Chem. Phys. Letters 130, 444 (1986).
7. V. Poncon, J. Kalisky, G. Boulon, R. Reisfeld, Chem. Phys. Letters 133, 363 (1987).
8. A. Van Die, G. Blasse and W. F. Van der Weg, J. Phys. C, Solid State Phys. 18, 3379 (1985).
9. G. Boulon, C. Garapon and A. Monteil, Advances in Laser Science II. Proceedings of the 1986 International Laser Science Conference: Seattle 21-24 October 1986--American Institute of Physics (1987).
10. A. Monteil, C. Garapon, and G. Boulon, Internal Report - Université Lyon I.
11. C. D. Brandle and R. L. Burns, J. of Crystal Growth, 26, 169 (1974).
12. A. Kahn, A. M. Lejus, M. Madsac, J. Thery, D. Vivien, J. C. Bernier, J. Appl. Phys. 52, 6864 (1981).
13. R. Moncorge, T. Benyattou, D. Vivien and A. M. Lejus, J. of Luminescence, 35, 199 (1986).
14. V. Viana, A. M. Lejus, D. Vivien, V. Poncon and G. Boulon, J. Solid State Chem. (to be published, 1987).
15. A. M. Ghazzuwi, J. Tyminski, R. C. Powell and J. C. Walling, Phys. Rev. B30, 7182 (1984).
16. R. C. Powell, Lin Xi, Xu Gang, G. J. Quarles and J. C. Walling, Phys. Rev. B 32, 2788 (1985).
17. R. E. Newham, R. Santoro, J. Pearson and C. Jansen, Am. Mineral, 449, 427 (1964).
18. M. Inokuti and H. Hirayama, J. Chem. Phys. 43, 1978 (1965).

PHYSICS AND SPECTROSCOPY OF LASER INSULATING CRYSTALS WITH DISORDERED STRUCTURE

A.A. Kaminskii

Institute of Crystallography
Academy of Sciences of the USSR
Leninsky prospekt 59
117333 Moscow, USSR

ABSTRACT

The systematization and analysis of the spectroscopic properties and temperature behavior of stimulated-emission parameters in activated condensed media, as well as the investigation of these properties in connection with the crystallo-chemistry of these compounds has resulted in the understanding of fundamental regularities of crystal-field disorder phenomena at Ln^{3+} sites in insulating laser crystals. In one of the discovered regularities, the structural-dynamic disorder is connected with the formulation of Ln-polyhedra statistics, the variety of which depends, in particular, on the concentration of the lasing-activator ions. Another regularity, the structural-static disorder is connected with the statistics of filling up the similar crystallographic positions by cations with different valency, and is generally independent from the content of laser-Ln^{3+} ions in the crystal. Experimental results among which are new low-temperature measurements on the spectroscopy of stimulated emission of Nd^{3+} ions on different disordered fluorine- and oxygen-containing crystals are presented to illustrate these objective principles. New data on record laser-wavelength tuning of octahedral Cr^{3+} ions in disordered $La_3Ga_5SiO_{14}$ single crystals at 300K are also presented.

I. INTRODUCTION

The tasks of investigating the spectroscopic properties of lasing compounds and of producing their systematization are most interesting in the whole problem of activated insulating crystals [1]. These tasks are especially connected with the improvement of excitation conditions and stimulated-emission (SE) processes of trivalent lanthanides (Ln^{3+}) and transition metal ions. In a number of cases they were promoted by the progress both in fundamental and applied directions of solid-state physics and quantum electronics. In accord with the findings of a number of scientific centers, in order to increase the effectiveness of the SE processes in the laser crystals doped by Ln^{3+} ions, one can go in three directions: the first is to use crystals with sensitizing ions (both Ln^{3+}, or transition metals and Ln^{3+}); the second, is to use crystals with high concentration (up to 100 at %) of lasing Ln^{3+} ions; the third is connected with the search and creation of crystals with disordered structure, where Ln^{3+} ions form activator centers with slightly different crystal fields. It should be noted that disordered laser crystals and laser glasses with Ln^{3+} activators have similar spectroscopic properties. In particular nowadays, the greatest efficiency (η~9%) of solid-state lasers with lamp pumping is achieved by using neodymium glass [2]; sensitized crystals are also present attractive characteristics (η~6.5% for $(Y,Er)_3Al_5O_{12}:Tm^{3+}$, Ho^{3+} [3]).

The ideas of first and second ways, based on the spectroscopic properties and physics of the processes occurring in the laser crystals, are quite clear and can be directly used for achieving some specific goals. The basic idea of the third way cannot be formulated as easily. Here, the situation is complicated by the great number of crystals and the insufficient knowledge of their structure. In different cases, e.g. for nonstoichiometric fluorides with CaF_2 and LnF_3 structure, the search has been going on for a long time [4,5]; likewise for the composite oxides, the largest class of media used for Ln^{3+} ions stimulated-emission excitation [1]. The success of some of these searches was due mainly to impirical or intuitive considerations. For this reason a lasing garnet with a disordered structure has not been found, despite the fact that this problem has been treated in numerous laboratories in the world for more than 20 years.

Recently, fundamental regularities have been uncovered and have shown how the crystal-field disorder at Ln^{3+} activator sites in laser crystals is created. The use and application of such principles has resulted not only in a wide variety of new types of lasing crystals with disordered structure [1,6-8], but also in the solution to the problem of disordered laser garnets [1,9,10]. These principles also explain the connection between spectral-laser and structural properties of some known disordered compounds with Nd^{3+} ions, which were obtained at different stages of the laser-crystal-physics development [5,11].

The main goal of the present work is to sum up the results of the investigations of the fundamental regularities which appear in the crystal-field disorder of Ln^{3+} and transition-metal ion sites in insulating laser crystals, used in the experiments on stimulated-emission spectroscopy.

The first part of this work deals with the results, which we have obtained in our investigations, going by the third above mentioned way. Here we shall discuss the physico-chemical nature of the structural disorder of different new and known classes of laser crystals, among them: cubic fluorides, trigonal crystals with Ca-gallogermanate structure, tetragonal crystals with scheelite and mililite structure, the first disordered garnet, and others.

In the second part of this paper the attention will be devoted to the stimulated emission of Cr^{3+} ions in disordered crystals.

First the main results of the work will be presented in concise form in order to show more clearly the nature of the discovered or perceived crystallo-chemical laws. Then, in order to illustrate their manifestations in some detail, the laser and structural properties of different disordered crystals will be presented.

II. STATISTICAL PHENOMENA IN DISORDERED-STRUCTURE CRYSTALS WITH Ln^{3+} ACTIVATORS

The results of numerous investigations [1,5,11,12] have shown that the crystal-field disorder phenomena at Ln^{3+} activator ions most strongly manifest themselves in crystals having either statistics of the structural elements (variety of Ln polyhedra), or statistics of filling up similar crystallographic positions in the lattice by cations with different valency. Such phenomenan result in the formation of a variety of Ln^{3+} activator centers having different crystal field parameters, and which are strongly connected with each other by electron-excitation-energy migration. Such collectives of Ln^{3+} centers and their properties can be described using the quasi-center concept [5,12,13]. A quasi-center is some conceivable formation, with the general properties of several elementary activator centers differing in structure, but having close Stark splitting of energy states. The structure of these centers and their variety will be determined by the statistical properties of the atomic structure of a given crystal, i.e., by the degree of disorder, which in some cases will depend on the concentration of Ln^{3+} quasi-centers. The crystals with disordered structures present inhomogeneously broadened bands in absorption and luminescence spectra (Fig. 1) and also usually broader SE lines at low temperatures. In crystals with ordered structures the picture is the opposite: all inter-Stark absorption (or luminescence) transitions of Ln^{3+} ions have homogeneously broadened lines *) and hence, narrower SE lines at low temperature. The pecularities of laser crystals with ordered and disordered structures are convincingly illustrated by the SE spectra of Nd^{3+} ions in the fluorine- and oxygen-containing crystals, shown in Figs. 2-4.

II.A. The Structural-Dynamic Disorder of the Crystal Field at Ln^{3+} Ions in Insulating Laser Crystals

The results on the disordered-laser-crystal structures and the data on the spectroscopic properties and SE indicate that the nature of crystal-field disorder at Ln^{3+} activators (in particular Nd^{3+}) is connected with two fundamental regularities. The first of these regularities, the structural-dynamic disorder, is manifest in a variety of nonstoichiometric $Me_{1-x}R_xF_{2+x}$ and $R_{1-x}Me_xF_{3-x}$ phases (where Me=Ca, Sr, Ba, Cd, Pb and R=Y, Sc, Ln) [1,5,12]. It is called dynamic because the degree of structural disorder will depend on the x value of the concentration of rare-earth and alkali-earth metals.

Some properties of the fluoride disordered nonstoichiometric laser crystals will be considered in detail below.

One can say that, with great probability, the structural-dynamic disorder determines the spectroscopic and laser properties of the Ln^{3+} ions in a number of oxygen-containing laser crystals (solid solutions) in binary form MeO_2-R_2O_3 systems (where Me=Hf, Zr and R=Y, Sc, Ln), in the

*At very low temperature inhomogeneous broadening of lines due to defects of real-crystal structure can be observed.

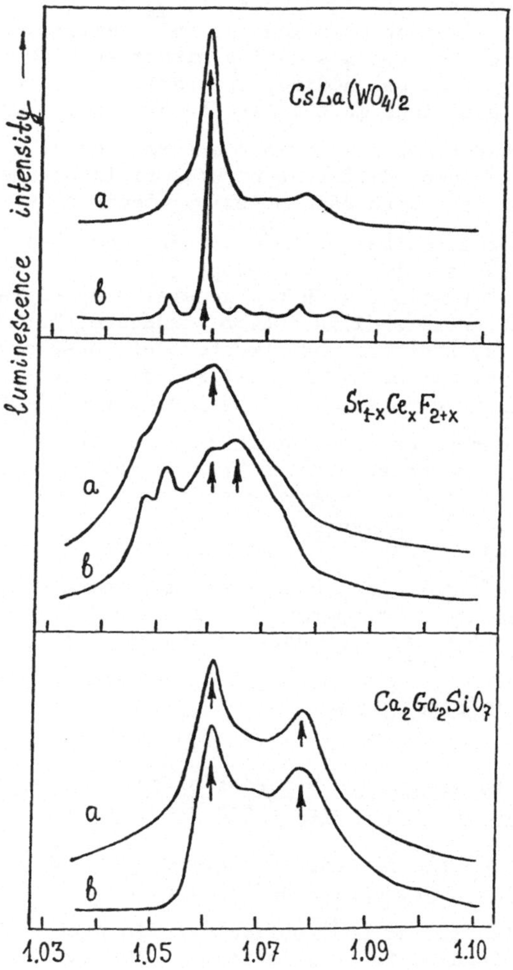

Fig. 1. Luminescence spectra of Nd^{3+} ions ($^4F_{3/2} \rightarrow {}^4I_{11/2}$ channel) at 300K (a) and 77K (b) in crystals with ordered $CsLa(WO_4)_2$ and two disordered $Sr_{1-x}Ce_xF_{2+x}$ and $Ca_2Ga_2SiO_7$ structure. Lasing lines are indicated by arrows.

crystals $YScO_3$ and $LnScO_3$ and in some fluorine- and oxygen-containing laser crystals as shown by Figs. 2 and 3, as well as Table 1.

II.B. The Structural-Static Disorder of the Crystal Field at Ln^{3+} Ions in Insulating Laser Crystals

Another fundamental type of the crystal-field disorder at the Ln^{3+} activators, called structural-static one, will determine the spectroscopic properties and character of the SE-temperature behaviour of the compounds formed in ternary and more complicated physico-chemical systems. In such crystals the construction of their polyhedra structure versus the Ln^{3+}-ion concentration does not take place, because the disorder is set in the origin of the structure and connected with the statistical introduction (filling) of different-valency cations of the compounds into the similar crystallographic positions. Data for representatives of some of these compounds, crystallized in different space groups, are shown by Figs. 3 and 4, as well as in Tables 2 and 3.

II.C. Preliminary Conclusions

From the figures and tables presented, it follows that independently from the type of laser-crystal disorder, whether it is structural-dynamic or structural-static, at low temperatures the SE excitation (E_{thr}) of Nd^{3+} ions is always higher than at 300K. When the same excitation is used and with the energy exceeding the threshold energy (in our cases $E_{exc}/E_{thr}=3$), the SE linewidth ($\Delta\nu_{SE}$) of Nd^{3+} ions in such cyrstals is always higher at low temperatures. These temperature changes of SE parameters are determined by the electron-excitation-energy migration weakening along with the temperature decrease over elementary Nd^{3+} ions centers "entering" into quasi-center. The same temperature behavior in lasing parameters takes place in numerous neodymium laser glasses (Fig. 2) [15,16], characterized by the maximum of the crystal-field disorder at Ln^{3+} (Nd^{3+}) ions among all known disordered solid laser materials.

III. LUMINESCENCE PROPERTIES AND STIMULATED EMISSION OF Nd^{3+} IONS IN CRYSTALS WITH DISORDERED STRUCTURE (EXAMPLES)

III.A. Structural-Dynamic Disorder

1. Cubic Nonstoichiometric $Me_{1-x}R_xF_{2+x}$ Fluorides with Nd^{3+} Ions (Me=Sr,Cd and R=Sc, La, Ce, Nd, Gd, Lu). In cubic MeF_2 single crystals with fluorite structure (space group O_h^5) Nd^{3+} ions at low concentrations ($C_{Nd} \leq 0.2$ wt. %) occupy positions at the centers of cubes formed by F^- anions (Fig. 5). When heterovalent substitution of Me^{2+} for R^{3+} occurs, the excess charge is compensated by the entering of the additional F^- ions either into the neighboring empty inter-sited tetragonal centers, or into the peripheral cubic centers. These structural formations, in the process of stimulated emission show themselves independently

Table 1. Stimulated-emission spectroscopic parameters of Nd^{3+} ions (laser channel $^4F_{3/2} \rightarrow {}^4I_{11/2}$) in fluoride and oxide crystals with structural dynamic disorderness of crystal field on activator ions

Crystal	T (K)	Laser wavelength*) (μm)	Threshold**) (J)	SE linewidth (cm^{-1})
$Ca_{1-x}Y_xF_{2+x}$	300	1.0540	7	3
		1.0632	5	10
	~110	1.0536	35	3***)
		1.0671	20	15
$Sr_{1-x}Ce_xF_{2+x}$	300	1.0590	14	13
	~110	1.0594	55	7***)
		1.0636	21	27
$Cd_{1-x}Gd_xF_{2+x}$	300	1.0672	3.5	8
	~110	1.0676	7	9
		1.0694	11	5***)
$Ba_{1-x}La_xF_{2+x}$	300	1.0534	2.7	5
	77	1.0538	15	17
		1.0580	50	30
$La_{1-x}Sr_xF_{3-x}$	300	1.0486	5.5	9
		1.0635	11	6
	~110	1.0636	25	10
$YScO_3$	300	1.0843	30	11
	~130	1.0774	50	11
		1.0837	100	5***)
$HfO_2-Y_2O_3$	300	1.0604	15	12
	~110	1.0615	26	15

*) The SE spectra were registered a photographic method (accuracy ±0.0002 μm, grating DFS-8 spectrograph and I-1070 type of IR-film).

**) The Nd^{3+}-ion concentration were about 2 wt.% for fluorides and (0.5 to 1) at.% for oxide crystals.

***) For $E_{exc}/E_{thr} \approx 1.5$.

Table 2. Stimulated-emission spectroscopic parameters of Nd^{3+} ions (laser channel $^4F_{3/2} \rightarrow ^4I_{11/2}$) in oxide crystals with structural-static disorderness of crystal field on activator ions

Crystal	T (K)	Laser wavelength*) (μm)	Threshold**) (J)	SE linewidth (cm^{-1})
$Ca_3(nb,Ga)_2Ga_3O_{12}$	300	1.0588	1.3	2.5
		1.0612	1.9	2.5
	~110	1.0583	2.6	6
		1.0605	5.5	6.3***)
		1.0648	2.6	6.5
$LaSr_2Ga_{11}O_{20}$	300	1.0572	6	7
		1.0706	8	7
	~110	1.0690	13	11
		1.0725	18	6***)
		1.0778	12	8
$La_3Ga_5SiO_{14}$	300	1.0640	3	12
		1.0670	3	10
	77	1.0675	4	20
$LaMgAl_{11}O_{19}$	300	1.0552	1.1	4
		1.0817	1.5	5
	77	1.0550	2	15****)
		1.0812	10	10****)
$Ca_2Ga_2SiO_7$	300	1.0610	1.6	23
		1.0788	3.4	18
	77	1.0609	10	45
		1.0780	11	32
$LiGd(moO_4)_2$	300	1.0599	1.3	7
	~110	1.0595	3	9
$NaY(MoO_4)_2$	300	1.0674	4	7
	~110	1.0663	5	13
$NaLa(MoO_4)_2$	300	1.0595	2	7
		1.0653	5	2***)
	~110	1.0650	8	20
$NaGd(MoO_4)_2$	300	1.0667	4	7
	~110	1.0663	5	13
$KLa(MoO_4)_2$	300	1.0587	3.5	5
	~110	1.0580	5	16
		1.0645	10	5***)

*) See notes for Table 1.

**) The Nd^{3+}-ion concentration were about 2 at.%.

***) See notes for Table 1.

****) For $E_{exc}/E_{thr} \approx 6$.

Table 3. Statistics of crystallographic-position filling in laser-crystalline hosts with disorderness of crystal field on Ln^{3+} (Nd^{3+}) ions

Host	Singony	Crystallographic position with heterovalent cations			
		Sites	Symmetry	Coordination number	Filling
$Ca_3(Nb,Ga)_2Ga_3O_{12}$	Cubic	16a	C_{3i}	6	$Ga^{3+}:Nb^{5+}=1:6$
$LaSr_2Ga_{11}O_{20}$	Monocl.	4i	C_s	7	$Sr^{2+}:La^{3+}=1:6$
$La_3Ga_5SiO_{14}$	Trigon.	2d	C_3	4	$Ga^{3+}:Si^{4+}=1:1$
$LaMgAl_{11}O_{19}$	Hexag.	4e*)	C_v	5*)	$Mg^{2+}:Al^{3+}=(?)$
$Ca_2Ga_2SiO_7$	Tetrag.	4e	C_s	4	$Ga^{3+}:Si^{4+}=1:1$
$LiGd(MoO_4)_2$	Tetrag.	4a	S_4	8	$Li^+:Gd^{3+}=1:1$
$NaY(MoO_4)_2$	Tetrag.	4a	S_4	8	$Na^+:Y^{3+}=1:1$
$NaLa(MoO_4)_2$	Tetrag.	4a	S_4	8	$Na^+:La^{3+}=1:1$
$NaGd(MoO_4)_2$	Tetrag.	4a	S_4	8	$Na^+:Gd^{3+}=1:1$
$KLa(MoO_4)_2$	Distored tetrag.	4a	S_4**)	8**)	$K^+:La^{3+}=1:1$

*) Should be refined, possible $4f_1$ sites with coordination number 4 also.

**) Should be refined.

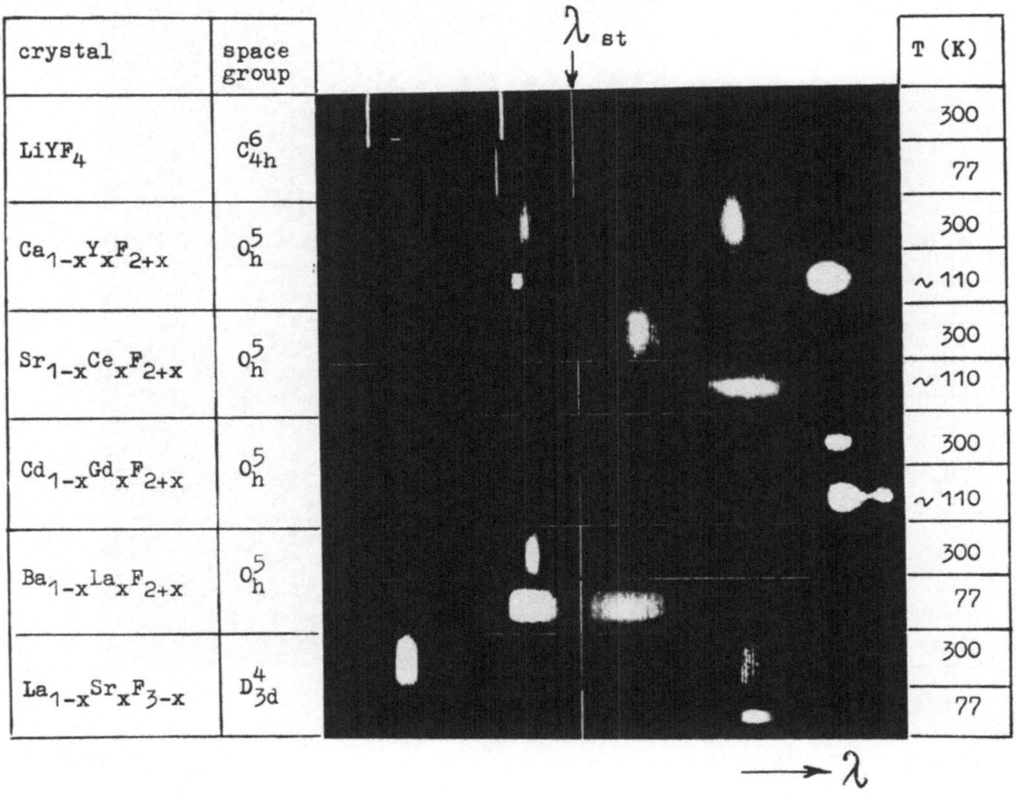

Fig. 2. Stimulated-emission spectra of Nd^{3+} ions ($^4F_{3/2} \rightarrow {}^4I_{11/2}$ channel) in disordered fluoride crystals and for comparison in $LiYF_4$ crystal with ordered structure. The standard line at $\lambda_{st}=1.0561$ µm is indicated by an arrow.

and are characterized by narrow laser lines at 300K [17,18], such as in $YAlO_3:Nd^{3+}$ and other ordered single laser crystals [5,11].

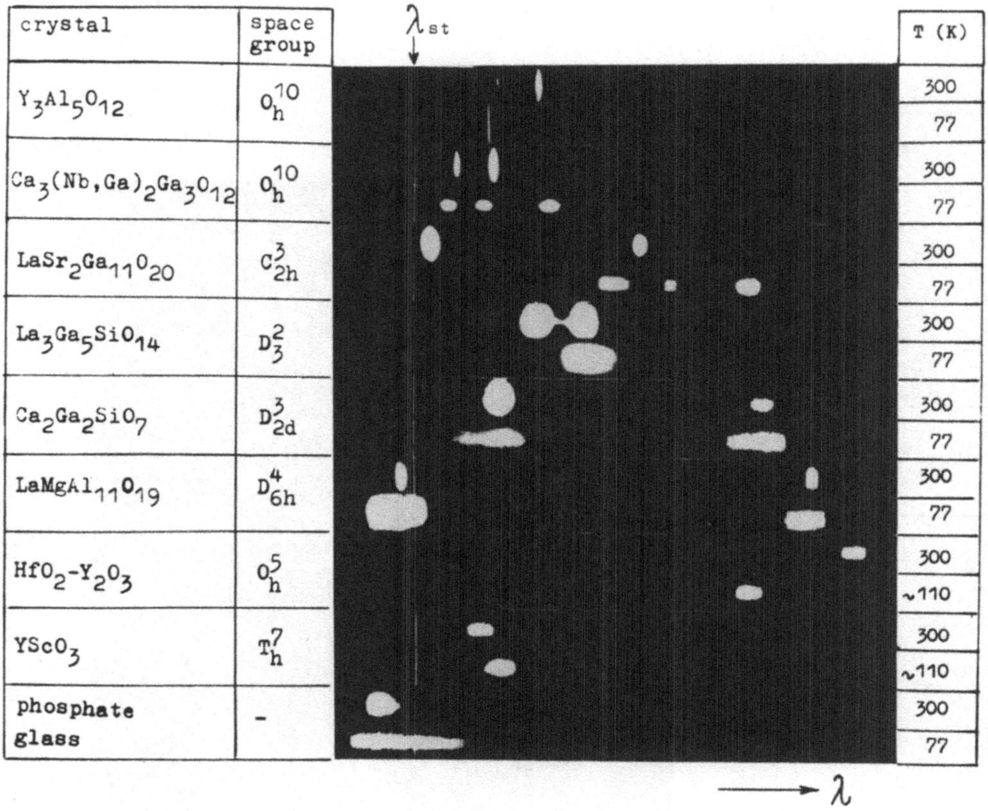
crystal
space group
λ_{st}
T (K)
$Y_3Al_5O_{12}$
O_h^{10}
$Ca_3(Nb,Ga)_2Ga_3O_{12}$
O_h^{10}
$LaSr_2Ga_{11}O_{20}$
C_{2h}^3
$La_3Ga_5SiO_{14}$
D_3^2
$Ca_2Ga_2SiO_7$
D_{2d}^3
$LaMgAl_{11}O_{19}$
D_{6h}^4
HfO_2-Y_2O_3
O_h^5
$YScO_3$
T_h^7
phosphate glass
-
300
77
300
77
300
77
300
77
300
77
300
77
300
~110
300
~110
300
77
λ

Fig. 3. Stimulated-emission spectra of Nd^{3+} ions ($^4F_{3/2} \rightarrow {}^4I_{11/2}$ channel) in disordered oxide crystals and for comparison in $Y_3Al_5O_{12}$ crystal with ordered structure and in phosphate glass. The standard line at λ_{st}=1.0561 μm is indicated by an arrow.

In the spectra of nonstoichiometric phases of $Me_{1-x}R_xF_{2+x}$ under increasing R^{3+} ion concentrations some regularities occur which cannot be explained merely by the formation of elementary activator centers (as in MeF_2 crystals) or by their structural combinations. The concentrational insensitivity of nonstoichiometric-$Me_{1-x}R_xF_{2+x}$-phases spectra, starting at $C_R \geq 7$ wt.%, is manifested by only slight changes in the internal construction of crystals without changing of their space group. What does result is the formation of aggregates of elementary centers. These new structural formations differ sharply from the above considered simple activator centers in CaF_2 type crystals. According to the investigations

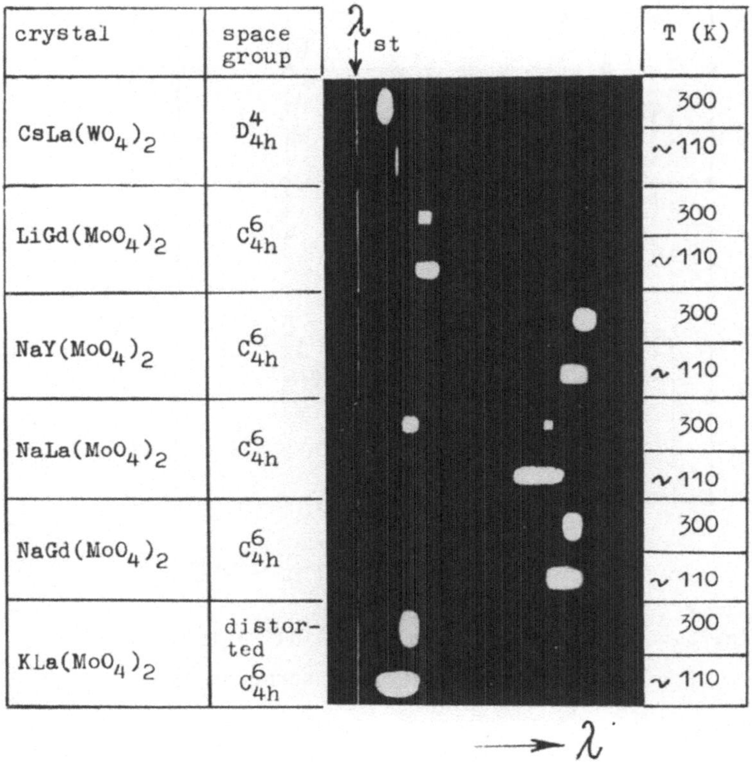

crystal
space group
λ_{st}
T (K)
$CsLa(WO_4)_2$
D_{4h}^4
300
~110
$LiGd(MoO_4)_2$
C_{4h}^6
300
~110
$NaY(MoO_4)_2$
C_{4h}^6
300
~110
$NaLa(MoO_4)_2$
C_{4h}^6
300
~110
$NaGd(MoO_4)_2$
C_{4h}^6
300
~110
$KLa(MoO_4)_2$
distorted C_{4h}^6
300
~110
λ

Fig. 4. Stimulated-emission spectra of Nd^{3+} ions ($^4F_{3/2} \rightarrow {}^4I_{11/2}$ channel) in disordered crystals with scheelite structure and (for comparison) in a $CsLa(WO_4)_2$ crystal with ordered structure. The standard line at λ_{st}=1.0561 μm is indicated by an arrow.

in nonstoichiometric $Me_{1-x}R_xF_{2+x}$, the Nd^{3+}-centers formation leads to complicated coordinational polyhedra (Fig.6) in which fluorine atoms occupy new positions in contrast with their usual 1/4, 1/4, 1/4 position in MeF_2 crystals. Here it should be noted that, in this case, the introduction of the additional F^- ions is not simple: the substitution of one ion of the initial structure by two (or more) ions takes place for each R^{3+} cation. These ions are distributed not in the centers of the

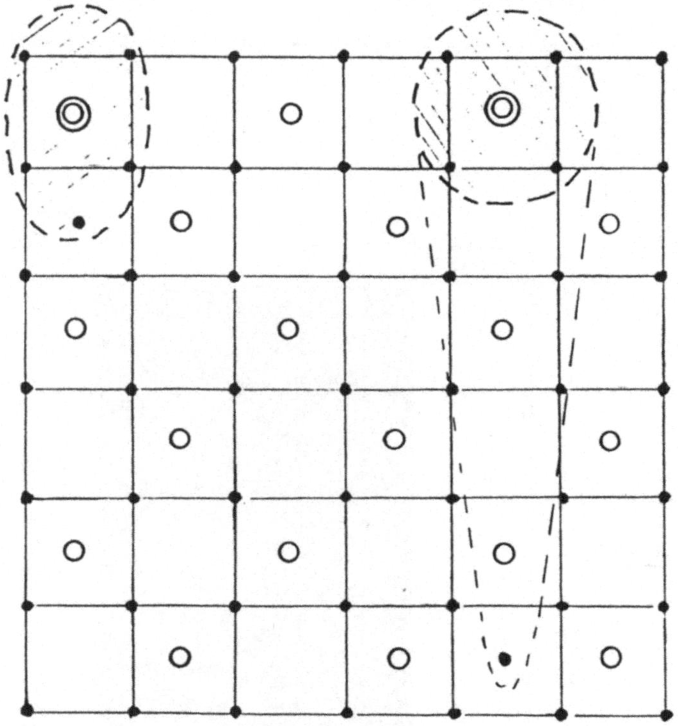

• - F^- ○ - Me^{2+} ◎ - Ln^{3+}

Fig. 5. Activator-center of Ln^{3+} (Nd^{3+}) ions in cubic (O_h^5) MeF_2 crystals (CaF_2 type structure).

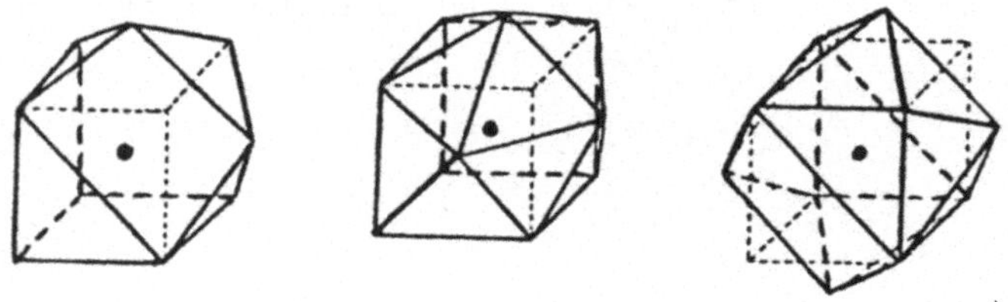

• - Me^{2+}, Sc^{3+}, Y^{3+} , Ln^{3+}

Fig. 6. Examples of probable coordination polyhedra in the structure of $Me_{1-x}R_xF_{2+x}$ crystals. Dashed lines show the basic cubes of pure MeF_2 with fluorite CaF_2 structure.

fluorite-structure-empty cubes, but are statistically shifted in their direction at some distance. As for the statistically probable polyhedron, it will be presented by ninevertex as a distorted cube [14]. The absorption and luminescence spectra of $Me_{1-x}R_xF_{2+x}:Nd^{3+}$ nonstoichiometric phases consist of broad slightly-structured bands (Fig. 7). The nature of inhomogeneous broadening is reflected also in SE spectra, showing all the effects typical of neodymium laser glasses [19] (Fig. 8).

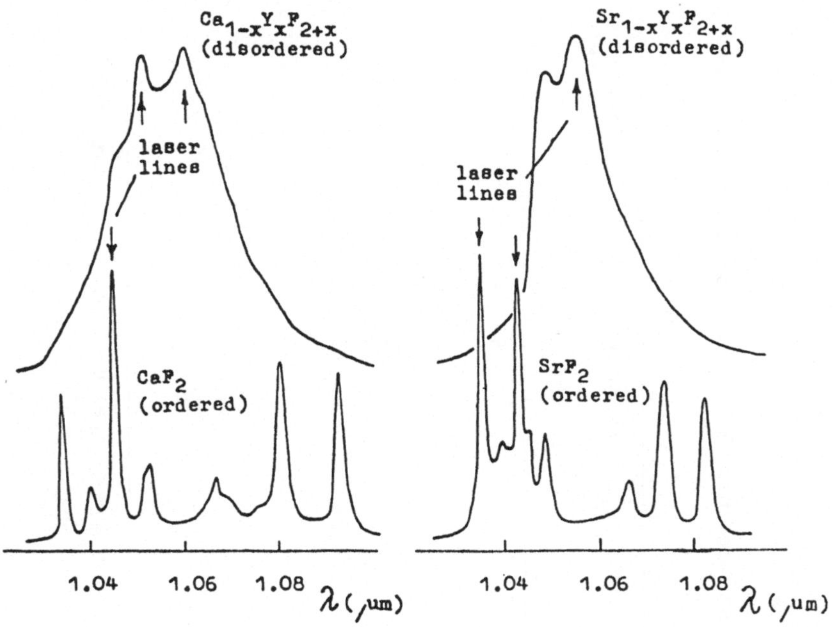

Fig. 7. Luminescence spectra of Nd^{3+} ions ($^4F_{3/2} \rightarrow {}^4I_{11/2}$ channel) in fluorides CaF_2 and SrF_2 with ordered structure (C_{Nd}=0.2 wt.%) and in disordered $Me_{1-x}R_xF_{2+x}$ compounds ($C_R \simeq 0.2$ wt.%) at 300K.

The possibility of chemical-composition variation without the disordered $Me_{1-x}R_xF_{2+x}$ fluorides-structure changing, and the creation on their base of new multi-component compounds, is very large. More than 30 such laser crystals are already known [1,5,11]. Data on seven new crystals, which we have studied recently, are listed in Table 4. Here there are indicated compositions, lasing wavelengths of Nd^{3+} ions, threshold of excitation, and SE linewidths.

As one can see, the values of peak cross-sections of the induced transitions (σ_e^p) in these crystals are approximately equal to or slightly higher than the cross-sections of the best neodymium laser glasses [19]. Also these crystals have the potential for growth scaling [20].

Table 4. Stimulated-emission parameters of disordered $Me_{1-x}R_xF_{2+x}$ fluorides with Nd^{3+} ions ($^4F_{3/2} \rightarrow {}^4I_{11/2}$ channel) at 300 K (C_R=7-10 wt.% and C_{Nd}=2 wt.%)

Crystal	Lasing wavelength (μm)	SE threshold*) (J)	SE linewidth**) (cm^{-1})	Peak cross-section of induced transition (10^{-20} cm^2)
$Cd_{1-x}Sc_xF_{2+x}$	1.0507	4	6	6.2
$Cd_{1-x}La_xF_{2+x}$	1.0668	5.5	15	5.3
$Cd_{1-x}Ce_xF_{2+x}$	1.0667	6	15	5.2
$Cd_{1-x}Nd_xF_{2+x}$	1.0666	8	15	5.2
$Cd_{1-x}Gd_xF_{2+x}$	1.0672	3.5	15	5.0
$Cd_{1-x}Lu_xF_{2+x}$	1.0652	7	20	4.5
$Sr_{1-x}Sc_xF_{2+x}$	1.0543 1.0605	2.5 3	10 6	5.5 5.3

*) At excitation of Xe flash-tube emission and optical confocal resonator with high Q-factor.

**) For $E_{exc}/E_{thr} \approx 3$.

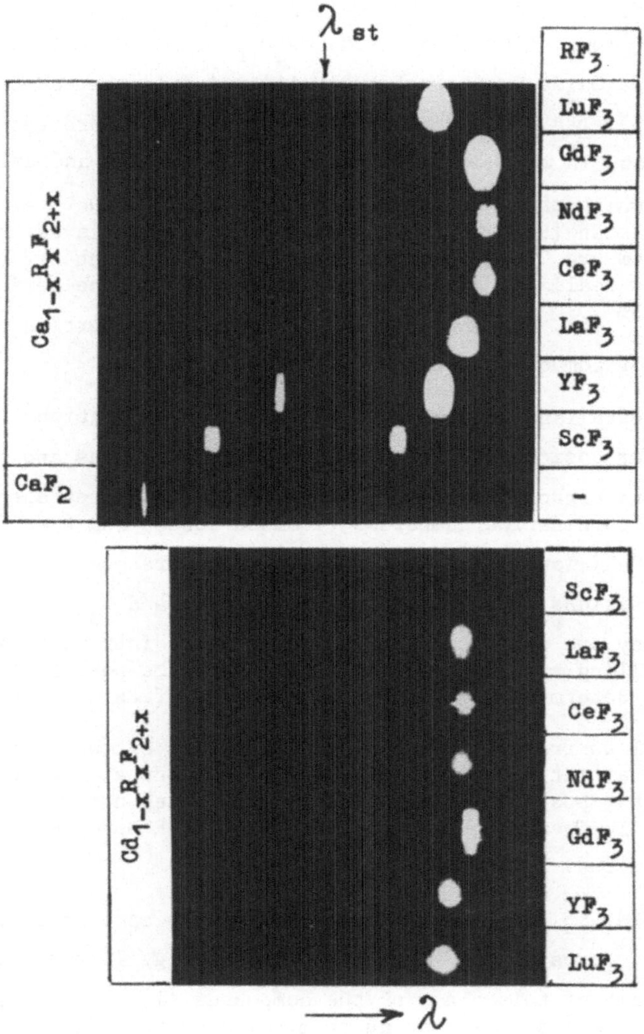

Fig. 8. Stimulated-emission spectra of Nd^{3+} ions ($^4F_{3/2}\rightarrow{}^4I_{11/2}$ channel) in cubic (O_h^5) disordered $Ca_{1-x}R_xF_{2+x}$, $Cd_{1-x}R_xF_{2+x}$ ($C_R\simeq7$ wt.%) and ordered CaF_2 single crystals at 300K. Standard line with λ_{st}=1.0561 μm is indicated by an arrow.

III.B. Structural-Static Disorder

The order type of disorder of the internal structure of laser crystals-structural-static disorderness was discovered by us while studying compounds formed in ternary-oxide systems. In such systems no polyhedra reformation of crystal structure takes place as the race-earth laser ion concentration is changed. As stated above, disorder here is initially set in the structure of crystals and it is connected with statistical filling of the similar crystallographic structural positions.

Now let us consider several classes of new laser crystals where structural-static disorder is realized.

1. Trigonal Crystals with Ca-gallogermanate Structure ($Ca_3Ga_2Ge_4O_{14}$ Type) with Nd^{3+} Ions. A trigonal $Ca_3Ga_2Ge_4O_{14}$ phase, discovered while investigating $Ca_3Ga_2Ge_4O_{12}$ garnet crystallization in the $CaO-Ga_2O_3-GeO_2$ system (Table 5), was unknown before as having a structure type with space group D_3^2; a group in which a large number of germanates and gallates are formed, among which are a number of compounds used for Nd^{3+} stimulated-emission excitation [6,20-26]. The results of complex investigations show that all of them are "pure" disordered crystals, in which structural-static disorderness is realized. The first laser crystal in the series is $La_3Ga_5SiO_{14}:Nd^{3+}$ [6], and on its example some main properties of the considered class of compounds can be explained.

In the crystalline structure with space group D_3^2 cations are distributed at four positions (Fig. 9). La^{3+} (or Nd^{3+}) ions and some fraction of Ga^{3+} ions are arranged between layers of Ga- and Si-tetrahedra in the cavities with coordination numbers 8 (Thomson cubes) and 6 (octahedra). The remaining Ga^{3+} ions in the $La_3Ga_5SiO_{14}$ structure occupy two types of tetrahedral positions with point symmetry C_2 (3f) and C_3 (2d); in the latter case they are statistically in a 1:1 ratio distributed with Si^{4+} ions. Such filling statistics of the tetrahedral 2d positions by heterovalent cations determine the structural disorder of this crystal.

As it follows from Table 6, in other crystals of this class of laser compounds show a statistical distribution of cations at different crystalloagraphic positions. All this is manifested not only in the absorption and luminescence spectra, but also in SE spectra of the activator Nd^{3+} ions.

For example, in Fig. 10 is shown luminescence spectra of a $La_3Ga_5SiO_{14}:Nd^{3+}$ crystal at 77 and 300K, and in Fig. 11 the SE spectra of laser crystals of this class of the compounds [1]. As it can be seen, SE lines are broader at 77K than at room temperature indicating that here we are dealing with disordered crystals. The same effect is observed for neodymium-doped laser glasses.

Before concluding, it is worthwile to note that the cross-sections of induced transitions (Table 7) are slightly higher than those of neodymium-doped glasses [27].

2. Tetragonal Crystals with Ga-Gehlenite Structure ($Ca_2Ga_2SiO_7:Nd^{3+}$). Recently we have grown new Nd^{3+}-doped Ga-Gehlenite $Ca_2Ga_2SiO_7$ using the Czochralski technique [7]. Precise structural analysis, spectral-luminescence, and stimulated-emission investigations show that for this crystal the case of structural-static disorder is realized. Let us briefly highlight the main results.

Fig. 12 shows a fragment of $Ca_2Ga_2SiO_7$ structure; such structure is typical for a large number of melilite-like compounds. For cations of tetragonal $Ca_2Ga_2SiO_7$ structure with space group D_{2d}^3 there are three types of crystallographic positions--they are Thomson cubes (4e), filled

Table 5. Disordered trigonal laser crystals with Ca-gallogermanate ($Ca_3Ga_2Ge_4O_{14}$) structure (D_3^2)*)

System	Compounds	Crystalline matrix for lasing Nd^{3+} ions
$AO-Ga_2O_3-GeO_2$	$A_3^{2+}Ga_2Ge_4O_{14}$	$Ca_3Ga_2Ge_4O_{14}$
(A=Ca, Sr)		$Sr_3Ga_2Ge_4O_{14}$
$La_2O_3-Ga_2O_3-BO_2$	$La_3Ga_5B^{4+}O_{14}$	$La_3Ga_5SiO_{14}$
(B=Si, Ge)		$La_3Ga_5GeO_{14}$
$La_2O_3-Ga_2O_3-B_2O_5$	$La_3Ga_{5.5}B^{5+}_{0.5}O_{14}$	$La_3Ga_{5.5}Nb_{0.5}O_{14}$
(B=Nb, Ta)		$La_3Ga_{5.5}Ta_{0.5}O_{14}$

*) History: they were found at crystallization of garnet $Ca_3Ga_2Ge_4O_{12}$ (O_h^{10}) in the $CaO-Ga_2O_3-GeO_2$ system (see all references in [1]).

Table 6. Cation distribution in the new disordered laser crystals with Ca-gallogermanate structure

Crystal	Crystallographic positions (oxygen coordination)			
	3e (8)	la (6)	2d (4)	3f (4)
$Ca_3Ga_2Ge_4O_{14}$	Ca^{2+}	$Ge^{4+}:Ga^{3+}=4:1$	Ge^{4+}	$Ge^{4+}:Ga^{3+}=2:3$
$Sr_3Ga_2Ge_4O_{14}$	Sr^{2+}	$Ge^{4+}:Ga^{3+}=3:2$	Ge^{4+}	$Ge^{4+}:Ga^{3+}=7:8$
$La_3Ga_5SiO_{14}$	La^{3+}	Ga^{3+}	$Ga^{3+}:Si^{4+}=1:1$	Ga^{3+}
$La_3Ga_5GeO_{14}$	La^{3+}	Ga^{3+}	$Ga^{3+}:Ge^{4+}=1:1$	Ga^{3+}
$Nd_3Ga_5SiO_{14}$	Nd^{3+}	Ga^{3+}	$Ga^{3+}:Si^{4+}=1:1$	Ga^{3+}
$La_3Ga_{5.5}Nb_{0.5}O_{14}$	La^{3+}	$Ga^{3+}:Nb^{5+}=1:1$	Ga^{3+}	Ga^{3+}
$La_3Ga_{5.5}Ta_{0.5}O_{14}$	La^{3+}	$Ga^{3+}:Ta^{5+}=1:1$	Ga^{3+}	Ga^{3+}

Table 7. Peak cross-sections at 300K for two intermanifold laser transitions of Nd^{3+} ions in disordered crystals with Ca-gallogermanate structure

Crystal	Peak cross-section (10^{-20} cm^2)*)	
	$^4F_{3/2} \rightarrow {}^4I_{11/2}$ (1.06 μm)	$^4F_{3/2} \rightarrow {}^4I_{13/2}$ (1.35 μm)
$Ca_3Ga_2Ge_4O_{14}$	5.6	1.5
$Sr_3Ga_2Ge_4O_{14}$	4.5	1.4
$La_3Ga_5SiO_{14}$	3.7	0.7
$La_3Ga_5GeO_{14}$	5.8	-
$La_3Ga_{5.5}Nb_{0.5}O_{14}$	6	1.5
$La_3Ga_{5.5}Ta_{0.5}O_{14}$	5.4	1.5

*) There are spectral and orientation dependences.

Table 8. Cation filling of crystallographic positions in disordered $Ca_2Ga_2SiO_7$ gehlenite

Position	Polyhedron	Filling
4e	dodecahedron	Ca^{2+}
2a	tetrahedron	Ga^{3+}
4e	tetrahedron	$Ga^{3+}:Si^{4+}=1:1$

Table 9. Spectroscopic characteristics of the intermanifold ($^4F_{3/2} \rightarrow {}^4I_{J'}$) luminescence channels in $Ca_2Ga_2SiO_7:Nd^{3+}$ crystals

Terminal J′-manifold	Channel wavelength (μm)	Probability of spontaneous emission (s^{-1})	Luminescence lifetime (μs)	Intermanifold branching ratio of luminescence
$^4I_{9/2}$	0.89	1064	270±10	0.304
$^4I_{11/2}$	1.07	1968		0.562
$^4I_{13/2}$	1.35	450		0.128
$^4I_{15/2}$	1.85	21		0.006

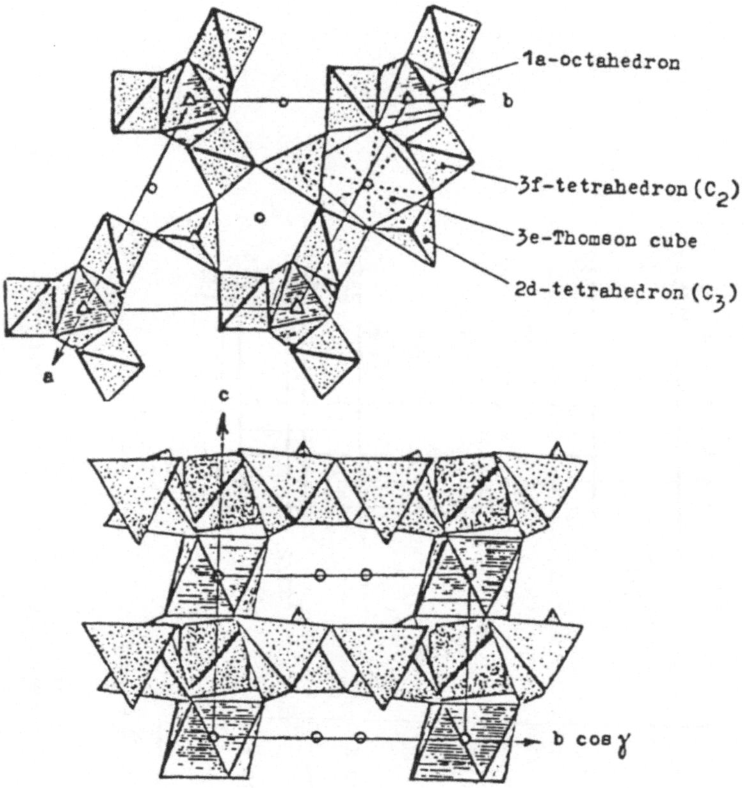

Fig. 9. Fragments of trigonal (D_3^2) Ca-gallogermanate crystal structure.

by Ca^{2+} ions, and two types of tetrahedra (2a and 4e). The former are totally occupied by Ga^{3+} ions, and the latter are statistically filled with heterovalent Ga^{3+} and Si^{4+} cations in a 1:1 ratio (Table 8) [28].

Spectroscopically this compound is very similar to the crystals with Ca-gallogermanate structure considered above. SE spectra (Fig. 13) show a broadening of generating lines at low temperature typical of disordered laser media. Disorder is also manifested in the luminescence spectra of Nd^{3+} ions (Fig. 14). The broad widths of the Raman-spectrum lines (Fig. 15) of a pure $Ca_2Ga_2SiO_7$ single crystal again indicates the structural disorder of this compound. [Tables 9 and 10 show data on luminescence-intensity characteristics of $Ca_2Ga_2SiO_7:Nd^{3+}$ crystals and their SE parameters [28]].

In our work [28] full concentration series of tetragonal melilite-like $Ca_{2-x}Nd_xGa_{2+x}Si_{1-x}O_7$ single crystals were grown, among them a self-activated $CaNdGa_3O_7$ crystal having a 100 at % concentration of Nd^{3+} ions, and all show a structural-static type disorder.

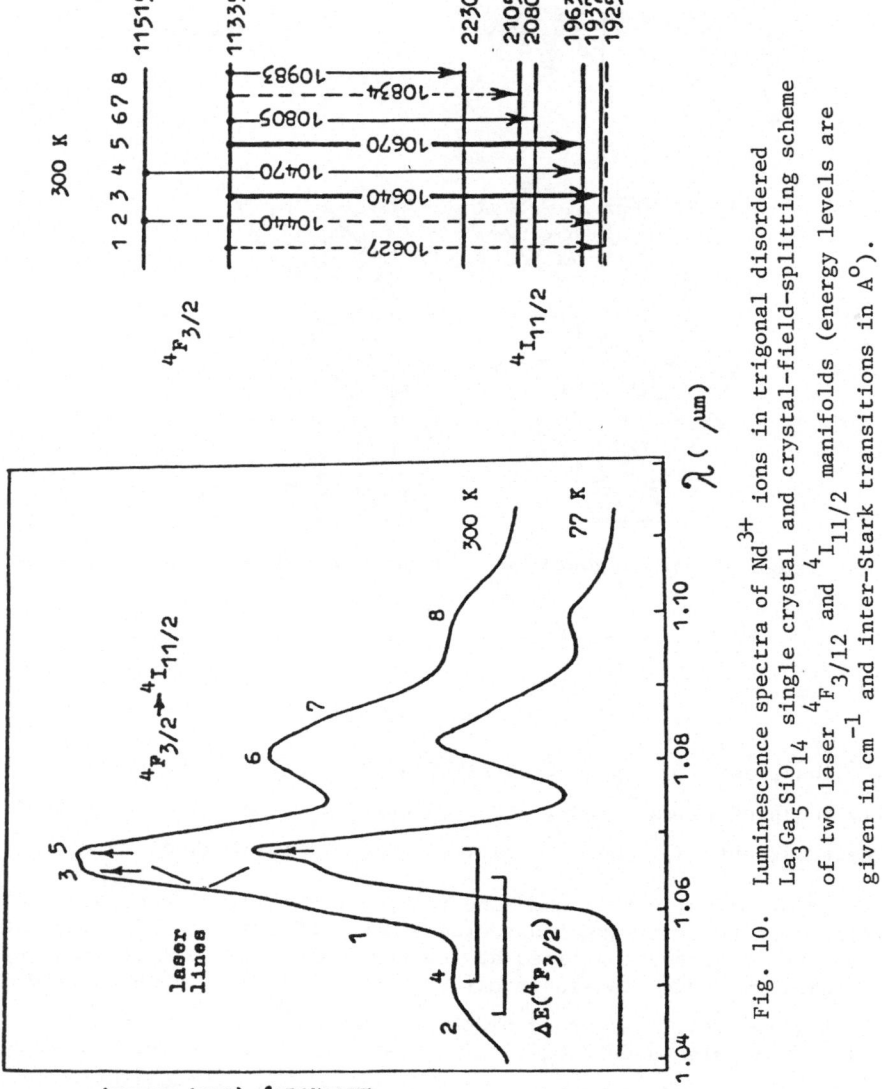

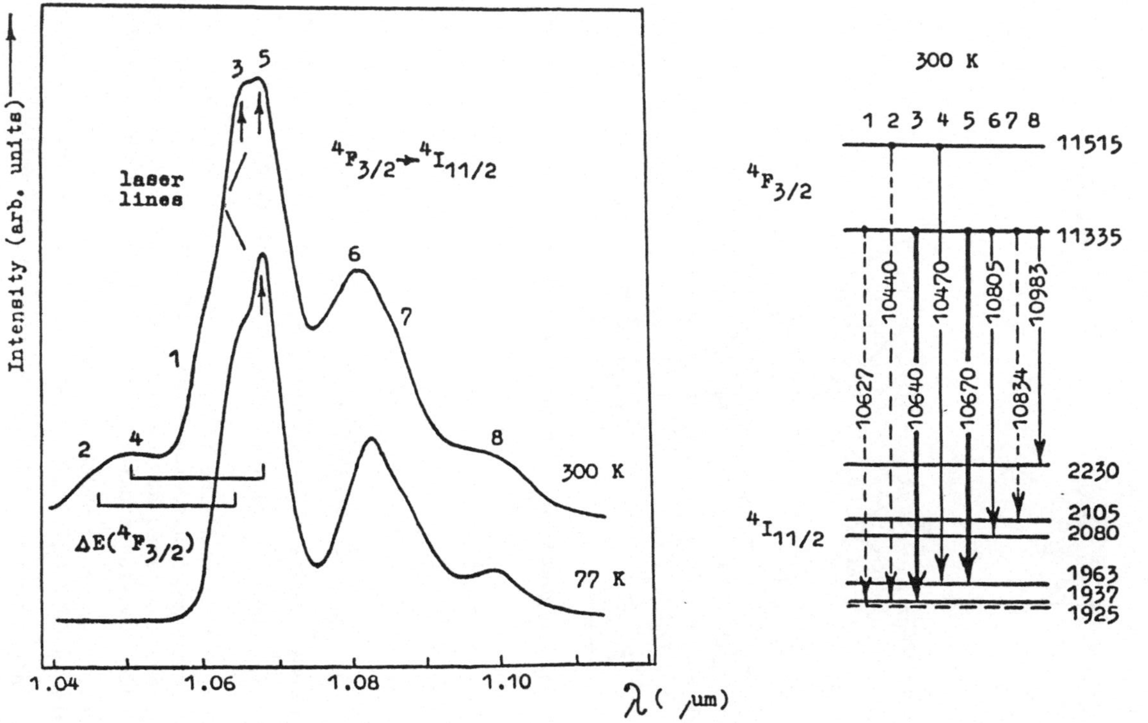

Fig. 10. Luminescence spectra of Nd^{3+} ions in trigonal disordered $La_3Ga_5SiO_{14}$ single crystal and crystal-field-splitting scheme of two laser $^4F_{3/12}$ and $^4I_{11/2}$ manifolds (energy levels are given in cm^{-1} and inter-Stark transitions in A^o).

Crystal	T(K)	x	C_{Nd} ($10^{20} cm^{-3}$)
$Ca_3Ga_2Ge_4O_{14}$	77		5,36
	300		
$Sr_3Ga_2Ge_4O_{14}$	77		5,02
	300		
$(La_{1-x}Nd)_3Ga_5SiO_{14}$	77	0,1	10,3
	300	0,01	1,02
		0,05	5,12
		0,1	10,3
		0,2	20,3
		0,4	41,5
		0,6	62,5
		1,0	105
$La_3Ga_5GeO_{14}$	77		5,12
	300		
$La_3Ga_{5.5}Nb_{0.5}O_{14}$	77		5,01
	300		
$La_3Ga_{5.5}Ta_{0.5}O_{14}$	77		5,01
	300		

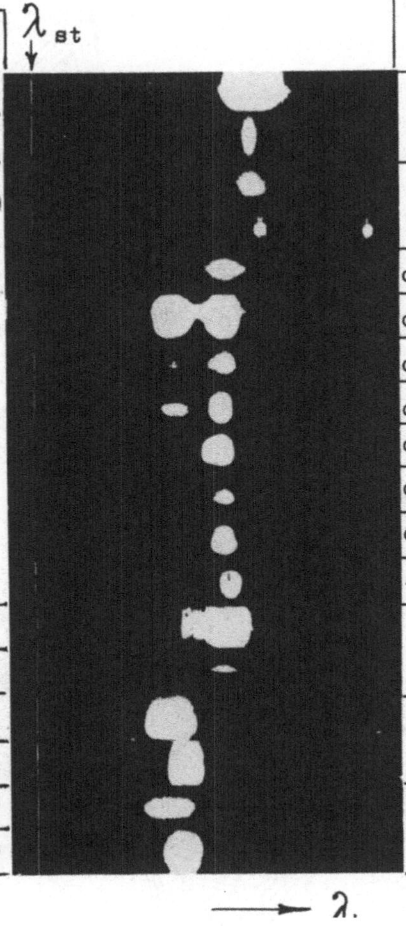

Fig. 11. Stimulated emissions spectra of Nd^{3+} ions ($^4F_{3/2} \rightarrow ^4I_{11/2}$ channel) in disordered trigonal crystals with Ca-gallogermanate structure. The standard line λ_{st}=1.0561 μm is indicated by an arrow.

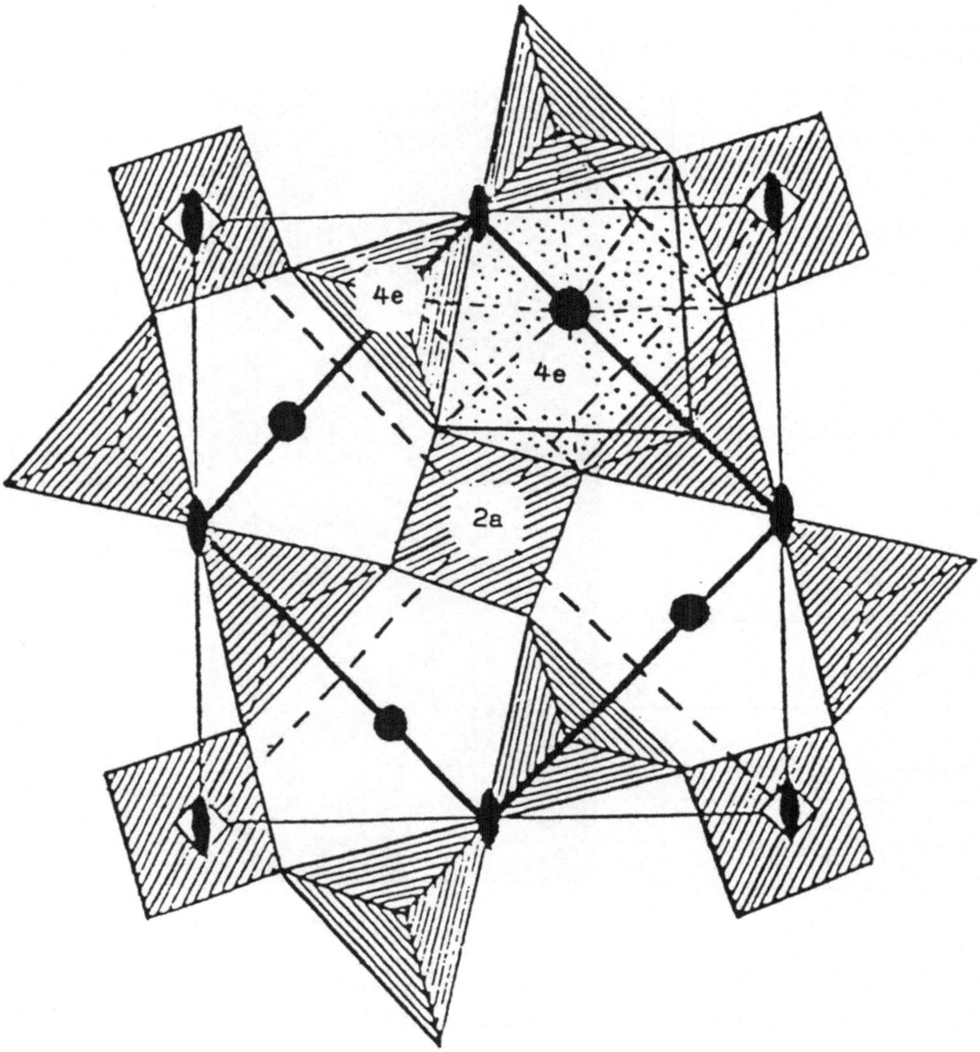

Fig. 12. Fragment of the melilite crystal structure.

Table 10. Some stimulated-emission parameters of $Ca_2Ga_2SiO_7:Nd^{3+}$ single crystals (C_{Nd} ≈ 2 at.%) at 300K

Parameter	Laser channel		
	$^4F_{3/2} \rightarrow {}^4I_{11/2}$		$^4F_{3/2} \rightarrow {}^4I_{13/2}$
SE wavelength (μm)	1.0610	1.0788	1.3365
Peak cross-section for ‖ c-axis (10^{-20} cm^2)	5.3	4.0	1.5
Peak cross-section for ⊥ c-axis (10^{-20} cm^{2})	2.8	1.5	1.3
SE threshold for ‖ c-axis (J)*)	1.6	3.4	5.5

*) At excitation of Xe flash-tube emission and optical confocal resonator with high Q-factor.

Table 11. Cation distribution in crystals with melilite structure

Crystal	Crystallographic positions (oxygen coordination)		
	4e (8)	2a (4)	4e (4)
$Ca_2Ga_2SiO_7$*)	Ca^{2+}	Ga^{3+}	$Ga^{3+}:Si^{4+}$=1:1
$CaNdGa_3O_7$*)	$Ca^{2+}:Nd^{3+}$=1:1	Ga^{3+}	Ga^{3+}
$Ba_2ZnGe_2O_7$**)	Ba^{2+}	Zn^{2+}(?)	Ge^{4+}(?)
$Ba_2MgGe_2O_7$***)	Ba^{2+}	Mg^{2+}(?)	Ge^{4+}(?)

*) Crystals belong to the $Ca_{2-x}Nd_xGa_{2+x}Si_{1-x}O_7$ system; $Ca_2Ga_2SiO_7$ is formed at x=0, and $CaNdGa_3O_7$ - at x=1.

**) Laser action of $Ba_2ZnGe_2O_7:Nd^{3+}$ was excited in [30].

***) Laser action of $Ba_2MgGe_2O_7:Nd^{3+}$ was excited in [29].

Heterovalent Ca^{2+} and Nd^{3+} cations in this compound, statistically in 1:1 ratio, fill still another crystallographic position of the structure-Thomson cubes (4e) (Table 11).

Finishing this part of the lecture, it should be noted that SE of Nd^{3+} ions in the melilite structural type of laser crystals for the first time was excited in $Ba_2MgGe_2O_7:Nd^{3+}$ [29] and $Ba_2ZnGe_2O_7:Nd^{3+}$ [30].

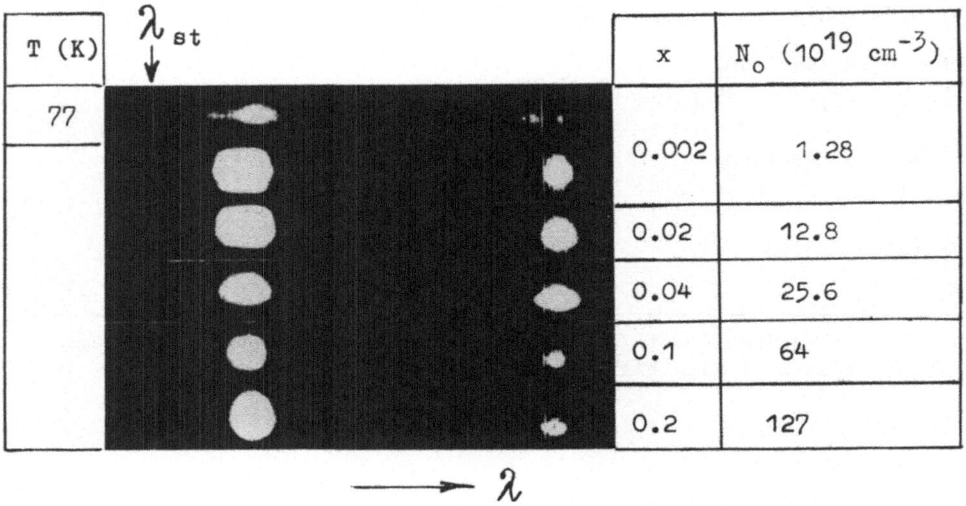

Fig. 13. Stimulated-emission spectra of Nd^{3+} ions (${}^4F_{3/2} \rightarrow {}^4I_{11/2}$ channel) in tetragonal $Ca_2Ga_2SiO_7$ crystals. Standard line λ_{st}=1.0561 μm is indicated by an arrow.

According to data in references 29 and 30 luminescence linewidths ($\simeq$ 40 cm^{-1}) at the wavelengths of the reported SE were attributed by me to the disordered laser media [5,11]. After detailed investigations of the family of $Ca_{2-x}Nd_xGa_{2+x}Si_{1-x}O_7$ crystals the question on the type of these two barium laser germanates again became interesting.

3. Monoclinic $LaSr_2Ga_{11}O_{20}:Nd^{3+}$ Crystals with Disordered Structure. In ref.[8] is presented evidence of the new laser $LaSr_2Ga_{11}O_{20}:Nd^{3+}$ crystal. Here we briefly discuss some results of an investigation of this inorganic material.

Detailed X-ray analysis shows that in the monoclinic (space group C^3_{2h}) crystal $LaSr_2Ga_{11}O_{20}$ structural-static disorder takes place. From Fig. 16 and Table 12 it can be seen that heterovalent Sr^{2+} and La^{3+} (or Nd^{3+})

cations of this compound occupy, in a 1:1 ratio, crystallographic positions 4i. The observed Nd^{3+} luminescence and SE are typical of disordered laser media (see Fig. 3 and Table 2).

4. Disordered $Ca_3(Nb,Ga)_2Ga_3O_{12}:Nd^{3+}$ Laser Garnet. In two recent works [9,10] complex investigations of cation-deficient $Ca_3(Nb,Ga)_2Ga_3O_{12}$ garnet with Nd^{3+} ions were carried out. Results of X-ray structural and luminescence-absorption measurements, as well as stimulated-emission and Raman-spectra analyses show that this crystal has a disordered structure. The structural-static disorder occurs via the statistical introduction of

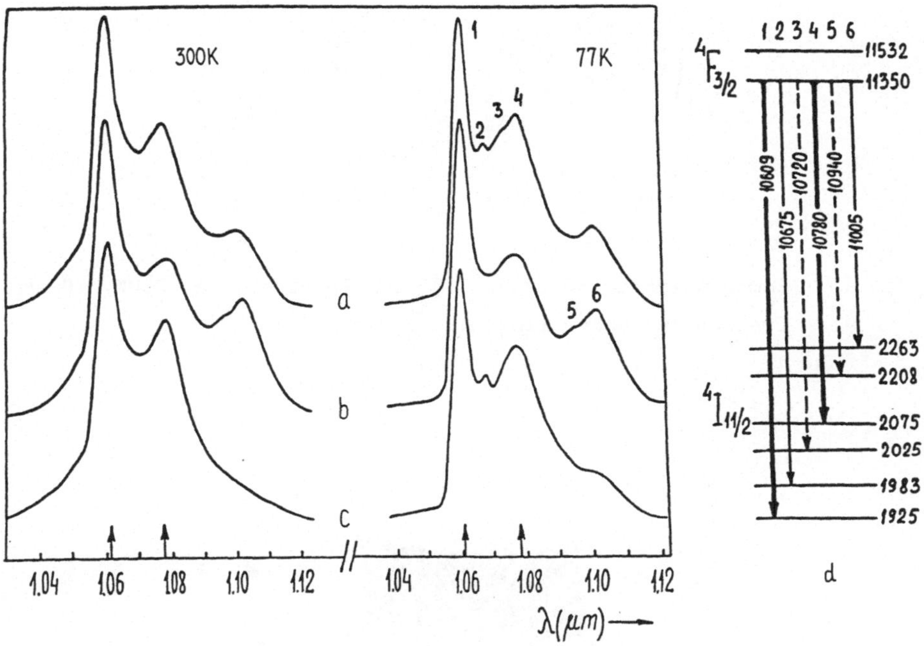

Fig. 14. Luminescence spectra of Nd^{3+} ions in $Ca_2Ga_2SiO_7$ (1c) in the principle ${}^4F_{3/2} \rightarrow {}^4I_{11/2}$ channel at 77 and 300K, and crystal-field splitting scheme of the ${}^4F_{3/2}$ and ${}^4I_{11/2}$ manifolds; (a) nonpolarized spectra; (b) π-polarization, and (c) σ-polarization. Induced transitions in the scheme are indicated by thick arrows, and transition wavelengths which need refinement are shown by dashed arrows. Lines in the spectra and transitions in the scheme have the same denotations. Lasing lines in the spectra are indicated by arrows.

heterovalent cations (Ga^{3+} and Nb^{5+}) into octahedral positions of the crystal approximately in the ratio 1:6 (Fig. 17 and Table 13). This is only a rough approximation. As for the exact distribution of cations, Table 13, that in the all cation positions it has certain concentration of vacancies and heterovalent cations.

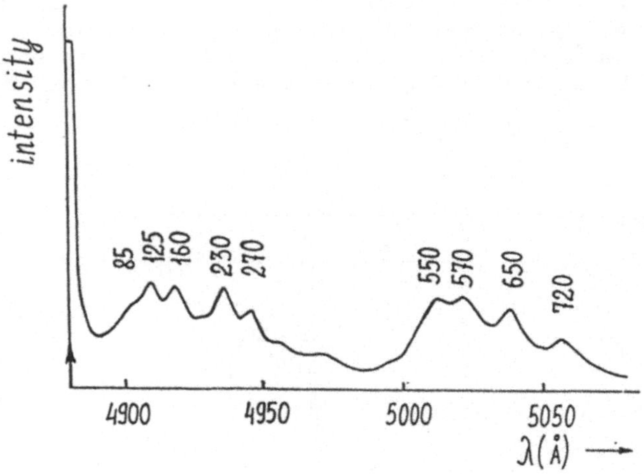

Fig. 15. Raman spectrum of $Ca_2Ga_2SiO_7$ single crystal at 300K. Peak frequencies above lines are given in cm^{-1}.

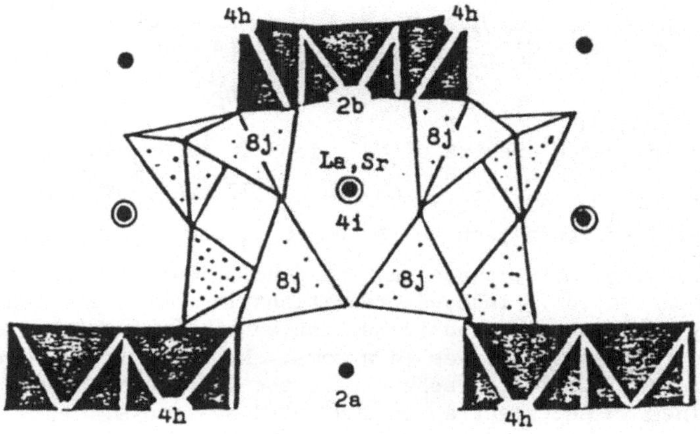

Fig. 16. Fragment of the $LaSr_2Ga_{11}O_{20}$ crystal structure.

Table 12. Cations filling of crystallographic positions in disordered $LaSr_2Ga_{11}O_{20}$ crystals

Position	Polyhedron	Filling
2a	octahedron	Sr^{2+}
4i	seven-vertex polyhedron*)	$Sr^{2+}La^{3+}$=1:1
2b	octahedron	Ga^{3+}
4h	octahedron	Ga^{3+}
8j	tetrahedron	Ga^{3+}
8j	tetrahedron	Ga^{3+}

*) One-cap trigonal prism.

Table 13. Cation filling of crystallographic positions in disordered $Ca_3(Nb,Ga)_2Ga_3O_{12}$ garnet

Position	Polyhedron	Filling		Exact filling*)
c	dodecahedron	Ca^{2+}		$Ca^{2+}_{0.983}\square$
a	octahedron	$Ga^{3+}:Nb^{5+}$	1:6	$Ca^{2+}_{0.035}Ga^{3+}_{0.138}Nb^{5+}_{0.81}\square$
d	tetrahedron	Ga^{3+}		$Ga^{3+}_{0.975}Nb^{5+}_{0.017}\square$

*) For composition $\square_{0.10}Ca_{3.02}Nb_{1.67}Ga_{3.2}O_{12}$.

Table 14. Spectroscopic characteristics of the intermanifold ($^4F_{3/2}\rightarrow{}^4I_{J'}$) luminescence in $Ca_3(Nb,Ga)_2Ga_3O_{12}:Nd^{3+}$ garnet

Terminal J′-manifold	Channel wavelength (μm)	Probability of spontaneous emission (s^{-1})	Luminescence lifetime (μs)	Intermanifold branching ratio of luminescence
$^4I_{9/2}$	0.89	1647		0.379
$^4I_{11/2}$	1.07	2227		0.512
$^4I_{13/2}$	1.35	454	220±10	0.104
$^4I_{15/2}$	1.85	23		0.005

Table 15. Some stimulated-emission parameters of $Ca_3(Nb,Ga)_2Ga_3O_{12}:Nd^{3+}$ single-crystal ($C_{Nd} \approx 2$ at.%) at 300K

Parameter	Laser channel		
	$^4F_{3/2} \rightarrow {}^4I_{11/2}$		$^4F_{3/2} \rightarrow {}^4I_{13/2}$
SE wavelength (μm)	1.0588	1.0612	1.3290
Peak cross-section (10^{-20} cm^2)	5.0	5.4	1.5
SE threshold (J)*)	1.3	0.9	5.3

*) At excitation of Xe flash-tube emission and optical confocal resonator with high Q-factor.

Fig. 18 shows SE spectra of this new garnet, and for a comparison are shown SE spectra of a number of garnets known to have an ordered structure. The lines of the latter in the low-temperature spectra are narrower than those at room temperature, while the SE line of the disordered garnet $Ca_3(Nb,Ga)_2Ga_3O_{12}:Nd^{3+}$ shows a broadening at low temperature.

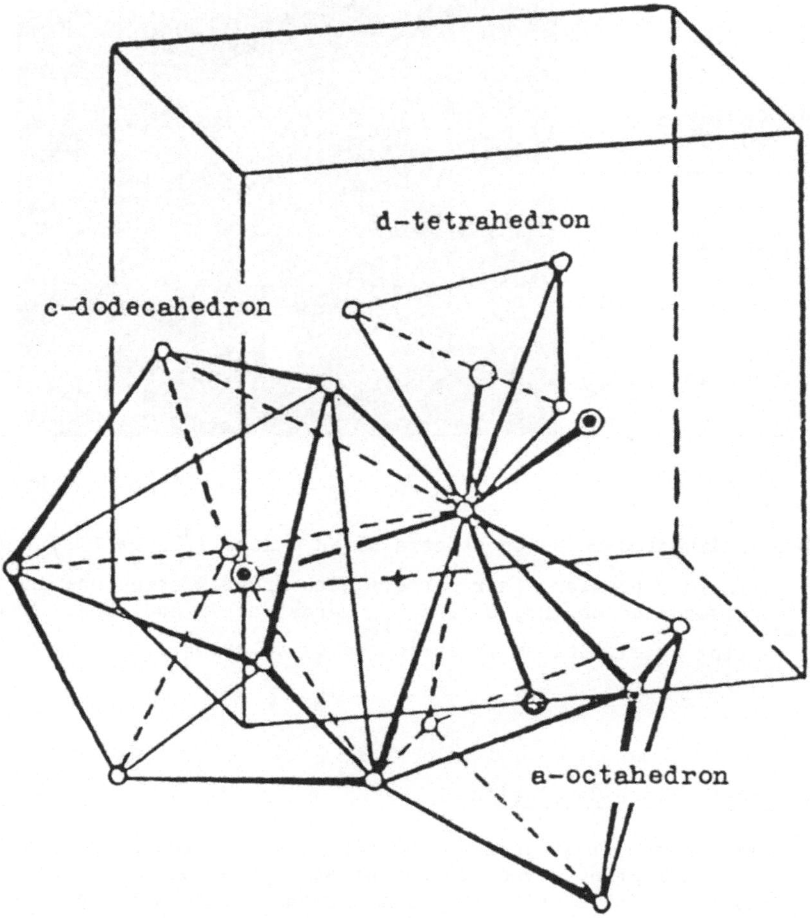

Fig. 17. Fragment of the garnet-crystal structure.

Fig. 19 shows Raman spectra of our disordered and one of an ordered garnet. Their difference is obvious.

At least some luminescence and stimulated-emission-intensity characteristics of the disordered calcium-niobium-gallium garnet with Nd^{3+} ions at 300K are listed in Table 14 and 15.

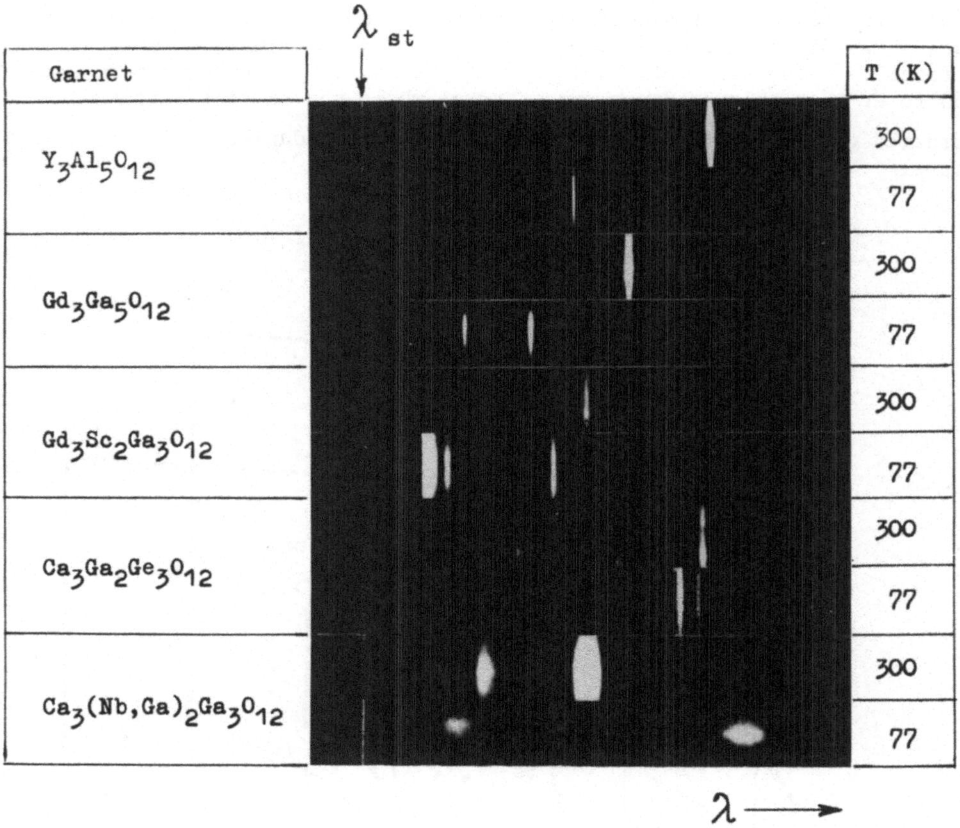

Fig. 18. Stimulated-emission spectra of Nd^{3+} ions ($^4F_{3/2} \rightarrow {}^4I_{11/2}$ channel) in four popular laser garnets with ordered structure and in disordered $Ca_3(Nb, Ga)_2Ga_3O_{12}$ garnet at 77 and 300K. Standard line λ_{st}=1.0561 μm is indicated by an arrow.

III.C. Intermediate Conclusion

Completing the first part of the paper, it is necessary to emphasize, that on the basis of observed new fundamental physicochemical properties of compound formation in ternary oxide systems, that is to say the entering of heterovalent cations into similar crystallographic positions, we are the first to obtain and identify several new types of the disordered laser crystals. This identification of heterovalent species in a similar crystalographic position gives substantial insight into the problem of disordered laser garnet, investigations of which have been a subject of interest in numerous laboratories all over the world during the last 20 years.

A variety of crystal-structure types, of course, does not exclude the possibility of crystal-field disorder of La^{3+} activators only on the basis of structural and phisicochemical objective laws. This phenomenon can be stipulated also by their simultaneous influence, which e.g., takes place (with different contribution) in some considered here disordered laser crystals.

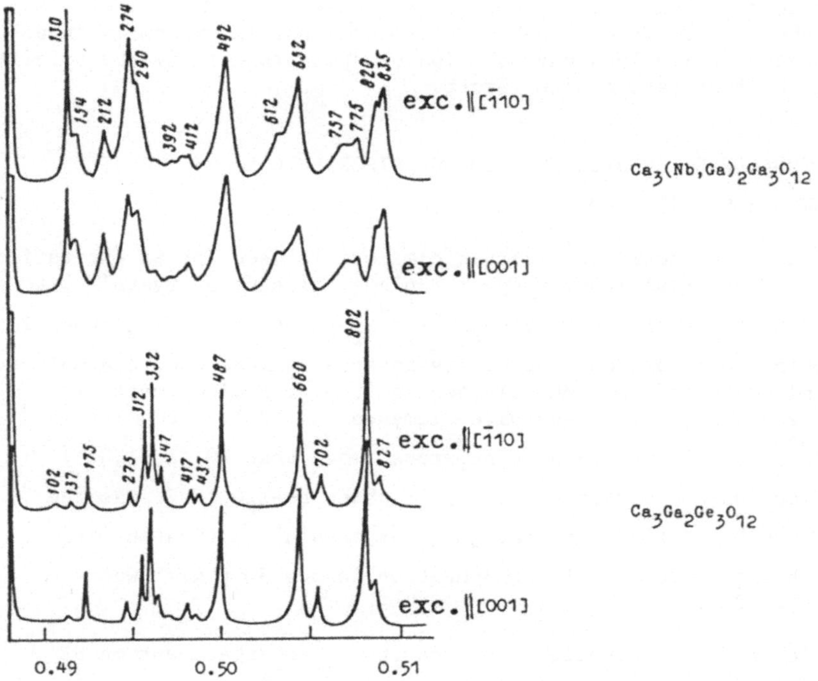

Fig. 19. Raman spectra of garnets with ordered ($Ca_3Ga_2Ge_3O_{12}$) and disordered (Ca (Nb, Ga)$_2$Ga$_3$O$_{12}$) structure at 300K. Peak frequencies above lines are given in cm^{-1}

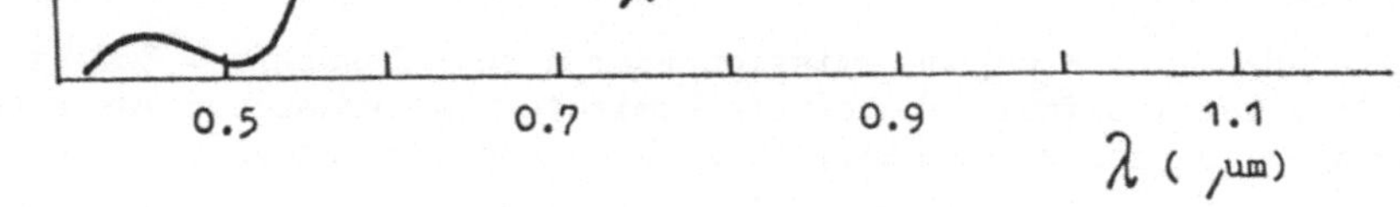

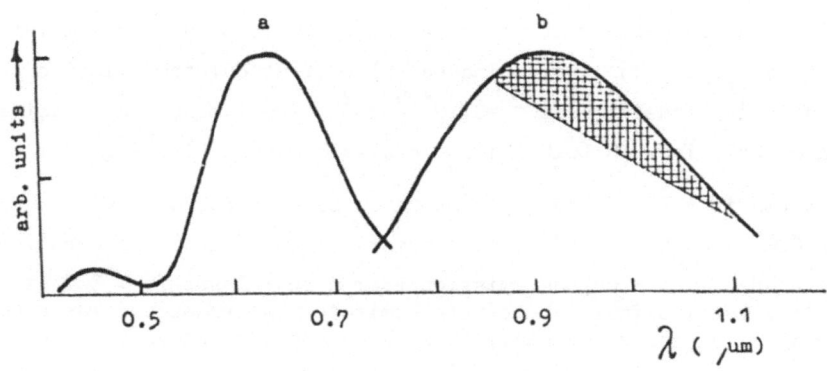

Fig. 20. Exciatation (a) and luminescence (b) spectra ($^4T_2 \rightarrow {}^4A_2$ channel) of Cr^{3+} ions in $La_3Ga_5SiO_{14}$ crystal at 300K. The shading indicates the continuous-tuning region for SE wavelength.

Special attention should be paid to the investigation of crystal-field disorder at transition-metal ions, which in comparison with Ln^{3+} activators characterized by medium crystal field and more strong covalent bonding and electron-phonon coupling. In this task the influence of disorderness on the properties of tunable SE of such ions is extremely interesting. An investigation into the solution of this problem will be briefly considered in the last part of the report.

IV. BROADLY TUNABLE STIMULATED EMISSION OF DISORDERED $La_3Ga_5SiO_{14}:Cr^{3+}$ SINGLE CRYSTALS [35]

Among a number of activated inorganic materials used for SE excitation of Ln and transition-metal ions, a special place belongs to crystals with Cr^{3+} ions, which at electron-vibrational ${}^4T_2 \rightarrow {}^4A_2$ transition. The specific features of this broad-band and reasonably intense luminescence channel permit the achievement of continuously tuning SE over a wide spectral region using resonators with dispersive elements [31,32]). Such laser-action from Cr^{3+} ions is realized in numerous (more than 10 [1,31,32]) oxygen- and fluorine-containing crystals, of which $BeAl_2O_4:Cr^{3+}$ was the first [33]. Possible methods of searching for crystals activated with Cr^{3+} ions which have a potential for operation in the SE regime was analyzed in ref. [34].

Here, briefly let us consider some results of an experiment on SE spectroscopy in the near infrared of a $La_3Ga_5SiO_{14}:Cr^{3+}$ single crystal [35]. The information on its luminescence is given in ref. [1]. As mentioned above, this crystal activated with Nd^{3+} ions was observed to show SE excitation [6].

The spectroscopic investigations show that Cr^{3+} ions in $La_3Ga_5SiO_{14}$ are in the weak octahedral crystal field ($Dq/B \simeq 2.1$) and the observed infrared luminescence is due to a spin allowed ${}^4T_2 \rightarrow {}^4A_2$ transition (Fig. 20). The room temperature luminescence lifetime of its initial ${}^4T_2$ state is $\tau_{lum} \simeq 6$ μs, and estimated value of stimulated-emission cross-section of said transition is $\sigma_e = 3.10^{-19}$ cm^2. The luminescence excitation spectrum of Cr^{3+} ions in $La_3Ga_5SiO_{14}$ is also shown in Fig. 20.

Reported in ref. 35 are experiments on SE spectroscopy of a 3cmx4cmx20cm $La_3Ga_5SiO_{14}:Cr^{3+}$ single crystal ($2x10^{-3}$ atoms of chromium to formula unit). The pulsed emission from a ruby laser ($\tau \simeq 300$ μs with spikes) in a longitudinal scheme in a semispherical resonator was used for pumping. The garnet crystal, in such a geometry the absorption of the lasing bars from $La_3Ga_5SiO_{14}:Cr^{3+}$ at 0.6943 μs was about 85%. The SE excitation threshold was not higher than 0.05 J when using an output flat mirror with 40% transmission. The laser action with a 60% transmittance output mirror was also obtained, which confirms a high σ_e value for the ${}^4T_2 \rightarrow {}^4A_2$ channel of the $La_3Ga_5SiO_{14}:Cr^{3+}$ single crystal. Using a special prism selector inside the resonator continuous broad-band tuning of SE was realized from 0.86 to 1.1 μm at 300K. In Table 16 are some spectral-lasing parameters of this $La_3Ga_5SiO_{14}:Cr^{3+}$ crystal, and several other Cr^{3+}-doped compounds.

Now some words about the atomic structure of $La_3Ga_5SiO_{14}$ doped with Cr^{3+} ions. Let us return to Fig. 9. Chromium ions in $La_3Ga_5SiO_{14}$ crystal occupy the la positions with site symmetry 32. The shortest Cr-Cr distances in this structure occur along the a-axis (8.168A°) and c-axis (5.095 A°), and the Cr-O distance is 2.00A°. Nearest neighbours of Cr^{3+} ions are connected with it through common oxygen ions, Ga^{3+} in oxygen 3f-tetrahedra (Cr-Ga distance is 3.189 A°) and La^{3+} ions in the Thomson cubes 3e (Cr-La is 3.420 A°). In the second coordination spheres are tetrahydral heterovalent Ga^{3+} and Si^{4+} ions occupying 2d crystallographic positions in the 1:1 ratio (Cr-Ga, Si is 5.282 A°) resulting, as mentioned above, in disorderness of the crystal field at Nd^{3+} ion sites. From what has been said and from what is shown in Fig. 9 it follows that in $La_3Ga_5SiO_{14}$ crystal the influence of the statistical filling of 2d positions by heterovalent Ga^{3+} and Si^{4+} cations on the fluctiation of the crystal field at Cr^{3+} ions, separated from them by two oxygen atoms, is sufficiently weaker than in the case of Nd^{3+} ions. Preliminary measurements (using site-selective laser spectroscopy) show that the vibronic band of the luminescence from $^4T_2 \rightarrow ^4A_2$ channel is generally homogenously broadened by its nature.

By this means, the laser action of the vibronic transition of Cr^{3+} ions in a $La_3Ga_5SiO_{14}$ single crystal having the Ca-gallogermanate structure was excited and investigated at 300K for the first time. In our experiments a record spectral interval of continuous tuning of SE for insulating laser crystals with octahedral Cr^{3+} ions was achieved.

CONCLUSIONS

The systematization and analysis of accumulated knowledge of the spectroscopic properties and SE of activated condensed media permit an understanding of the phenomenon of crystal-field disorderness at Nd^{3+} and Cr^{3+} ions in the insulating compounds. This new understanding addresses some searching problems in laser-crystal physics, and introduces an entirely new class of disordered solid-state-laser crystals.

I should like to add also, that all results presented here were obtained on the basis of data of our fundamental investigations of the properties and processes occuring in the activated laser crystals, as well as on the basis of physico-chemical and growth studies in search of new compounds.

As for a deeper understanding of the physics and spectroscopy of laser disordered crystals, I am looking forward with optimism. Undoubtedly in the future scientists of different centers from many countries will get more interesting results than those I explained in my report.

Table 16. Spectral and stimulated-emission parameters of $La_3Ga_5SiO_{14}$ and other oxygen- and fluoride-containing crystals with Cr^{3+} ions ($^4T_2 \rightarrow {}^4A_2$ channel) at 300K

Crystal	Spectral range of continuous tuning of SE (μm)	Luminescence lifetime (μs)	SE cross-section (10^{-20} cm^2)	$\Delta E^{*)}$ (cm^{-1})	Reference**)
$La_3Ga_5SiO_{14}$	0.86 -1.1	~ 6	~30	-1800	[35]
$KZnF_3$	0.78 -0.867	180	1.3	-1000	[36]
$Gd_3Sc_2Ga_3O_{12}$	0.742-0.842	115	~ 1	50	[37]
$Be_3Al_2(SiO_3)_6$	0.728-0.809	70	1.8	400	[38]
$BeAl_2O_4$	0.7 -0.815	~260	~ 3	800	[39]

*) $\Delta E = E(^4T_2) - E(^2E)$.

**) Only pioneer journal publications are indicated.

ACKNOWLEDGEMENTS

The author expresses his thanks to all the co-authors of past publications dealing with the problem of disordered laser crystals.

REFERENCES

1. A.A. Kaminskii, L.K. Aminov, V.L. Ermolaev, A.A. Kornienko, V.B. Kravchenko, B.Z. Malkin, B.V. Mill, Yu. E. Perlin, A.G. Petrosyan, K.K. Pukhov, V.P. Sakun, S.E. Sarkisov, E.B. Sveshnikova, G.A. Skripko, N.V. Starostin, and A.P. Shkadarevich, Fizika i Spektroskopia Lasernykh Kristallov, Nauka, Moskva 1986.
2. A.A. Mak, V.A. Fromzel, A.A. Shcherbakov, V.M. Volynskii, V.E. Gavrilov, V.A. Gerasimov, V.M. Gradov, V.I. Zhitsov, G.I. Kramskii, A.G. Mursin, and L.K. Sukhareva, Optiko-mechanicheskaya promyshlennost, N1, 58 (1983).
3. R. Beck and K. Gurs, J. Appl. Phys., 46, 5224 (1975).
4. A.A. Kaminskii, Dokl. Akad. Nauk SSSR, 211, 811 (1973).
5. A.A. Kaminskii, Lasernye Kristally, Nauka, Moskva 1975.
6. A.A. Kaminskii, S.E. Sarkisov, B.V. Mill, and G.G. Khodzhabagyan, Dokl. Akad. Nauk SSSR, 274, 93 (1982).
7. A.A. Kaminskii, E.L. Belokovena, B.V. Mill, S.A. Tamasyan, and K. Kurbanov, Izv. Akad. Nauk SSSR, Ser. neorgan. Materials, 22, 1138 (1986).
8. A.A. Kaminskii, B.V. Mill, E.L. Belokoneva, A.V. Butashin, S.E. Sarkisov, K. Kurbanov, and G.G. Khodzhabagyan, Izv. Akad. Nauk SSSR, Ser. neorgan. Materials, 22, 1869 (1986).
9. A.A. Kaminskii, E.L. Belokoneva, A.V. Butashin, S.E. Sarkisov, and O.N. Nikolskaya, Izv. Akad. Nauk SSSR, Ser. neorgan. Materials, 21, 2093 (1985).
10. A.A. Kaminskii, E.L. Belokoneva, A.V. Butashin, K. Kurbanov, A.A. Markosyan, B.V. Mill, and O.K. Nikolskaya, Izv. Akad. Nauk SSSR, Ser. neorgan. Materials, 22, 1061 (1986).
11. A.A. Kaminskii, Laser Crystals, Their Physics and Properties, Springer-Verlag, Berlin/Heidelberg/New York 1981.
12. V.M. Garmash, A.A. Kaminskii, M.I. Polyakov, S.E. Sarkisov, and A.A. Filimonov, Physica Status Solidi (a), 75, K111 (1983).
13. A.A. Kaminskii, Zh.eksper. teor. Fiz., 58, 407 (1970).
14. V.A. Aleksandrov and L.S. Garashina, Dokl. Akad. Nauk SSSR, 189, 307 (1969).
15. M.N. Tolstoi and V.N. Shapavalenko, Optika i spektroskopia, 23, 648 (1967).
16. A.A. Kaminskii, Zh. eksper. teor. Fiz., 54, 727 (1968).
17. A.A. Kaminskii, L.S. Kornienko, L.V. Makarenko, A.M. Prokhorov, and M.M. Fursikov, Zh. eksper. teor. Fiz., 46, 386 (1964).
18. A.A. Kaminskii, L.S. Kornienko, and A.M. Prokhorov, Zh. eksper. teor. Fiz., 48, 476 (1985).
19. N.E. Alekseev, V.P. Gapontsev, M.E. Zhabotinskii, V.B. Kravchenko, and Yu. P. Rudnitskii, Lazernye Fosfatnye Stekla, Nauka, Moskva 1980.
20. A.A. Kaminskii, B.V. Mill, G.G. Khodzhabagyan, A.F. Konstantinova, A.I. Okorochkov, and I.M. Silvestrova, Physica Status Solidi (a), 80, 387 (1983).
21. A.A. Kaminskii, I.M. Silvestrova, S.E. Sarkisov, and G.A. Denisenko, Physica Status Solidi (a), 80, 607 (1983).
22. A.A. Kaminskii, E.L. Belokoneva, B.V. Mill, Yu. V. Pisarevskii, S.E. Sarkisov, I.M. Silvestrova, A.V. Butashin, and G.G. Khodzhabagyan, Physica Status Solidi (a), 86, 345 (1984).
23. A.A. Kaminskii, S.E. Sarkisov, B.V. Mill, G.G. Khodzhabagyan, Izv. Alad. Nauk SSSR, Ser. neorgan. Materials, 18, 1396 (1982).

24. A.A. Kaminskii, B.V. Mill, E.L. Belokoneva, G.G. Khodzhabagyan, Izv. Adak. Nauk SSSR, Ser. neorgan. Materials, 19, 1762 (1983).
25. A.A. Kaminskii, B.V. Mill, E.L. Belokoneva, S.E. Sarkisov, T. Yu. Pastukhova, and G.G. Khodzhabagyan, Izv. Akad. Nauk SSSR, Ser. neorgan. Materials, 20, 2058 (1984).
26. A.A. Kaminskii, K. Kurbanov, A.A. Markosyan, B.V. Mill, S.E. Sarkisov, and G.G. Khodzhabagyan, Izv. Akad. Nauk SSSR, Ser. neorgan. Materials, 21, 1970 (1985).
27. R. Reisfeld and C.K. Jorgensen, Lasers and Excited States of Rare Earths, Springer-Verlag, Berlin/Heidelberg/New York 1977.
28. A.A. Kaminskii, E.L. Belokoneva, B.V. Mill, S.E. Sarkisov, and K. Kurbanov, Physica Status Solidi (a), 97, 279 (1986).
29. M. Alam, K.H. Gooen, B. DiBartolo, A. Linz, E. Sharp, L. Gillespie, and G. Janney, J. Appl. Phys., 39, 4728 (1968).
30. D.J. Horowitz, L.F. Gillespie, J.E. Miller, E.J. Sharp, J. Appl. Phys., 43, 3527 (1972).
31. Tunable Solid State Lasers, Editors: P. Hammerling, A.B. Budgor, and A. Pinto, Springer-Verlag, Berlin/Heidelberg/New York/Tokyo 1985.
32. Tunable Solid State Lasers for Remote Sensing, Editors: R.L. Byer, E.K. Gustafson, and R. Trebino, Springer-Verlag, Berlin/Heidelberg/New York/Tokyo 1985.
33. J.C. Walling, O.G. Peterson, H.P. Jenssen, R.C. Morris, and E.W. O'Dell, IEEE J. Quantum Electronics, 16, 1302 (1980).
34. P.T. Kenyon, L. Andrews, B. McCollum, and A. Lempicki, IEEE J. Quantum Electronics, 18, 1189 (1982).
35. A.A. Kaminskii, A.P. Shkadarevich, B.V. Mill, V.G. Koptev, and A.A. Demidovich, Izv. Akad. Nauk SSSR, Ser. neorgan. Materials, 23, (1987).
36. U. Brauch and U. Durr, Optics Communications, 49, 61 (1984).
37. B. Struve, G. Huber, V.V. Laptev, I.A. Shcherbakov, and E.V. Zharikov, Appl. Phys. B., 28, 235 (1982).
38. M.L. Shnad and J.C. Walling, IEEE J. Quantum Electronics, 18, 1829 (1982).
39. J.C. Walling, H.P. Jenssen, R.C. Morris, E.W. O'Dell, and O.G. Peterson Optics Letters, 4, 182 (1979).

FRACTAL BEHAVIOUR OF POROUS GLASSES

Renata Reisfeld*

Department of Inorganic and Analytical Chemistry
The Hebrew University of Jerusalem
Jerusalem 91904 Israel

ABSTRACT

Fractal geometry provides a measure of randomness and permits characterization of porous materials and systems having rough surfaces. Fractal objects show dilation symmetry implying that the essential geometric features are invariant to scale changes such as magnification factors. Theoretical background and experimental methods for studying fractal structure such as low-angle neutron and X-ray scattering, adsorption and energy transfer between donor and acceptor molecules embedded in fractal structures are presented. It will be shown that results obtained today by various methods provide different answers about the fractal dimensionality. A proposition is made for unifying the different methods.

INTRODUCTION

Fractal geometry describes objects which show dilation symmetry, the geometrical features of which are invariant with scale changes. In other words the same picture should be obtained at very low or very high magnifications. Fractal objects may be very small with dimensions of molecules, 1 - 10 Å, or very large, many meters or even kilometers, like coast lines or mountain ranges.

Fractal geometry [1] provides a quantitative measure of randomness and irregularity of surfaces and thus permits characterization of random systems such as polymers [2], colloidal aggregates [3], rough surfaces [4,5] and porous materials [6,7].

The fractal dimension expresses the degree of irregularity which is $1 \leq D < 2$ for lines and $2 \leq D < 3$ for surfaces.

At least three different dimensions are required to define a fractal [8,9]. The first is the Euclidean dimension, d, in which the structure is embedded. The second is called the fractal dimension, D [1,10]. This describes the dependence of the number of sites N(R) on the distance R, through the relation $N(R) = R^D$. The third dimension is the spectral or fracton dimension $\tilde{d}$ [8,9], which governs the random walk and relaxation processes and determines the density of states of the

* Enrique Berman Professor of Solar Energy

structure. The spectral dimension has been previously discussed in electron-spin relaxation studies in proteins [11,12] and triplet-triplet annihilation studies in mixed molecular crystals [13,14].

In a series of papers, Alexander, Orbach and collaborators have developed a general theory for the excitations, or "fractons" in fractal systems [8,15-18]. They introduced a fractal dimension, $\tilde{d}$, related to the static Hausdorff dimension D through the relation [8]

$$\tilde{d} = \frac{2D}{2 - \theta} \tag{1}$$

where θ is the diffusion constant scaling exponent.

It can be shown that the fracton dimension, d, is always smaller than D and that, in general, the following double inequality applies [19]

$$d \geq D \geq \tilde{d} \tag{2}$$

where d is the dimension of the Euclidian space.

Fracton theory applies in principle to glassy and disordered materials because in these systems there is a characteristic length scale larger than the lattice parameters. Anomalies in the density of the vibrational states of glasses can be eventually understood within the fracton theory. This has been done for the interpretation of low-temperature thermal conductivity and heat capacity data of epoxy resin [20]. Results for the low frequency vibrations in vitreous silica, obtained by inelastic neutron scattering [21], can also be interpreted within this theory.

Studies of one-step electronic energy transfer have been discussed, both theoretically [22] and experimentally [6,23-25], in terms of the fractal dimension.

Porous materials and aggregates frequently display self-similarity. Their fractal structure is associated with power-law density-density correlation function. Within some units the system is well-described by a fractal or Hausdorff dimension.

In a general way the fractal or Hausdorff dimension D is defined by

$$N(r) \propto (r/r_o)^D \tag{3}$$

where N(r) is the quantity obtained by measuring a fractal medium with a gauge r_o. For instance N(r) can be the number of particles of radius r_o which lie within a sphere of radius r centered on an arbitrary particle.

Depending on the characteric size of the fractal object different experimental methods have to be used. For an experimentalist the determination of the fractal dimension will be connected with some observable lengths' scale.

The main experimental methods used for studying fractal geometry are:

1. Adsorption of molecules on surfaces

In this method the surfaces are covered by a monolayer of molecules of adsorbent with cross-sectional area σ or radius r. The relation between the number of moles per gram of adsorbent m, σ or r, may be expressed by [23,26,27]

$$m \propto \sigma^{-D/2}$$

or $m \propto r^{-D}$

This is based on a general finding that a simple scaling law relates the size of the molecule (the yardstick σ) to the apparent monolayer value, m

$$m = k\sigma^{-D/2} \tag{4}$$

where k is a units constant and D/2 is a characteristic exponent which carries information on the degree of irregularity which is higher the higher D is. Equation (4) indicates simply that for a given surface fewer molecules are needed to form a monolayer as the size of the molecule increases. From equation (5):

$$A = N\sigma m \tag{5}$$

where A is the apparent or effective surface area obtained from a monolayer value m (mol of adsorbent/g adsorbate) of a molecule with cross sectional area σ and N is Avogadro's number, one obtains

$$A = k\sigma^{(2-D)/2} \tag{6}$$

In this relation D has been developed to deal with the connection between geometric parameters such as length, area and the size of the yardstick used to determine these parameters. When only physical adsorption takes place D has the meaning of dimension.

The intrinsic property of a fractal surface D, can be also obtained from equation (7)

$$n = cR^{-D} \tag{7}$$

Here a hypothetical experiment determines the number n of spheres of radius R needed to cover a surface with a monolayer [4,10 p.30] where c and D, the fractal dimensions, are quantities which do not depend on R but are characteristic of the particular surface. The meaningful values of the fractal dimension D for a surface lie in the interval $2 < D < 3$. When $D = 2$, the surface is smooth, while fractal for $D > 2$.

Experimentally the fractal dimensions c and D are obtained from the power-law expressed in equation (7).

When chemisorption takes place the limits of $2 \leq D < 3$ do not hold since in chemical interactions the selectivity of adsorption is governed not only by the fractal geometry but by the chemical heterogeneity of the surface [28].

2. Small-angle scattering of X-rays or neutrons

Small-angle scattering of X-rays or neutrons [29-31] is a useful technique for learning about the structures of porous materials on a scale from about 5 Å through 2000 Å. These lengths range from distances slightly larger than the diameters of single atoms to distances which are almost large enough to be resolved in an ordinary optical microscope. Since this interval of lengths includes much of the structure in porous solids small-angle scattering can play an important role in the study of porosity. In this method the decrease of scattered electromagnetic radiation intensity is measured as a function of scattering angle [5,29].

Experimentally X-rays or neutrons, at angle θ with respect to the incident beam, are formed into a fine beam, usually by slits, and fall on the sample. A small fraction of the beam is scattered (i.e. reemitted) in directions different from that of the incident beam at angle θ and is often defined by additional slits and registered by a detector.

The intensity of the scattered beam and the way in which the intensity varies with the scattering angle θ are determined by the structure of the sample. One therefore can obtain information about the sample structure by analysis of the scattered intensity measured at a series of scattering angles. The scattering angle θ can be conveniently described by the quantity [32]

$$q\ell = 4\pi \ell/\lambda \sin(\theta/2) \tag{8}$$

where q is the momentum transfer, λ is the wavelength of the X-rays or neutrons employed in the experiment and ℓ the characteristic scattering length. When θ does not exceed a few degrees,

$$q\ell = 2\pi \ell \theta/\lambda \tag{9}$$

In a scattering measurement the scattered intensity I(q) is measured for several values of q. From an analysis of these data an attempt is made to learn about the structure of the sample on the scale of lengths accessible to small angle scattering [29].

X-rays are scattered by electrons [32] while the atomic nuclei are responsible for the type of neutron scattering [33]. The force exerted by a nucleus on a neutron during the scattering process is described by a quantity which is known as the scattering length and which depends on the particular isotope which causes the scattering. Although by a suitable choice of the isotopes in the sample, certain features of the sample structure can be emphasized or masked, in the scattering studies of porous solids X-ray and neutron scattering give essentially the same results.

Since the scattering at small angles is not affected by atomic-scale structure, an approximation known as the "two-phase" approximation can be used [34] in the analysis of most small-angle scattering curves. In this approximation the sample is considered to consist of two phases each of which has a constant density on a scale of lengths from about 5 Å to 2000 Å. In porous solids one of these phases is the solid material and the other phase consists of empty pores. In the two-phase approximation the intensity of the small-angle X-ray scattering is proportional to the square of the difference of the electron densities of the two phases [32]. For neutrons the intensity scattered at small angles is proportional to the square of the difference of the scattering-length densities [33]. With a few exceptions the boundary separating the two phases can be considered to be infinitely sharp.

The scattering intensity is given by [5]

$$I(q) = \pi c\delta^2 I_e \Gamma(5-D)\sin[\pi(D-1)/2]\, q^{-(6-D)} \tag{10}$$

where δ is the difference between the electron or scattering length densities of the two phases, I_e is the scattering intensity per electron or per unit scattering length (for X-rays or neutrons, respectively), and Γ(5-D) is a gamma function.

According to (10), when ql >> 1, I(q) is proportional to $q^{-(6-D)}$. Thus, when (10) describes the scattering, a plot of the logarithm of q will be a straight line, and the fractal dimension D can be calculated from the slope of this line.

The X-ray and neutron small-angle scattering curves for two kinds of controlled-pore glasses, Vycor(R) 7930 "Thirsty Glass" (Corning Glass Works), and, Electro-Nucleonics (USA) cpg-10 in which, according to the manufacturer, the average pore sizes were 75, 170, 500, 700, 1000 and 2000 Å, were obtained by Schmidt [29] whose curves for Vycor(R) 7930 are in good agreement with the neutron system at Oak Ridge National Laboratory where data could be recorded for smaller q than was possible with the X-ray apparatus. With X-rays, however, it was possible to measure the scattering at somewhat larger q values than those accessible with the neutron system.

There are some features of the Vycor(R) 7930 scattering curve which should be pointed out. The pronounced maximum at about $q = 0.25\ A^{-1}$ suggests that there is a rather highly ordered structure in the glass. More evidence for this ordering comes from the small shoulder near $q = 0.05\ A^{-1}$. This shoulder is easily visible in the X-ray scattering curve, although it is harder to detect in the neutron scattering curve, because the inner and outer parts of the neutron curve were recorded separately, and the shoulder falls in the region where the two sets of the scattering data overlapped when they were combined.

The power-law (linear) region in the outer part of the scattering curve is large enough to permit a reliable evaluation of the exponent of the power-law scattering. From this exponent, calculation of the fractal dimension D of the pore surfaces gives 2.40 ± 0.06. During the last few years, several people have discussed [35,36] small-angle scattering determinations of the fractal dimension of Vycor(R) 7930. The value of D obtained [20] agrees with that of Sinha et al. [36] in a very recent small-angle scattering investigation of Vycor(R) 7930. The data which Schmidt and Sinha et al. have obtained suggest that the pore surfaces in Vycor(R) 7930 are fractal on the scale of lengths corresponding to the interval of q for which the scattering curves follow a power law. A rough, order-of-magnitude estimate of these lengths can be obtained from the condition that the scattering at a given value of q is associated with a distance π/q. Since the curve is linear for q between approximately 0.06 and $0.2\ A^{-1}$, it is concluded that the pore surfaces have a fractal dimension 2.40 ± 0.06 on a length scale from about 50 Å down to at least 15 Å.

The scattering curves for the cpg-10 controlled-pore glasses with pore sizes of 75 and 170 Å have a form similar to that obtained for Vycor(R) 7930. However, because the ratio of the q values corresponding to the positions of the maxima in these curves is not equal to the reciprocal of the ratio 75/170 of their pore sizes, the scattering curves for the controlled-pore glasses cannot be combined into a single curve. For all members of this series, the exponents of the power-law scattering in the outer parts of the curves give a fractal dimension D = 2.20 ± 0.05. Though this value of D is obtained for all of the Electron-Nucleonics controlled-pore glasses, it is different from the fractal dimension of the pore surfaces of Vycor(R) 7930.

The interpretation of scattering curves for silica gels obtained by the sol-gel method by fractal geometry was also made by Schaefer and Keefer [7]. Scattering curves of polymer-like fractals were prepared by hydrolysis followed by polycondensation of TEOS (tetraethoxysilane). Depending on the ratio of water to the silicon and to the type of catalyzer used different fractal geometries are obtained from the scattering experiments.

Similarly to the work of Schmidt [29] a power-law dependence of the form

$$I(q) \sim q^{-x} \qquad (11)$$

was used in order to interpret the scattering curves of various silica gels [7].

According to [7] the Porod exponent x depends on the origin of the scattering.

For so-called mass fractals (i.e. polymer-like structures) the exponent is simply D, the fractal dimension which relates the size R of the object to its mass M,

$$M \sim R^{D} \qquad (12)$$

For a polymer-like fractal object with a 1-dimensional backbone, $1 \leq D \leq 3$, depending on the degree of branching and folding. For a sheet-like fractal object, $2 \leq D \leq 3$, where D is greater than two for branched and tortuous structures.

For scattering from 3-dimensional objects with fractal surfaces [5]

$$x = 6 - D_s \qquad (13)$$

where D_s is the fractal dimension of the surface ($2 \leq D_s \leq 3$). $D_s = 2$ represents a classical smooth surface. Finally, for fractally porous [4,5,28] materials, $x = 7 - \gamma$, where γ is the exponent describing the distribution P(r), of pores of radius r

$$P(r) \sim r^{-\gamma} \qquad (14)$$

The Porod plots of scattering data for silicate polymerized under various conditions are as follows [7];

Two step TEOS acid catalyzed has a slope of	-1.9
Two-step base catalyzed	-2.0
Single-step base catalyzed W = 1	-2.8
Single-step base catalyzed W = 2	-3.3
Ludox SM	-4.0

where W is the stoichiometric ratio between concentration of water to concentration of TEOS.

3. Energy transfer in fractal structures

Since porous materials, silica gel surfaces [37], Vycor glass [6] and sandstone rocks [38] have fractal character, the dynamics of excitations is expected to be influenced by the self-similar nature of these structures.

Energy transfer from donors to randomly distributed acceptors has been considered to depend on the fractal dimension of the host [22]. In general the energy transfer in a regular system is reflected in the survival probability function of the donor which is given by [39]:

$$d/dt(P_n(t)) = -\{A_R+X_n+\Sigma_{n'\neq n} W_{nn'}\}.P_n(t)+\Sigma_{n'=n} W_{n'n}.P_{n'}(t) \qquad (15)$$

where:

$P_n(t)$ - probability that donor n is in the excited state at time t

A_R^{-1} - radiative lifetime of the donor

X_n - total transfer rate to acceptor species or traps

$W_{n'n}$ - energy transfer rate from donor n to donor n'

$W_{nn'}$ - back transfer form donor n' to donor n.

If the donor ion or molecule is fluorescent the integrated intensity at time t is proportional to the number of excited donors and the survival probability of donors is expressed as:

$$P(t) = \exp[-A_R.t].f(t) \qquad (16)$$

$$P(t) = \frac{N_D(t)}{N_D(0)} \qquad (17)$$

where $N_D(0)$ is the number of excited donors present at the time the excitation source is turned off and f(t) denotes the fraction of excited donors that would be present were the excited state to have an infinite radiative lifetime.

An exact solution for f(t) based on the rate equation is possible in two limiting cases in the Euclidian space:
-A- No donor-donor transfer
-B- Infinitely rapid donor-donor transfer

In the former case we have:

$$f(t) = \prod_{\ell}(1-C_A+C\exp[-X_{o\ell}t]) \qquad (18)$$

which is a generalization of a result first obtained by Inokuti and Hirayama [40] to all values of C_A. $X_{o\ell}$ denotes the transfer rate from a donor at site 0 to an acceptor at site ℓ. C_A is the probability that the site is occupied by the acceptor.

In the other extreme it is assumed that all donors interact with each other rapidly so that for t>0 all donors have equal probability of being excited.

In such a case f(t) becomes:

$$f(t) = \exp[-C_{A\ell}\Sigma X_{o\ell}t] \qquad (19)$$

In the intermediate region a diffusion model first proposed by Yokota and Tanimoto [41] is appropriate:

$$\partial/\partial t\ (N_D(r,t)) = D.\nabla^2 N_D(r,t) - \Sigma_j X(r-r_j)\ .\ (N_D(r,t) \qquad (20)$$

where:

D = diffusion constant (for donor-donor transfer)
$X(r-r_j)$ = Transfer rate from donor at site r to acceptor at site r_j.

In the limit of long times the diffusion model predicts that f(t) varies as:

$$f(t) \sim \exp[-\Lambda_D t]$$

where: $\Lambda_D = 4\pi D N_A a_S$

and a_S is defined:

$$a_S = (\alpha/D)^{1/(S-2)}.(S-2)^{-2/(S-2)}.\Gamma(1-1/(S-2)).\Gamma(1+1/(S-2))^{-1} \qquad (21)$$

where the subscript S is the order of multipolar interaction; S = 6 for dipole-dipole interaction of the form α/r^S.

Experimentally the diffusion constant can be obtained from the semilog plot of survival probability [39].

The diffusion of energy on fractal structures was recently approached by continuous time random walk (CTRW) method by Blumen et al. [42,43].

Two limiting cases were considered for energy transfer on fractal structures, the direct, and the indirect, multistep, energy transfer. In the direct transfer case an excited donor transfers its energy directly to randomly distributed acceptors. One is then interested in the ensemble averaged decay law of the excited donor. The decay law can be expressed exactly for acceptors randomly distributed in a fractal if the back-transfer to the donor is unimportant. The indirect transfer problem, where the excitation randomly hops in many steps over the donor system until encountering an acceptor, is more complicated.

In this second mechanism the energy transfer is governed by a multistep transfer where the excitation migrates among the fractal sites (donor-donor migration) until encountering an acceptor. For the indirect case the survival probability is determined from the number of distinct sites visited by the excitation in acceptor-free fractals and the short time behavior and the long time tails of the survival probability are governed by the spectral dimension d. The interpolation between these two limits is done by introducing a scaling law. Thus, for each transfer mechanism a different fractal dimension is decisive.

For energy transfer on fractals it is necessary to calculate the survival probability of the excitation in the framework of random walk theory in terms of the number of distinct sites visited. Unlike in the situation in regular lattices where the dimension d governs both mechanisms, in fractal structures the direct transfer depends on the fractal dimension D, while the multistep transfer is given in terms of the spectral dimension $\tilde{d}$.

When the donor and the acceptor ions present at low concentrations are embedded in a fractal the survival probability P(t) of the excited state energy of the donor, at time t is given by [22]

$$P(t) = \exp[-\gamma(t/\tau)^{\beta} - (t/\tau)] \text{ for direct energy transfer} \tag{22}$$

where

$$\beta = D/s \tag{23}$$

τ is the radiative decay lifetime of the donor and s is the order of the multipolar EET rate w(R) R^{-s}. The proportionality factor γ is

$$\gamma = \chi_A(d/D)\Gamma(1-D/s)(R_o/a)^D \tag{24}$$

where χ_A is the fraction of the fractal sites occupied by the acceptors ($\chi_A << 1$), a is an (average) size of a unit cell, and R_o is the "critical" radius for EET, corresponding to $w(R_o)\tau - 1$. As is apparent from eqs. (22) and (23), direct unistep EET on a fractal is determined by the fractal dimensionality.

For multipolar interactions [22]:

$$f(t) = \exp(-\chi pAt^{D/s}) \tag{25}$$

where A is time-independent and p is the probability of finding the acceptor in the fractal site. Eq. (25) is an extension of a known result for Euclidean dimension d to the fractal dimension D. Recently it has been used in analyzing time resolved measurements of energy transfer from rhodamine B to malachite green molecules doped into a porous glass. These measurements resulted in a fractal dimension d = 1.74 for the void structure in the porous glass [6].

EXPERIMENTAL MEASUREMENTS OF ENERGY TRANSFER ON FRACTALS

The first experiment of energy transfer on fractals was performed on a porous glass [6,25]. In this experiment time-resolved picosecond spectroscopy was utilized to study energy transfer from rhodamine B to malachite green incorporated in porous glass Vycor(R) 7930.

Equation (22) connecting the survival probability with the fractal dimension D was applied for interpretation of the results.

This Vycor glass which is prepared by leaching of borosilicate glass constitutes a porous open structure (effective surface area A = 200 m^2/g; density ρ = 1.5g cm^{-3}). The average pore diameter is r_v = 40 ± 3 Å. The volume fraction of the pores is 0.28, which exceeds the critical volume fraction ϕ_c = 0.16 ± 0.02 for continuous percolation in three dimensions. The pores form topologically connected, intermeshed, random paths in the porous glass, whose stochastic geometry characteristics are amenable to description in terms of a self-similar structure. Disks of Corning Vycor 7930 porous glass (14.5 mm diameter and 1.4 mm thickness) were cut from a glass rod [6]. The disks were dipped for 48 h into methanol solutions containing rhodamine B or rhodamine B and malachite green and subsequently dried for 48 h in a vacuum at 30°C until a pressure of 10^{-4} Torr was attained. The effective concentrations $\bar{C}_D$ of rhodamine B and $\bar{C}_A$ of malachite green (per unit volume of the glass) were determined from the change of the dye content of the solutions. These effective concentrations were varied in the range $\bar{C}_D$ = (3 x 10^{-5} - 6 x 10^{-5})M and $\bar{C}_A$ = (1.9 x 10^{-4} - 5.1 x 10^{-4})M. The porous glass was uniformly doped, as indicated by optical absorption measurements as well as by optical bleaching experiments conducted at high laser intensities. EET in the malachite green + rhodamine B system is expected to be induced via dipole-dipole coupling (s = 6). The high fluorescence quantum yield of rhodamine B (Y = 0.9 - 1.0) and the low fluorescence quantum yield of malachite green (Y $\simeq$ 10^{-4} in methanol and Y $\simeq$ 10^{-2} in the porous glass) enabled the interrogation of the time evolution of the donor fluorescence without interference from the fluorescence of the acceptor. The EET for this donor-acceptor pair is characterized by a critical radius of R_o = 90 Å, as determined in solution. Accordingly, R_o somewhat exceeds r_v, and the unistep EET process monitors the site distribution of the fractal over a microscopic scale length of $\sim r_v$, which is considerably shorter than the characteristic length of the fractal object.

The decay dynamics of the rhodamine B + malachite green system was simulated by the convolution of the survival probability, eq. (22) in the form I(t) = F(t) $\otimes$ P(t). With use of the value of τ obtained from $I_o(t)$, (3.25 ± 0.1 ns), the best values of β and γ were extracted by a multidimensional mean least-squares fit analysis of I(t). To test the reliability of the experimental procedure for the determination of β, some data was obtained on EET between rhodamine B and malachite green in methanol solution. The donor concentration in solution was C_D = (10^{-5} - 2 x 10^{-5}) M, while the acceptor concentration was C_A = (10^{-4} - 2.2 x 10^{-4})M. From the exponential decay curve of rhodamine B, τ = 2.5 ± 0.1 nsec was obtained. The decay curve for rhodamine B + malachite green resulted in the value of β = 0.50 ± 0.05, which is in perfect agreement with the Forster result β = 1/2 for dipole-dipole EET in three-dimensional space (s = 6 and d = 3). The values of γ were found to be independent of C_D and were proportional to C_A within the experimental uncertainty of ± 5%. From the Forster relation $\gamma = 4\pi^{3/2}R_o^3C_A/3$, together with the experimental value of γ = 0.75 ± 0.08 for C_A = 2.2 x 10^{-4}M, R_o = 91 ± 5 A was obtained for the rhodamine B + malachite pair.

This value of β was found to be independent of $\bar{C}_D$ over the narrow (effective) concentration range used, demonstrating that a unistep direct EET is involved. Eq. (23), together with s = 6, result in D = 1.74 ± 0.12 for the fractal dimension of the porous glass. γ was found to be linear within an experimental uncertainty of ± 20% on $\bar{C}_A$ in the (effective) concentration range employed herein, which is in accord with eq. (24). The experimental values of γ were utilized for extraction and estimation of the fraction of the fractal sites occupied by the acceptor molecules. Eq. (24), together with the approximate value a $\simeq$ 6 Å, which corresponds to the molecular radius, results in $X_A = 4.0 \times 10^{-3}$. For $C_A = 5.1 \times 10^{-4}$ M, we get $\gamma = 3.9 \pm 0.1$, which gives $X_A = 1.6 \times 10^{-2}$ for this doped porous glass. From this a posteriori analysis of the experimental data was inferred that the glasses used were indeed lightly doped (e.g., X_A = 1.6% and X_D - 0.2% for the fraction of occupied donor sites in the glass studied), whereupon $X_D \ll X_A \ll 1$ and multistep energy diffusion between the donor centers is negligible.

Kopelman [44] describes the theory of heterogeneous kinetics on triplet exciton fusion of naphthalene on porous glass of the same type used in [6] and finds that the effective spectral dimension of the glass,1.1, is consistent with the fractal dimension of 1.7.

Levitz and Drake [45] have used the method suggested in [6] for studying energy transfer on porous silicas. Here rhodamine 6G was used as a donor and malachite green as acceptor. The silicas used were referred to as Si-40, Si-60, Si-100 and Si-500, having measured Brunauer-Emmett-Teller (BET) surface areas (S_{BET}) of 768, 391, 281 and 68 m^2/g, mean pore radii (R_p) of 18, 35, 60 and 200 Å, and pore volumes (V_p) of 0.66, 0.924, 1.25 and 0.61 cm^3/g, respectively. Their decay curves were fitted to eq. 22. The silicas 60, 100 and 500 were fitted by D of nearly 2 while silica 40 could be fitted to D close to 3.

The results of [45] are inconsistent with recent results of Rojanski et al. [46] where energy transfer between rhodamine B and malachite green adsorbed on varying pore size silica was measured. Here a gradual decrease in average pore sizes causes a gradual increase in the fractal dimension. By applying single parameter fit, eq. (26), of survival probability it was found that the fractal dimension D increases gradually from the value of D = 2.02 for silica with average pore size 5000 Å up to D = 2.82 for silica with apparent pore size of 60 A. In their work Rojanski et al have analyzed the survival probability as a function of only a single parameter using the formula

$$P(t) = \exp\{-\frac{N}{S}\frac{2\pi r^2}{D}\left[\Gamma\left(1-\frac{D}{6}\right)\left(\frac{R_o}{r}\right)^D\right]\left(\frac{t}{\tau}\right)^{D/6} - \frac{t}{\tau}\} \qquad (26)$$

where N/S is the surface concentration of the acceptor molecule and r is the radius of nitrogen which was used as a yardstick in adsorption measurements of the surface area S [46].

By performing a fit to eq. 25 where D was the only parameter to be fitted it was found [46] that D increases gradually with the decrease of average pore size. For silica of average pore sizes of 5000 A, D was 2.02 as compared with D = 2.05 for non-porous aerosil, and for silica of apparent pore size of 60 Å, D was equal to 2.82. Values between 2.02 and 2.82 of D were obtained for gradual decrease of the following average pore sizes 2500, 1000, 500, 200 and 100 Å.

Yang et al [47] studied energy transfer between rhodamine B and malachite green and have performed a computer simulation for one-step dipolar energy transfer and attribute the energy transfer in porous Vycor to an excluded volume effect.

The inconsistency of the results obtained by different laboratories may be connected with the preparation procedure of porous silica. Porous solids with different structures can be prepared from gels. Depending on the details of the preparation procedures three distinct structures may be produced. These are the non-fractal, mass-fractal and the surface-fractal [7]. They will behave differently towards scattering and adsorption measurements as well as towards energy tranfer experiments.

As far as eq. (22) goes, the evaluation of fractal dimension D is strongly dependent on parameter R_o, the critical radius, and the concentration of the acceptor molecules. Changing these parameters can alter the fractal dimensions obtained by fitting the survival probability curves.

The similarity of eq. (21) to that of Inokuti and Hirayama [40] dealing with survival probabilities of the donor in 3-dimensional systems is striking.

Inokuti and Hirayama [40]

$$P(t) = \exp[-t/\tau - C_A 4/3\pi R_o^3 \times \Gamma(1-d/s) \,.\, (t/\tau)^{d/s}] \tag{27}$$

Fractal - direct energy transfer [46]

$$P(t) = \exp[-t/\tau - \theta d/D(R_o/R)^D \times \Gamma(1-D/s) \,.\, (t/\tau)^{D/s}] \tag{28}$$

where θ is the degree of surface coverage of the acceptor molecule, d is the Euclidean dimension (2 for adsorption on a surface), R_o is the critical radius and a is the average size of the unit cell (not very well defined for a glass).

One has to be careful in the interpretation of the results especially due to the fact that the availability of a large number of unequal sites can result in a functional behavior identical in both equations.

The numerous results obtained recently in our laboratory on energy transfer between inorganic ions in glasses [48-50] provide a large number of survival probability curves. By analyzing these curves using the Klafter Blumen model [43] we notice that the Euclidean dimension d = 3 is obtained for all cases. However, by changing the parameters R_o and C the curves may be fitted to a slightly lower dimension of 2.8. The measurements of these quantities in porous glasses is not trivial and the critical radius R_o for the solution does not necessarily correspond exactly to an amorphous medium.

SUGGESTIONS FOR FURTHER INVESTIGATIONS

We plan now to perform a systematic study of energy transfer in systems of controlled pore distribution in which the critical radius and the distance between molecules will be determined by independent spectroscopic measurements.

"Sol-gel" technology refers to a low-temperature technique for preparing single-component or multicomponent oxide glasses. The process involves the hydrolysis and condensation of a solution of inorganic/organic alkoxides or other inorganic compounds in an organic or aqueous solvent.

In our recent studies we have shown that the pore sizes and their distribution depend strongly on the temperature of gelation, the amount of catalyzer and the nature of the glass precursors, TMOS or TEOS, (tetramethoxysilane or tetraethoxysilane) [51-55]. This is also consistent with the various scattering curves obtained by Schaefer and Keefer [7].

Their interpretation of scattering curves to fractal geometry was applied to silica-gels, both polymeric and colloidal. Porous solids with different structures can be prepared from the gels.

Polymer-like silicates can be synthesized in two ways. If TEOS (tetraethyoxysilane) is polymerized under base-catalyzed conditions with substoichiometric $[H_2O]/[Si]$ ratio, their scattering curves with a slope of 2.8 indicates a mass-fractal object. Presumably this is a densely crosslinked polymer molecule. The structures cross over smoothly from mass fractals to surface fractals near W = 2 which is the stoichiometric water ratio.

Polymers can also be synthesized if the polymerization is carried out in two stages [2,7]. In the first stage, small 5 A prepolymers are grown under water-starved, acid-catalyzed conditions [7]. In the second stage, these prepolymers are linked under either base or acid-catalyzed conditions to yield the final polymers. Regardless of the details of the second-stage polymerization, we always observe Porod slopes near -2 indicating polymer-like, mass-fractal structures. By studying the evolution of the scattering curves with dilution, it was shown that the base-catalyzed systems are more high branched than their acid-catalyzed counterparts [2]. These numerous different structures demonstrate that the materials scientist can exploit chemical and physical growth phenomena in precursor solutions to control the properties of solid materials.

A systematic study of fractal dimensions should be performed simultaneously by the 3 methods; energy transfer, small-angle scattering and adsorption. This can be achieved by preparing sol-gel glasses doped by dyes in which the critical radius R_o and the concentration of donor and acceptor molecules are kept constant but the fractal dimensions are gradually changed by changing the gelation conditions. The use of synchrotron radiation may increase the sensitivity of small-angle scattering data.

Experiments performed in our laboratory [56] on computer analysis of decay curves of manganese(II) where manganese(II) acts as a donor in energy transfer to neodymium(III) in fluoride glasses, show an excellent agreement with D = 3. These dimensions are not surprising for glasses obtained by melting. We expect that when the same donor and accceptor ion will be introduced into porous glasses obtained by various gelation procedures, fractal dimensions of less than 3 may be observed.

ACKNOWLEDGEMENTS

The author is grateful to Mrs. E. Greenberg for her invaluable help in preparation of the manuscript and to Dr. Marek Eyal and Professor David Avnir for numerous discussions.

REFERENCES

1. B.B. Mandelbrot, "Fractals: Form, Chance and Dimension", W.H. Freeman, San Francisco, 1977.
2. D.W. Schaefer and K.D. Keefer, Phys. Rev. Lett. 53, 1383 (1984).
3. D.W. Schaefer, J.E. Martin, P. Wiltzius and D.S. Cannell, Phys. Rev. Lett. 52, 2371 (1984).
4. P. Pfeifer and D. Avnir, J. Chem. Phys. 79, 3558, 3566 (1983).
5. H.D. Bale and P.W. Schmidt, Phys. Rev. Lett. 53, 596 (1984).
6. U. Even, K. Rademann, J. Jortner, N. Manor and R. Reisfeld, Phys. Rev. Lett. 52, 2164 (1984).
7. D.W. Schaefer and K.D. Keefer, Structure of random silicates: Polymers, colloids and porous solids, in "Fractals in Physics", Eds. L. Pietronero and E. Tosatti, North-Holland, Amsterdam - Oxford - New York - Tokyo, 1986. Part II, p 39.
8. S. Alexander and R. Orbach, J. Phys. Lett. (Paris) 43, L625 (1982).
9. R. Rammal and G. Toulouse, J. Phys. Lett. 44, L-13 (1983).
10. B.B. Mandelbrot, "The Fractal Geometry of Nature", W.H. Freeman, San Francisco, 1982.
11. J.P. Allen, J.J. Colvin, D.G. Stimson, C.P. Flynn and H.J. Stapleton, J. Biophys. 38, 299 (1982)

12. H.J. Stapleton, J.P.Allen, C.P.Flynn, D.G. Stimson and S.R. Kurz, Phys. Rev. Lett. 45, 1456 (1980).
13. (a) P.W. Kylmko and R. Kopelman, J. Phys. Chem. 87, 4565 (1983).
(b) P. Argyrakis and R. Kopelman, Phys. Rev. B: Condens. Matter 29, 511 (1984).
(c) P. Argyrakis and R. Kopelman, J. Chem. Phys. 81, 1015 (1984).
14. (a) P. Evesque, J. Phys. (Les Ulis, Fr) 44, 1217 (1983).
(b) P. Evesque and J. Duran, J. Chem. Phys. 80, 3016 (1984).
15. S. Alexander, C. Laermans, R. Orbach and H.M. Rosenberg, Phys. Rev. B 28, 4615 (1983).
16. P.F. Tua, S.J. Puttermann and R. Orbach, Phys. Lett. 98A, 357 (1983).
17. B. Berrida, R. Orbach and K.-W. Yu, Phys. Rev. B 29, 6645 (1984).
18. A. Aharony, S. Alexander, O. Entin-Wohlman and R. Orbach, Phys. Rev. B. 31, 2565 (1985).
19. R. Rammal and G. Toulouse, J. Phys. Lett. (Paris) 43, L625 (1982).
20. S. Kelham and H.M. Rosenberg, J. Phys. C 14, 1737 (1981).
21. U. Buchenau, N. Nucker and A.J. Dianoux, Phys. Rev. Lett. 53, 2316 (1984).
22. J. Klafter and A. Blumen, J. Chem. Phys. 80, 875 (1984).
23. D. Rojanski, D. Huppert, H.D. Bale, Xie Dacai, P.W. Schmidt, D. Farin, A. Seri-Levy and D. Avnir, Phys. Rev. Lett. 23, 2505 (1986).
24. U. Even, K. Rademann, J. Jortner, N. Manor and R. Reisfeld, Phys. Rev. Lett. 58, 285 (1987).
25. U. Even, K. Rademann, J. Jortner, N. Manor and R. Reisfeld, J. Luminescence 31/32, 634 (1984).
26. D. Avnir, D. Farin and P. Pfeifer, Nature 308, 261 (1984).
27. V.R. Kaufman and D. Avnir, Langmuir 2, 717 (1986).
28. D. Farin and D. Avnir, The fractal nature of molecule-surface chemical activities and physical interactions in porous materials. Proc. IUPAC Symp. Characterization of Porous Solids, FRG, April 1987. K.K. Unger et al. Eds. Elsevier, Amsterdam, 1987.
29. P.W. Schmidt, Small-angle scattering studies of porous solids. Proc. IUPAC Symp. Characterization of Porous Solids, FRG, April 1987. K.K. Unger et al. Eds. Elsevier, Amsterdam, 1987.
30. A. Guinier, G. Fournet, C.B. Walker and K.L. Yudowitch, "Small-Angle Scattering of X-Rays". Wiley, New York, 1955.
31. O. Glatter and O. Kratky, "Small-Angle X-Ray Scattering". Academic Press, New York, 1982.
32. Ref. 30, pp. 3-4.
33. Ref. 31, Chapter 6, especially pp.197-199.
34. Ref. 30, pp. 5-30.
35. D.W. Schaefer, B.C. Bunker and J.P. Wilcoxon, Phys. Rev. Lett. 58, 284 (1987).
36. S.K. Sinha, J.M. Drake, P. Levitz and G. Grest, unpublished research and paper RRI presented at the meeting of the American Physical Society, New York, March 20, 1987.
37. D. Avnir, D. Farin and P. Pfeifer, J. Chem. Phys. 79, 3566 (1983).
38. A.J. Katz and A.H. Thompson, Phys. Rev. Lett. 54, 1325 (1985).
39. D.L. Huber, Dynamics of incoherent transfer, in "Laser Specroscopy of Solids", Eds. W.M. Yen and P.M. Selzer, Topics in Applied Physics, Vol. 49, Springer-Verlag, Berlin, Heidelberg, New York, 1981. p 83.
40. M. Inokuti and F. Hirayama, J. Chem. Phys. 43, 1978 (1985).
41. M. Yokota and I. Tanimoto, J. Phys. Soc. (JPN) 22, 779 (1967).
42. A. Blumen, J. Klafter, B.S. White and G. Zumofen, Phys. Rev. Lett. 53, 1301 (1984).
43. J. Klafter, A. Blumen and G. Zumofen, J. Luminescence 31,32, 627 (1984).
44. R. Kopelman, J. Statistical Phys. 42, 185 (1986).

45. P. Levitz and J.M. Drake, Phys. Rev. Lett. 58, 686 (1987).
46. D. Pines-Rojanski, D. Huppert and D. Avnir, Pore size effects on the fractal distribution of adsorbed molecules as revealed by electronic energy transfer on silica surfaces. Submitted to Chem. Phys. Lett. 1987.
47. C.L. Yang, P. Evesque and M.A. El-Sayed, J. Phys. Chem. 89, 3442 (1985).
48. R. Reisfeld, M. Eyal, C.K. Jørgensen and C. Jacoboni, Chem. Phys. Lett. 129, 392 (1986).
49. M. Eyal, R. Reisfeld, C.K. Jørgensen and C. Jacoboni, Chem. Phys. Lett. 129, 550 (1986).
50. R. Reisfeld and M. Eyal, Acta Physica Polonia, to be published 1987.
51. D. Levy, R. Reisfeld and D. Avnir, Chem. Phys. Lett. 109, 593 (1984).
52. R. Reisfeld, M. Eyal and R. Gvishi, Spectroscopic behaviour of fluorescein and its di(mercury acetate) adduct in glasses. Chem. Phys. Lett., (1987) in publication.
53. D. Brusilovsky and R. Reisfeld, Comparison of gelation mechanisms of TEOS and TMOS by spectroscopic measurements. Chem. Phys. Lett. submitted.
54. D. Avnir, D. Levy and R. Reisfeld, J. Phys. Chem. 88, 5956 (1984).
55. D. Avnir, V.R. Kaufman and R. Reisfeld, J. Noncryst. Solids 74, 395 (1985).
56. M. Eyal and R. Reisfeld, to be published in Chem. Phys. Lett.

PHOTOCHEMICAL HOLE BURNING

W. Richter

Physikalisches Institut, Universitat Bayreuth
D-8500 Bayreuth, F.R.G.

ABSTRACT

Photochemical hole buring (PHB) is a laser spectroscopy method for measuring quasihomogeneous linewidths in the presence of inhomogenous line broadening (for a recent survey see 1). The holes, which are burnt with laser irradiation, appear as small indentations in the absorption spectra of dye molecules which are doped into a polymer or glass in minute concentrations.

A hole burning experiment can be used to measure a series of experimental parameters characterizing both the matrix (host) and the dopant (guest). From the intensity and width of the zero-phonon line as a function of light intensity and time, one can obtain information on molecular parameters, e.g. Debye-Waller factor, photochemical quantum yield, Debye temperatures etc. From experimental data of the phonon sideband one can get information on parameters such as characteristic phonon frequencies (local mode frequencies), distribution of Debye-Waller factors and distribution functions of the electron-phonon coupling strength. Because of the narrow linewidth of the zero-phonon part of the photochemical hole, the PHB method yields a gain in resolution of the order of 10^3 to 10^5. It can be used to detect small perturbations of the system by a variation of external parameters, giving rise to line shifts and broadenings. Two examples will show the variety of this spectroscopic technique in getting quantitative results.

The black reaction from the photoproduct state to the initial ground state, measured by the hole area, occurs on logarithmic time scales. This behavior can be well described by assuming the same distribution of parameters for the two-level-systems (TLS), which are assumed for describing the specific heat of the material. The so calculated relaxation rates describe the dynamic behavior of the glassy matrix on time scales covering a wide dynamic range.

The second example shows pressure-induced matrix changes. The PHB technique allows to study extremely small strain-induced shifts of the homogeneous optical lines. If one assumes a R^{-6}-law, governing the molecular interactions responsible for the solvent shift, one can calculate the hydrostatic compressibility values of different matrices. These optically determined values agree with the mechanical ones within 10-20%.

Besides the many well documented spectroscopic applications of the PHB, it may offer interesting future developments for the understanding of biological processes and for high-density data storage.

REFERENCES

1) J. Friedrich and d. Haarer in "Optical Spectroscopy of Glasses", ed. I. Zschokke, Reidel Publ. Comp. 1986

SHORT SEMINARS
(Titles Only)

LUMINESCENCE OF WIDE BAND GAP SEMIMAGNETIC SEMICONDUCTORS (H. E. Gumlich)

STRUCTURAL TRANSFORMATIONS IN AMORPHOUS $Ge_{1-x}Sn_xSe_2$ (L. McNeil)

THE EXISTENCE OF BANDS IN AMORPHOUS CARBON FROM (e,2e) SPECTROSCOPY (A. Ritter)

OPTICAL PROPERTIES OF HIGHLY EXCITED II-VI SEMICONDUCTORS (E. Swoboda)

SPECTROSCOPY OF DISORDERED $CdS_{1-x}Se_x$ MIXED CRYSTALS UNDER HIGH EXCITATION (F. A. Majumder)

APPLICATION OF AMORPHOUS SEMICONDUCTORS IN OPTICAL RECORDING (P. Scholte)

DEPENDENCE OF THE OPTICAL SPECTRUM OF Mn^{2+} IMPURITIES IN FLUORIDES ON THE $Mn^{2+}-F^-$ DISTANCE (M. Moreno)

FLUORESCENCE LINE NARROWING IN URANYL GLASS (J. Thorne)

FLUORESCENCE BEHAVIOR OF TRIVALENT CROMIUM IN A SERIES OF STRUCTURALLY DIFFERENT GLASSES (D. B. Hollis)

DIFFUSION AND CROSS-RELAXATION AMONG Tb IONS IN YF_3 (J. Collins)

STUDIES OF Cr^{3+} IN MULLITE CERAMICS (R. Knutson)

CRYSTAL STRUCTURE AND SPECTROSCOPIC PROPERTIES OF DISORDERED $Ca_3(Nb,Ga)_2Ga_3O_{12}$ (S. E. Sarkisov)

THE SPECTROSCOPY OF ANTHRACENE-ACRIDINE MIXED CRYSTALS (O. Morawski)

RADIATIVE AND NON-RADIATIVE ENERGY TRANSFER BETWEEN Cr AND Nd IN GSGG (G. Armagan)

INFRARED ABSORPTION MEASUREMENTS IN HIGH T_C SUPERCONDUCTORS (R. Zamboni)

AGING AND RELAXATION OF THERMO-REMANENT MAGNETIZATION IN A SPIN GLASS (E. Vincent)

DYNAMIC SPECIFIC HEAT OF SPIN GLASSES (G. Szamel)

SOFT X-ray SPECTROSCOPY OF METALLIC SOLIDS (J. Pelka)

RAMAN SCATTERING OF GLASSY AND MOLTEN $ZnCl_2$ (S. Magazù)

SPECIAL TOPIC

FUNDAMENTAL INTERACTIONS OF ELEMENTARY PARTICLES: DISCOVERIES AND PERSPECTIVES

G. Costa

Dipartimento di Fisica dell'Università, Padova, Italy

Istituto Nazionale di Fisica Nucleare, Sezione di Padova

Italy

ABSTRACT

Recent progress in the unification of the fundamental interactions among elementary constituents of matter is reviewed. A great achievement has been the unification of electromagnetic and weak interactions, which implies the existence of heavy intermediate bosons ($W^{\pm}$ and Z°), discovered in 1983 at the proton-antiproton collider at CERN, Geneva. The peculiar properties of strong interactions can be understood in a theoretical framework, called quantum chromodynamics, in which the relevant quanta are the quarks and the gluons. Some progress has been made in the next step of unifying also strong and electroweak interactions in a theory of Grand Unification. A characteristic feature is the prediction of the proton decay, for which extensive experimental searches have put already stringent limits. Finally, recent ideas about the important role that gravitation might have in elementary particles are briefly discussed.

I. INTRODUCTION

A constant trend in the development of physics has been the idea of explaining the variety of phenomena in terms of fundamental interactions among elementary constituents. The choice of the basic constituents changed from time to time, but this point of view still dominates the present research in Particle Physics. The possibility of investigating smaller and smaller physical systems (atoms, nuclei, nucleons) revealed, in fact, different levels of compositness. The present picture of the elementary particles and their interactions is the result of the new theoretical ideas and experimental discoveries in the last 20 years.

Two main ingredients have been essential in the realization of this picture: the intensive use of Quantum Field Theory (QFT), which represents one of the greatest achievements of the theoretical physics of this century, and the extensive exploitation of symmetry.

QFT combines the laws of Quantum Mechanics with those of Special Relativity: new important features emerge from this combination. The interactions between two particles are not at-a-distance, but are transmitted by a field; in other words, the interactions are mediated by the quanta of the field. In fact, even the constituents of matter are described by fields: they are identified with the quanta of these fields. In the following, we shall adopt the usual nomenclature:

MATTER consists of quanta with half-integer ($\frac{1}{2}$) spin, i.e. fermions,
RADIATION consists of quanta with integer (generally 1) spin, i.e. bosons.
With the generic name of "particles" we refer both to fermions and bosons.

Another important consequence of QFT is that for each particle (with a definite mass m, spin s, charge q) there is a corresponding antiparticle (with the same m and s, and opposite charge -q).

About 25 years ago, the "elementary" fermions were identified with 4 particles: proton (P), neutron (N), electron (e) and neutrino (ν). They are stable and, maybe with the exception of the elusive neutrino, they have non-vanishing masses.

In the present picture:

i) Matter consists of leptons and quarks; the former interact only weakly, the latter both strongly and weakly. Three kinds of charged leptons are known (e^-, μ^-, τ^-), each of which with a corresponding neutrino (ν_e, ν_μ, ν_τ); there is evidence for 6 kinds of quarks (u=up, d=down, c=charm, s=strange, b=bottom, t=top), but no experimental signature of the top has been found up to now. For each particle there is the corresponding antiparticle.

ii) Radiation consists of photons, intermediate weak bosons ($W^{\pm}$,Z°), gluons and gravitons. Gravitons have spin 2, while all the other quanta have spin 1.

Matter interacts through radiation; all the interactions can be grouped into 4 categories, the main distinctive features of which are summarized in Table 1.

Each type of interaction plays a dominant role in a specific domain of physics. The electromagnetic interactions characterize the properties of atoms and the structure of ordinary matter; the strong interactions determine the structure of atomic nuclei; the weak interactions play an essential role in thermonuclear processes, which are the main source of energy in the sun. All these interactions are essential at the level of elementary particles. Gravitational interactions are the dominant ones at macroscopic and astronomical scales, but their role in elementary particle physics was considered unessential until recently.

As a final remark, we should recall that the range r of the interaction is related to the mass of the exchanged boson by the uncertainty relation

$$r \approx c\Delta t \approx \frac{\hbar}{mc} \; ; \qquad (1)$$

Table 1. Fundamental Interactions

Type of interaction	Relevant system or process	Elementary fermions involved	Strength ($\hbar=c=1$)	Range (cm)	Quanta of radiation
EM	atoms	charged	$\alpha=\frac{e^2}{4\pi}\cong\frac{1}{137}$	∞	photons (γ)
Weak	β-decay	leptons and quarks	$\alpha_W=\frac{G_F\, m_p^2}{4\pi}\cong 10^{-5}$	$2.5\text{x}10^{-16}$	$W^{\pm}$, $Z°$
Strong	nuclei	quarks	$\alpha_s=\frac{g^2}{4\pi}\cong 1$	$1.4\text{x}10^{-13}$	gluons
Gravitational	solar system	all	$\alpha_G=\frac{G_N\, m_p^2}{4\pi}$ $\cong 6\text{x}10^{-39}$	∞	gravitons

II. EM AND WEAK INTERACTIONS AND THEIR UNIFICATION

Quantum Electrodynamics (QED) represents the best known and most complete example of quantum field theory. It has been tested with very high accuracy, and all its predictions are in excellent agreement with experiments.

It is also a good example for illustrating the important role of symmetry.

In QED a system of electrons interacting through the EM field is described by a Lagrangian which can be split into 3 terms:

$$L=L^{\circ}_{\text{electrons}}+L^{\circ}_{\text{EM}}+L_{\text{int}}. \tag{2}$$

The first term, which describes a system of free electrons, is invariant under global (or rigid) gauge transformations, i.e. phase transformations of the electron field

$$\psi(x)\to e^{i\alpha}\,\psi(x), \tag{3}$$

where $e^{i\alpha}$ is a generic element of the Abelian group U(1).

Now suppose that we let the global transformation (3) turn into a local gauge transformation

$$\psi(x)\to e^{i\alpha(x)}\,\psi(x), \tag{4}$$

i.e. that we require that the unobservable phase of the field $\psi(x)$ is changed independently at each space-time point by an arbitrary function $\alpha(x)$. This is a very strong constraint: $L^{\circ}_{\text{electrons}}$ is no longer invariant by itself, but the invariance now requires the inclusion of the term L_{int}, which describes the interaction of the electrons with the EM potential $A_\mu(x)$. The transformation of $A_\mu(x)$ which goes together with (4) is given by

$$A_\mu(x)\to A_\mu(x)-\partial_\mu\,\alpha(x) \tag{5}$$

The term L°_{EM}, which corresponds to non-interacting EM field, is then added to complete the description of the system.

The important point we would like to stress is that the requirement of local gauge invariance implies the introduction of a (massless) vector field A_{μ}, which describes the photon.

Next, let us consider the weak interactions, and compare a weak process, like the neutron decay ($N \to P\, e^- \bar{\nu}_e$) or, better, the related process of neutrino absorption ($\nu_e\, N \to e^- P$), with the analogous EM process $e^- P \to e^- P$. If one assumes that the weak interactions are mediated by an intermediate boson $W^{\pm}$, the two processes are represented by the graphs of Figure 1. Replacing the massless photon by a massive boson $W^{\pm}$ ($m_W \approx 10 \div 10^2$ GeV), one gets:

i) the range is reduced from infinity to $1/m_W \approx 10^{-15} \div 10^{-16}$ cm.;
ii) the strength is reduced from α to $\alpha_W \approx \frac{\alpha}{m_W^2} \approx (10^{-2} \div 10^{-4})\, \alpha$.

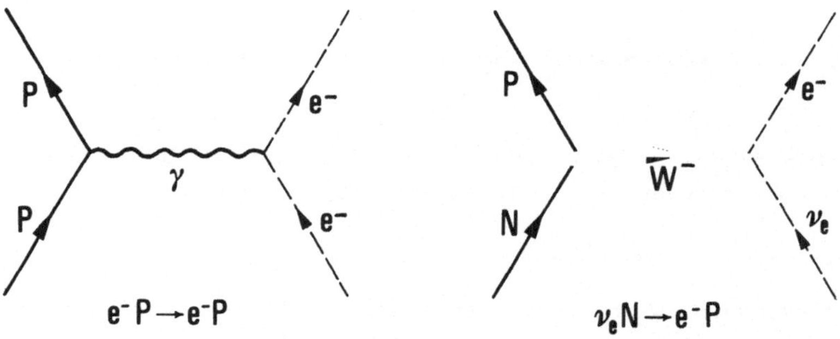

Figure 1. Comparison between an EM and a weak process.

On the basis of these analogies (in spite of the striking apparent differences!) it was proposed [1] that the EM and the weak interactions are related by a gauge symmetry principle.

The group U(1) of QED is extended to a (non-Abelian) group of transformations with 4 generators:

$$SU(2) \times U(1) \tag{6}$$

This structure was suggested by the properties of the currents: besides the EM current (coupled to γ) there must be a charged weak current (coupled to $W^{\pm}$). Moreover, the symmetry SU(2) implies that there is also a neutral weak current, coupled to the neutral counterpart (Z°) of $W^{\pm}$.

The difference between a charged current process and a neutral current one can be understood comparing the processes represented in Figure 2.

In the first process, $\nu_\mu N \to \mu^- P$, a neutrino (ν_μ) is absorbed by a neutron and it is transformed into the charged lepton μ^-. Instead, the second process represents the elastic scattering of a ν_μ by a proton (beams of neutrinos of ν_μ type are available for experiments). The elastic neutrino scattering is difficult to detect, but one can look for a similar process

$$\nu_\mu + P \to \nu_\mu + P + \pi^\circ, \tag{7}$$

which is easier to detect. Events of this type were discovered in a big bubble chamber at CERN in 1973 [2], providing evidence for the weak neutral current. By now the existence of such current is well established, and its specific properties have been determined to be those predicted by the symmetry SU(2)xU(1).

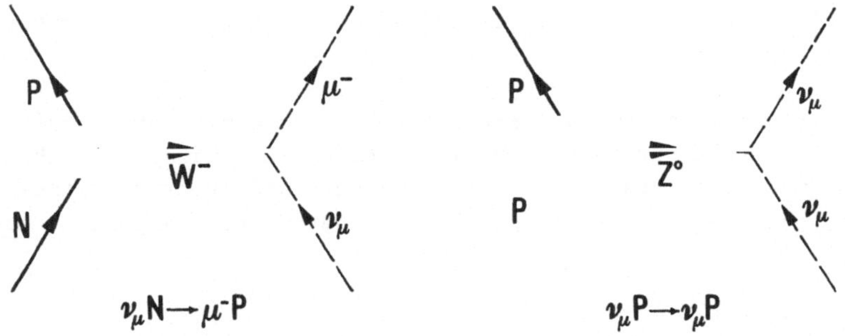

Figure 2. Weak processes mediated by charged and neutral vector bosons.

We have mentioned a few of the ingredients which should be present in a unifying picture of the electro-weak interactions. However, this picture does not appear, at first sight, to be consistent with the general framework of gauge field theories, which requires that the bosons mediating the interactions (usually denoted by gauge bosons) are strictly massless. While the physical photon is indeed massless and it is described in terms of 2 degrees of freedom (2 transverse polarizations), the weak ($W^\pm$, Z°) bosons must be massive to reproduce the characteristic features of weak interactions. Each of the weak bosons is then described in terms of 3 degrees of freedom (2 transverse and 1 longitudinal polarizations).

The problem of conciliating the opposite gauge invariance conditions and the phenomenological requirements is solved making use of the concept of spontaneous symmetry breaking. This concept, first introduced in solid state physics, gives explanation to well-known phenomena, such as spontaneous magnetization and superconductivity.

The main point is that, even if the Lagrangian of a system, and therefore its equations of motion, have a given symmetry, the ground state of the system can have a lower symmetry.

The electro-weak Lagrangian has the gauge symmetry SU(2)xU(1). The ground state has, in general, the same symmetry but, below a certain energy scale (or, equivalently, below a certain temperature), it presents a lower symmetry:

$$SU(2)\times U(1) \to U(1)_Q \tag{8}$$

The residual symmetry $U(1)_Q$ is just the gauge symmetry of QED, which is preserved at all temperatures. This reduced symmetry does not prevent that the 3 vector bosons ($W^{\pm}$ and Z°) acquire mass, while the photon remains massless.

This mechanism, discovered by Higgs in 1964 [3], requires the presence of two doublets of scalar fields (the so-called Higgs fields), 3 components of which are "eaten" by the massless vector fields $W^{\pm}$ and Z°: then they acquire the additional longitudinal components that are necessary for massive vector fields.

The values of the masses are determined by a parameter, which is the analogue of the "order parameter" in a phase transition. From the knowledge of the weak (charged and neutral) currents and their couplings at low energies, one can infer the value of this parameter and predict the masses of $W^{\pm}$ and Z°:

$$m_{W^{\pm}} \approx 81 \text{ GeV}$$

$$m_{Z°} \approx 92 \text{ GeV}$$

After the construction of the $\bar{P}P$ collider at CERN, in which antiproton-proton collisions are produced at the C.M. energy of 540 GeV, both $W^{\pm}$ and Z° were discovered in 1983 [4]. The reactions which lead to this discovery are the following:

$$\bar{P}+P \to W^{\pm} + \ldots, \quad W^{\pm} \to e^{\pm} + \overset{(-)}{\nu}_e \tag{9}$$

$$\bar{P}+P \to Z° + \ldots, \quad Z° \to e^+e^-,\ \mu^+\mu^- \tag{10}$$

The analysis of these reactions allowed the determination of the properties of the heavy weak bosons, which are those predicted by the electroweak theory. In particular, the experimental values obtained for the masses

$m_{W^\pm} = (81\pm2)$ GeV

$m_{Z^\circ} = (92\pm2)$ GeV

are in very good agreement with the expected values.

As a conclusion of this section, one can say that the first step in unification has been achieved.

Above the scale of about 100 GeV, the appropriate QFT is a gauge theory based on the symmetry SU(2) x U(1); it contains 4 massless vector bosons which mediate the electroweak interactions. Going down in the energy scale, the symmetry is spontaneously broken, and electromagnetic and weak interactions become different and manifest their specific characteristics.

III. STRONG INTERACTIONS

Before going to the next step of unification, it is convenient to discuss the main features of strong interactions.

We know that strong interactions are responsible for nuclear forces; in fact, they bind nucleons (protons and neutrons) in the atomic nuclei. According to Yukawa's hypothesis, strong interactions among nucleons are mediated by scalar (spin zero) bosons. This hypothesis was formulated by Yukawa in 1935 [5]; the pion ($\pi^\pm,\pi^\circ$), discovered in 1947 [6], appeared to be the expected particle ! Its mass was found to be $m_\pi \cong 140$ MeV, in agreement, according to eq. (1), with the range of the interactions $r_\pi \cong 1.4 \times 10^{-13}$ cm.

For a few years, the world of strongly interacting particles was limited to nucleons and pions. But, with the advent of high energy accelerators, a large variety of other particles was discovered during the decades 1950s and 1960s. They are named hadrons (for strongly interacting particles), and they are separated into baryons and mesons, in correspondence with half-integer and integer spin values, respectively. Among the baryons, some look like excited states of nucleons, while others require the introduction of a new quantum number, called "strangeness", not possessed by nucleons. The situations is analogous for the mesons: some appear to be of the same type of the pions, while others, called strange mesons, possess the extra quantum number. The large number of particles with similar properties gave clear indication that they were different states of compound systems of more elementary constituents.

The present tables of particles [7] show the spectra of baryons and mesons, which consist of hundreds of states with increasing mass and spin (S=1/2, 3/2, 5/2 ... for baryons, and S=0,1,2,3,... for mesons).

A peculiar regularity was soon manifested in the spectrum of these states: both baryons and mesons can be grouped into multiplets, the components of each multiplet being very close in mass, and having the same spin. Mesons appear only in singlets and octets; baryons only in singlets, octets and decuplets.

These multiplets correspond to the lowest representations of the unitary group SU(3), which was the symmetry proposed independently by Gell--Mann and Ne'eman in 1960 [8].

In order to give a hint on this type of symmetry, it is instructive to consider an example, i.e. the case of the (S=3/2) decuplet which is represented in Figure 3. This example will be useful also for clarifying other distinctive features of the spectra of baryons and mesons.

It should be pointed out that the 10th particle, denoted by Ω^-, was not known when the symmetry SU(3) was proposed. In fact, the Ω^- was predicted by Gell-Mann in 1961 [9], and discovered at the same time (1963) at the Brookhaven National Laboratories (USA) and at CERN [10].

This discovery provided a test of SU(3), which appeared to be a (approximate) symmetry of strong interactions, describing rather well the properties of the spectra of ordinary and "strange" baryons and mesons.

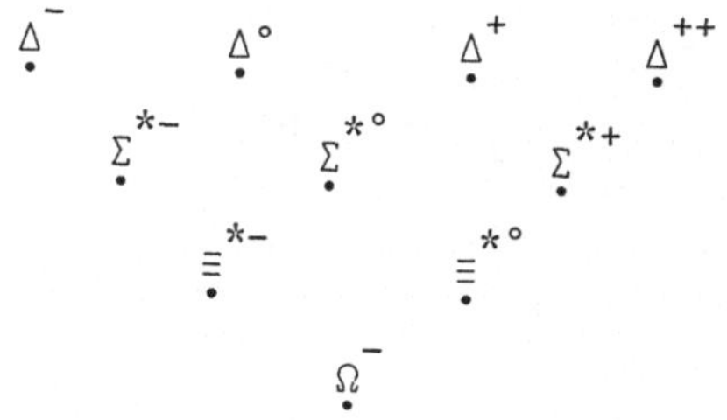

Figure 3. The Δ(S=3/2) decuplet.

However, even if the situation appeared to be rather satisfactory, the picture presented a few puzzles, the solution of which required the introduction of new ideas in particle physics. In the following, we shall discuss briefly the puzzles and the solutions which were proposed.

<u>First Puzzle</u>: Why <u>only</u> the representations <u>1</u>, <u>8</u> and <u>10</u> of SU(3) seem to be relevant for the classification of baryons and mesons ?

The solution of this puzzle, proposed independently in 1964 by Gell--Mann and Zweig [11], was based on the introductions of <u>quarks</u> as elementary constituents of particles. Taking into account that the fundamental representation of SU(3) is 3 (and its conjugate $\bar{3}$), they introduced a <u>triplet</u> of quarks (and a corresponding triplet of anti-quarks, assigned to $\bar{3}$), so that one could say:

<u>baryons</u> are made of 3 quarks

$$3 \times 3 \times 3 = 1 + 8 + 10 \; ;$$

<u>mesons</u> are made by a quark-antiquark pair

$$3 \times \bar{3} = 1 + 8.$$

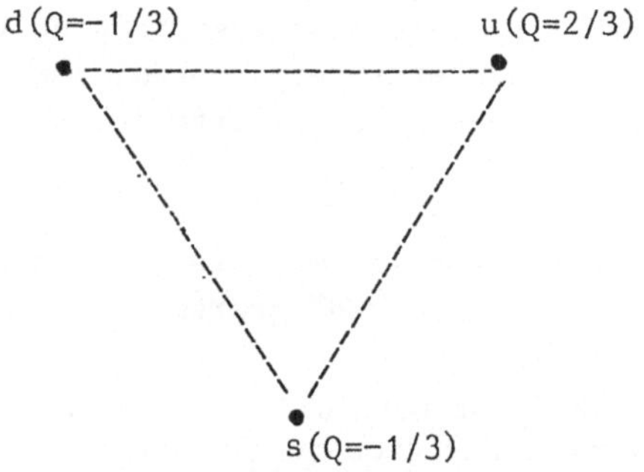

Figure 4. Triplet of quarks.

We point out that quarks are peculiar particles: they are fermions with spin 1/2, and fractional charges Q=2/3, -1/3. They are denoted by u,d,s as indicated in Figure 4.

The s-quark carries "strangeness" i.e. it has a non-vanishing quantum number (S=-1), while the two others have S=0. In terms of its constituents quarks, the Δ-decouplet appears as in Figure 5.

Second Puzzle: How can there be 3 identical quarks in the same state?

This question arises from the observation that the state Δ^{++} (with $S=3/2$, $S_z=3/2$), which is a component of the decuplet represented in Figure 5, would be composed of 3 identical quarks of u-type, each in the same state of spin $S_z=3/2$ and orbital momentum $\ell=0$: which seems in clear contradiction with the Pauli exclusion principle.

The solution of this puzzle is based on the hypothesis that each quark possesses extra quantum numbers [12], which correspond to those of a second SU(3). To make a distinction between the two symmetries one uses whimsical names: flavour and colour. With respect to the first symmetry, denoted by $SU(3)_f$, the quarks belong to a triplet (each component has a different flavour: u, d, s); with respect to the second one, deno-

(ddd) (ddu) (duu) (uuu)

(sdd) (sdu) (suu)

(ssd) (ssu)

(sss)

Figure 5. The Δ decuplet in terms of its constituent quarks.

ted by $SU(3)_c$, each quark q belongs to another triplet (each component has a different colour: q_1, q_2, q_3, where e.g. 1=red, 2=white, 3=green). The state quoted above becomes then compatible with the Pauli principle:

$$\Delta^{++}(u_1^{\uparrow}\ u_2^{\uparrow}\ u_3^{\uparrow}).$$

At this point one should enquire what is the evidence for the existence of quarks and moreover of "coloured" quarks.

Quarks have been looked for with different experimental methods, but no experimental evidence exists for free quarks and for free particles carrying fractional charges. However, there is indirect evidence of quarks (i.e. fermions with spin 1/2 and electric charges 2/3, -1/3) inside the nucleons. This evidence come from a Rutherford kind of experiments performed at SLAC (Stanford Linear Accelerator Center) in the early 70's, through deep inelastic scattering of 20 GeV electrons by proton targets, as represented in Figure 6 (for an updated review see [13]).

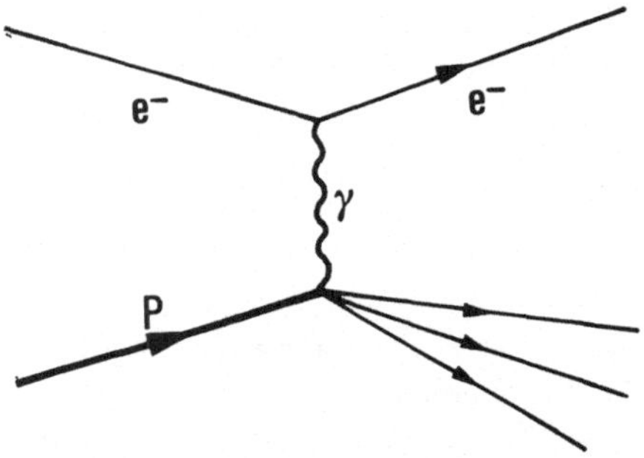

Figure 6. Deep inelastic electron-proton scattering.

By high momentum transfer scattering one can investigate (at very short distances) the structure of the proton: this is the analogue of the Rutherford's experiments, in which the structure of atoms were investigated by analysing the scattering of α-particles. In fact, in these experiments the nucleons appear to be composite objects made of point-like constituents carrying spin 1/2 and fractional charges.

There is also experimental evidence for colour. At least two facts indicate that each type of quark (u,d,s) comes in three different internal states. The first is the measurement of the cross section of the process $e^+e^- \to$ hadrons. Theoretically, one obtains the expression (neglecting higher order corrections)

$$R=\frac{\sigma(e^+e^- \to \text{hadrons})}{\sigma(e^+e^- \to \mu^+\mu^-)} = 2\left(\frac{N}{3}\right) \tag{11}$$

where N=number of colours. The experimental value for R [7], taken in the appropriate energy range, shows that the preferred number is N=3. The second is the comparison of the experimental value of the decay width for $\pi^{\circ}\to\gamma+\gamma$, which is [7]

$$\Gamma_{\pi^{\circ}\to\gamma\gamma} = 7.95 \pm 0.55 \text{ eV},$$

with the theoretically estimate

$$\Gamma_{\pi^{\circ}\to\gamma\gamma} = 7.9 \left(\frac{N}{3}\right)^2 \text{ eV}. \tag{12}$$

Also, in this case, the favoured case is N=3.

So far, the picture looksrather nice, but there is still a further puzzle to solve.

Third Puzzle: Why there are no free quarks ? An alternative, more mathematical, way of formulating the question is based on the observation that the physical states appear only as singlets (i.e. in the representation 1) of $SU(3)_c$: why only colour singlets?

Once more gauge symmetry is advocated to solve the problem. The global $SU(3)_c$ symmetry is promoted to a local gauge symmetry: one gets in this way a new type of field theory: Quantum chromodynamics (QCD) [14]. The theory contains 8 vector bosons (corresponding to the 8 generators of SU(3)) which are massless: they are called gluons: they mediate the interactions among coloured quarks.

It is very instructive to point out what is the main difference between QCD and QED.

In QED, photons are neutral; the vacuum polarization produces a screening of the electric charge.The electric charge appears smaller and smaller at large distances (low momentum transfers) and it increases at short distances (high momenta). The neutral photons do not modify this feature.

In QCD, gluons carry colour as quark do, so that they interact directly among themselves. This fact changes dramatically the situation: the effective coupling decreases by decreasing the distance, while it increases beyond any limit by going to larger and larger distances. In concise words,at short distances one gets asymptotic freedom, while at larger distances confinement becomes operating.
These properties were recognized by analysing the evolution equations of the effective couplings [15]; however, while the theory is under control at short distances, it becomes very difficult, from the technical point of view, to deal with the large distance behaviour: for this reason, we are still lacking a rigorous proof of confinement.

As a conclusion of this section, we have to point out that also strong interactions are described by a gauge field theory. It is based on the gauge symmetry $SU(3)_c$, which remains unbroken down to low energy scales. We have reached by now a Standard Model for strong and electroweak interactions based on the gauge symmetry

$$SU(3) \times SU(2) \times U(1). \tag{13}$$

IV. GRAND UNIFICATION

Before going beyond the "Standard Theory", we have to look at the present picture of the matter constituents. The situation is summarized in Table 2.

Table 2. Constituents of matter *

	SINGLETS	TRIPLETS			of $SU(3)_c$
DOUBLETS	ν_e	u	u	u	1st GENERATION
	e^-	d	d	d	(light)
SINGLETS of SU(2)	e^+	$\bar{u},\bar{d}$	$\bar{u},\bar{d}$	$\bar{u},\bar{d}$	
	ν_μ	c	c	c	2nd GENERATION
	μ^-	s	s	s	(heavier)
	μ^+	$\bar{c},\bar{s}$	$\bar{c},\bar{s}$	$\bar{c},\bar{s}$	
	ν_τ	t	t	t	3rd GENERATION
	τ^-	b	b	b	(much heavier)
	τ^-	$\bar{t},\bar{b}$	$\bar{t},\bar{b}$	$\bar{t},\bar{b}$	

*We denote by $\bar{x}$ the antiparticle of x. We consider particles and antiparticles of a definite helicity; for a given neutrino there exists the corresponding antineutrino which has, however, the opposite helicity.

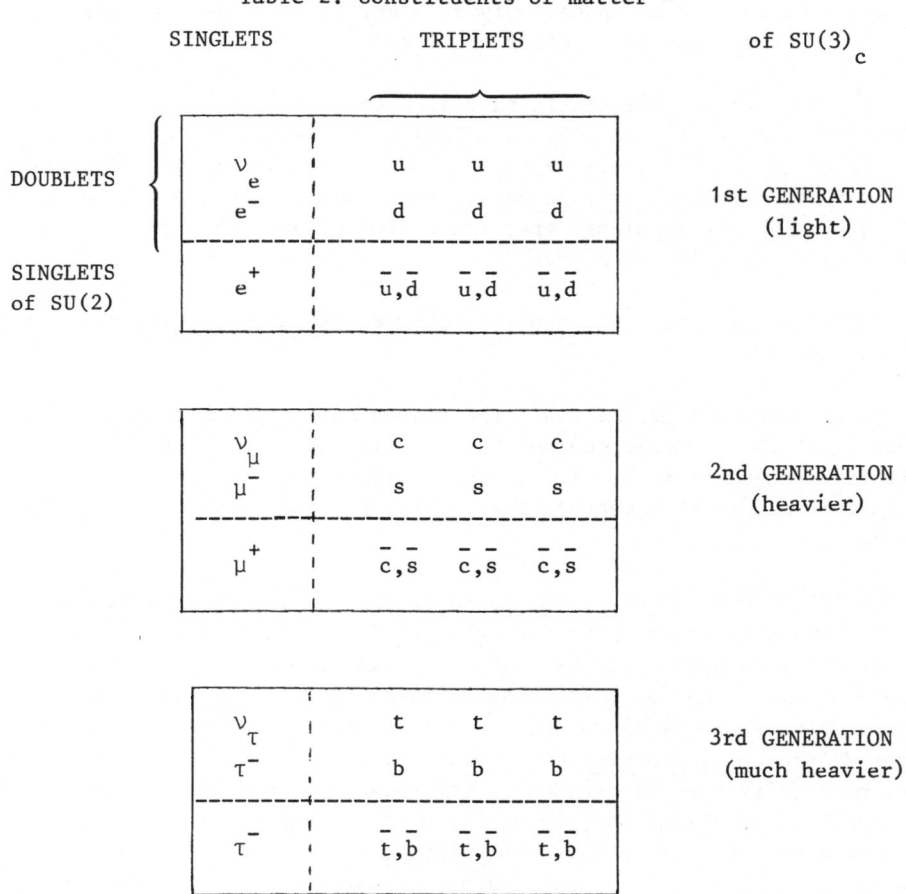

In the previous sections, we have made reference to the situation before 1974, until when there was evidence of only 3 "flavours" of quarks (u,d,s,) and of two lepton pairs: (e, ν_e) and (μ, ν_μ). Two more types of quarks were discovered in the next few years, denoted by c for "charm" and b for "beauty" or "bottom", and there are reasons (at least theoretical prejudices) for a sixth type of quark: t for "truth" or "top".

The quark c was first introduced for theoretical reasons, in order to give an explanation to a peculiar feature of the weak neutral current [16]. A new type of hadron, requiring a new quantum number, i.e. "charm", was discovered in 1974. It was discovered independently in two different laboratories, Stanford and Brookhaven, and it was observed immediately afterwards at Frascati [17].

It received two different names and since then it is denoted by J/ψ. All the experimental information about it is in agreement with the interpretation of the J/ψ (which is a meson with spin 1) as a bound state of a quark-antiquark pair to the type $c\bar{c}$.

The existence of the quark b was inferred after the discovery at the Fermi National Laboratory, [18], of a new type of heavy meson, denoted by Υ, which had to carry a new quantum number ("beauty") and could be interpreted as a bound state of a quark-antiquark pair: $b\bar{b}$.

The 6th quark, the "top", is believed to be much heavier than the others, probably heavier than 50 GeV; experimental searches are under way.

On the side of leptons, a charged one, denoted by τ, was discovered at SLAC in 1976 [19]; it is very similar to e and μ, except for the fact that it is heavier. It is believed that it has a neutral counterpart, of the neutrino type: ν_τ.

Looking at the previous table, one realizes that quarks and leptons appear to be grouped into "families" or "generations", each containing two types of quarks, and two types of leptons. There is no theoretical argument to fix the number of generation equal to $n_G=3$; present phenomenological implication allow both values $n_G=3$ and 4, while values $n_G>4$ would be disfavoured.

As we know from the previous sections, the particles listed in Table 2 (which are fermions with spin S=1/2) interact among themselves through the exchange of vector (i.e. spin S=1) bosons. Specifically, the electroweak interactions are mediated by photons and heavy vector bosons ($Z°$ and $W^\pm$), while the strong interactions are mediated by gluons. All this is described in a satisfactory way in the frame of the Standard Model based on the gauge symmetry SU(3)xSU(2)xU(1).

However, even if the construction of the Standard Model represents a great achievement since it unifies a variety of different phenomena, it cannot be considered as the ultimate theory of elementary particles. In particular, it contains too many free (unrelated) parameters which can be determined only by experiments: first of all, the three independent coupling constants which give the strengths of the interactions between fermions and gauge bosons, and then all the masses of quarks and leptons.

Keeping the good features and results of the Standard Model, one is lead to go beyond it looking for a complete unification which provides some answers to the still open questions:

a) why there are three different coupling constants?
b) why leptons have integer charges (0,±1) and quarks fractional ones (± 1/3, ± 2/3)?
c) why there is repetition of the generations of fermions?

The idea was to build a grand unified field theory based on a simple group containing SU(3)xSU(2)xU(1) as a subgroup. The minimal realization of GUT (Grand Unified Theory) is based on the group SU(5) [20]:

$$SU(5) \supset SU(3)\times SU(2)\times U(1). \tag{14}$$

The main outcomes of the SU(5) model are the following:

1) Each generation of fermions is fitted nicely into the representation (multiplet) $15=\bar{5}+10$ of SU(5), according to the scheme

$$\begin{array}{ccccccc} \bar{5} = & (1,2) & + & (\bar{3},1) & & \\ & \begin{pmatrix} \nu_e \\ e^- \end{pmatrix} & + & (\bar{d}_1\bar{d}_2\bar{d}_3) & & \\ 10 = & (3,2) & + & (\bar{3},1) & + & (1,1) \\ & \begin{pmatrix} u_1u_2u_3 \\ d_1d_2d_3 \end{pmatrix} & + & (\bar{u}_1\bar{u}_2\bar{u}_3) & + & (e^+) \end{array}$$

2) By group properties it is required that the sum of the electric charges vanishes separately for each (irreducible) representation. Then, in the case of the $\bar{5}$ multiplet, one gets

$$Q_{e^-} + 3Q_{\bar{d}} = 0,$$

i.e.

$$Q_d = - Q_{\bar{d}} = \frac{1}{3} Q_{e^-} .$$

Since the quark u is in the same doublet as d(with respect to SU(2)), it follows in units $|Q_{e^-}|=1$:

$$Q_d = - \frac{1}{3}, \quad Q_u=Q_d + 1 = \frac{2}{3} .$$

This explains why the charge of the electron and the charge of the proton (P=uud) add exactly to zero.

3) The Standard Model contains three independent couplings $\alpha_1, \alpha_2, \alpha_3$ related to the three factors U(1), SU(2), SU(3) of the gauge group SU(3)xSU(2)xU(1). They depend on the energy scale μ according to the equation [15,21]

$$\frac{1}{\alpha_i(\mu)} = \frac{1}{\alpha_{GUT}} + b_i \ln \left(\frac{M_{GUT}}{\mu}\right)^2 \quad (i=1,2,3)$$

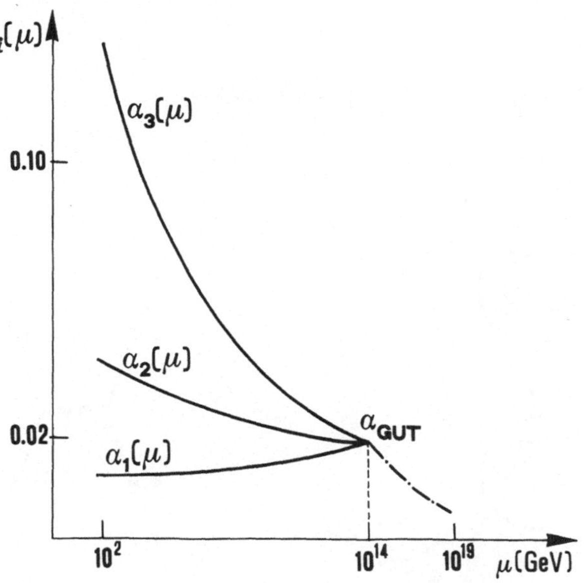

Figure 7. Behaviour of the couplings $\alpha_i(\mu)$ versus the energy scale μ

If one requires that the three couplings meet with the same value α_{GUT} at a scale $\mu=M_{GUT}$ (scale of Grand Unification), one can predict the value of M_{GUT} by extrapolating from the low-energy values of the α_i. The situation is represented in Figure 7. Since the rate of approach to Grand Unification is very slow (the curves of $\alpha_i(\mu)$ have a logarithmic behaviour), one has to go well beyond the electroweak scale 10^2 GeV before reaching unification for values of M_{GUT} in the range 10^{14}-10^{15} GeV [22].

Moreover, there is a very important prediction of GUT which is related to the higher symmetry involved. Going from SU(3)xSU(2)xU(1) to SU(5), the number of gauge vector bosons is increased from 12 to 24. The 12 extra bosons are very massive (mass of the order of M_{GUT}), and possess very peculiar quantum numbers: which are those of either a lepton-quark or a quark-quark pair. Since quarks and leptons belong to the same multiplets, they can be transformed among themselves by exchange of these super-heavy bosons. These transitions would manifest as physical processes violating both baryon adn lepton numbers. Even the proton, which was believed to be absolutely stable, would decay into lighter particles!

The process of proton decay is represented in Figure 8. Since the effective coupling is of the order of

$$G_B \approx \frac{\alpha_{GUT}}{M^2_{GUT}} ,$$

it turns out to be very small, and the predicted lifetime τ_p is very long [22]:

$$\tau_p \approx \left(\frac{M_{GUT}}{10^{15}\text{GeV}}\right)^4 \times 10^{30}\text{years} \approx 10^{30}\text{years}.$$

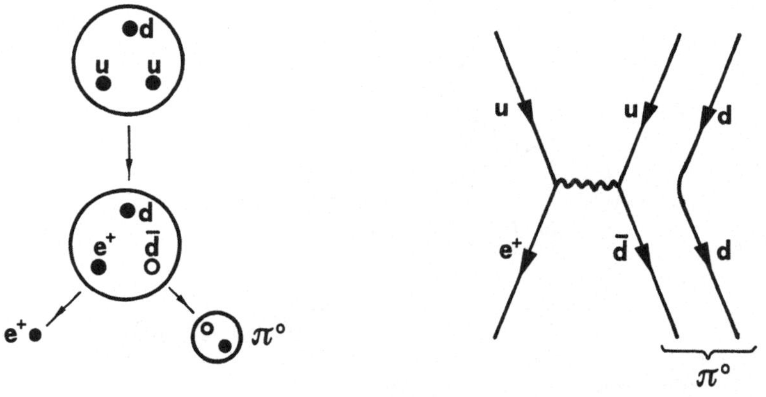

Figure 8. The proton decay

How is it possible to detect such a long lifetime?
(Compare with the age of the universe, which is of the order of 10^{10} years). The solution is very simple: one has to make observations on a sample that contains a sufficiently large number of protons. A detector of the weight of 100 tons contains about 10^{32} nucleons (protons and neutrons); according to the above value predicted for the lifetime, one would detect about 100 events for year.

Experiments using very big detectors have been performed in underground laboratories, in order to reduce the cosmic ray background; specifically in the tunnel under Mount Blanc between France and Italy, and in various mines in USA, India and Japan. A few candidates have been found; however, they are not statistically significant to exclude ths possibility that they are due to the background. They provide a lower limit for the proton lifetime (in the specific decay mode $P \to e^+\pi^\circ$) [23]:

$$(P \to e^+\pi^\circ) \gtrsim 10^{32} \text{ years.}$$

This experimental bound would rule out the theoretical prediction. The question is whether the idea of Grand Unification has to be abandoned. We have to point out that the above prediction was obtained in the frame of a specific GUT, i.e. SU(5) model. Probably this minimal model is too simple-minded; may be other kinds of particles exist between the electroweak and grand-unification scales, so that Grand Unification would be reached through intermediate steps*. From this point of view, the question of proton decay is still open. Experiments with bigger detectors have been planned, in particular in the underground laboratory under the Gran Sasso mountain in central Italy; hopefully these experiments will clarify the situation.

V. PERSPECTIVES

In the previous sections we have reviewed the main ideas which lead to the formulation of a quantum field theory for the electroweak and strong interactions. The standard model, based on the gauge group SU(3)xSU(2)xU(1), reproduces very well the properties of these interactions, but it cannot be considered more than a successful step toward the construction of the complete theory of fundamental interactions.

Several ideas for the construction of theories beyond the Standard Model have been proposed, first of all the idea of Grand Unification. The failure of the simplest model based on the gauge group SU(5) does not mean that the road of Grand Unification has to be abandoned: maybe the situation is not so simple, and intermediate stages of unification have to be introduced before reaching the goal of a complete unification.

From the experimental point of view, there is a great interest in exploring the energy region beyond the SM scale. The discovery of new phenomena would give some knowledge on this unknown land and some restrictions on the theoretical possibilities. A big effort is being made for building larger and larger accelerators, which will investigate higher energy regions. A list is given in Table 3. Underground laboratories are also in construction: experiments will be performed to look for feeble effects, and rare phenomena, such as proton decay which would not be detectable in the accelerators laboratories.

On the theoretical side, there is a big effort in the investigation of the main ingredients for a complete theory which unifies all interactions, including gravity besides strong and electroweak ones.

---

* For a discussion on non-minimal GUT's, see ref. [24].

Table 3. New high-energy colliders

Colliding Particles	Beam Energy (GeV)	Total C.M. Energy (GeV)	Luminosity ($10^{30}cm^{-2}s^{-1}$)	Name and site	Expected starting date
e^+e^-	30+30	60	20	TRISTAN KEK, Tokyo	End of 1986
e^+e^-	50+50	100	6	SLC SLAC, Stanford	Spring 1987
e^+e^-	50+50	100	16	LEP, CERN Geneva	Early 1989
$\bar{p}p$	315+315	630	4	ACOL/Sp$\bar{p}$S CERN, Geneva	End of 1989
$\bar{p}p$	1000+1000	2000	1	TEVATRON, FNAL, Chicago	End of 1986
ep	30+820	314	15	HERA, DESY Hamburg	Spring 1990
pp	400+3000	2200	100	UNK Serpukhov	1993
pp	20000+20000	40000	1000	SSC, USA (site to be decided)	1994

Gravitational interactions are well understood at macroscopic and astronomical scales, in terms of Einstein's general relativity. They are not well understood at microscopic scales: though they are expected to be negligible up to scales of energy of order 10^{19} GeV (i.e. down to distances of order 10^{-33} cm!), they are believed to play a very important role in unification. The unification of gravity with the other interactions seems to require a much bigger symmetry: supersymmetry [25], which is invariance under the interchange between fermions and bosons. Recent developments show that this field of research is very promising. A possible scheme of unification is sketched in Figure 9. The last step, boldly denoted by TOE (Theory of Everything!), might require a drastic change in the formulation of quantum field theory: the fundamental objects would be string-like rather than point-like, and a possible TOE would be realized by the theory of superstrings [26]. These new ideas have very important implications for cosmology; also, there is a hint to understand the question, left unsolved by GUT's, about the number of fermion generations.

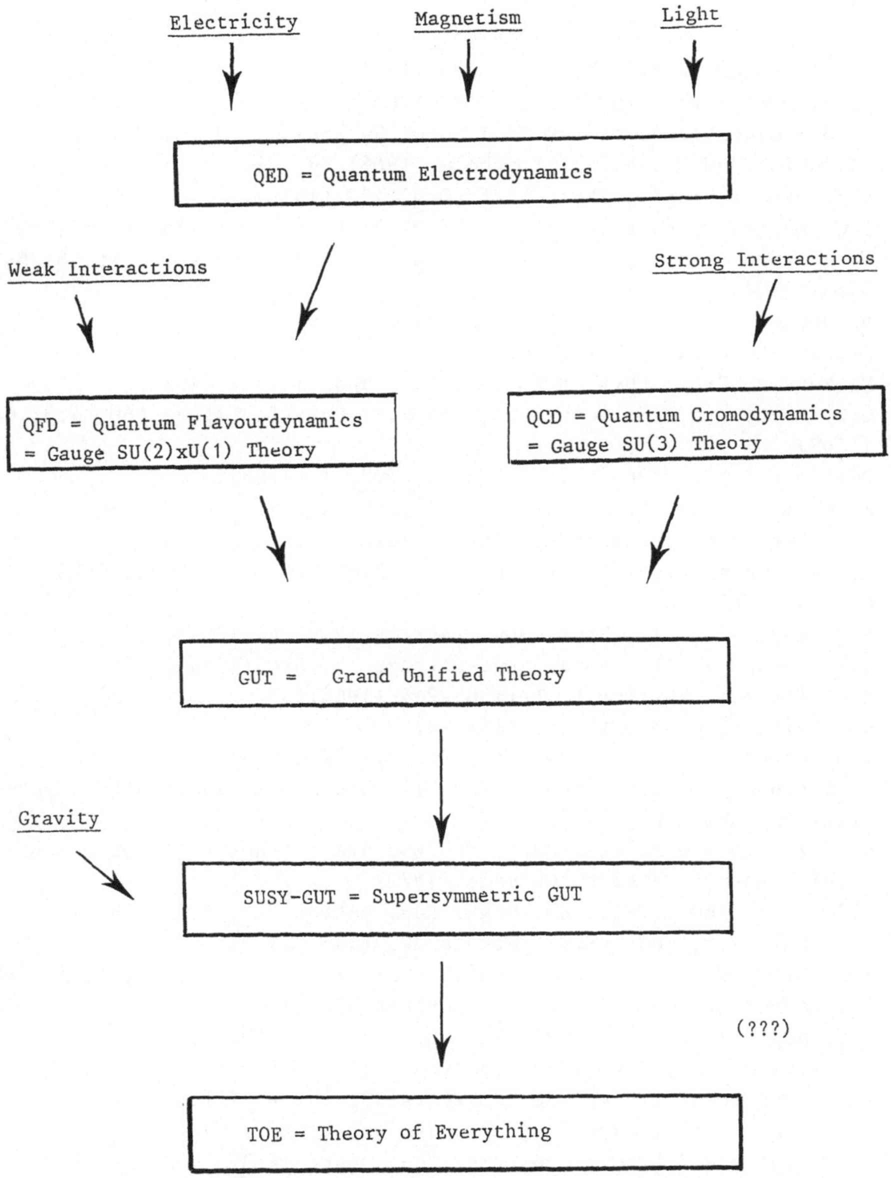

Figure 9. Scheme of unification of the fundamental interactions.

In the unification process, as represented in Figure 9, a larger symmetry is reached in each step, and the energy scales become higher and higher. It might be that the highest steps in unification cannot be tested in experiments performed in laboratories on earth, but only by observation of new phenomena, looking into far-away space-time distances in the Universe. Then, progress in understanding the ultimate constituents of matter would be bound to progress in astrophysics and cosmology.

REFERENCES

1. S.L. Glashow, Nucl. Phys. 22, 579 (1961).
   S. Weinberg, Phys. Rev. Lett. 19, 1269 (1967).
   A. Salam, Proceed. of the VIII Nobel Symposium, ed. N. Svartholm (Almquist and Wiksells, Stockholm, 1968), p. 367.
2. F.J. Hasert et al., Phys. Lett. B46, 138 (1973).
3. P.W. Higgs, Phys. Lett. 12, 132 (1964); Phys. Rev. 145, 1156 (1966).
4. G. Arnison et al., Phys. Lett. B122, 103 (1983); Phys. Lett. B126, 398 (1983);
   M. Banner et al., Phys. Lett. B122, 476 (1983);
   P. Bagnaia et al., Phys. Lett. B129, 130 (1983).
5. H. Yukawa, Proc. Phys. Math. Soc. of Japan 17, 48 (1935).
6. C.M. Lattes, G.P.S. Occhialini and C.F. Powell, Nature 160, 453, 486 (1947).
7. Particle Data Group, Phys. Letters 170B, 1 (1986).
8. M. Gell-Mann and Y. Ne'eman, The Eightfold Way, W.A. Benjamin, New York (1964), and references therein.
9. M. Gell-Mann, Proc. Intern. Conf. on High Energy Physics, CERN, Geneva (1962), p. 805.
10. V.E. Barnes et al., Phys. Rev. Letters 12, 204 (1964);
    G.S. Abrams et al., Phys. Rev. Letters 13, 670 (1964).
11. M. Gell-Mann, Physics Letters 8, 214 (1964);
    G. Zweig, CERN report (unpublished).
12. O.W. Greenberg, Phys. Rev. Letters 13, 122 (1964).
13. J. Steinberg, Proceedings of the XII SLAC Summer Institute on Particle Physics (July 1984).
14. H. Fritzsch and M. Gell-Mann, Proceedings of the XVI Intern. Conf. on High Energy Physics, Chicago (1972).
15. D.J. Gross and F. Wilczek, Phys. Rev. Letters 30, 1343 (1973);
    H.D. Politzer, Phys. Rev. Letters 30, 1346 (1973).
16. S.L. Glashow, J. Iliopoulos and L. Maiani, Phys. Rev. D2, 1285 (1970).
17. J.J. Aubert et al., Phys. Rev. Letters 33, 1404 (1974);
    J.E. Augustin et al., Phys. Rev. Letters 33, 1406 (1974);
    C. Bacci et al., Phys. Rev. Letters 33, 1408 (1974).
18. S.W. Herb et al., Phys. Rev. Letters 39, 252 (1977).
19. M.L. Perl et al., Phys. Letters 63B, 466 (1976).
20. H. Georgi and S.L. Glashow, Phys. Rev. Letters 32, 438 (1974).
21. H. Georgi, H.R. Quinn and S. Weinberg, Phys. Rev. Letters 33, 451 (1974).
22. A.J. Buras, J. Ellis, M.K. Gaillard and D.V. Nanopoulos, Nuclear Phys. B135, 66 (1978).
23. R.M. Bionta et al., Phys. Rev. Letters 51 (1983) 27;
    H.S. Park et al., Phys. Rev. Letters 54 (1985) 54.
24. G. Costa and F. Zwirner, Rivista del Nuovo Cimento 9, no.3 (1986)
25. J. Wess and J. Bagger, Supersymmetry and Supergravity, Princeton University Press, Princeton, N.J. (1982).
26. M.B. Green, J.H. Schwarz and E. Witten, Superstring Theory, Cambridge University Press, Cambridge (1987).

SUMMARY OF THE MEETING

A. Ritter

Department of Physics
Virginia Tech
Blackburg, VA 24061

L. McNeil

Department of Physics
University of North Carolina at Chapel Hill
Chapel Hill, NC 27514

I. INTRODUCTION

The eighth course of the International School of Atomic and Molecular Spectoscopy has examined the structure and dynamic properties of disordered solids with emphasis on the processes which occur in these materials. The field is very wide and rich, and only a fraction of it could be covered in two weeks. In order to summarize the subjects that were covered, four themes which appeared in various degrees throughout the school were identified. First is the obvious need to define disorder and how ordered and disordered systems differ, i.e., how the static structure of a disordered solid can be characterized. Second, one would like to understand the transition from the ordered to the disordered state (or vice-versa), including both the structural and electronic evolution. Dimensionality, the third theme, played an extremely significant role in the discussion of the static and dynamic processes of disordered solids. Finally, the effect of disorder on the dynamic properties of a disordered system was a recurrent subject which threaded through many of the lectures at the school.

II. DEFINITIONS OF DISORDER

The static structure of disordered solids can be divided into two classifications: compositional and topological disorder. It is possible to have both types of disorder in a single system, but this possibility was not considered at the school since the physics of the separate cases proved sufficiently challenging.

Compositionally disordered systems have an underlying lattice on which two (or more) entities are placed at random. An example might be donor or acceptor impurities in a silicon lattice. Other examples which were discussed included semiconductor alloys, magnetic systems with different interactions between different sites, and systems with two isotopes on different sites giving rise to variable hyperfine splitting. These relatively "simple" disordered systems demonstrated a rich variety of phenomena, including mobility edges, frustration, percolation, and metastability.

In a topologically disordered system there is no underlying lattice, which makes the theoretical treatment more difficult. A simplifying feature which makes the problem tractable is the existence of short range order. Systems with topological disorder exhibit the same variety of phenomena as is seen in compositionally disordered solids. An example of extreme topological disorder is provided by fractals, and these fascinating objects received considerable theoretical and experimental attention at the school, as will be described below.

III. DISORDER AND TRANSITIONS

The nature of the transition between an ordered and a disordered state of matter is a subtle problem. The concept of symmetry breaking is a useful approach to understanding phase transitions. For example, in the process of going from gas to liquid to crystalline solid the number of symmetry operations which leave the system invariant is successively reduced. Many systems cool from a liquid into a glass phase rather than into the crystalline phase, and the precise nature of this transition is still unclear. The glass phase is metastable since it is not the true ground state of the system. Why nature chooses this sidetrack and what symmetry operation is associated with the transition from liquid to glass is a fascinating problem. In order to gain insight into this question, the magnetic analog (called the spin glass) of the structural glass phase was discussed.

When a system undergoes a structural transition from the ordered to the disordered state, the electronic properties also change. Solid state physicists raised on the concepts of the importance of translational symmetry and Bloch's theorem can be surprised to find that the electronic structure of the ordered and disordered states are not as different as might be expected. The band widths do not change appreciably and in semiconductors. A sharp boundary occurs between disorder-induced localized states and the extended states in the bulk of the bands. The nature of this "mobility edge" was demonstrated by scaling theories and experimental evidence for its existence was presented.

IV. DISORDER AND DIMENSIONALITY

In ordered systems the dimensionality of the structure in which interactions occur is a primary determinant of the system's behavior. This is most abundantly illustrated in the phenomena of phase transitions and critical phenomena. When we observe novel behavior in disordere systems it is natural to investigate the question of the dimensionality of such systems. If a method whereby a disordered structure could be assigned a dimension different from the Euclidean one in which it is embedded could be developed and if this could be done in a coherent way, then perhaps the behavior of such a structure could be analyzed in detail and new phenomena predicted.

One such method explored at some length during the school was the use of fractal geometry to describe the relevant structures. The aesthetic beauty of fractals and the analytic techniques for determining their (fractional) dimensions were demonstrated by several lecturers, and results were presented for theoretical descriptions of random walks on fractal structures and for experimental tests of such behavior using energy transfer probes.

Several objections to the use of fractals in describing disordered systems were raised by the participants. Fractal geometry necessarily implies a filamentary structure with lower density than that of the corresponding Euclidean structure (indeed, the density of a fractal approaches infinity). This requirement is not met by bulk glasses, which have densities which differ little from those of their crystalline analogs. It is met by percolation clusters within a Euclidean matrix and b porous solids such as Vycor glass. Even in such structures, however, the requirement of dilation symmetry cannot be met in the strict mathematical sense and the physical length scales are bound by the interatomic distance and the size of the sample (admittedly, a range of 10^8). Some kinds of behavior predicted by theoretical analyses (such as certain types of time dependence in diffusive transport) may be unobservable if they involve correlation lengths which are larger than the sample size.

Experimental evidence was presented which appeared to support the fractal model for certain systems, but the range of length scales sampled by the probes used was not necessarily large enough to reveal dilational symmetry. The utility of non-integar dimensionality to describe the behavior of disordered systems remains an open question.

V. DISORDER AND DYNAMICS

The dynamics of disordered systems remains one of the more difficult questions in the field. In a system without translational symmetry, even the description of the elementary excitations (such as those corresponding to phonons in crystalline solids) remains unclear. The concept of fractons was introduced by some lecturers to describe the density of states of vibrational modes in fractal structures, but its predictive value is not obvious and in any case it does not necessarily give insight into non-fractal structures. An understanding of disordered dynamics is vital to progress in the study of non-radiative transitions and of systems in which a distribution of lifetimes leads to diffusive transport and non-exponential decay. The relationship between the disordered structure and such dynamical processes is not yet well understood. Two-level systems characteristic of disorder were invoked to explain the temperature dependence of certain processes, but the nature of these states and their relationship to the disordered structure remain mysterious.

VI. CONCLUSION

The purpose of this school was to bring together scientists working in diverse fields involving various types of disordered solids. In this effort it appears to have succeeded, having allowed the participants to see the connections among the phenomena of disorder in insulators, semiconductors, and metals. Related behavior was described for systems with electronic and with magnetic interactions, and different types of experimental probes and theoretical formulations were found to reveal complementary information. It is hoped that the participants will be able to use these new insights to better understand the disordered systems which are the focus of their own studies.

The participants of a course on Disordered Solids: Structures and Properties, held June 15-29, 1987, in Erice, Sicily.

PARTICIPANTS

1. E. M. Amrhein
University of Puerto Rico
Mayaguez Campus
Department of Physics
Mayaguez
Puerto Rico 00708

2. G. Armagan
Department of Physics
Boston College
Chestnut Hill, MA 02167
U.S.A.

3. R. Balda
Department of Physics
E. S. Ingenieros Industriales
C/Alameda de Urquijo s/n
Bilbao (48103)
Spain

4. F. Bertinelli
Dipartimento di Chimica Fisica ed Inorganica
Università degli Studi di Bologna
Viale del Risorgimento 4
40316 Bologna
Italy

5. M. A. Boukenter
Université Claude Bernard-Lyon I
Laboratoire de Physico-Chimie des Materiaux Luminescents
Unité Associée au CNRS n. 442
Batiment 205
43, Bd. du 11 Novembre 1918
69622 Villeurbanne, Cedex
France

6. G. Boulon
Université Claude Bernard-Lyon I
Laboratoire de Physico-Chimie des Materiaux Luminescents
Unité Associée au CNRS n. 442
Batiment 205
43, Bd. du 11 Novembre 1918
69622 Villeurbanne, Cedex
France

7. E. M. Buchberger
Institut fur Physikalische Chemie der Universitat Wien
IX, Wahringerstrasse 42
A-1090 Wien
Austria

8. G. Careri
Università degli Studi di Roma
"La Sapienza"
Dipartimento di Fisica
Piazzale Moro, 2
00185 Roma
Italy

9. J. Collins*
Department of Physics
Boston College
Chestnut Hill, MA 02167
U.S.A.

10. G. Costa
Università degli Studi
Istituto di Fisica Galileo Galilei
Via F. Marzolo 8
35100 Padova
Italy

11. R. Cowley
Department of Physics
James Clerk Maxwell Building
The King's Building
University of Edinburgh
Edinburgh, EH93JZ
U.K.

12. R. Degli Esposti
Istituto di Fotochimica e Radiazioni d'Alta Energia
(FRAE) del C.N.R.
Via de' Castagnoli, 1
40126 Bologna
Italy

13. B. Di Bartolo
Department of Physics
Boston College
Chestnut Hill, MA 02167
U.S.A.

14. D. L. Di Bartolo
Westbridge School
333 Market Street
Brighton, MA 02135
U.S.A.

15. S. J. Di Bartolo**
Westbridge School
333 Market Street
Brighton, MA 02135
U.S.A.

*Scientific Secretary of the School
**Administrative Secretary of the School

16. H. Domes
Lehrstuhl fur Experimentalphysik IV
Universität Bayreuth
Postf. 101251
D-8580 Bayreuth
F. R. of Germany

17. J. M. Dutta
Department of Physics
P.O. Box 19708
North Carolina Central University
Durham, NC 27707
U.S.A.

18. P. Evesque
Laboratoire d'Optique de la Matière Condensée, Tour 13
Université Paris VI,
4 Place Jussieu
75231 Paris Cedex 05
France

19. J. Fernandez
Department of Physics
E. S. Ingenieros Industriales
C/Alameda de Urquijo s/n
Bilbao (48013)
Spain

20. A. Ferreira
Departemento de Fisica
Universidade de Aveiro
Aveiro
Portugal

21. L. Flight
Department of Physics
University of Leicester
Leicester, LEI 7RH
U.K.

22. H. E. Gumlich
Technische Universität Berlin
Institut für Festkorperphysik III
Sekr PN 4-1
Hadenbergstrasse 36
D-1000 Berlin
F.R. of Germany

23. D. B. Hollis
The University of Sheffield
Department of Ceramics, Glasses and Polymers
Northumberland Road
Sheffield S10 2TZ
U.K.

24. C. R. Jones
Department of Physics
P.O. Box 19708
North Carolina Central University
Durham, NC 27707
U.S.A.

25. A. Kaminskii
Institute of Crystallography
Academy of Sciences of USSR
Moscow, 117333
U.S.S.R.

26. H. Kaul
Lehrstuhl fur Experimentalphysik IV
Universität Bayreuth
Postf. 101251
D-8580 Bayreuth
F.R. of Germany

27. C. Klingshirn
Physicalisches Institut
Johann Wolfgang Goethe Universität
Robert Mayer Strasse 2-4
600 Frankfurt am Main-1
F.R. of Germany

28. R. Knutson
University of Georgia
Department of Physics and Astronomy
Athens, Georgia 30602
U.S.A.

29. A. La Francesca
44 Chester Road
Belmont, MA 02178
U.S.A.

30. S. Magazù
Istituto di Fisica dell' Università
Via dei Verdi
98100 Messina
Italy

31. D. Majolino
I.T.S. del C.N.R.
Istituto di Fisica dell' Università
Via dei Verdi
98100 Messina
Italy

32. F. A. Majumder
Physicalisches Institut
Johann Wolfgang Goethe Universität
Robert Mayer Strasse 2-4
600 Frankfurt am Main-1
F.R. of Germany

33. R. N. Mantegna
Istituto di Fisica dell' Università
Via Archirafi 36
90132 Palermo
Italy

34. J. F. Marcerou
Université Claude Bernard-Lyon I
Laboratoire de Physico-Chimie des Materiaux Luminescents
Unité Associée au CNRS n. 442
Batiment 205
43, Bd. du 11 Novembre 1918
69622 Villeurbanne, Cedex
France

35. L. E. McNeil
Department of Physics
University of North Carolina at Chapel Hill
Chapel Hill, NC 27514
U.S.A.

36. O. Morawski
Institute of Physics
Polish Academy of Sciences
Al. Lotnikow 32/46
02-668 Warsaw
Poland

37. M. Moreno
Departmento de Optica y Estractura de la Materia
Facultad de Ciencias
Universitad de Santander
Avda. de los Castros, s/n
39005 Santander
Spain

38. R. Murri
Dipartimento di Fisica
Università di Bari
Via Amendola 173
70126 Bari
Italy

39. I. Oppenheim
Department of Chemistry
M.I.T.
Cambridge, MA 02139
U.S.A.

40. P. Ostoja
Consiglio Nazionale delle Ricerche
Istituto di Chimica e Tecnologia dei Materiali e dei
Componenti per l'Elettronica (LAMEL)
Via de' Castagnoli, 1
40126 Bologna
Italy

41. G. Ozen
Department of Physics
Boston College
Chestnut Hill, MA 02167
U.S.A.

42. J. B. Pelka
Institute of Physics
Polish Academy of Sciences
Al. Lotnikow 32/46
02-668 Warsaw
Poland

43. R. Reisfeld
Department of Inorganic and Analytical Chemistry
The Hebrew University of Jerusalem
Jerusalem
Israel

44. W. Richter
Universitat Bayreuth
Postf. 101251
D-8580 Bayreuth
F.R. of Germany

45. A. L. Ritter
Department of Physics
Virginia Tech
Blacksburg, Virgnia 24061
U.S.A.

46. J. L. Rousset
Université Claude Bernard-Lyon I
Laboratoire de Physico-Chimie des Materiaux Luminescents
Unité Associée au CNRS n. 442
Batiment 205
43, Bd. du 11 Novembre 1918
69622 Villeurbanne, Cedex
France

47. S. Sarkisov
Institute of Crystallography
Academy of Sciences of USSR
Moscow, 117333
U.S.S.R.

48. L. Schiavulli
Dipartimento di Fisica
Università degli Studi di Bari
Via Amendola 173
70126 Bari
Italy

49. E. Schiff
Department of Physics
Syracuse University
Syracuse, NY 13244
U.S.A.

50. P.M.L.O. Scholte
Philips Research Laboratory
P.O. Box 80000
5600 JA Eindhoven
The Netherlands

51. D. Sherrington
Department of Physics
Imperial College of Science and Technology
SW7 2AZ, London
U.K.

52. R. Silbey
Department of Chemistry
M.I.T.
Cambridge, MA 02139
U.S.A.

53. A. Soldi
Department of Physics
P.O. Box 19708
North Carolina Central University
Durham, NC 27707
U.S.A.

54. B. Spagnolo
Departimento di Energetica ed Applicazioni della Fisica
Università degli Studi di Palermo
Viale della Scienze
90128 Palermo
Italy

55. H. E. Swoboda
Physicalisches Institut
Johann Wolfgang Goethe Universität
Robert Mayer Strasse 2-4
600 Frankfurt am Main-1
F.R. of Germany

56. G. Szamel
Uniwersytet Warszawski
Instytut Fizyski Teoretycznej
ul. Hoza 69
Warszawa (00-681)
Poland

57. J. R. C. Thorne
Inorganic Chemistry Laboratory
South Parks Road
Oxford Oxl 3QR
U.K.

58. N. G. van Kampen
University of Utrecht
Institute of Theoretical Physics
Princetonplein 5
Utrecht
The Netherlands

59. E. Vincent
Service de Physique du Solides et de Resonance Magnetique
C.E.N. Saclay
F-91191 GIF CEDEX
France

60. W. M. Yen
Department of Physics and Astronomy
University of Georgia
Athens, Georgia 30602
U.S.A.

61. R. Zallen
Department of Physics
Virginia Tech
Blacksburg, Virginia 24061
U.S.A.

62. R. Zamboni
Consiglio Nazionale delle Ricerche
Istituto di Spectroscopia Moleculare
Via de' Castagnoli 1
40126 Bologna
Italy

63. Xinxiong Zhang
P.O. Box 1209
Xiamen University
Xiamen, Fujian
P.R. of China

INDEX